Mathematics for Industry

Volume 5

For further volumes:
http://www.springer.com/series/13254

Aims & Scope

The meaning of "Mathematics for Industry" (sometimes abbreviated as MI or MfI) is different from that of "Mathematics in Industry" (or of "Industrial Mathematics"). The latter is restrictive: it tends to be identified with the actual mathematics that specifically arises in the daily management and operation of manufacturing. The former, however, denotes a new research field in mathematics that may serve as a foundation for creating future technologies. This concept was born from the integration and reorganization of pure and applied mathematics in the present day into a fluid and versatile form capable of stimulating awareness of the importance of mathematics in industry, as well as responding to the needs of industrial technologies. The history of this integration and reorganization indicates that this basic idea will someday find increasing utility. Mathematics can be a key technology in modern society.

The series aims to promote this trend by (1) providing comprehensive content on applications of mathematics, especially to industry technologies via various types of scientific research, (2) introducing basic, useful, necessary and crucial knowledge for several applications through concrete subjects, and (3) introducing new research results and developments for applications of mathematics in the real world. These points may provide the basis for opening a new mathematics-oriented technological world and even new research fields of mathematics.

Ryuei Nishii · Shin-ichiro Ei
Miyuki Koiso · Hiroyuki Ochiai
Kanzo Okada · Shingo Saito
Tomoyuki Shirai
Editors

A Mathematical Approach to Research Problems of Science and Technology

Theoretical Basis and Developments in Mathematical Modeling

Springer

Editors
Ryuei Nishii
Kyushu University
Fukuoka
Japan

Shin-ichiro Ei
Department of Mathematics
Hokkaido University
Sapporo
Japan

Miyuki Koiso
Hiroyuki Ochiai
Kanzo Okada
Tomoyuki Shirai
Institute of Mathematics for Industry
Kyushu University
Fukuoka
Japan

Shingo Saito
Faculty of Arts and Science
Kyushu University
Fukuoka
Japan

ISSN 2198-350X ISSN 2198-3518 (electronic)
ISBN 978-4-431-55059-4 ISBN 978-4-431-55060-0 (eBook)
DOI 10.1007/978-4-431-55060-0
Springer Tokyo Heidelberg New York Dordrecht London

Library of Congress Control Number: 2014943753

Printed on acid-free paper

Springer is part of Springer Science+Business Media (www.springer.com)

Preface

The Institute of Mathematics for Industry (IMI), Kyushu University, is a relatively new institute that will soon mark the fourth anniversary of its founding. In April 2013, the IMI received official recognition from the Ministry of Education, Culture, Sports, Science and Technology as Japan's second Joint Usage/Research Center in mathematics following the Research Institute for Mathematical Sciences of Kyoto University. Given the title of Center for Collaborative Research in Advanced and Fundamental Mathematics for Industry, the breadth of IMI activities is widening in collaboration with the research community as a motivating force. The IMI is also the third national mathematical science research institute in Japan if we include the Institute of Statistical Mathematics under the Research Organization of Information and Systems. Members of the IMI are researchers in what would traditionally be called industrial mathematics, (theoretical) applied mathematics, and pure mathematics, and they are divided about evenly among these fields. In addition, most of these members are currently engaged in joint research with industry while also being responsible for educating students majoring in mathematics including those in Master's degree courses and Ph.D. programs. For this reason, I think it would be fair to say that this book has a feel different from a typical compilation on mathematical modeling.

This book is based on a Japanese-language version prepared exactly one year ago, but its text has been revised and enhanced while adding contributions from new members in IMI. Instead of summarizing its contents, I will here quote from the preface to the Japanese edition.

"This book has been achieved through the cooperation of IMI members as well as researchers in industry who have made time to give keynote addresses at research gatherings sponsored either solely or jointly by IMI or to speak at the IMI Colloquium held regularly on the third Wednesday of every month. The themes covered in the book were selected according to the specialties and interests of each author, with attention given to one or more problems within each theme. The idea here was to create a guide for solving those problems through mathematical modeling. The world of applied mathematics and industrial mathematics is, of course, quite vast, and only a few themes from that world are taken up in this book. The purpose of the book, however, is to introduce those fields of mathematics—even if only a small portion of that world—that are now contributing to other scientific fields and to industry and that have the potential of contributing in the future. The readers that we have in mind begin with undergraduate students and graduate students with an interest in mathematics and mathematical science, followed by individuals in industry and finally researchers/faculty members in various fields including mathematics. It is with this order in mind that an editorial policy was established. In particular, the authors were asked to prepare their manuscripts assuming that readers would have a level of knowledge typical of second- and third-year undergraduate students majoring in mathematics for the Japanese standard.

"Plans for publishing this book go back to the preparatory stage in the founding of IMI, but it has been a matter of 'easier said than done'. Nevertheless, an editorial committee for preparing the book was established in IMI in April of last year with Prof. Ryuei Nishii taking on the responsibilities of chairman. The result of this committee's efforts was a book consisting of 36 chapters.

"Although I cannot say for sure that the contents of this book have completely satisfied our objective here, I sincerely hope that it finds its way into the hands of many readers. The publishing of this book is, in a way, an experimental endeavor, and we plan to use the results that we have achieved here as a basis for enhancing the content of next year's edition."

All of us at the IMI would be greatly pleased if this book created through the process described above were to breathe new life, if even slightly, into the research of industrial mathematics and mathematical modeling. At the same time, we look forward to the frank opinions and comments offered by reviewers and readers of this book.

March 2014

Masato Wakayama
Director, Institute of Mathematics for Industry
Kyushu University
Fukuoka, Japan

Preface

The purpose of this book is to introduce those fields of mathematics that are contributing to other fields and to industry and that have the potential of contributing in the future. The readers targeted by the book are upper-level undergraduate students, graduate students, and corporate individuals. The six people responsible for editing the Basic Volume and Applied Volume of the book and the person in charge of overall editing made up the Editorial Committee, and all members of the Institute of Mathematics for Industry (IMI) and individuals from industry and academia having a deep relationship with IMI made contributions. The members of the Editorial Committee are listed below.

Algebra: Takayuki Ochiai (IMI, Kyushu University)
Geometry: Miyuki Koiso (IMI, Kyushu University)
Analysis: Shin-ichiro Ei (IMI, Kyushu University, now at Hokkaido University)
Probability and Statistics: Tomoyuki Shirai (IMI, Kyushu University)
Applied Mathematics: Ryuei Nishii (IMI, Kyushu University)
Application of Mathematics: Kanzo Okada (IMI, Kyushu University)
Overall Editing: Shingo Saito (Faculty of Arts and Science, Kyushu University)

On reading the submitted manuscripts, we could not help but be reminded of the wide dynamic range of mathematics and the great potential for its application to other fields and industry. We feel confident that the goal of publishing this book—to help others become more knowledgeable about the great possibilities of mathematical modeling—will be achieved, and it is our hope that this book and its individual articles will prove useful in a variety of situations and scenarios. Finally, we would like to extend our deep appreciation to those in industry and academia who took time from their busy schedules to prepare manuscripts for this book.

March 2014 Ryuei Nishii

Contents

Part III Analysis

Part IV Probability and Statistics

Part I
Algebra

Mathematics: As an Infrastructure of Technology and Science

Hiroyuki Ochiai

Abstract One of the roles of mathematics is to serve as a language to describe science and technology. The terminology is often common over several branches of science and technology. In this chapter, we describe several basic notions with the emphasis on what is the point of a definition and what are key properties. The objects are taken from set theory, groups and algebras.

Keywords Group · Lie algebra · Exponential map · Spherical linear interpolation · Unit quaternion

1 Sets and Functions

In mathematics, the objects under discussion are described by *sets*, and the relations are described by *functions*.

1.1 Sets: Two Methods of Description

There are two ways to describe sets:

- List all the elements. This method is called *extension*.
- Specify the conditions that are satisfied by all the elements in the sets. This method is called *intension*.

H. Ochiai (✉)
Institute of Mathematics for Industry, Kyushu University, 744, Motooka, Nishiku, Fukuoka 819-0395, Japan
e-mail: ochiai@imi.kyushu-u.ac.jp

R. Nishii et al. (eds.), *A Mathematical Approach to Research Problems of Science and Technology*, Mathematics for Industry 5,
DOI: 10.1007/978-4-431-55060-0_1,

Example 1 A surface is recognized as a subset of the space $\mathbb{R}^3$. The point cloud is an extensional definition of this set. Another extensional definition of a surface is a parametric description:

$$x = x(u, v), y = y(u, v), z = z(u, v).$$

An intensional definition of a surface is

$$f(x, y, z) = 0$$

by using an implicit function f in (x, y, z).

Example 2 Solving a system of equations can be understood as rewriting a set described in an intensional definition into an extensional definition. Practical examples of this are linear or nonlinear programming and numerical problem solving. A more abstract example is Fermat's Last Theorem: The intensionally defined set $\{(x, y, z, n) \in \mathbb{Z}^4 \mid x^n + y^n = z^n, x, y, z > 0, n > 2\}$ is an empty set which is a most trivial example of a set in an extensional method!

Example 3 Given a data set $D = \{(x_i, y_i) \mid i = 1, 2, \ldots, N\}$, finding a correlation f such that $f(x_i, y_i) = 0$ (or nearly equal zero with noise) can be understood as the converse procedure of the above example: a set given by an extensional method is embedded in a set given by an intensional method $D \subset \{(x, y) \mid f(x, y) = 0\}$.

In mathematics and in an application, the translations from intension to extension and vice versa are often useful. The ideal-variety correspondence in algebraic geometry is one example [2]. It is easy to judge an element to be in a subset given by an intensional method. It is straightforward to give all the elements in a subset given by an extensional method. As one of the applications of these observations, in order to show the inclusion $S_1 \subset S_2$, we consider the following: if S_1 is given in extension while S_2 is given in intension, then it is easy to check whether $S_1 \subset S_2$.

1.2 Function

1.2.1 Function and Map

A map f: $X \to Y$ is a correspondence, denoted by $x \mapsto f(x)$, from a set X to a set Y. If Y consists of numbers, then f is also called a function.

Example 4 A vector $\mathbf{a} \in \mathbb{R}^n$ can be considered as a map from $\{1, 2, \ldots, n\}$ to $\mathbb{R}$. Data consisting of N points in $\mathbb{R}^3$ can be considered as a map $\{1, 2, \ldots, N\} \to \mathbb{R}^3$ and also as a map $\{1, 2, \ldots 3N\} \to \mathbb{R}$. A series $\{a_n\}$ is considered to be a map $\mathbb{N} \to \mathbb{R}$.

1.2.2 Average

The average (or *mean*) of data $a_1, \ldots, a_n$ is defined to be

$$(a_1 + \cdots + a_n)/n.$$

This operation usually requires the operation called *addition* satisfying the commutative and associative law and the scalar $(1/n)$ multiple operation. This implies that the data should be in a (convex subset of) vector space in order for the average to make sense. In other words, if the data are not contained in a vector space, then the definition of average is not trivial. For example, if we have weather records like (fine, fine, rain, cloudy, rain, fine, cloudy), we can not obtain the average of these data. Of course, if we assign the numbers 3, 2, 1, 0 to fine, cloudy, rain, snow, then we can obtain an average of the translated scores, but we need to consider whether this assignment makes sense.

If the data are expressed in terms of a set of numbers (like vectors), we could naively take the average by means of the coordinates of the data, but this procedure may not make sense or may even be not well defined. For example, if the given data are on a curved manifold, such as a sphere, then the naive average is often located outside of the given manifold. This phenomenon suggests a modification of the notion of line; a *geodesic* and a connection in differential geometry is an alternative to lines and linear interpolation and extrapolation.

On the other hand, a smooth short curve and piece of a surface are well approximated by a segment (a piece of a line) and a face like a triangle, respectively, so the linear approximation works well in such a case. The set of data obtained by motion capture is considered to be a low-dimensional submanifold of a higher (like 10^4) dimensional spaces and is non-linear, but it is locally well controlled by a linear combination. This is the idea of *blendshape*, which is commonly used in computer graphics.

1.3 Injective, Surjective

1.3.1 Definition

Definition 1 We define several notions on a map $f : X \to Y$.

- The map f is called *surjective* if for any $y \in Y$, there exists $x \in X$ such that $y = f(x)$.
- The map f is called *injective* if $f(x) \neq f(x')$ for any $x, x' \in X$ with $x \neq x'$.
- The map f is called *bijective* if f is surjective and injective.

In general, injectivity and surjectivity are independent. However, in the following cases, these notions are accidentally equivalent:

- In the case X and Y are finite sets and the numbers of elements are same.
- In the case X and Y are finite-dimensional vector space with the same dimension, and f is a linear map.
- In particular, A is a square matrix and f is a multiplication map $v \mapsto Av$ on the set of column vectors.

These cases are often treated seriously in the first course of linear algebra, especially when eigenvalues and eigenvectors are introduced. So, one might have the wrong impression that injective implies surjective and vice versa, even in a general situation.

1.3.2 Bijective and Inverse

Suppose a map $f: X \to Y$ is bijective. Then there exists a map $g : Y \to X$ such that $g(f(x)) = x$ and $f(g(y)) = y$ for all $x \in X$ and $y \in Y$. This g is called the *inverse* of f and is denoted by f^{-1}. In mathematics, once we have a bijective f, then we immediately get f^{-1} and we often identify X with Y by using f. This is a powerful way of thinking. On the other hand, in a practical setting, even if we know f is bijective, the actual construction of f^{-1} is often non-trivial. This fact is a key to *one-way* function, which are often important in cryptography. Moreover, even if we have a bijection between X and Y, the information read in picture X and that in picture Y are often different. We explain these phenomena by means of an example from the logarithm and the Fourier transform, respectively.

1.3.3 Logarithm

For a real variable, the reason the logarithm $y = \log x$ is defined to be an inverse function of the exponential function $y = e^x$ is based on the continuity and monotonicity of the exponential function, and on the completeness of real numbers. The numerical computation supported by software is effective because $\log x$ is close enough to $\log x'$ if x is so to x'. On the other hand, in the case of the discrete logarithm, once we have fixed a and n, the exponential function $\mathbb{Z} \ni x \mapsto (a^x \mod n) \in \mathbb{Z}/n\mathbb{Z}$ does not have continuity on x. Therefore, the analysis of the discrete logarithm as the inverse function cannot be reduced to elementary calculus.

1.3.4 Fourier Transform

A Fourier transformation is a linear bijection on an appropriate function space, e.g., the space of square integrable functions (L^2). Nevertheless, the information given in the original space variables and that in the transformed frequency variables are different in a practical situation, e.g., an approximation by a finite number of terms.

1.4 Generalization of Map

A generalization of a map is a *correspondence*. This is a subset of the direct product $X \times Y$. Once we are given a map $f : X \to Y$, the graph of f is defined to be $\{(x, f(x)) \mid x \in X\}$ as a subset of $X \times Y$. On the other hand, if a subset $C \subset X \times Y$ satisfies the condition that the fiber $C_x := \{y \in Y \mid (x, y) \in C\}$ consists of one element for each $x \in X$, then such a C comes from a map. For a general correspondence C, the fiber C_x can be an empty set or can consist of more than one element. The correspondence is effective in many situation, e.g., to describe an integral transform like the X-ray transform, or Radon transform.

Other generalization of the notions of sets and maps are *category* and *functor*.

1.5 Topology

There are several notions to describe near or far; e.g., topology, metric, norm, inner product. The energy function, the error term, and penalty also play the same role in some cases.

1.5.1 Definition

A *distance* is a non-negative symmetric map $d : X \times X \to \mathbb{R}$ with a triangle inequality: $d(x, z) \leq d(x, y) + d(y, z)$ for all $x, y, z \in X$. A metric space (X, d) is a set X with a distance d. A generalization of a metric space is a *topological space*, by definition, a space with topology. Topology is defined by the set of open sets, which is closed under the union and the finite intersection, while the metric uses a single real-valued function. So topology can express a more complicated structure. A *norm* is a special case of distance, and it has a scalar multiple structure. A norm only has only a length, but an inner product measures both length and angle. In particular, there is the notion of orthogonality in an inner product space. Orthogonal basis and an orthogonal projection only make sense only in an inner product space, whereas basis and a projection do make sense in a vector space.

1.5.2 Complete

A metric space is called *complete* if every Cauchy sequence converges. We can find and specify an element in a complete space by an approximating series. The expression $\pi = 3.14\ldots$ is an example. This idea is also effective to give a function, curve, surface and data, and is the basis of construction principle.

A complete norm space is called a *Banach space*, and a complete inner product space is called a *Hilbert space*.

Example 5 Suppose $p > 1$. For a series $\{a_n\}$, we define the l^p norm

$$\|\{a_n\}\|_p := \sqrt[p]{\sum_{n\geq 1} |a_n|^p},$$

and we denote by l^p the set of all series $\{a_n\}$ with $\|\{a_n\}\|_p < \infty$. Then l^p is the Banach space for $p > 1$, and l^p is a Hilbert space if and only if $p = 2$. Note that for a finite-dimensional vector space, topology does not depend on a norm, whereas it does depend on a norm for an infinite-dimensional vector space [8].

2 Algebra

In this section, we introduce several algebras. A group is used for the description of motion and symmetry. A typical example is a group consisting of matrices. A set of numbers and that of polynomials are rings.

2.1 Definition of Groups (Static)

A set G and a binary operation $G \times G \to G$ is called a *group* if they satisfy the associative law $(g_1g_2)g_3 = g_1(g_2g_3)$, the existence of the unit (identity) and the existence of the inverse of every element. We often call G a group without explicitly mentioning the operation (produce) we choose.

Example 6 A set of regular matrices is a group, by the multiplication of matrices.

2.1.1 Abelian Group

A group satisfying the commutative law $g_1g_2 = g_2g_1$ is called an *abelian group* (commutative group). We often denote the operation by the addition for an abelian group.

Example 7 With the addition, $\mathbb{Z}$, $\mathbb{Z}/n\mathbb{Z}$ are examples of abelian group. The set $\{z \in \mathbb{C} \mid |z| = 1\}$ of complex numbers with absolute value 1 is an abelian group, which is called a unitary group $U(1)$ of size one.

Example 8 The set of points on an elliptic curve turns out to be an abelian group by a non-trivial definition of "addition".

2.1.2 Homomorphism

A map $f : G \to H$ between two groups G and H is called a (group) *homomorphism* if it satisfies $f(g_1 g_2) = f(g_1)f(g_2)$. If a group homomorphism is bijective, then it is called an *isomorphism*.

Example 9 An exponential function $y = e^x$ gives a group isomorphism from the additive group $\mathbb{R}$ to the multiplicative group $\mathbb{R}_{>0} := \{t \in \mathbb{R} \mid t > 0\}$.

Note that the exponential function $\exp(A) = \sum_{n=0}^{\infty} \frac{1}{n!} A^n$ of matrices is not a group homomorphism.

Example 10 The Weierstrass $\wp$ function gives a group isomorphism

$$(\wp, \wp') : \mathbb{C}/(\mathbb{Z}\omega_1 + \mathbb{Z}\omega_2) \longrightarrow \{(x, y) \in \mathbb{C}^2 \mid y^2 = 4x^3 - g_2 x - g_3\} \cup \{\infty\} \quad (1)$$

Note that

- $\mathbb{C}/(\mathbb{Z}\omega_1 + \mathbb{Z}\omega_2)$ is an intensional description, whereas $\{(x, y) \in \mathbb{C}^2 \mid y^2 = 4x^3 - g_2 x - g_3\} \cup \{\infty\}$ is an extensional description.
- By a natural topology, this map is also a homeomorphism.
- The operation on $\mathbb{C}$ divided by the lattice $\mathbb{Z}\omega_1 + \mathbb{Z}\omega_2$ is the addition of complex numbers, so it is easy to calculate the n sum $z + z + \cdots + z$. On the other hand, the operation on elliptic curve $y^2 = 4x^3 - g_2 x - g_3$ is non-trivial, so it is not easy to calculate the n sum $P + P + \cdots + P$.
- By this isomorphism, the set of rational points on elliptic curve bijectively corresponds to a subset of $\mathbb{C}/(\mathbb{Z}\omega_1 + \mathbb{Z}\omega_2)$. However, the non-triviality of $\wp$ function reflects to the difficulty of the logarithm on an elliptic curve.

2.2 A Multiplicative Group of a Ring

2.2.1 Definition of a Ring

A set R with two operations, called an addition and a multiplication, is called a *ring* if R is an abelian group with an addition, the multiplication satisfies the associative law, and the two operations satisfies the distribution law such as $(a+b)c = ab+ac$. Here the last expression means $(ab)+(ac)$, of course. We often assume the existence of the unit 1 of the multiplication. If the multiplication satisfies the commutative law $ab = ba$, then R is called a *commutative ring*. (Note that we never call it an abelian ring.) We also remark that the existence of the unit 0 of the addition and the commutativity of the addition $a + b = b + a$ have already been assumed for a ring.

Example 11 The set $\mathbb{C}[x_1, \ldots, x_n]$ of polynomials is a commutative ring with unit.

A non-zero element a in a ring does not necessarily have its inverse (with respect to the multiplicative operation; $ab = ba = 1$). An element of a ring is called *invertible* if it has an inverse. We denote by $R^\times$ the set of invertible elements, then $R^\times$ is a group.

Example 12 An invertible element of the ring $M(n, \mathbb{R})$ of real square matrices of size n is a regular matrix. We denote $M(n, \mathbb{R})^\times$ by $GL(n, \mathbb{R})$, which is called a general linear group.

A ring is called a *field* if every non-zero element is invertible.

Example 13 The set $\mathbb{Q}, \mathbb{R}, \mathbb{C}$ (respectively) of rational, real, complex (respectively) numbers is field. The set $\mathbb{H} = \mathbb{R} + \mathbb{R}i + \mathbb{R}j + \mathbb{R}k$ of quaternions is also a field, and is a typical example of a non-commutative field [5]. We remark that in some literatures, especially on Galois theory, a field has been already assumed to be commutative.

An algebra is defined to be a vector space with a ring structure. There are a lot of variety of algebraic systems depending on its purpose; e.g., monoid, semigroup, Lie algebra, Jordan algebra.

2.3 Transformation Group, a Dynamic Definition of Group

A group G and a set X with a binary operation $G \times X \to X$ is called an *action* if it satisfies $(g_1 g_2)x = g_1(g_2 x)$.

Example 14 A coordinate transformation group and a fractional linear transformation group give a natural action.

Example 15 Galois group acts on the set of roots of a polynomial equation.

The definition of an action is equivalent to a group homomorphism $G \to Aut(X)$, where $Aut(X)$ is a group consisting of all bijection on the set X. An action is called a *representation* if X is a vector space and the action $x \mapsto gx$ is a linear map. In general, an action is not a representation, but the action naturally induces a representation of G on a function space on X.

2.4 Homogeneous Space

2.4.1 Definition

An action of G on X is called *transitive* if for any two elements x, y in X there exists a $g \in G$ with $y = gx$. In such a case, X is called a homogeneous space of G. For example, the rotation group $SO(3)$ acts transitively on the sphere S^2. All the elements on a homogeneous space are equivalent.

2.4.2 Isotropy Subgroup

We fix an element x in a homogeneous space X of G. Then the subgroup $G_x = \{g \in G \mid gx = x\}$ consisting of elements fixing the element x is called a *isotropy subgroup* (or stabilizer) at x. Then the homogeneous space X has the expression $X = G/G_x$. For example, $S^2 = SO(3)/SO(2)$. The expression $X = G/G_x$ is useful to understand the symmetry of X, introduce coordinates and a metric on X, and the comparison with other spaces. Remark that the isotropy subgroup G_x and the expression G/G_x depend on the choice of the base point x.

2.4.3 Principal Homogeneous Space

A homogeneous space X of G is called a *principal* homogeneous space if for any $x, y \in X$ there exists a unique $g \in G$ with $y = gx$. In this case, the action is called *transitive*. As a set, X can be identified with G. However, G has the special point "the identity", but X does not.

Example 16 The three-dimensional vector space $\mathbb{R}^3$ is an additive group. Our living three-dimensional space is a homogeneous space of this translation group $\mathbb{R}^3$, and our living space does not have a specific base point (the origin of the coordinates). Most of geometric properties of figures in spaces does not depend on the choice of the origin.

Example 17 Let X be the set of all triangles (on the fixed plane) whose barycenters are located at the origin. Then X is a principal homogeneous space of the general linear group $GL(2, \mathbb{R})$.

In computer graphics (see, e.g., [1, 4]) a continuous deformation of a triangulated figure is described by the interpolation not in terms of the coordinates of the vertices of triangles but in terms of the corresponding elements in $GL(2, \mathbb{R})$.

Example 18 Consider a single rigid body, such as a camera. The set of possible positions and direction of this rigid body is a principal homogeneous space of the motion group.

Example 19 For a plane cubic curve X, if we fix a point x_0 on X, then we can define the additive group operation on X so that x_0 is the identity. This is called an *elliptic curve* E. The group structure on X depends on x_0. The curve X is a principal homogeneous space of E.

2.5 Lie Group

2.5.1 Definition

A *Lie group* is a manifold with a compatible group structure. If one is not familiar with manifolds, one may consider a Lie group to be a group consisting of matrices. We can use both calculus and linear algebra to investigate Lie groups.

2.5.2 Exponential Map

The exponential map exp (Example 9) gives a map from the Lie algebra to the Lie group.

Example 20 For example, the Lie algebra of the general linear group $GL(n, \mathbb{R})$ (Example 12) is $\mathfrak{gl}(n, \mathbb{R}) = M(n, \mathbb{R})$, and the exponential map is $\exp : M(n, \mathbb{R}) \to GL(n, \mathbb{R})$. This map is neither injective nor surjective [3].

Example 21 The set

$$\mathfrak{so}(3) = \{A \in M(3, \mathbb{R}) \mid {}^tA = -A\} \tag{2}$$

of skew-symmetric matrices of size three is the Lie algebra of the three-dimensional rotation group $SO(3) = \{g \in M(3, \mathbb{R}) \mid {}^tgg = I_3, \det g = 1\}$, and the exponential map $\exp : \mathfrak{so}(3) \ni A \mapsto \exp(A) \in SO(3)$ is surjective.

These exponential maps exp are not group homomorphisms. In the case of two-dimensional rotation group $SO(2)$, the exponential map is a group homomorphism, but it is exceptional.

One of the definition of the Lie algebra of a Lie group is the tangent space of the Lie group at the identity. In particular, the Lie algebra is a vector space, and the linear interpolation makes sense in a Lie algebra. For example, a linear interpolation of two rotation matrices may not be a rotation matrix, but the exponential image of a linear interpolation of the corresponding two elements in the Lie algebra makes sense.

2.5.3 Dual Numbers

The Lie algebra is a linear approximation of a Lie group. We introduce a formal variable ε, and consider the ring $\mathbb{R}[\varepsilon]$ by adding ε to $\mathbb{R}$ and dividing by the ideal (ε^2), then we obtain the quotient ring $\mathbb{R}[\varepsilon]$. As an actual computation, we deal with a number-like expression $a + b\varepsilon$ and replace ε^2 by 0.

Example 22 Let us compute the Lie algebra of $G = SO(3)$. We consider the tangent space of G at the identity. We put $g = I + \varepsilon A$ with $A \in M(3, \mathbb{R})$. Examine the condition $g \in SO(3)$. By the rule $\varepsilon^2 = 0$, we obtain ${}^tgg = (I + \varepsilon {}^tA)(I + \varepsilon A) = I + \varepsilon({}^tA + A)$. We see that $g \in SO(3)$ is equivalent to ${}^tA + A = O$. This proves the formula (2).

A linear approximation of functions and manifolds usually looses some information. It is significant that the Lie algebra has all the higher order (local) information in Lie algebra. The key of this recovery is encoded in Lie bracket $[A, B] = AB - BA$ and the exponential map. Remark that the global information of Lie groups, such as a connectivity π_0 and the fundamental group π_1 can not be read off from the Lie algebra (c.f., Sect. 2.6).

2.5.4 Semidirect Product

Let G be a group, H a subgroup of G, and K a normal subgroup of G. If a map $H \times K \ni (h,k) \mapsto hk \in G$ is bijective, then G is called a *semidirect product* group of H and K and we denote $G = H \ltimes K$. In other words, for any $h \in H$ and $k \in K$, if we consider $k' = h^{-1}kh$, then $kh = hk'$ and $k' \in K$. Note that the role of H and K is not symmetric.

Example 23 We consider the three-dimensional vector space, and regard $H = SO(3)$, $K = \mathbb{R}^3$ as a set of rotations and translations, respectively. The motion group G is a semidirect product group $SO(3) \ltimes \mathbb{R}^3$, and is the set of orientation-preserving congruences of the three-dimensional space. If we rotate, translate and rotate reversely, then we obtain a translation. This paraphrases the fact that $K = \mathbb{R}^3$ is a normal subgroup of G.

2.6 *Unit Quaternion*

We explain the *spherical linear interpolation* used in the control of characters and cameras in computer graphics [7]. This is considered to be a method of interpolation in a rotation group $SO(3)$.

A unit quaternion is by definition a quaternion with its norm 1, and we denote by $\mathbb{H}^1$ the set of unit quaternions. There are several notations $U(1,\mathbb{H})$, $Sp(1)$ for it. Explicitly,

$$\mathbb{H}^1 = \{a + bi + cj + dk \in \mathbb{H} \mid a^2 + b^2 + c^2 + d^2 = 1\}. \tag{3}$$

Then $\mathbb{H}^1$ is homeomorphic to the three-dimensional sphere S^3 as a manifold, and is a compact connected Lie group. We regard $\mathbb{H}$ as a two-dimensional complex (right) vector space; $\mathbb{H} = \mathbb{C} + j\mathbb{C}$. Consider the matrix expression of the left multiplication of an element $g = a + bi + cj + dk = \alpha + j\beta \in \mathbb{H}$ with this basis $1, j$ then we obtain $\begin{pmatrix} \alpha & -\bar{\beta} \\ \beta & \bar{\alpha} \end{pmatrix}$ The set of matrices of this form is a special unitary group $SU(2) = \{A \in M(2,\mathbb{C}) \mid {}^t\bar{A}A = I_2, \det(A) = 1\}$. $\mathbb{H}^1 \cong SU(2)$. This is an example of accidental isomorphisms of low-dimensional Lie groups [6].

The conjugate action of $q \in U(1,\mathbb{H})$ is defined by $\varphi_q(z) = qzq^{-1}$ $(z \in \mathbb{H})$. Since the quaternion is non-commutative field, this map is non-trivial. We put the imaginary part of quaternions by $\mathrm{Im}\mathbb{H} = \mathbb{R}i + \mathbb{R}j + \mathbb{R}k$. If $z \in \mathrm{Im}\mathbb{H}$, then $\varphi_q(z) \in \mathrm{Im}\mathbb{H}$. We naturally identify $\mathrm{Im}\mathbb{H} = \mathbb{R}^3$, and then we obtain a map

$$\varpi : \mathbb{H}^1 \ni q \mapsto \varphi_q \in SO(3). \tag{4}$$

This map ϖ is surjective, a group homomorphism, and its kernel consists of two elements; $\ker \varpi = \{\pm 1\}$. The map ϖ is also a universal covering map of $SO(3)$

Consider the interpolation of two elements $g_1, g_2 \in SO(3)$ For example, we consider a problem to give a smooth and natural motion of a camera from one starting position to another ending position specified by g_1, g_2. Put q_1, q_2 be a lift of g_1, g_2 by a map ϖ. That is, $\varphi_{q_1} = g_1, \varphi_{q_2} = g_2$. If we can interpolate q_1, q_2 in $\mathbb{H}^1$, then its image by ϖ gives an interpolation of g_1, g_2 in $SO(3)$. There are several possibilities of interpolations in $\mathbb{H}^1$.

- A geodesic (shortest path) in a Riemannian manifold S^3.
- The image of the exponential map of the linear interpolation in the Lie algebra $\mathrm{Im}\mathbb{H}$.
- Spherical linear interpolation (slerp) [9].

We will explain these methods are same. We first recall the exponential map, the Lie algebra and the Lie group in this setting. For a θ with $-\pi < \theta < \pi$ and an $\mathbf{n} \in \mathrm{Im}\mathbb{H}$ with $|\mathbf{n}| = 1$, we obtain $\exp(\theta\mathbf{n}) = (\cos\theta) + (\sin\theta)\mathbf{n}$ and $\exp(\theta\mathbf{n}) \in \mathbb{H}^1$. On the other hand, every unit quaternion other than -1 can be expressed in this form. This gives a geometric meaning of the conjugate action of a unit quaternion. For $q = \exp(\theta\mathbf{n})$, the map φ_q gives a rotation with the axis $\mathbf{n}$ and the angle 2θ.

On the other hand, for $q_0, q_1 \in \mathbb{H}^1$, we denote by ϕ the angle of these two elements in $\mathbb{R}^4$. Then the spherical linear interpolation is, by definition, for $0 \le t \le 1$,

$$\mathrm{slerp}(q_0, q_1, t) = \frac{\sin(1-t)\phi}{\sin\phi} q_0 + \frac{\sin t\phi}{\sin\phi} q_1.$$

The slerp satisfies the following properties:

$$\begin{aligned}
\mathrm{slerp}(q_0, q_1, 0) &= q_0, \\
\mathrm{slerp}(q_0, q_1, 1) &= q_1, \\
\mathrm{slerp}(q_0, q_1, t) &\in \mathbb{H}^1, \\
\mathrm{slerp}(q_0, q_1, t) &= \mathrm{slerp}(1, q_1 q_0^{-1}, t) q_0, && (5) \\
\mathrm{slerp}(1, \exp(\theta\mathbf{n}), t) &= \exp(t\theta\mathbf{n}). && (6)
\end{aligned}$$

The property (5) shows the covariance, and the property (6) shows that it is the image by the exponential map of the linear interpolation in the Lie algebra. These two properties characterize slerp.

References

1. M. Alexa, D. Cohen-Or, D. Levin, As-rigid-as-possible shape interpolation, in *Proceedings of ACM SIGGRAPH* (2000), pp. 157–164
2. D.A. Cox, J. Little, D. O'Shea Ideals, Varieties, and Algorithms, *An Introduction to Computational Algebraic Geometry and Commutative Algebra*, 3rd edn. (Springer, New York, 2007)

3. S. Kaji, S. Hirose, H. Ochiai, K. Anjyo, A Lie theoretic parameterization of affine transformation, in *Mathematical Progress in Expressive Image Synthesis*, MI Lecture Note, vol. 50, (Kyushu University, 2013), pp. 134–140
4. S. Kaji, S. Hirose, S. Sakata, Y. Mizoguchi, K. Anjyo, Mathematical analysis on affine maps for 2D shape interpolation. in *Proceedings of SCA2012* (2012), pp. 71–76
5. M. Koecher, R. Remmert, *Hamilton's Quaternions, in Numbers*, (Springer, New York, 1991)
6. G. Matsuda, S. Kaji, H. Ochiai, Anti-commutative dual complex numbers and 2D rigid transformation, in *Mathematical Progress in Expressive Image Synthesis*, MI Lecture Note, vol. 50, (Kyushu University, 2013), pp. 128–133
7. H. Ochiai, K. Anjyo, Mathematical Description of Motion and Deformation—From Basics to Graphics Applications—, SIGGRAPH Asia 2013 Course, http://portal.acm.org, (Revised course notes are also available at http://mcg.imi.kyushu-u.ac.jp/english/index.php) (2013)
8. F. Reinhardt, H. Soeder, G. Falk, in *dtv-Atlas zur Mathematik, Deutscher Taschenbuch* (Springer, New York, 1978)
9. J. Vince, in *Quaternions for Computer Graphics* (Springer, New York, 2011)

Remarks on Quantum Interaction Models by Lie Theory and Modular Forms via Non-commutative Harmonic Oscillators

Masato Wakayama

Abstract As typically the *quantum Rabi model*, particular attention has been paid recently to studying the spectrum of self-adjoint operators with non-commutative coefficients, not only in mathematics but also in theoretical/experimental physics, e.g. aiming at an application to quantum information processing. The *non-commutative harmonic oscillator* (NcHO) is a self-adjoint operator, which is a generalization of the harmonic oscillator, having an interaction term. The Rabi model is shown to be obtained by a second order element of the universal enveloping algebra of the Lie algebra $\mathfrak{sl}_2$, which is arising from NcHO through the oscillator representation. Precisely, an equivalent picture of the model is obtained as a confluent Heun equation derived from the Heun operator defined by that element via another representation. Though the spectrum of NcHO is not fully known, it has a rich structure. In fact, one finds interesting arithmetics/geometry described by e.g. elliptic curves, modular forms in the study of the spectral zeta function of NcHO. In this article, we draw this picture, which may give a better understanding of interacting quantum models.

Keywords Eichler integral · Heun ODE · Non-commutative harmonic oscillator · Oscillator representation · Rabi model · Spectral zeta function · Universal enveloping algebra · Zeta regularization.

1 Introduction

The *non-commutative harmonic oscillator* Q (NcHO) is a parity-preserving (or possessing $\mathbb{Z}_2$ symmetry) differential operator introduced in [28, 29] as

M. Wakayama (✉)
Institute of Mathematics for Industry, Kyushu University, 744, Motooka, Nishiku, Fukuoka 819-0395, Japan
e-mail: wakayama@imi.kyushu-u.ac.jp

R. Nishii et al. (eds.), *A Mathematical Approach to Research Problems of Science and Technology*, Mathematics for Industry 5,
DOI: 10.1007/978-4-431-55060-0_2, © Springer Japan 2014

$$Q := A\Big(-\frac{1}{2}\frac{d^2}{dx^2}+\frac{1}{2}x^2\Big)+B\Big(x\frac{d}{dx}+\frac{1}{2}\Big),$$

where A is positive definite symmetric and B is skew-symmetric ($A, B \in \mathrm{Mat}_2(\mathbb{R})$). We assume the Hermitian matrix $A + iB$ is positive definite, i.e. $\det(A) > \mathrm{pf}(B)^2$. The former requirement arises from the formal self-adjointness of Q relative to the inner product on $\mathbb{C}^2 \otimes L^2(\mathbb{R})$. The latter condition guarantees that the eigenvalues of Q are all positive and form a discrete set with finite multiplicity. As a normalized form we may take $A = \begin{bmatrix}\alpha & 0\\ 0 & \beta\end{bmatrix}$ and $B = \begin{bmatrix}0 & -1\\ 1 & 0\end{bmatrix}$ (The assumption is $\alpha\beta > 1$). It should be noted that, when $\alpha = \beta$, Q is unitarily equivalent to a couple of quantum harmonic oscillators, whence the eigenvalues are easily calculated as $\{\sqrt{\alpha^2-1}(n+\frac{1}{2}) \mid n \in \mathbb{Z}_{\geq 0}\}$ having multiplicity 2 ([28], I). Actually, when $\alpha = \beta$, behind Q, there exists a structure corresponding to the tensor product of the 2-dimensional trivial representation and the oscillator representation (e.g. [10]) of the Lie algebra $\mathfrak{sl}_2$. The clarification of the spectrum in the general $\alpha \neq \beta$ case is, however, considered to be highly non-trivial. Indeed, while the spectrum is well described theoretically by using certain continued fractions [28, 29] and also by Heun's ordinary differential equation (the second order Fuchsian differential equation with four regular singular points in a complex domain, [23, 24, 34]), and in particular there are some results related to the estimate of upper bound of the lowest eigenvalue and distribution of eigenvalues [9, 13, 22, 30], only very little information is available in reality when $\alpha \neq \beta$ (see [26] and references therein. Figure 1 represents a numerical graph for the spectrum for the ratio β/α. Note that the eigenvalue curves are continuous w.r.t. the parameter β/α [22]).

In fact, only quite recently, it was proved that the multiplicity of each eigenvalue is always less than or equal to 2 by the monodromy representation of Heun's equations [34], and the ground state is simple (and even) [8] using the criterion given in [35] (see also [9]). Therefore, in spite of many studies, the spectral description of the NcHO is still incomplete (see [27] for an overview of the recent progress). One of the difficulties to obtain the eigenfunctions and eigenvalues is, representation theoretically, the apparent lack of an operator which commute with Q (second conserved quantity) besides the Casimir operator, the image of the generator of the center $\mathcal{Z}\mathcal{U}(\mathfrak{sl}_2)$ of the universal enveloping algebra of the Lie algebra $\mathfrak{sl}_2$. (Moreover, it has been shown that there is no annihilation/creation operators associated to NcHO when $\alpha \neq \beta$ [27].)

Recently, however, particular attention has been paid to studying the spectrum of self-adjoint operators with non-commutative coefficients, in other words, interacting quantum systems, like the *quantum Rabi model* [3, 20, 21, 31, 38], the *Jaynes-Cumming (JC) model* etc., not only in mathematics (e.g. [7]) but also in theoretical physics and experimental physics founded e.g. in the book by Haroche and Raimond [6] (also [39]). For instance, the quantum Rabi model [31] is known to be the simplest model used in quantum optics to describe interaction of light and matter beyond the harmonic oscillator, and the JC model is the widely studied rotating-wave approximation of the Rabi model (see e.g. [4]).

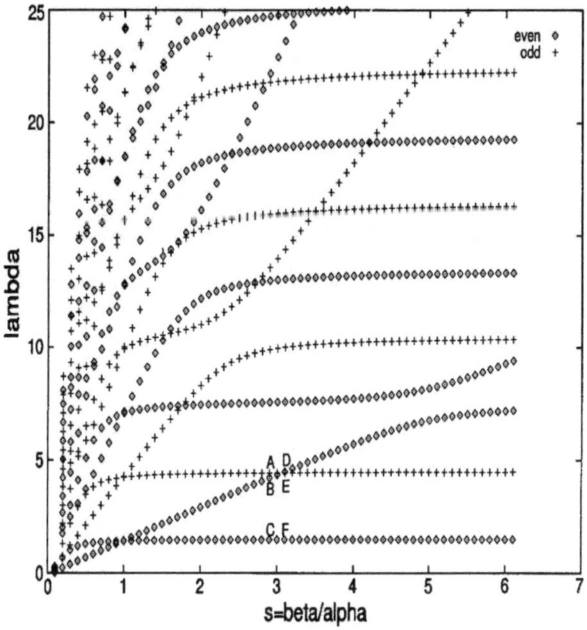

Fig. 1 Approximate N-th eigenvalues $\hat{\lambda}_N$ of Q [22]

The quantum Rabi model is defined by the Hamiltonian

$$H_{\text{Rabi}}/\hbar = \omega a^\dagger a + \Delta\sigma_z + g\sigma_x(a^\dagger + a).$$

Here $a = (x + \partial_x)/\sqrt{2}$ (resp. $a^\dagger = (x - \partial_x)/\sqrt{2}$) is the annihilation (resp. creation) operator for a bosonic mode of frequency ω, $\sigma_x = \begin{bmatrix} 0 & 1 \\ 1 & 0 \end{bmatrix}$, $\sigma_y = \begin{bmatrix} 0 & -i \\ i & 0 \end{bmatrix}$, $\sigma_z = \begin{bmatrix} 1 & 0 \\ 0 & -1 \end{bmatrix}$ are the Pauli matrices for the two-level system, 2Δ is the energy difference between the two levels, and g denotes the coupling strength between the two-level system and the bosonic mode. The Rabi model considers a two-level atom coupled to a quantized, single-mode harmonic oscillator (in the case of light, this could be a photon in a cavity, as in Fig. 2 [33]). Introduced over 70 years ago [31], its applications range from quantum optics, magnetic resonance to solid state and molecular physics. Very recently, the model applies to a variety of physical systems, including cavity quantum electrodynamics, the interaction between light and trapped ions or quantum dots, and the interaction between microwaves and superconducting qubits.

Although this model has had an impressive impact on many fields of physics [6], only recently (in 2011) could this model be declared solved by D. Braak [3]. It is now pointed out [33] that as physicists gain intuition for Braak's mathematical solution, it is very much expected that the results could have implications for further theoretical

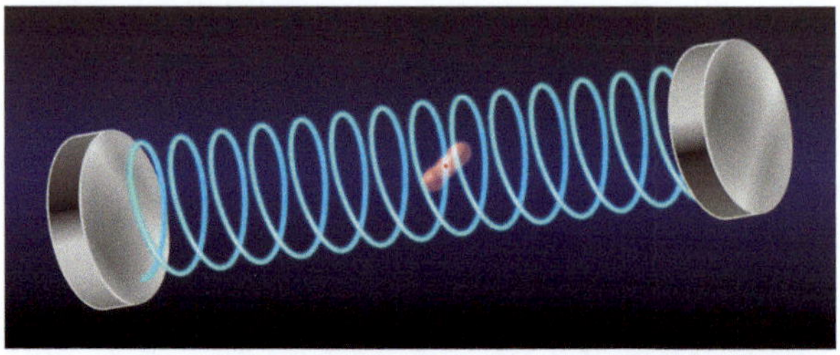

Fig. 2 Courtesy of APS/Alan Stonebraker in [33]: The Rabi model describes the simplest interaction between quantum light and matter. The model considers a two-level atom coupled to a quantized, single-mode harmonic oscillator

and experimental work that explores the interaction between light and matter, from weak to extremely strong interactions.

The NcHO has been similarly expected to provide one of these Hamiltonians describing such quantum interacting systems. In this article, we will observe that the quantum Rabi model is obtained by a second order element $\mathscr{R}$ of the universal enveloping algebra $\mathscr{U}(\mathfrak{sl}_2)$, which is arising from the NcHO through the oscillator representation π' of the Lie algebra $\mathfrak{sl}_2$, and a confluence procedure for Heun's equation under another representation π'_a. Roughly speaking, the quantum Rabi model can be obtained by a confluence process by a "certain rescaling" of the NcHO through their respective Heun's pictures:

$$\begin{array}{ccccc} \text{NcHO} & \xleftarrow{\ \pi'\ } & \underset{\mathscr{U}(\mathfrak{sl}_2)}{\overset{\mathscr{R}}{\cap}} & \xrightarrow[\pi'_a(\cong\varpi_a)]{\mathscr{L}_a} & \text{Heun ODE} \\ & & & & \Big\Downarrow \text{confluence process} \\ & & & & \text{Confluent Heun ODE} \sim \text{Rabi model.} \end{array}$$

2 Number Theoretic Structure of NcHO

Although the explicit eigenvalues of Q are not known, the spectrum of Q possesses a very rich mathematical structure. Denote the (repeated) eigenvalues of Q by $0 < \lambda_1 \leq \lambda_2 \leq \lambda_3 \leq \cdots (\to \infty)$. Define the spectral zeta function of Q by

$$\zeta_Q(s) = \sum_{n=1}^{\infty} \lambda_n^{-s}.$$

This series is absolutely convergent and defines a holomorphic function in s in the region $\mathrm{Re}(s) > 1$. The function $\zeta_Q(s)$ is analytically continued to the whole complex plane $\mathbb{C}$ as a single-valued meromorphic function that is holomorphic, except for a simple pole at $s = 1$ with residue $\frac{\sqrt{\alpha+\beta}}{\sqrt{\alpha\beta(\alpha\beta-1)}}$ [11]. It is notable that $\zeta_Q(s)$ has 'trivial zeros' at $s = 0, -2, -4, \ldots$. When $\alpha = \beta(> 1)$, $\zeta_Q(s)$ is identified (by a elementary holomorphic factor) with the Riemann zeta function $\zeta(s)$.

Similarly to the Apéry numbers which were introduced in 1978 by R. Apéry for proving the irrationality of $\zeta(2)$ and $\zeta(3)$ (see, e.g. [2]), *Apéry-like numbers* have been introduced in [12] for the description of the special values $\zeta_Q(2)$ and $\zeta_Q(3)$. These Apéry-like numbers $J_2(n)$ and $J_3(n)$ share with many of the properties of the original Apéry numbers, e.g. recurrence equations, congruence properties, etc (see [12, 18], also [25]). Actually, the Apéry-like numbers $J_2(n)$ for $\zeta_Q(2)$ obtain a remarkable modular form interpretation, as that shown by F. Beukers [2] in the case of the Apéry numbers. We have shown in [14] that the differential equation satisfied by the generating function $w_2(t)$ of $J_2(n)$ is the Picard-Fuchs equation for the universal family of elliptic curves equipped with rational 4-torsion: $\Omega_{AL} w_2(t) = 0$. The parameter t of this family is regarded as a modular function for the congruence subgroup $\Gamma_0(4)(\cong \Gamma(2)) \subset SL_2(\mathbb{Z})$. Moreover, one observes ([14]) that $w_2(t)$ is considered as a $\Gamma_0(4)$ meromorphic modular form of weight 1 in the variable τ as the classical Legendre modular function $t(\tau) = -\frac{\theta_4(\tau)^2}{\theta_4(\tau)^4}$. We also remark that the modular form $w_2(t)$ can be found at #19 in the list of [37].

The formulas of the special values $\zeta_Q(k)$ for the general cases $k \geq 4$ are much more complicated than those of $k = 2, 3$. Thus, we will focus only on the *first anomaly* $R_{k,1}(x)$ (in the terminology by Kimoto) which expresses the 1st order difference (in a suitable sense) of $\zeta_Q(k)$ from $\zeta(k)$ with respect to the parameters α, β [15, 16]. The first anomaly $R_{k,1}(x)$ for $x = 1/\sqrt{\alpha\beta - 1}$ describes the special value $\zeta_Q(k)$ partly. (When $k = 2, 3$, $R_{k,1}(x)$ possesses full information of each special value.) The Taylor expansion of $R_{k,1}(x)$ in x yields k-th Apéry-like numbers $J_k(n)$. Then, remarkably, one can show that the generating function $w_k(t)$ of $J_k(n)$ satisfies an inhomogeneous differential equation whose homogeneous part is given by the same Fuchsian differential operator which annihilates $w_2(t)$ as $\Omega_{AL} w_k(t) = w_{k-2}(t)$.

In order to solve this differential equation for $w_4(t)$, it is necessary to "integrate twice" a certain explicitly given modular form. Then one can prove that the generating function $w_4(t)$ can be expressed as a differential of a *residual modular form* multiplied by a modular form (a product and quotient of theta functions) for $\Gamma(2)$. The notion of residual modular forms is a generalization of the Eichler (or automorphic) integral. Note that the Abelian integrals and the Eisenstein series $E_2(\tau)$ of weight 2 for $SL_2(\mathbb{Z})$ are special examples of the Eichler integral. The name "residual" comes from the following two facts.

- Eichler's integral possesses an "integral constant" given by a polynomial in τ, which is known as a period function and computed as residues of the integral when one performs the inverse Mellin transform of L-function of the corresponding modular form.
- To obtain another meaningful expression of such Eichler's integral, we define *differential Eisenstein series* by a derivative of the analytic continuation of generalized Eisenstein series (e.g. [1]) at negative integer points.

We remark that the "residual part" of a differential Eisenstein series is in general given by a rational function in τ, whence it can not be handled in a framework of Eichler integrals. Moreover, one should note that only $w_4(t)$ one can give its explicit expression by a sum of two such differential Eisenstein series [16]. Furthermore, to understand the structure, especially the dimension of a space of residual modular forms, it is important to consider the Eichler cohomology groups [5] associated with several $\Gamma(2)$-modules made by a set of certain functions on the Poincaré upper half plane, such as the space (field) of rational functions $\mathbb{C}(\tau)$, the space of holomorphic/meromorphic functions with some decay condition at the infinity (cusps), etc. In the course of this analysis, we focus on a particular subgroup of the Eichler cohomology group, which we call a periodic cohomology, for the explicit determination of the space of residual modular forms which contains $w_4(t)$. We leave the detailed discussion about this arithmetic study of NcHO to the paper [16] (also [15]).

3 Quantum Rabi and Jaynes-Cummings Models

The Hamiltonian of the Rabi model ($\hbar = 1$) reads

$$H_{\text{Rabi}} = \omega a^\dagger a + \Delta\sigma_z + g(\sigma^+ + \sigma^-)(a^\dagger + a),$$

with $\sigma^\pm = (\sigma_x \pm i\sigma_y)/2$. It is regarded as exactly solvable only since the work of [3]. The simpler related model defines the JC Hamiltonian

$$H_{\text{JC}} = \omega a^\dagger a + \Delta\sigma_z + g(\sigma^+ a + \sigma^- a^\dagger).$$

It is known to be integrable, even in the Dicke version with n two-state atoms. Actually, unlike in the Rabi case, the operator

$$\mathscr{I} := a^\dagger a + \frac{1}{2}(\sigma_z + 1)$$

commutes with the Hamiltonian H_{JC} and leads to the solvability of the JC-model. The conservation (i.e. invariance w.r.t. H_{JC}) of $\mathscr{I}$ signifies that the state space decomposes into an infinite sum of two-dimensional invariant subspaces. Each eigenstate of H_{JC} is then labeled by (the eigenstates of $\mathscr{I}$) $0, 1, 2, \ldots$ with a two-valued index, e.g. $+$ and $-$, denoting a basis vector in the two-dimensional subspace which belongs to the eigenspace of $\mathscr{I}$. Representation theoretically, the conserved quantity $\mathscr{I}$ generates a continuous $U(1)$ symmetry of the JC-model which is broken down to $\mathbb{Z}_2$ in the Rabi model due to the presence of the term $(\sigma^+ + \sigma^-)(a^\dagger + a)$. This residual $\mathbb{Z}_2$ symmetry, usually called parity, leads to a decomposition of the state space into just two subspaces $\mathscr{H}_\pm$, each with infinite dimension. Hence the Rabi model shares a similar situation as NcHO.

By [3], one knows that the spectrum of $H_{\rm Rabi}$ consists of two parts, the regular and the exceptional (degenerate) spectrum. Almost all eigenvalues are regular and given by the zeros of the transcendental functions $G_\pm(x)$ in the variable x. The functions $G_\pm(x)$ are defined through the power series in the coupling constant g as

$$G_\pm(x) = \sum_{n=0}^{\infty} K_n(x)\Big[1 \mp \frac{\Delta}{x - n\omega}\Big]\Big(\frac{g}{\omega}\Big)^n,$$

where the coefficients $K_n(x)$ are defined recursively,

$$nK_n(x) = f_{n-1}(x)K_{n-1}(x) - K_{n-2}(x),$$

with the initial condition $K_0 = 1,\ K_1(x) = f_0(x)$, and

$$f_n(x) = \frac{2g}{\omega} + \frac{1}{2g}\Big(n\omega - x + \frac{\Delta^2}{x - n\omega}\Big).$$

The function $G_\pm(x)$ is meromorphic in x having simple poles at $x = 0, \omega, 2\omega, \ldots$ (essentially the eigenvalues of harmonic oscillator). Then the regular energy spectrum of the Rabi model in each invariant subspace $\mathscr{H}_\pm$ with parity $\pm$ is given by the zeros of $G_\pm(x)$: for all zeros $x_n^\pm$ of $G_\pm(x)$, the nth eigenenergy with parity $\pm$ reads $E_n^\pm = x_n^\pm - g^2/\omega$. All exceptional eigenvalues E have the form $E_n^{deg} = n\omega - g^2/\omega$, and the necessary and sufficient condition for the occurrence of the eigenvalue E_n^{deg} reads $K_n(n\omega) = 0$, which furnishes a condition on the model parameters g and $|\Delta|$. Actually, we have the following interesting result due to Kus [20].

We will assume $\omega = 1$ (without loss of generality) in the sequel of this article.

Lemma 1 *Let $P_k^{(n)}(x, y)$ be the polynomial of two variables defined by the following recursion formula:*

$$\begin{aligned} P_0^{(n)} &= 1, \quad P_1^{(n)} = x + y - 1, \\ P_k^{(n)} &= (kx + y - k^2)P_{k-1}^{(n)} - k(k-1)(n-k+1)xP_{k-2}^{(n)} \end{aligned}$$

If $P_n^{(n)}((2g)^2, \Delta^2) = 0$, then there exist two linearly independent eigenfucntions $\psi_n^\pm$ ("positive and negative parity") of $H_{\rm Rabi}$ corresponding to the eigenvalue $E_n^{deg} = n - g^2$, that is, the multiplicity of E_n^{deg} is 2.

Remark 1 The eigenfunctions $\psi_n^\pm$ are constructed in [20]. Also, if $0 < \Delta < 1$ there exist exactly n distinct positive roots of $P_n^{(n)}(x, \Delta^2)$ as a polynomial in x [20]. Similar polynomials for the NcHO [34] should be well formulated as $P_k^{(n)}$.

The analysis on the Rabi model above have been extensively using the Bargmann representation of bosonic operators which is realized by the following Bargmann transform $\mathscr{B}$ (from real coordinate x to complex variable z).

$$(\mathscr{B}f)(x) = \sqrt{2}\int_{-\infty}^{\infty} f(x)e^{2\pi xz - \pi x^2 - \frac{\pi}{2}z^2}\mathrm{d}x.$$

Here the Bargmann space is by definition a Hilbert space of entire functions equipped with the inner product

$$(f|g) = \frac{1}{\pi}\int_{\mathbb{C}} \overline{f(z)}g(z)e^{-|z|^2}\mathrm{d}(\mathrm{Re}(z))\mathrm{d}(\mathrm{Im}(z)).$$

The main advantage is simply due to the fact; $a^\dagger = (x - \partial_x)/\sqrt{2} \to z$ and $a = (x + \partial_x)/\sqrt{2} \to \partial_z$. This makes the Rabi model to be a matrix-valued first order differential operator. The same situation, however, does not appear for NcHO. This explains one of the reasons that the analysis for NcHO is rather difficult and very likely richer.

Now we consider the spectral (Hurwitz type) zeta function of the Rabi model as

$$\zeta_{\mathrm{Rabi}}(s, z) = \sum_{\lambda\in\mathrm{Spec}(H_{\mathrm{Rabi}})} (z - \lambda)^{-s},$$

where the sum runs over all eigenvalues λ of the Rabi model (counted with multiplicity). As in the case of the spectral zeta function $\zeta_Q(s)$ of NcHO, one can easily prove that the sum converges absolutely and uniformly on compacts in the right hall plane $\mathrm{Re}(s) > 1$ so that it defines an analytic function in this region. Then, as in the case of $\zeta_Q(s)$, one can naturally expect that $\zeta_{\mathrm{Rabi}}(s, z)$ has a meromorphic continuation to the whole complex plane $\mathbb{C}$, in particular meromorphic at $s = 0$. If we may assume that $\zeta_{\mathrm{Rabi}}(s, z)$ is holomorphic at $s = 0$, we define the zeta regularized product by

$$\prod_{\lambda\in\mathrm{Spec}(H_{\mathrm{Rabi}})} (z - \lambda) := \exp\left(-\frac{\mathrm{d}}{\mathrm{d}s}\zeta_{\mathrm{Rabi}}(0, z)\right).$$

(Notice that a zeta regularized product is identified with a usual product when the defining series is finite. Moreover, even if $\zeta_{\mathrm{Rabi}}(s, z)$ is not holomorphic at $s = 0$, one may still define the zeta regularized product similarly. See [17, 19] and the references therein for zeta regularizations.) It is known that the function $\prod_{\lambda\in\mathrm{Spec}(H_{\mathrm{Rabi}})}(z - \lambda)$ is an entire function whose zeros are exactly given by the λ's. Then, the following claim follows naturally from the results of Braak [3] and Kus [20] above.

Conjecture 1 *Let* $\prod_{E_n^{\mathrm{deg}}\in\mathrm{Spec}(H_{\mathrm{Rabi}})}(z - n)$ *be the zeta regularized product defined by the series* $\sum_{E_n^{\mathrm{deg}}\in\mathrm{Spec}(H_{\mathrm{Rabi}})}(z - E_n^{\mathrm{deg}})^{-s}$, *where* $E_n^{\mathrm{deg}} = n - g^2$ *denotes the doubly degenerate eigenvalue of the Rabi Hamiltonian. Then, there is a non-zero entire function* $C(z)$ *such that the following holds.*

$$\prod_{\lambda \in \mathrm{Spec}(H_{\mathrm{Rabi}})} (z - \lambda - g^2) = C(z)\Gamma(z)^{-2} G_+(z) G_-(z) \prod_{E_n^{\mathrm{deg}} \in \mathrm{Spec}(H_{\mathrm{Rabi}})} (z - n)^2.$$

Remark 2 It is important to study common zeros of the polynomials $P_n^{(n)}(x, \Delta^2)$. Also, the proof should be done by an analytic continuation of the (Hurwitz) spectral zeta function of the Rabi model in an explicit manner.

4 Lie Algebraic Description

To draw the picture more precisely we recall the representation theoretic setting. Let H, E and F be the standard generators of $\mathfrak{sl}_2$ defined by

$$H = \begin{bmatrix} 1 & 0 \\ 0 & -1 \end{bmatrix}, \quad E = \begin{bmatrix} 0 & 1 \\ 0 & 0 \end{bmatrix}, \quad F = \begin{bmatrix} 0 & 0 \\ 1 & 0 \end{bmatrix}.$$

They satisfy the commutation relations

$$[H, E] = 2E, \ [H, F] = -2F, \ [E, F] = H.$$

For the triplet $(\kappa, \varepsilon, \nu) \in \mathbb{R}^3_{>0}$, define a second order element $\mathscr{R}$ of the universal enveloping algebra $\mathscr{U}(\mathfrak{sl}_2)$ of $\mathfrak{sl}_2$ by

$$\mathscr{R} := \frac{2}{\sinh 2\kappa} \Big\{ \Big[(\sinh 2\kappa)(E - F) - (\cosh 2\kappa) H + \nu \Big] (H - \nu) + (\varepsilon \nu)^2 \Big\}.$$

Let us consider the representation $(\pi', \mathbb{C}[y])$ of $\mathfrak{sl}_2$ given by

$$\pi'(H) = y\partial_y + 1/2, \ \pi'(E) = y^2/2, \ \pi'(F) = -\partial_y^2/2.$$

Define an inner product on $\mathbb{C}[y]$ by $(f, g)_F = \sqrt{\pi}(f(\partial_y)\bar{g}(y))|_{y=0}$ $(f, g \in \mathbb{C}[y])$. Then $(y^m, y^n)_F = \delta_{m,n}\sqrt{\pi}n!$. If we denote by $\overline{\mathbb{C}[y]}$ the completion of $\mathbb{C}[y]$ w.r.t. this inner product, then it is shown that the representation $(\pi', \overline{\mathbb{C}[y]})$ is unitarily equivalent to the oscillator representation of $\mathfrak{sl}_2$ realized on the Hilbert spaces $L^2(\mathbb{R})$.

The following lemma follows immediately from [23] (Corollary 9 with Lemma 8), which translates the eigenvalue problem of Q into a single differential equation.

Lemma 2 *Assume $\alpha \neq \beta$ ($\alpha\beta > 1$). Determine the triplet $(\kappa, \varepsilon, \nu) \in \mathbb{R}^3_{>0}$ by the formulas*

$$\cosh \kappa = \sqrt{\frac{\alpha\beta}{\alpha\beta - 1}}, \quad \sinh \kappa = \frac{1}{\sqrt{\alpha\beta - 1}}, \quad \varepsilon = \left| \frac{\alpha - \beta}{\alpha + \beta} \right|, \quad \nu = \frac{\alpha + \beta}{2\sqrt{\alpha\beta(\alpha\beta - 1)}} \lambda.$$

Then the eigenvalue problem $Q\varphi = \lambda\varphi$ ($\varphi \in L^2(\mathbb{R}, \mathbb{C}^2)$) *is equivalent to the equation* $\pi'(\mathscr{R})u = 0$ ($u \in \overline{\mathbb{C}[y]}$).

Remark 3 Notice that $\pi'(\mathscr{R})$ is a third order differential operator. The correspondence $\varphi \leftrightarrow u$ in the lemma above can be given explicitly. Remark also that the recurrence equation (or its corresponding continued fraction) in [28] is equivalent to this third order differential equation.

4.1 Intertwiners Arising from Laplace Transforms

In order to obtain a complex analytic picture of the equation $\pi'(\mathscr{R})u = 0$ in Lemma 2 and observe a connection between NcHO and the Rabi model through Heun ODE, we introduce two representations of $\mathfrak{sl}_2$.

Let $a \in \mathbb{N}$. Define first the operator T_a acting on the space of Laurent polynomials $\mathbb{C}[y, y^{-1}]$ (or $y^2\mathbb{C}[y]$) by

$$T_a := -\frac{1}{2}\partial_y^2 + \frac{(a-1)(a-2)}{2} \cdot \frac{1}{y^2}.$$

Define a modified Laplace transform $\mathscr{L}_a$ by

$$(\mathscr{L}_a u)(z) := \int_0^\infty u(yz)e^{-\frac{y^2}{2}} y^{a-1} dy.$$

Then, one finds that

$$(\mathscr{L}_a T_a u)(z) = \left(-\frac{1}{2z}\partial_z + \frac{a-1}{2z^2}\right)(\mathscr{L}_a u)(z) + \frac{1}{2z}u'(0)\delta_{a,1} - \frac{a-1}{2z^2}u(0)\delta_{a,2},$$

where $\delta_{a,k} = 1$ when $k = a$ and 0 otherwise. This can be true whenever $u(0)$, $u'(0)$ and $(\mathscr{L}_a u)(z)$ exist.

We now define a representation π_a' of $\mathfrak{sl}_2$ on $y^{a-1}\mathbb{C}[y]$ by

$$\pi_a'(H) = \pi'(H), \ \pi_a'(E) = \pi'(E), \ \pi_a'(F) = T_a = \pi'(F) + \frac{(a-1)(a-2)}{2} \cdot \frac{1}{y^2}.$$

Moreover, introduce another representation of $\mathfrak{sl}_2$ on $\mathbb{C}[z, z^{-1}]$ by

$$\varpi_a(H) = z\partial_z + \frac{1}{2}, \ \varpi_a(E) = \frac{1}{2}z^2(z\partial_z + a), \ \varpi_a(F) = -\frac{1}{2z}\partial_z + \frac{a-1}{2z^2}.$$

Then one easily verifies the following.

Lemma 3 *Let $a \neq 1, 2$. Then one has $\mathscr{L}_a \pi'_a(X) = \varpi_a(X)\mathscr{L}_a$ $(X \in \mathfrak{sl}_2)$. Furthermore, when $a = 1$ (resp. $a = 2$) the restriction of $\mathscr{L}_1$ (resp. $\mathscr{L}_2$) to the space of even (resp. odd) functions turns out to be an intertwiner between two representations $\pi'(= \pi'_1)$ (resp. $= \pi'_2$)) and ϖ_1 (resp. ϖ_2).*

Remark 4 Observe that $\mathscr{L}_a$ defines an isometry. For instance, assume $a = 1$. If $u(y) = \sum_{n=0}^{N} u_n y^n \in \mathbb{C}[y]$ then $(\mathscr{L}_1 u)(z) = \frac{1}{\sqrt{2}} \sum_{n=0}^{N} u_n \Gamma(\frac{n+1}{2})(\sqrt{2}z)^n$. Moreover, if one defines the inner product in z-space such that $\{z^n \mid n \in \mathbb{N}\}$ forms an orthogonal basis and $(z^n, z^n)_1 = \frac{2\Gamma(\frac{n}{2}+1)}{\Gamma\left(\frac{n+1}{2}\right)}$, then $\mathscr{L}_1$ is an isometry. The others are similar.

Since $\varpi_a(E)z^{-a} = 0$, one has the following second equivalence: the representation $(\pi'_a,\ y^{2-a}\mathbb{C}[y^2])$ can be considered as the Langlans quotient of the representations $(\varpi_a,\ \mathbb{C}[z^2, z^{-2}])$ or $(\varpi_a,\ z\mathbb{C}[z^2, z^{-2}])$ depending on the parity of a.

Lemma 4 *The operator $\mathscr{L}_a$ gives the equivalence of irreducible modules of $\mathfrak{sl}_2$:*

$$(\pi'_a,\ y^{a-1}\mathbb{C}[y^2]) \cong (\varpi_a,\ z^{a-1}\mathbb{C}[z^2]),$$
$$(\pi'_a,\ y^{2-a}\mathbb{C}[y^2]) \cong (\varpi_a,\ z^{a}\mathbb{C}[z^2, z^{-2}]/z^{-a}\mathbb{C}[z^{-2}]).$$

Moreover, the Casimir operator $Z_C := 4EF + H^2 - 2H \in \mathscr{Z}\mathscr{U}(\mathfrak{sl}_2)$ takes the value $(a-1)(a-2) - \frac{3}{4}$ in both representations $(\pi'_a,\ y^{a-1}\mathbb{C}[y^2])$ and $(\pi'_a,\ y^{2-a}\mathbb{C}[y^2])$.

Remark 5 There is a symmetry $a \leftrightarrow 3 - a$ for π'_a. Actually, when $a \notin \mathbb{Z}$, there is a equivalence between two representations π'_a and π'_{3-a} in a suitable setting.

4.2 Heun Differential Operators

In this section, we follow the results from [34]. Recall the operator $\mathscr{R} \in \mathscr{U}(\mathfrak{sl}_2)$. Then, one observes

$$\begin{aligned}\varpi_a(\mathscr{R}) =& \Big\{(z^2 + z^{-2} - 2\coth 2\kappa)\left(\theta_z + \frac{1}{2}\right) \\ &+ (a - \frac{1}{2})(z^2 - z^{-2}) + \frac{2\nu}{\sinh 2\kappa}\Big\}(\theta_z + \frac{1}{2} - \nu) + \frac{2(\varepsilon\nu)^2}{\sinh 2\kappa},\end{aligned}$$

where $\theta_z = z\partial_z$. Hence, conjugating by z^{a-1} one obtains the following lemma.

Lemma 5 *For each integer a one has*

$$z^{-a+1}\varpi_a(\mathscr{R})z^{a-1} = \Big\{(z^2+z^{-2}-2\coth 2\kappa)(\theta_z+a-\frac{1}{2}) + (a-\frac{1}{2})(z^2-z^{-2}) + \frac{2\nu}{\sinh 2\kappa}\Big\}(\theta_z+a-\frac{1}{2}-\nu) + \frac{2(\varepsilon\nu)^2}{\sinh 2\kappa}.$$

Furthermore, notice that the operators $\varpi_a(H)$, $\varpi_a(E)$ and $\varpi_a(F)$ are invariant under the symmetry $z \to -z$. This implies that the $\varpi_a(\mathscr{R})$ can be expressed in terms of the variable z^2. We therefore put $w := z^2\coth\kappa$. Using $z\partial_z = 2w\partial_w$ and the relations

$$\begin{aligned} z^2+z^{-2}-2\coth 2\kappa &= (\tanh\kappa)w^{-1}(w-1)(w-\coth^2\kappa),\\ z^2-z^{-2} &= (\tanh\kappa)w^{-1}(w^2-\coth^2\kappa),\\ 2/\sinh 2\kappa &= (\tanh\kappa)(\coth^2\kappa-1), \end{aligned}$$

factoring out the leading coefficient of $\varpi_a(\mathscr{R})$ in its expression one obtains

Proposition 1 *The following relation holds:*

$$z^{-a+1}\varpi_a(\mathscr{R})z^{a-1} = 4(\tanh\kappa)\,w(w-1)(w-\coth^2\kappa)H^a(w,\partial_w),$$

where $H^a(w,\partial_w)$ is the Heun differential operator given as follows:

$$H^a(w,\partial_w) = \frac{d^2}{dw^2} + \left(\frac{3-2\nu+2a}{4w} + \frac{-1-2\nu+2a}{4(w-1)} + \frac{-1+2\nu+2a}{4(w-\coth^2\kappa)}\right)\frac{d}{dw} + \frac{\frac{1}{2}(a-\frac{1}{2})(a-\frac{1}{2}-\nu)w-q_a}{w(w-1)(w-\coth^2\kappa)}.$$

Here the accessory parameter q_a is given by

$$q_a = \Big\{-(a-\frac{1}{2}-\nu)^2+(\varepsilon\nu)^2\Big\}(\coth^2\kappa-1) - 2\left(a-\frac{1}{2}\right)\left(a-\frac{1}{2}-\nu\right).$$

5 Heun Operators' Description for NcHO

The equivalence between the spectral problem of Q and the existence/non-existence of holomorphic solutions of Heun ODE's in a certain complex domain is described in [23] for odd parity and in [34] for even parity. The proof follows from the following quasi-intertwining property of the operator $\mathscr{L}_j$ resulted from Lemma 3 and the realization of the representation ϖ_j.

Proposition 2 *The element $\mathscr{R} \in \mathscr{U}(\mathfrak{sl}_2)$ satisfies the following equations:*

$$(\mathscr{L}_1\pi'(\mathscr{R})u)(z) = \varpi_1(\mathscr{R})(\mathscr{L}_1u)(z) + (\nu - \tfrac{3}{2})u'(0)z^{-1},$$
$$(\mathscr{L}_2\pi'(\mathscr{R})u)(z) = \varpi_2(\mathscr{R})(\mathscr{L}_2u)(z) - (\nu - \tfrac{1}{2})u(0)z^{-2}.$$

In particular, the eigenvalue problem $Q\varphi = \lambda\varphi$ for the even and odd case is respectively equivalent to the equation

$$\varpi_1(\mathscr{R})(\mathscr{L}_1u)(z) = 0 \text{ (the even case) } \quad \text{and} \quad \varpi_2(\mathscr{R})(\mathscr{L}_2u)(z) = 0 \text{ (the odd case).}$$

Noting that, for instance the even case,

$$\varpi_1(\mathscr{R}) = 4(\tanh\kappa)\, w(w-1)(w-\alpha\beta)H_\lambda^+(w, \partial_w),$$

one has the following by Proposition 2 ([34]). The odd case was obtained in [23].

Theorem 1 *There exist linear bijections:*

$$\text{Even}: \{\varphi \in L^2(\mathbb{R}, \mathbb{C}^2) \mid Q\varphi = \lambda\varphi,\ \varphi(-x) = \varphi(x)\} \xrightarrow{\sim} \{f \in \mathscr{O}(\Omega) \mid H_\lambda^+ f = 0\},$$
$$\text{Odd}: \{\varphi \in L^2(\mathbb{R}, \mathbb{C}^2) \mid Q\varphi = \lambda\varphi,\ \varphi(-x) = -\varphi(x)\} \xrightarrow{\sim} \{f \in \mathscr{O}(\Omega) \mid H_\lambda^- f = 0\},$$

where Ω is a simply-connected domain in $\mathbb{C}$ (w-space) such that $0, 1 \in \Omega$ while $\alpha\beta \notin \Omega$, $\mathscr{O}(\Omega)$ denotes the set of holomorphic functions on Ω, and $H_\lambda^\pm = H_\lambda^+(w, \partial_w)$ are the Heun ordinary differential operators given respectively by

$$H_\lambda^+(w, \partial_w) := \frac{\mathrm{d}^2}{\mathrm{d}w^2} + \left(\frac{\frac{1}{2} - p}{w} + \frac{-\frac{1}{2} - p}{w-1} + \frac{p+1}{w - \alpha\beta}\right)\frac{\mathrm{d}}{\mathrm{d}w} + \frac{-\frac{1}{2}(p+\frac{1}{2})w - q^+}{w(w-1)(w-\alpha\beta)},$$
$$H_\lambda^-(w, \partial_w) := \frac{\mathrm{d}^2}{\mathrm{d}w^2} + \left(\frac{1-p}{w} + \frac{-p}{w-1} + \frac{p+\frac{3}{2}}{w-\alpha\beta}\right)\frac{\mathrm{d}}{\mathrm{d}w} + \frac{-\frac{3}{2}pw - q^-}{w(w-1)(w-\alpha\beta)}.$$

Here $p = \frac{2\nu-3}{4}$ with $\nu = \frac{\alpha+\beta}{2\sqrt{\alpha\beta(\alpha\beta-1)}}\lambda$. The accessory parameters $q^\pm = q^\pm(\lambda, \alpha, \beta)$ can be explicitly expressed by the parameters α, β and eigenvalue λ [34].

Remark 6 The modified Laplace transform $\hat{u}(= \mathscr{L}_2u)$ in [23] defines the intertwiner when restricting to the space of odd functions but does not for the even case.

6 Capturing the Rabi Model by $\mathscr{R}$

In this section, employing the standard confluence process of Heun equations, we observe that the Rabi model can be obtained from $\mathscr{R} \in \mathscr{U}(\mathfrak{sl}_2)$ by a suitable choice of a triple $(\kappa, \varepsilon, \nu) \in \mathbb{R}^3$. In the sequel, we assume $a \in \mathbb{R}$, not necessarily an integer.

6.1 Confluent Heun's Equation Derived from Rabi's Model

The Schrödinger equation $H_{\text{Rabi}}\varphi = E\varphi$ of the quantum Rabi model is reduced to the following second order differential equation (see e.g. [34, 38]):

$$\frac{\mathrm{d}^2 f}{\mathrm{d}z^2} + p(z)\frac{\mathrm{d}f}{\mathrm{d}z} + q(z)f = 0,$$

where

$$p(z) = \frac{(1-2E-2g^2)z - g}{z^2 - g^2}, \quad q(z) = \frac{-g^2z^2 + gz + E^2 - g^2 - \Delta^2}{z^2 - g^2}.$$

Write $f(z) = e^{-gz}\phi(x)$, where $x = (g+z)/2g$. Substituting f into the equation above, one finds that the function ϕ satisfies the following confluent Heun equation (by a similar calculation in [38]). Actually, one has $H_1^{\text{Rabi}}\phi = 0$, where

$$H_1^{\text{Rabi}} := \frac{\mathrm{d}^2}{\mathrm{d}x^2} + \left\{ -4g^2 + \frac{1-(E+g^2)}{x} + \frac{1-(E+g^2+1)}{x-1} \right\}\frac{\mathrm{d}}{\mathrm{d}x} + \frac{4g^2(E+g^2)x + \mu}{x(x-1)}$$

with the accessory parameter $\mu = (E+g^2)^2 - 4g^2(E+g^2) - \Delta^2$.

Remark 7 Setting $f(z) = e^{gz}\phi(x)$, where $x = (g-z)/2g$, one obtains another equation $H_2^{\text{Rabi}}\phi = 0$. Here

$$H_2^{\text{Rabi}} := \frac{\mathrm{d}^2}{\mathrm{d}x^2} + \left\{ -4g^2 + \frac{1-(E+g^2+1)}{x} + \frac{1-(E+g^2)}{x-1} \right\}\frac{\mathrm{d}}{\mathrm{d}x} + \frac{4g^2(E+g^2-1)x + \mu}{x(x-1)}.$$

6.2 Confluence Process of the Heun Equation

Put $t = \coth^2\kappa (> 1)$. The Heun operator $H^a(w, \partial_w)$ derived from $\varpi_a(\mathscr{R})$ is given by

$$\begin{aligned} H^a(w, \partial_w) =& \frac{\mathrm{d}^2}{\mathrm{d}w^2} + \left(\frac{3-2\nu+2a}{4w} + \frac{-1-2\nu+2a}{4(w-1)} + \frac{-1+2\nu+2a}{4(w-t)} \right)\frac{\mathrm{d}}{\mathrm{d}w} \\ &+ \frac{\frac{1}{2}(a-\frac{1}{2})(a-\frac{1}{2}-\nu)w - q_a}{w(w-1)(w-t)}. \end{aligned}$$

The corresponding generalized Riemann scheme [32] is expressed as

$$\begin{pmatrix} 1 & 1 & 1 & 1 & \\ 0 & 1 & t & \infty & ; w \ q_a \\ 0 & 0 & 0 & a-\frac{1}{2} & \\ \frac{1+2\nu-2a}{4} & \frac{5+2\nu-2a}{4} & \frac{5-2\nu-2a}{4} & \frac{-1-2\nu+2a}{4} & \end{pmatrix}.$$

Here the first line indicates the s-rank of each singularity. Replace a (resp. ν) by $a+p$ (resp. $\nu+p$) in the expression of $H^a(w,\partial_w)$ above. It follows then that

$$A := \frac{1}{4}(-1-2\nu+2a),\ B := a+p+\frac{1}{2},\ C := \frac{1}{4}(3-2\nu+2a) = 1+A,\ D := A.$$

Write it as

$$\begin{aligned} w(w-1)(w-t)H^a(w,\partial_w) &= w(w-1)(w-t)\partial_w^2 \\ &\quad + \Big[C(w-1)(w-t) + Dw(w-t) \\ &\qquad + (A+B+1-C-D)w(w-1)\Big]\partial_w + ABw - q_a. \end{aligned}$$

Consider a confluence process of the singular points at $w=t$ and ∞ (Table 3.1.2 in [32]). The process is given as $t := \rho^{-1}$, $B := r\rho^{-1}$ and $\rho \to 0$ (equivalently $p \to \infty$):

$$\begin{aligned} &-\lim_{\rho\to 0} w(w-1)(w-t)\rho H^a(w,\partial_w) \\ &= w(w-1)\partial_w^2 + \big[C(w-1) + Dw - rw(w-1)\big] - rAw + \lim_{\rho\to 0}\rho q_a. \end{aligned}$$

Now we take $\varepsilon = k\rho$ for some constraint k. Then one has a confluent Heun equation.

$$\frac{d^2\phi}{dw^2} + \Big[-r + \frac{1+A}{w} + \frac{A}{w-1}\Big]\frac{d\phi}{dw} + \frac{-rAw-(2A)^2-4A+k^2}{w(w-1)}\phi = 0.$$

Notice that $w=\infty$ is an irregular singularity with s-rank 2 (see e.g. [32]). Compare this with the confluent Heun operator H_1^{Rabi} for the Rabi model. Then, taking $r = 4g^2$, $A = -(E+g^2)$ with a suitable choice of k in this equation gives the latter.

Remark 8 Let $K = \begin{bmatrix} 0 & 1 \\ 1 & 0 \end{bmatrix}$. Similarly to Lemma 2, one can show that the eigenvalue problem $KQK\varphi = \lambda\varphi$ is equivalent to the equation $\pi'(\tilde{\mathscr{R}})u = 0$, where

$$\tilde{\mathscr{R}} := \frac{2}{\sinh 2\kappa}\Big\{(H-\nu)\Big[(\sinh 2\kappa)(E-F) - (\cosh 2\kappa)H + \nu\Big] + (\varepsilon\nu)^2\Big\} \in \mathscr{U}(\mathfrak{sl}_2).$$

Then, a confluence procedure for $\varpi_a(\tilde{\mathscr{R}})$ similarly to that of $\varpi_a(\mathscr{R})$ yields H_2^{Rabi} in Remark 7. Moreover, one can find an element $\mathscr{K}$ (resp. $\tilde{\mathscr{K}}$) $\in \mathscr{U}(\mathfrak{sl}_2)$ of order two such that $\varpi_a(\mathscr{K})$ (resp. $\varpi_a(\tilde{\mathscr{K}})$) essentially provides H_1^{Rabi} (reps. H_2^{Rabi}) [36].

7 Conclusion

So far, even well-developed Lie theory has never contributed the spectral problems of quantum interaction models in a definite way. One of the simplest reasons is obviously the absence of the creation/annihilation operators (see [27]). Hence the observation in this article might provide a new insight. Also, probably, as many of physicists may think there is an important questions: What are the detailed meaning of the exact solvability of [3], if it really differs from integrability? At the same time, according to [33], no second operator—integral of motion—exists. Therefore we should explore a fundamentally new category of exact solvability. As we have seen, the NcHO can be a "mother" of the Rabi model through the confluence procedure, whence one may expect to obtain an unified understanding of some sort of quantum interaction models using (in general, much higher $\mathbb{R}$-rank) Lie groups/algebras.

References

1. B.C. Berndt, Generalized Eisenstein series and modified Dedekind sums. J. Reine Angew. Math. **272**, 182–193 (1974)
2. F. Beukers, in *Irrationality of π^2, periods of an elliptic curve and $\Gamma_1(5)$*, Diophantine approximations and transcendental numbers (Luminy 1982), Progr. Math. vol. 31 (Birkhäuser, Boston, 1983), pp. 47–66
3. D. Braak, On the integrability of the Rabi model. Phys. Rev. Lett. **107**, 100401–100404 (2011)
4. J. Casanova, G. Romero, I. Lizuain, J.J. Garca-Ripoll, E. Solano, Deep strong coupling regime of the Jaynes-Cummings model. Phys. Rev. Lett. **105**, 263603 (2010)
5. R.C. Gunning, The Eichler cohomology groups and automorphic forms. Trans. Am. Math. Soc. **100**, 44–62 (1961)
6. S. Haroche, J.M. Raimond, *Exploring Quantum. Atoms, Cavities, and Photons* (Oxford University Press, Oxford, 2008)
7. M. Hirokawa, F. Hiroshima, Absence of energy level crossing for the ground state energy of the Rabi model. Comm. Stoch. Anal. (To appear)
8. F. Hiroshima, I. Sasaki, Spectral Analysis of Non-commutative Harmonic Oscillators: The Lowest Eigenvalue and No Crossing. J. Math. Anal. Appl. **105**, 595–609 (2014)
9. F. Hiroshima, I. Sasaki, Multiplicity of the lowest eigenvalue of non-commutative harmonic oscillators. Kyushu J. Math. **67**, 355–366 (2013)
10. R. Howe, E.C. Tan, in *Non-abelian Harmonic Analysis. Applications of $SL(2,\mathbb{R})$* (Springer, Berlin, 1992)
11. T. Ichinose, M. Wakayama, Zeta functions for the spectrum of the non-commutative harmonic oscillators. Commun. Math. Phys. **258**, 697–739 (2005)
12. T. Ichinose, M. Wakayama, Special values of the spectral zeta function of the non-commutative harmonic oscillator and confluent Heun equations. Kyushu J. Math. **59**, 39–100 (2005)

13. T. Ichinose, M. Wakayama, On the spectral zeta function for the noncommutative harmonic oscillator. Rep. Math. Phys. **59**, 421–432 (2007)
14. K. Kimoto, M. Wakayama, Elliptic curves arising from the spectral zeta function for non-commutative harmonic oscillators and $\Gamma_0(4)$-modular forms, in *Proceedings of the Conference on L-functions*, ed. by L. Weng, M. Kaneko (World Scientific, Singapore, 2007) pp. 201–218
15. K. Kimoto, M. Wakayama, Residual modular forms and Eichler cohomology groups arising from non-commutative harmonic oscillators (2014) (preprint)
16. K. Kimoto, M. Wakayama, Spectrum of non-commutative harmonic oscillators and residual modular forms, in *Noncommutative Geometry and Physics* ed. by G. Dito, H. Moriyoshi, T. Natsume, S. Watamura (World Scientific, Singapore, 2012), pp. 237–267
17. K. Kimoto, M. Wakayama, Remarks on zeta regularized products. Int. Math. Res. Not. **17**, 855–875 (2004)
18. K. Kimoto, M. Wakayama, Apéry-like numbers arising from special values of spectral zeta functions for non-commutative harmonic oscillators. Kyushu J. Math. **60**, 383–404 (2006)
19. N. Kurokawa, M. Wakayama, Zeta regularizations. Acta Appl. Math. **81**, 147–166 (2004)
20. M. Kus, On the spectrum of a two-level system. J. Math. Phys. **26**, 2792–2795 (1985)
21. A. Moroz, On the spectrum of a class of quantum models. Lett. J. Exploring Frontiers Phys. **100**, 60010–60015 (2012)
22. K. Nagatou, M.T. Nakao, M. Wakayama, Verified numerical computations for eigen-values of non-commutative harmonic oscillators. Numer. Funct. Anal. Optim. **23**, 633–650 (2002)
23. H. Ochiai, Non-commutative harmonic oscillators and Fuchsian ordinary differential operators. Commun. Math. Phys. **217**, 357–373 (2001)
24. H. Ochiai, Non-commutative harmonic oscillators and the connection problem for the Heun differential equation. Lett. Math. Phys. **70**, 133–139 (2004)
25. H. Ochiai, A special value of the spectral zeta function of the non-commutative harmonic oscillators. Ramanujan J. **15**, 31–36 (2008)
26. A. Parmeggiani, in *Spectral Theory of Non-commutative Harmonic Oscillators: An Introduction*. Lecture Notes in Mathematics, vol. 1992. (Springer, Berlin, 2010)
27. A. Parmeggiani, Non-commutative harmonic oscillators and related problems. Milan J. Math. (2014). doi:10.1007/s00032-014-0220-z
28. A. Parmeggiani, M. Wakayama, Non-commutative harmonic oscillators-I, II, Corrigenda and remarks to I. Forum. Math. **14**, 539–604, 669–690 (2002) [ibid **15**, 955–963 (2003)]
29. A. Parmeggiani, M. Wakayama, Oscillator representations and systems of ordinary differential equations. Proc. Nat. Acad. Sci. USA **98**, 26–30 (2001)
30. A. Parmeggiani, On the spectrum of certain non-commutative harmonic oscillators and semi-classical analysis. Commun. Math. Phys. **279**, 285–308 (2008)
31. I.I. Rabi, Space quantization in a gyrating magnetic field. Phys. Rev. **51**, 652–654 (1937)
32. S.Y. Slavyanov, W. Lay, *A Unified Theory Based on Singularities*, Oxford Mathematical Monographs (Oxford University Press, Oxford, 2000)
33. E. Solano, Viewpoint: the dialogue between quantum light and matter. Physics **4**, 68 (2011)
34. M. Wakayama, Equivalence between the eigenvalue problem of non-commutative harmonic oscillators and existence of holomorphic solutions of Heun differential equations, eigenstates degeneration and the Rabi model, Kyushu University (2014) (preprint)
35. M. Wakayama, Simplicity of the lowest eigenvalue of non-commutative harmonic oscillators and the Riemann scheme of a certain Heun's differential equation. Proc. Jpn. Acad. Ser. A **89**, 69–73 (2013)
36. M. Wakayama, T. Yamazaki, The quantum Rabi model and Lie algebra representations of $\mathfrak{sl}_2$, Kyushu University (2014) (preprint)
37. D. Zagier, in *Integral Solutions of Apéry-Like Recurrence Equations, Groups and Symmetries*, CRM Proceedings and Lecture Notes, vol. 47 (American Mathematical Society, Providence 2009), pp. 349–366

38. H. Zhong, Q. Xie, M.T. Batchelor, C. Lee, Analytical eigenstates for the quantum Rabi model (arXiv:1305.6782v2 [quant-ph]) (2013) (preprint)
39. X. Zhu, S. Saito, A. Kemp, K. Kakuyanagi, S. Karimoto, H. Nakano, W.J. Munro, Y. Tokura, M.S. Everitt, K. Nemoto, M. Kasu, N. Mizuochi, K. Semba, Coherent coupling of a superconducting flux qubit to an electron spin ensemble in diamond. Nature **478**, 221–224 (2011)

Introduction to Public-Key Cryptography

Tsuyoshi Takagi

Abstract Cryptography was once considered to be a means of maintaining secrecy of communications only in military affairs and diplomacy. However, today, modern cryptography is used for various purposes in familiar circumstances. Public-key cryptography is a key technology of modern society; it is used for personal authentication, electronic commerce on the Internet, copyright protection of DVDs, and so on. In particular, the RSA public-key cryptosystem, which was proposed more than 30 years ago, has become the de facto standard of cryptographic software since the spread of the Internet in the 1990s. Another technology, called elliptic curve cryptography, was proposed in 1985. It can perform arithmetic processing at high speed, and since the beginning of the 2000s, it has been implemented in devices such as DVD players and personal digital assistants. Pairing-based cryptography, first proposed in 2000, can be incorporated in security technologies that are not practical with the previous public-key cryptographies. It is actively studied by various organizations around the world. In this chapter, we explain the basic mathematics and security evaluations of public-key cryptography.

Keywords Bilinear pairing · Public-key cryptography · Discrete logarithm problem · Elliptic curve · Factoring

1 Introduction

Public-key cryptography is one of the key technologies for maintaining information security in the world today. For example, secure socket layer (SSL), an encryption system incorporating public-key cryptography, is used for securely transmitting secret data such as credit card numbers.

T. Takagi (✉)
Institute of Mathematics for Industry, Kyushu University, 744, Motooka, Nishiku, Fukuoka 819-0395, Japan
e-mail: takagi@imi.kyushu-u.ac.jp

R. Nishii et al. (eds.), *A Mathematical Approach to Research Problems of Science and Technology*, Mathematics for Industry 5,
DOI: 10.1007/978-4-431-55060-0_3,

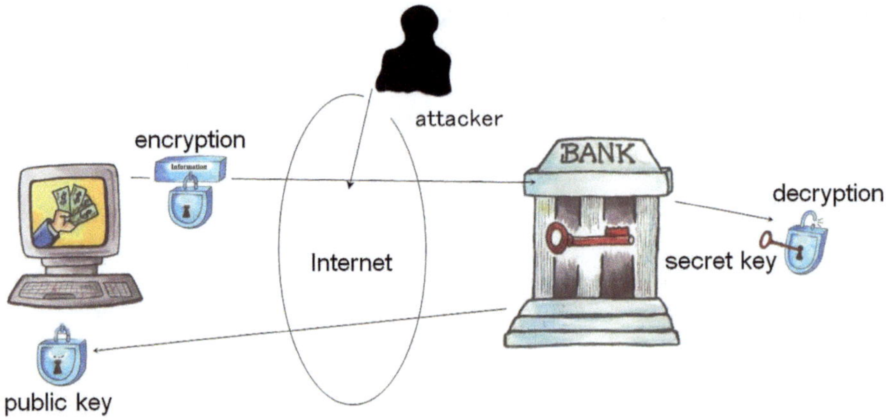

Fig. 1 Internet banking using public-key cryptography

Figure 1 sketches the public-key cryptography used in internet banking. Two different keys are generated, namely a secret key and the corresponding public key. The secret key is securely stored in the receiver (server), and the public key is visible to the whole network. If we want to send a credit-card number to the server, we encrypt it with the public key on the network, and send its ciphertext to the server via the Internet. Thanks to the encryption, potential attackers cannot learn the credit-card number. At the same time, the server, which has the secret key, is able to decrypt the ciphertext. Note that only the public key is required to encrypt messages, and thus, public-key cryptography does not need to transmit the secret key of the server.

The most frequently used public-key cryptography is the RSA cryptosystem, whose security is based on the hardness of the factoring problem [21]. On the other hand, elliptic curve cryptography (ECC) relies on the intractability of the discrete logarithm problem on elliptic curves over finite fields [16, 18]. ECC has equivalent security to that of an RSA cryptosystem, but with a shorter key size. Recently, pairing-based cryptography (PBC) has attracted the attention of researchers and organizations concerned with cryptography. PBC enables many novel cryptographic protocols, such as ID-based encryption [4, 22], which cannot be efficiently constructed using RSA or ECC. In particular, PBC involves a pairing that reduces the discrete logarithm problem on elliptic curves to one over finite fields.

In this chapter, we give a short overview of the development of public-key cryptography, starting with the RSA cryptosystem and moving on to ECC and PBC. We explain how the basic mathematical structure influences the security and efficiency of public-key cryptography.

2 RSA Cryptosystem

The RSA cryptosystem was proposed by Rivest, Shamir, and Adleman in 1977 [21]. In this cryptosystem, a one-way function with a trapdoor is constructed by using the basic mathematical property of the divisibility of integers.

For given integers a, b, there are uniquely determined integers q, r such that $a = bq + r$ and $0 \leq r < b$. The integers q, r are called the quotient q and remainder r of dividing a by b, and we write $r = a \bmod b$. The set $\mathbf{Z}_k = \{0, 1, 2, ..., k-1\}$ of all remainders by an integer k is called the residue ring modulo k. If the greatest common divisor of two integers is one, the two numbers are called relatively prime. We denote by $\mathbf{Z}_k^\times$ the set of all integers in the residue ring $\mathbf{Z}_k$ that are relatively prime to k, and $\mathbf{Z}_k^\times$ forms a group, which is called the multiplicative group modulo k.

The security of the RSA cryptosystem is based on the computational difficulty of factoring integers. The cryptosystem is constructed on the multiplicative group $\mathbf{Z}_n^\times$, where n is the product of two integers p, q of the same size. In the following, we explain the construction of the RSA cryptosystem.

[Key Generation] Let $n = pq$, where p, q are two distinct primes of the same bit-length. Generate integers (e, d) such that $ed = 1 \bmod (p-1)(q-1)$. The public key is (e, n), and the secret key is d.

[Encryption] We choose the a representative of the residue ring $\mathbf{Z}_n$ as $\mathbf{Z}_n = \{0, 1, 2, ..., n-1\}$. We encrypt a message $m \in \mathbf{Z}_n$ with the public key (e, n) by computing $c = m^e \bmod n$. Then c is the ciphertext of m.

[Decryption] For the ciphertext c, we decrypt m with the secret key d by computing $m = c^d \bmod n$.

Here, the order of the multiplicative group $\mathbf{Z}_n^\times$ is $(p-1)(q-1)$, and there is an integer k such that $ed = 1+k(p-1)(q-1)$. Therefore, as a result of Euler's theorem, we can decrypt m by computing $c^d = m^{ed} = m^{1+k(p-1)(q-1)} = m \bmod n$ for $\gcd(m, n) = 1$. In the case of $\gcd(m, n) \neq 1$, we have $m = 0 \bmod p$, $m = 0 \bmod q$, or $m = 0$, and the message can be recovered with the same decryption.

2.1 *Security of the RSA Cryptosystem*

The security of the RSA cryptosystem is based on the hardness of the factorization problem. If the public key $n = pq$ can be easily factored, the secret key d can be easily computed by $d = e^{-1} \bmod (p-1)(q-1)$, and the cryptosystem would be completely broken. Here, the asymptotically fastest algorithm for factoring n is the number field sieve (NFS) [17], which requires a subexponential time in the bit length of n, i.e., $O(\exp(((64/9)^{1/3} + o(1))(\log n)^{1/3}(\log\log n)^{2/3}))$, where $o(1) \to 0$ for $n \to \infty$.

The secure bit length of n in the RSA cryptosystem depends on the speed of the factoring algorithm used against it as well as the capabilities of software and hardware technology of computers. Currently, it is considered infeasible to factor more than

1024 bits, but a longer key size will be needed in the near future. In this regard, the CRYPTREC project has reported on secure key sizes for the RSA cryptosystem [10].

Next, we explain the security of the RSA cryptosystem in the sense of one-wayness. In the following, we denote by RSA_ℓ the set of all public keys of an RSA cryptosystem of ℓ bits. Let $\mathbf{N}$ and $\mathbf{R}$ be the set of natural and real numbers, respectively. A function $\epsilon(\ell) : \mathbf{N} \to \mathbf{R}$ is called negligible, for any integer $\alpha > 0$, there exists an integer $\ell_\alpha > 0$ that satisfies $\epsilon(\ell) < 1/\ell^\alpha$ for ℓ with $\ell > \ell_\alpha$.

Let us consider an algorithm $\mathcal{A}$ that computes a message m from a randomly chosen ciphertext c and public key (e, n) in RSA_ℓ. Given any polynomial-time algorithm $\mathcal{A}$ for the input size ℓ, if the probability

$$Pr\left[\begin{array}{c}(e,n) \leftarrow RSA_\ell,\ m \leftarrow \mathbf{Z}_n, \\ c \leftarrow m^e \bmod n : \mathcal{A}(e,n,c) = m\end{array}\right] < \epsilon(\ell)$$

is negligible, the RSA cryptosystem is called secure in the sense of one-wayness.

The one-wayness will be compromised if we can compute the eth root $m = c^{1/e} \bmod n$ from the public key (e, n) and ciphertext c. However, it is an open problem as to whether the one-wayness of the RSA cryptosystem can be compromised without having to factor the public key n [7].

3 Elliptic Curve Cryptography

Elliptic curve cryptography was independently proposed by Miller [18] and Koblitz [16] in 1985. This cryptosystem is constructed by using the elliptic curve over finite field.

3.1 Addition Formulae of Elliptic Curves

For a prime number $p > 3$, the elliptic curve over a finite field $GF(p)$ is

$$E(a,b,p) := \{(x,y) \in GF(p) \times GF(p) \mid y^2 = x^3 + ax + b\} \cup \{\infty\} \qquad (1)$$

where $a, b \in GF(p)$ satisfies $4a^3 + 27b^2 \neq 0$ and ∞ is the point at infinity. Equation (1) is called the Weierstrass form of the elliptic curve. $E(a,b,p)$ forms an additive group with the zero element ∞, and the inverse of $P = (x, y)$ is given by $-P = (x, -y)$. From Hasse-Weil's theorem, the order of $E(a,b,p)$ is $\#E(a,b,p) = p + 1 - t$, with $|t| \leq 2\sqrt{p}$, where t is the trace of the Frobenius map of $E(a,b,p)$. Namely, the order of $E(a,b,p)$ is approximately as large as p.

Given two points $P_1 = (x_1, y_1)$, $P_2 = (x_2, y_2)$ on elliptic curves $E(a,b,p)$ which are different from ∞, the addition $P_1 + P_2 = (x', y')$ can be computed by

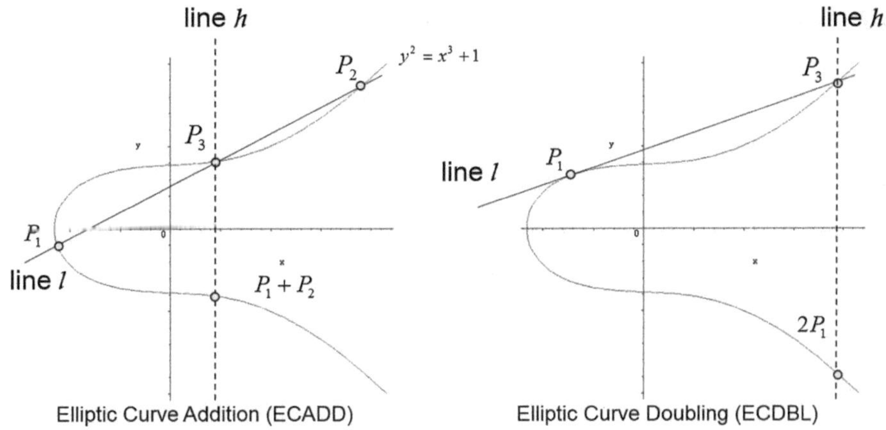

Fig. 2 Addition and doubling on elliptic curves

$$x' = \lambda^2 - x_1 - x_2, \quad y' = \lambda(x_1 - x') - y_1,$$

$$\lambda = \begin{cases} (y_2 - y_1)/(x_2 - x_1) \text{ for } P_1 \neq \pm P_2 \\ (3x_1^2 + a)/(2y_1) \text{ for } P_1 = P_2. \end{cases} \tag{2}$$

Addition $P_1 + P_2$, $(P_1 \neq \pm P_2)$ and doubling $2P_1$ on an elliptic curve $E(a, b, p)$ are, respectively, denoted by ECADD and ECDBL.

Figure 1 illustrates ECADD and ECDBL on the elliptic curve $y^2 = x^3 + 1$ defined over a real field. Regarding ECADD, let l be the line that passes both P_1 and P_2. There is a third point P_3 through line l on elliptic curve. Let h be the line through P_3 and the point at infinity ∞. Then h is the vertical line to the x-axis from P_3, and the symmetric point of P_3 to the x-axis on elliptic curve becomes the resulting addition $P_1 + P_2$. In ECDBL, let P_3 be another point on the elliptic curve that is on the tangent line l on point P_1. The doubling $2P_1$ is another crossing point of h that passes through P_3 and the point at infinity ∞.

We usually use elliptic curves of prime order in ECC in order to avoid subgroup attacks. The order $\#E$ of an elliptic curve $E(a, b, p)$ can be efficiently counted for given a, b, p of Eq. (1) by Schoof's algorithm [2, Chap. VII]. For a given order $\#E(a, b, p)$ and characteristic p, we can efficiently find the coefficients $a, b \in GF(p)$ by performing complex multiplication [2, Chap. VIII]. It is possible for many users to use one fixed elliptic curve as a set of system parameters in ECC. For example, SECG (http://www.secg.org/) recommends elliptic curves that are secure against known attacks (Fig. 2).

3.2 Elliptic Curve Cryptography

In the following, we explain about ElGamal encryption based on elliptic curve.

[System Parameter] Given a prime number $p > 3$, generate an elliptic curve $E(a, b, p)$ defined by $a, b \in GF(p)$ of prime order $\#E = \ell$. Let G be a generator of $E(a, b, p)$. All users share a, b, p, ℓ, G as system parameters.

[Key Generation] Let $s \in \mathbf{Z}_\ell$ be the secret key of a user, and let $Q = sG$ be the corresponding public key.

[Encryption] A message M is chosen as a point M on the elliptic curve $E(a, b, p)$. Using the system parameter G, a random integer $r \in \mathbf{Z}_\ell$, and public key Q, we encrypt the message M by computing $C_1 = rG \in E(a, b, p)$, $C_2 = rQ + M \in E(a, b, p)$. Then (C_1, C_2) is the ciphertext of M.

[Decryption] For ciphertext (C_1, C_2), we can decrypt the message M by using secret key s to compute $M = C_2 - sC_1 \in E(a, b, p)$.

Here we can uniquely decrypt the message M due to relationship $C_2 - sC_1 = (rQ + M) - s(rG) = rsG + M - rsG = M$.

The security of ECC relies on the intractability of the discrete logarithm problem on $E(a, b, p)$, wherein one tries to compute the secret key s of $Q = sG$ from the public key Q and system parameter G. No algorithm for solving the discrete logarithm problem on $E(a, b, p)$ in subexponential time in the bit length of p has been found so far. The fastest algorithm currently known is Pollard's ρ method [20], which requires $O(\sqrt{p})$. It is estimated that an ECC of 160 bits p is as secure as an RSA cryptosystem of 1024 bits n. This means that the key length of ECC can be made much shorter than that of RSA and that ECC is suitable for embedded devices that have small memories.

On the other hand, the public key $Q = (x, y)$ used for ECC is generated by $Q = sG$ from the random secret key $s \in \mathbf{Z}_\ell$ and the system parameter G, and thus, $Q = (x, y)$ is a randomly distributed point on the elliptic curve $E(a, b, p)$. An example of a public key $Q = (x, y)$ of 160 bits in hexadecimal is

$x =$ `4A96B568 8EF57328 46646989 68C38BB9 13CBFC82`

$y =$ `23A62855 3168947D 59DCC912 04235137 7AC5FB32`

In this example, the curve parameters are chosen as $p = 2^{160} - 2^{31} - 1, a = -3$, and $b =$ `1C97BEFC 54BD7A8B 65ACF89F 81D4D4AD C565FA45`. The y-coordinate of $Q = (x, y)$ on $E(a, b, p)$ can be computed as $y = (x^3 + ax + b)^{1/2}$, and thus, it is possible to use only the x-coordinate for the public key.

4 Pairing-Based Cryptography

Sakai, Ohgishi, and Kasahara presented a cryptosystem based on pairing at the 2000 Symposium on Cryptography and Information Security, Japan [22]. Since then, many cryptographic pairing-based protocols have been proposed that would have been difficult to construct by using RSA or ECC. These include ID-based encryption [4], keyword searchable encryption [5], and efficient broadcast encryption [6]. Here, we explain the algorithm for computing pairing and ID-based encryption.

4.1 Pairing

We deal with Tate pairing over supersingular elliptic curves defined on a finite field of characteristic $p > 3$ [8]. The details on how to construct pairings over elliptic curves can be found in [11, 25].

For $b = 0, 1$ we define an elliptic curve over a finite field $GF(p)$ as follows:

$$E^b(p) = \{(x, y) \in GF(p) \times GF(p) \mid y^2 = x^3 + (1 - b)x + b\} \cup \{\infty\}.$$

The trace of the Frobenius map of $E^b(p)$ is 0, and the order of the elliptic curve becomes $\#E^b(p) = p + 1$, if $p \equiv 3, 2 \bmod 4$, respectively, holds for $b = 0, 1$. Such elliptic curves are called supersingular, and the following assumes that $E^b(p)$ is supersingular. Let ℓ be a prime number with $\gcd(\ell, p) = 1$ and $\ell | \#E^b(p)$. We usually choose $\#E^b(p)$ to be as large as ℓ in cryptography. From $\ell | (p^2 - 1)$, the extension field $GF(p^2)$ contains the ℓth primitive root of 1. Denote by $E^b(p)[\ell]$ the subgroup of $E^b(p)$ of order ℓ. The Tate pairing is a non-degenerate bilinear pairing map e defined by

$$e : E^b(p)[\ell] \times E^b(p^2)/\ell E^b(p^2) \to GF(p^2)^\times / \left(GF(p^2)^\times\right)^\ell .$$

For a point $P \in E^b(p)$, we define the function $f_P^{(\ell)}(x, y)$ whose divisor $(f_P^{(\ell)})$ is equivalent to $\ell(P) - \ell(\infty)$. The Tate pairing is computed by $e(P, R) = f_P^{(\ell)}(R)$ for a point $R = (x, y) \in E^b(p^2)/\ell E^b(p^2)$.

We choose the basis of $GF(p^2)$ over $GF(p)$ to be $\{1, i\}$ for $p \equiv 3 \bmod 4$, where $i^2 = -1$. In the following, we will consider an elliptic curve with this fixed basis (we can similarly discuss an elliptic curve $E^1(p)$ using a different basis). The distortion map is defined by $\psi(x, y) = (-x, iy) \in E^0(p^2)$ for a point $Q = (x, y) \in E^0(p)$. There exists a point $Q \in E^0(p)[\ell]$ that satisfies $R = \psi(Q)$ for $R \in E^0(p^2)/\ell E^0(p^2)$, and thus, we can define a Tate pairing $e(P, \psi(Q))$ for the points $P, Q \in E^0(p)[\ell]$. In order to decide the value of Tate pairing uniquely in $GF(p^2)^\times / \left(GF(p^2)^\times\right)^\ell$, we define the reduced Tate pairing as $\hat{e}(P, Q) = e(P, \psi(Q))^{(p^2-1)/\ell}$ for $P, Q \in E^0(p)$. The reduced Tate pairing over $E^0(p)$ satisfies the bilinearity condition $\hat{e}(aP, Q) = \hat{e}(P, aQ) = \hat{e}(P, Q)^a$ for an integer a, and thus, it is a non-degenerate symmetric bilinear pairing map.

Next, let us explain the Miller algorithm, which is an efficient algorithm for computing Tate pairings [19]. Let l be a line that passes through points P_1, P_2, and let h be a line through P_3 and the point at infinity. These lines l, h were used for the addition formulae of elliptic curves (Fig. 1) described in the previous section. Let g_l, g_h be linear equations over $GF(p)$ corresponding to lines l, h. For points $P_1, P_2 \in E^b(p)$, function $f_P^{(\ell)}$ has the following relationship [3, Chap. IX]:

$$f_{P_1+P_2}^{(\ell)} = f_{P_1}^{(\ell)} f_{P_2}^{(\ell)} \frac{g_l}{g_h}. \tag{3}$$

The pairing $\hat{e}(P, Q) = f_P^{(\ell)}(\psi(Q))^(p^2 - 1)/\ell$ can be computed by calling this relationship $\mathcal{O}(\log p)$ times by using a binary expansion of $\ell = \sum_{i=0}^{t-1} \ell[i]2^i$, $\ell[t-1] = 1, \ell[i] \in \{0, 1\}$ for $i = 0, 1, ..., t-2$. Algorithm 1 is one for computing the reduced Tate pairing.

%queryPlease check and confirm the inserted opening brace in sentence starting with ?Algorithm 1 is one for computing the reduced Tate pairing..., and amend if necessary.

Algorithm 1: Computation of Tate Pairing for $E^0(p)$

Input: $P = (x_p, y_p), Q = (x_q, y_q) \in E^0(p)[\ell], \ell = \sum_{i=0}^{t-1} \ell[i]2^i, \ell[t-1] = 1$
Output: $\hat{e}(P, Q) \in GF(p^2)^\times/(GF(p^2)^\times)^\ell$
1. $f \leftarrow 1$ and $V \leftarrow P$
2. **for** $i \leftarrow t-2$ **to** 0 **do**
   2.1. Set the lines l and h for ECDBL(T)
   2.2. $f \leftarrow f^2 \frac{g_l(\psi(Q))}{g_h(\psi(Q))}$ in $GF(p^2)$
   2.3. $T \leftarrow$ ECDBL(T) in $E^0(p)$
   2.4. **if** $\ell[i] = 1$ **do**
   2.5. Set the lines l and h for ECADD(T, P)
   2.6. $f \leftarrow f \frac{g_l(\psi(Q))}{g_h(\psi(Q))}$ in $GF(p^2)$
   2.7. $T \leftarrow$ ECADD(T, P) in $E^0(p)$
3. **return** $T^{(p^2-1)/\ell}$

Each step of the second loop in Algorithm 1 can be implemented by arithmetic operations (addition, multiplication, inverse) in a finite field $GF(p)$ and requires $\mathcal{O}((\log p)^2)$ bit operations. The final exponentiation $f^{(p^2-1)/\ell}$ can be computed by in $\mathcal{O}(\log p)$ multiplications of the finite field $GF(p^2)$ by using the binary expansion of $(p^2-1)/\ell$. The total computation cost of the Miller algorithm is $\mathcal{O}((\log p)^3)$ bit operations. Moreover, there are several speed-up methods, such as eliminating the denominator $g_h(\psi(Q))$, reducing the Hamming weight of the binary expansion of ℓ, and using the Jacobian coordinate [15]. In recent implementations, bilinear pairing maps of 80- to 128-bit security can be computed as efficiently as the decryption of an RSA cryptosystem having the same security level [14, 26].

4.2 ID-Based Encryption

The bilinearity of the reduced Tate pairing $\hat{e}$ provides us with an ID-based encryption that cannot be efficiently constructed with an RSA cryptosystem or ECC [4, 22].

[System Parameter] Let p be a prime number, and let $b = 0, 1$. Let P be a generator of the subgroup of prime order ℓ in an elliptic curve $E^b(p)$. We generate a master key $s \in \mathbf{Z}_\ell$, and set $Q = sP$. All users share p, b, ℓ, P, Q as system parameters.

[Key Generation] The user ID is embedded in a point $Q_{ID} \in E^b(p)[\ell]$, and the secret key is computed by $S_{ID} = sQ_{ID}$. It might be possible to delete the master key s after generating the secret key of all users.

[Encryption] Let m be a message $m \in GF(p^2)$. Using the system parameters P, Q, a random integer $r \in \mathbf{Z}_\ell$, and public key Q_{ID}, we encrypt the message m by computing $C_1 = rP \in E^b(p)[\ell]$, $c_2 = m\hat{e}(Q_{ID}, Q)^r \in GF(p^2)$. Then (C_1, c_2) is the ciphertext of message m.

[Decryption] For the ciphertext (C_1, c_2), we can decrypt m with the secret key $S_{ID} \in E^b(p)[\ell]$ by computing $m = c_2\hat{e}(S_{ID}, C_1)^{-1} \in GF(p^2)$.

The bilinearity of $\hat{e}$ ensures the following relationship:

$$c_2\hat{e}(S_{ID}, C_1)^{-1} = m\hat{e}(Q_{ID}, Q)^r\hat{e}(S_{ID}, C_1)^{-1} = m\hat{e}(Q_{ID}, P)^{rs}\hat{e}(Q_{ID}, P)^{-rs} = m,$$

and thus, the message m can be uniquely decrypted.

ID-based encryption generates a secret key S_{ID} for each user ID, namely Q_{ID}. Therefore, the x-coordinate of the public key Q_{ID} can be a freely chosen bit string (for example, takagi@imi.kyushu-u.ac.jp). On the other hand, if we want to realize an ID-based encryption by using ECC, we have to solve a discrete logarithm problem on $E^b(p)[\ell]$ in order to get the secret key s from the system parameter G and $Q_{ID} = sG$.

The security of ID-based encryption is based on the difficulty of the discrete logarithm problem on both the finite field $GF(p^2)$ and elliptic curve $E^b(p)$. Indeed, the problem of finding the master key s from system parameters P and Q wherein $Q = sP$ is equivalent to the discrete logarithm problem on $E^b(p)$. As we stated in the previous section, this problem requires an exponential time $O(\sqrt{\ell})$, and the size of ℓ must be at least 160 bits. Moreover, for any $R \in E^b(p)[\ell]$, we have $\hat{e}(R, Q) = \hat{e}(R, sP) = \hat{e}(R, P)^s$. We can recover the master key s, if the discrete logarithm problem for $\hat{e}(R, Q)$ and $\hat{e}(R, P)$ over $GF(p^2)$ can be efficiently computed. The asymptotically fastest algorithm for solving the discrete logarithm problem on $GF(p^2)$ is the number field sieve (NFS) over finite fields [23]. The asymptotic speed of NFS is estimated to be subexponential in time: $\mathcal{O}(\exp\left(((64/9)^{1/3} + o(1)\right) (\log p^2)^{1/3}(\log\log p^2)^{2/3}\right)$. This asymptotic complexity is as large as that of the number field sieve used for factoring integers, and thus, the length of p^2 should be as large as that of the public key of the RSA cryptosystem.

Many large-scale experiments on PBC (including ECC) have attempted to estimate the maximum bit length for which the discrete logarithm problem can be solved in actual computational environments. The current world records on solving the discrete logarithm problem are 112 bits for an elliptic curve over finite field $GF(p)$, 532 bits for a finite field $GF(p)$, and 923 bits for a finite field $GF(3^n)$ of characteristic three [13]. In light of these figures, we can determine the secure key length of PBC in practical environments by considering the computational limits that attackers are likely to have.

5 Conclusion

This chapter described the development of public-key cryptography, one of the core technologies in the field of information security. In particular, we explained the construction and security of the RSA cryptosystem, ECC, and pairing-based cryptography. The development of PBC has been relatively quick. The first international conference on PBC was held in 2007 [24]. We expect that research on pairing-based cryptography will continue to advance at a rapid pace.

References

1. J.-L. Beuchat, N. Brisebarre, J. Detrey, E. Okamoto, M. Shirase, T. Takagi, Algorithms and arithmetic operators for computing the η_T pairing in characteristic three. IEEE Trans. Comput. **57**(11), 1454–1468 (2008)
2. I. Blake, G. Seroussi, N. Smart, in *Elliptic Curves in Cryptography*, London Mathematical Society Lecture Note Series, vol 265 (Cambridge University Press, Cambridge, 1999)
3. I. Blake, G. Seroussi, N. Smart (eds.), in *Advances in Elliptic Curve Cryptography*, London Mathematical Society Lecture Note Series, vol 317 (Cambridge University Press, Cambridge, 2005)
4. D. Boneh, M. Franklin, Identity based encryption from the Weil pairing. SIAM J. Comput. **32**(3), 586–615 (2003)
5. D. Boneh, G. Di Crescenzo, R. Ostrovsky, G. Persiano, Public key encryption with keyword search, in *Proceedings of EUROCRYPT 2004*. LNCS, vol. 3027 (Springer, Heidelberg, 2004), pp. 506–522
6. D. Boneh, C. Gentry, B. Waters, Collusion resistant broadcast encryption with short ciphertexts and private keys, in *Proceedings of CRYPTO 2005*. LNCS, vol. 3621 (Springer, 2005), pp. 258–275
7. D. Boneh, R. Venkatesan, Breaking RSA may not be equivalent to factoring, *Proceedings of EUROCRYPT'98*. LNCS, vol. 1233 (Springer, 1998), pp. 59–71
8. X. Boyen, L. Martin, in *Identity-Based Cryptography Standard (IBCS) #1: Supersingular Curve Implementations of the BF and BB1 Cryptosystems, RFC 5091 (Informational), December 2007*, http://www.ietf.org/rfc/rfc5091.txt
9. H. Cohen, A. Miyaji, T. Ono, in *Efficient Elliptic Curve Exponentiation Using Mixed Coordinates, ASIACRYPT 1998*. LNCS, vol. 1514 (Springer, 1998), pp. 51–65
10. Cryptography Research and Evaluation Committees, http://www.cryptrec.jp/
11. D. Freeman, M. Scott, E. Teske, A taxonomy of pairing-friendly elliptic curves. J. Cryptol. **23**(2), 224–280 (2010)

12. D. Hanerson, A. Menezes, S. Vanstone, *Guide to Elliptic Curve Cryptography* (Springer, Berlin, 2003)
13. T. Hayashi, T. Shimoyama, N. Shinohara, T. Takagi, in *Breaking Pairing-Based Cryptosystems Using η_T Pairing Over $GF(3^{97})$, ASIACRYPT 2012*. LNCS, vol. 7658 (Springer, 2012), pp. 43–60
14. T. Iyama, S. Kiyomoto, K. Fukushima, T. Tanaka, T. Takagi, in *IEICE Transaction on Implementation of Pairing Based Cryptosystem on Mobile Phones*, vol. J95-A, no. 7 (2012), pp. 579–587 (in Japanese)
15. T. Izu, T. Takagi, in *Efficient Computations of the Tate Pairing for the Large MOV Degrees, ICISC 2002*. LNCS, vol. 2513 (2002), pp. 283–297
16. N. Koblitz, Elliptic curve cryptosystems. Math. Comput. **48**, 203–209 (1987)
17. A.K. Lenstra, H.W. Lenstra Jr., (eds.), in *The Development of the Number Field Sieve*, Lecture Notes in Mathematics, vol 1554 (Springer, Berlin, 1993)
18. V. Miller, in *Use of Elliptic Curves in Cryptography, CRYPTO 1985*. LNCS, vol. 218 (Springer, 1985), pp. 417–426
19. V. Miller, The Weil pairing, and its efficient calculation. J. Cryptol. **17**(4), 235–261 (2004)
20. J. Pollard, A Monte Carlo method for factorization. BIT Numer. Math. **15**(3), 331–334 (1975)
21. R. Rivest, A. Shamir, L. Adleman, A method for obtaining digital signatures and public-key cryptosystems. Commun. ACM **21**(2), 120–126 (1978)
22. R. Sakai, K. Ohgishi, M. Kasahara, Cryptosystems based on pairing, in *The 2000 Symposium on Cryptography and Information, Security, SCIS2000-C20*, 2000
23. O. Schirokauer, Discrete logarithms and local units. Philos. Trans. Royal Soc. A **345**(1676), 409–424 (1993)
24. T. Takagi, T. Okamoto, E. Okamoto, T. Okamoto (Eds.), in *Pairing-Based Cryptography—Pairing 2007*. LNCS, vol. 4575 (Springer, 2007)
25. T. Yasuda, T. Takagi, K. Sakurai, in *Application of Scalar Multiplication of Edwards Curves to Pairing-Based Cryptography, IWSEC 2012*. LNCS, vol. 7631 (Springer, 2012), pp. 19–36
26. M. Yoshitomi, T. Takagi, S. Kiyomoto, T. Tanaka, in *IEICE Transaction on Efficient Implementation of the Pairing on Mobilephones using BREW*, vol. E91-D, no.5 (2008), pp. 1330–1337

Code-Based Public-Key Encryption

Kirill Morozov

Abstract We present a short survey of public-key encryption (PKE) schemes based on hardness of general decoding. Such the schemes are believed to be resistant even against attacks using quantum computers, which makes them candidates for the so-called post-quantum cryptography. First, we briefly introduce the state-of-the-art in the area of code-based PKE. Then, we describe the McEliece PKE, two major attacks against this scheme and the proposed parameters. Finally, we survey recent results on the variants of this PKE which are proven to be indistinguishable under chosen plaintext and chosen ciphertext attacks.

Keywords Goppa codes · General decoding · McEliece public-key encryption · Recommended parameter sets · Provable security

1 Introduction

The first public-key encryption (PKE) scheme based on error-correcting codes was introduced by McEliece in 1978 [28]. This scheme used Goppa codes [16, 26], a subclass of alternant codes, which has the following useful features: (1) The lower bound on its minimal distance (and hence the number of correctable errors) is known; (2) Hardness of recovering the decoding algorithm from a proper representation of the code—the meaning of this property will be explained later. In this section, we focus on irreducible Goppa codes over $\mathbb{F}_2$, and only note that working with codes over larger fields may help to decrease the key size [7]. There were many attempts to use different classes of codes in the McEliece-type public-key encryption schemes, but

K. Morozov (✉)
Institute of Mathematics for Industry, Kyushu University, 744, Motooka, Nishi-ku, Fukuoka 819-0395, Japan
e-mail: morozov@imi.kyushu-u.ac.jp

R. Nishii et al. (eds.), *A Mathematical Approach to Research Problems of Science and Technology*, Mathematics for Industry 5,
DOI: 10.1007/978-4-431-55060-0_4,

some of them turned out to be insecure, while others are currently under evaluation—we leave this topic out of scope of this survey.

Another famous code-based PKE scheme was introduced by Niederrieter in 1986 [31]. Originally, Generalized Reed-Solomon codes were proposed to be used, however this construction was shown insecure by Sidelnikov and Shestakov [42]. Nonetheless, when Goppa codes are employed, the security of Niederreiter PKE is equivalent to that of McEliece PKE as shown by Li et al. [24]. For details, we refer the reader to the surveys [12, 33].

The main advantage of the above code-based PKE's is that there is no efficient attack on this system using quantum computers [5, 35]. This makes them candidates for post-quantum PKE [35]. Although at this moment, quantum computers exist only as early prototypes, it is important to consider secure alternatives to currently used cryptographic systems (such as RSA [38]) which are not quantum-tolerant [35]. Another important advantage of code-based PKE is their fast encryption and decryption algorithms, that admit implementation even for embedded and memory-constraint devices, see e.g., [11, 17, 44]. The main disadvantage of both McEliece and Niederreiter PKE is their relatively large public key size.

Recent research on code-based encryption proceeds in the following main directions[1]:

- Attacks on underlying assumptions: decoding attacks [1, 6], structural attacks [13, 41].
- Study on compact keys [3, 29].
- Alternatives to Goppa codes [36].
- Efficient and compact implementations: [11, 17, 44].
- Advanced cryptographic protocols for code-based PKE [10, 18, 27].

The rest of this presentation will be focused on McEliece PKE.

2 Background

2.1 Notation

Let J be an ordered subset as follows: $\{j_1, \dots, j_m\} = J \subseteq \{1, \dots, n\}$, then we denote a vector $(x_{j_1}, \dots, x_{j_m}) \in \mathbb{F}_2^m$ by x_J. Similarly, we denote by M_J the submatrix of a $(k \times n)$ matrix M consisting of the columns which correspond to the indexes of J. A concatenation of vectors $x \in \mathbb{F}_2^{n_0}$ and $y \in \mathbb{F}_2^{n_1}$ is written as $(x|y) \in \mathbb{F}_2^{n_0+n_1}$. For $x, y \in \mathbb{F}_2^m$, $x + y$ denotes a bitwise exclusive-or. We denote by $x \leftarrow_R \mathcal{X}$ a uniformly random selection of an element from its domain $\mathcal{X}$.

[1] Note that this collection of references is by no means comprehensive—it only contains some of the representative works on the topics in question.

2.2 Elements of Coding Theory

2.2.1 Linear Codes

A binary (n, k)-code $\mathcal{C}$ is a k-dimensional subspace of the vector space $\mathbb{F}_2^n$; n and k are called the *length* and the *dimension* of the code, respectively. We call $\mathcal{C}$ an (n, k, d)-code, if its so-called *minimum distance* is $d := \min_{\substack{x,y \in \mathcal{C} \\ x \neq y}} d_H(x, y)$, where d_H denotes the Hamming distance (i.e., the number of positions where x and y differ). The distance of $x \in \mathbb{F}_2^n$ to the zero-vector 0^n denoted by $w_H(x) := d_H(x, 0^n)$ is called the *weight* of x. We will write $\mathbf{0}$ to represent the zero-vector 0^n, omitting n which will be clear from the context.

For the relevant topics in coding theory we refer the reader to [26, 39].

2.2.2 Goppa Codes

In this subsection, we will follow the presentation of [12]. Let us first define a binary irreducible Goppa code of length n. Set $m = \log_2 n$. Let t be an integer—in fact, it will be an upper bound on the number of errors, which the code can correct.

Let $g(X) = \sum_{i=0}^{t} g_i X^i \in \mathbb{F}_{2^m}[X]$ be a monic polynomial of degree t called the *Goppa polynomial* and $L = (\gamma_0, \ldots, \gamma_{n-1}) \in \mathbb{F}_{2^m}^n$ called the *support*, which is a tuple of n distinct elements such that $g(\gamma_i) \neq 0$, for all $0 \leq i \leq n-1$.

For any vector $y \in \mathbb{F}_2^n$, define the *syndrome* of y by $S_y(X) := \sum_{i=0}^{n-1} \frac{y_i}{X - \gamma_i} \mod g(X)$, where y_i denotes the i-th bit of y.

Definition 1 The binary Goppa code $\mathcal{G}(L, g(X))$ is the set of all vectors $y \in \mathbb{F}_2^n$ such that the identity $S_y(X) = 0$ holds in the polynomial ring $\mathbb{F}_{2^m}[X]$.

If $g(X)$ is irreducible over $\mathbb{F}_{2^m}$, then $\mathcal{G}(L, g(X))$ is called an *irreducible binary Goppa code.*

Parity-Check and Generator Matrices. A parity-check matrix of $\mathcal{G}(L, g(X))$ can be written as: $H = XYZ$, where

$$X = \begin{pmatrix} g_t & 0 & 0 & \cdots & 0 \\ g_{t-1} & g_t & 0 & \cdots & 0 \\ \vdots & \vdots & \vdots & \ddots & \vdots \\ g_1 & g_2 & g_3 & \cdots & g_t \end{pmatrix}, \quad Y = \begin{pmatrix} 1 & 1 & \cdots & 1 \\ \gamma_0 & \gamma_1 & \cdots & \gamma_{n-1} \\ \vdots & \vdots & \ddots & \vdots \\ \gamma_0^{t-1} & \gamma_1^{t-1} & \cdots & \gamma_{n-1}^{t-1} \end{pmatrix},$$

and $Z = Diag_n(g(\gamma_0)^{-1}, g(\gamma_1)^{-1}, \ldots, g(\gamma_{n-1})^{-1})$, where $Diag_n(\cdot)$ denotes the diagonal matrix of size n with the arguments being the elements of the main diagonal. Since X is a $k \times k$ invertible matrix, when multiplying the parity-check matrix by it, we obtain an equivalent representation of the same code. Therefore, one may omit X, and compute $H = YZ$.

We have

$$y \in \mathcal{G}(L, g(X)) \text{ if and only if } Hy^T = 0. \tag{1}$$

The entries of H are elements of the extension field $\mathbb{F}_{2^m}$ over $\mathbb{F}_2$. In order to obtain the binary form of H, we use a representation of $\mathbb{F}_{2^m}$ as a vector space over $\mathbb{F}_2$. Then, we write H as a $mt \times n$ matrix over $\mathbb{F}_2$.

Now, we need to compute the generator matrix G of the code $\mathcal{G}$. It follows by (1) that the Goppa code consists of the vectors belonging to the kernel of H. Therefore, the generator matrix G can be represented by the basis vectors of such the kernel.

Since H is an $mt \times n$ matrix, G is $k \times n$ with $k \geq n - mt$, defining the (n, k) Goppa code.

Error Correction. For any codeword $y \in \mathcal{G}(L, g(X)) \setminus \mathbf{0}$, the following relation holds [12]: $w_H(y) \geq 2 \deg g(X) + 1$. We have $\deg g(X) = t$, therefore the code $\mathcal{G}$ corrects up to t errors.

As the decoding algorithm $\mathsf{Dec}_{\mathcal{G}}$ used for decryption, we will employ the algorithm by Patterson [34]. We refer the reader to [12] for further details.

2.3 Security Assumptions

Definition 2 (General Decoding (G-SD) Problem) Input: $G \leftarrow_R \mathbb{F}_2^{k \times n}$, $c \leftarrow_R \mathbb{F}_2^n$ and $0 < t \in \mathbb{N}$.
Decide: If there exists $x \in \mathbb{F}_2^k$ such that $e = xG + c$ and $w_H(e) \leq t$.

This problem was shown to be NP-complete by Berlekamp et al. [4].

The following two definitions refer to the quantities defined in Sect. 3. No efficient (polynomial-time) algorithm is known for solving them when using recommended parameters [6, 12, 14].

Definition 3 (McEliece Problem) Input: A McEliece public key (G^{pub}, t), where $G^{pub} \in \mathbb{F}_2^{k \times n}$, $0 < t \in \mathbb{N}$;
and a McEliece ciphertext $c \in \mathbb{F}_2^n$.
Output: $m \in \mathbb{F}_2^k$ such that $d_H(mG^{pub}, c) = t$.

Definition 4 (Goppa Code Distinguishing (GD) Problem) Input: $G \in \mathbb{F}_2^{k \times n}$.
Decide: Is G a parity-check matrix of an (n, k) irreducible Goppa code, or of a random (n, k)-code?

An important step toward solving the GD problem was made by Faugère et al. [13] by introducing a distinguisher for *high rate* Goppa codes,[2] however, this distinguisher does not work for typical parameters of the McEliece PKE. Nonetheless, it shows that the assumption on hardness of the GD problem must be used with extra care.

[2] Such the codes are not typically used for public-key encryption, but rather for constructing code-based digital signatures [9].

3 McEliece Public-Key Encryption

The McEliece PKE scheme consists of the following triplet of algorithms $(\mathcal{K}, \mathcal{E}, \mathcal{D})$:

- System parameters: $n, k, t \in \mathbb{N}$.
- Key generation algorithm $\mathcal{K}$: On input n, k, t, generate the following matrices:
  - $G \in \mathbb{F}_2^{k \times n}$—the generator matrix of an irreducible binary Goppa code correcting up to t errors. Its decoding algorithm is denoted as $\mathsf{Dec}_{\mathcal{G}}$.
  - $S \in \mathbb{F}_2^{k \times k}$—a random non-singular matrix.
  - $P \in \mathbb{F}_2^{n \times n}$—a random permutation matrix (of size n).
  - $G^{pub} = SGP \in \mathbb{F}_2^{k \times n}$.

  Output the public key $pk = (G^{pub}, t)$ and the secret key $sk = (S, G, P, \mathsf{Dec}_{\mathcal{G}})$.

- Encryption algorithm $\mathcal{E}$: On input a plaintext $m \in \mathbb{F}_2^k$ and the public key pk, choose a vector $e \in \mathbb{F}_2^n$ of weight t at random, and output the ciphertext

$$c = mG^{pub} + e.$$

- Decryption algorithm $\mathcal{D}$: On input c and the secret key sk, calculate:
  - $cP^{-1} = mSG + eP^{-1}$.
  - $mSG = \mathsf{Dec}_{\mathcal{G}}(cP^{-1})$.
  - Let $J \subseteq \{1, \ldots, n\}$ be such that G_J is invertible. Output $m = (mSG)_J (G_J)^{-1} S^{-1}$.

It is easy to check that the decryption algorithm correctly recovers the plaintext: Since in the first step of decryption, the permuted error vector eP^{-1} is again of weight t, the decoding algorithm $\mathsf{Dec}_{\mathcal{G}}$ successfully corrects these errors in the next step.

3.1 Security Analysis

Let us discuss two major types of attacks against McEliece PKE.

Decoding Attack. For the parameter sizes related to McEliece PKE, the best algorithm is the *information-set decoding* [1, 6, 8, 22, 23, 30, 37, 43]. The time complexity of this algorithm is subexponential and for the relevant parameters can be (conservatively) lower bounded by the following expression [12]: $O(n^3)2^{-t \log_2(1-k/n)}$. For precise estimation, one may use the lower bounds [30, 37].

Structural Attack. If we employ the irreducible binary Goppa codes, then up to date, there is no efficient algorithm which can extract the secret key from the public key in the McEliece (or the Niederreiter) scheme as long as the so-called weak keys [25] are avoided. Moreover, there is no algorithm which can efficiently distinguish the

Table 1 Examples of Parameter Sets for the McEliece PKE (Niebuhr et al. [30])

Equivalent security (bits)	85	112	129
Code length n	1652	2440	2798
Code dimension k	1203	1877	2088
Weight of error vector t	42	50	62
Public key size (Kbytes)	66	129	181

matrices defined by the McEliece public keys and the same size generator matrices of random codes, for the typical parameters. The time complexity of the currently best algorithm [41] is still subexponential. Roughly speaking, this algorithm works as follows: Enumerate Goppa polynomials and verify whether each corresponding code and the code generated by G^{pub} are "permutation equivalent" or not by using the *support splitting algorithm* [41], which results in an algorithm running in time $n^t(1 + o(1))$.

For details on the attacks described above and their respective countermeasures, we refer the reader to the surveys in [12, 19, 33, 40].

Some recommended parameter sets along with their estimated security levels provided in [30] are given in Table 1.

4 Provable Security of Code-Based PKE

4.1 Chosen Plaintext Security

The *semantic security* (also called *indistinguishability under chosen plaintext attacks or IND-CPA*) defined by Goldwasser and Micali [15] is a security notion for public-key encryption. Its intuitive meaning is that a ciphertext does not leak any useful information about the plaintext except for its length. More precisely, suppose that the attacker is allowed to pick any pair of plaintexts. Then given a ciphertext, she must not be able to find out, which one was encrypted.

Nojima et al. [32] show that the McEliece encryption with a random padding of the plaintext (which is multi-bit) is IND-CPA secure under hardness of the learning parities with noise (LPN) problem[3] and GD problem. A little more formally, the Randomized McEliece encryption is constructed in the same way as described above, except that the ciphertext $c = (r|m)G^{pub} + e$, where $r \leftarrow_R \{0, 1\}^{k_0}$, $m \in \{0, 1\}^{k_1}$, $k = k_0 + k_1$. Particular choices of k_0 and k_1 are discussed in [32]. For typical parameters, taking $k_1 \approx k/4$ results in secure encryption. A similar padding will provide IND-CPA security for Niederreiter PKE as well [32].

[3] See e.g. [20] for a formal definition of LPN problem—it is similar to G-SD problem except that in the error vector e, each bit has Bernoulli distribution with fixed p, $0 < p < 0.5$.

4.2 Chosen Ciphertext Security

In some cases, a limited access to *decryption* algorithm may be available to an attacker. It may sound somewhat counter-intuitive, but consider, for instance, a mailing service that automatically decrypts the received correspondence. Then the adversary, who gains access to the output of such a service for some time, may be modeled by the above oracle. This model is to capture the property that the result of decryption must not reveal any additional information (for instance, nothing about the secret key), apart from the decrypted messages themselves.

Public-key encryption is *indistinguishable under the adaptive chosen ciphertetxt attack (IND-CCA2)*, if in the IND-CPA scenario described in the previous subsection, the attacker is allowed to access the decryption algorithm (but not the secret key). Naturally, the attacker is allowed to request decryption of any ciphertext, except those to be distinguished.

An IND-CCA2 conversion for the McEliece PKE in the random oracle model was presented by Kobara and Imai [21]. The random oracle model [2] assumes cryptographic hash functions to behave like random functions, hereby simplifying security proofs. Recently, IND-CCA2 conversions that do not use random oracles were presented for McEliece PKE [10] and for Niederreiter PKE [27].

5 Conclusion

We presented a summary of the state-of-the-art in code-based PKE, with a focus on the McEliece PKE scheme which is based on error-correcting codes by Goppa. We briefly described major attacks on this scheme, secure parameters sets, and conversions enhancing its security.

Current research trends in PKE based on error-correcting codes include studies on compact keys and fast implementations, as well as on related cryptographic protocols.

References

1. A. Becker, A. Joux, A. May, A. Meurer, Decoding Random Binary Linear Codes in $2^{n/20}$: How $1 + 1 = 0$ Improves Information Set Decoding. *EUROCRYPT 2012*. LNCS, vol. 7237 (Springer, Heidelberg, 2009), pp. 520–536
2. M. Bellare, P. Rogaway, Random Oracles are Practical: A Paradigm for Designing Efficient Protocols, in *ACM Conference on Computer and Communications Security* 1993, pp. 62–73, ACM (1993)
3. T. Berger, P. Cayrel, P. Gaborit, A. Otmani, Reducing Key Length of the McEliece Cryptosystem. *AFRICACRYPT 2009*. LNCS, vol. 5580 (Springer, Heidelberg, 2009), pp. 77–97
4. E. Berlekamp, R. McEliece, H. van Tilborg, On the inherent intractability of certain coding problems. IEEE Trans. Inf. Theory **24**, 384–386 (1978)
5. D.J. Bernstein, Grover vs. McEliece, *PQCrypto 2010*. LNCS, vol. 6061 (Springer, Heidelberg, 2010), pp. 73–80

6. D.J. Bernstein, T. Lange, C. Peters, Smaller Decoding Exponents: Ball-Collision Decoding, *CRYPTO 2011*. LNCS, vol. 6841 (Springer, Heidelberg, 2011), pp. 743–760
7. D.J. Bernstein, T. Lange, D.J. Peters, Wild McEliece, *Selected Areas in Cryptography 2010*. LNCS, vol. 6544 (Springer, Heidelberg, 2010), pp. 143–158
8. A. Canteaut, F. Chabaud, A new algorithm for finding minimum-weight words in a linear code: application to primitive narrow-sense bch-codes of length 511. IEEE Trans. Inf. Theory **44**, 367–378 (1998)
9. N. Courtois, M. Finiasz, N. Sendrier, How to Achieve a McEliece-Based Digital Signature Scheme, *ASIACRYPT 2001*. LNCS, vol. 2248 (Springer, Heidelberg, 2001), pp. 157–174
10. N. Döttling, R. Dowsley, J. Müller-Quade, A.C.A. Nascimento, A CCA2 secure variant of the McEliece cryptosystem. IEEE Trans. Inf. Theory **58**(10), 6672–6680 (2012)
11. T. Eisenbarth, T. Güeysu, S. Heyse, C. Paar, MicroEliece: McEliece for Embedded Devices, *CHES 2009*. LNCS, vol. 5747 (Springer, Heidelberg, 2009), pp. 49–64
12. D. Engelbert, R. Overbeck, A. Schmidt, A summary of McEliece-Type cryptosystems and their security, J. Math. Cryptol. **1**, 151–199 Walter de Gruyter (2007)
13. J. Faugère, A. Gauthier-Umaña, V. Otmani, L. Perret, J. Tillich, A distinguisher for high rate McEliece cryptosystems, in *Information Theory Workshop 2011*, pp. 282–286 (2011)
14. M. Finiasz, N. Sendrier, Security Bounds for the Design of Code-Based Cryptosystems, *ASIACRYPT 2009*, LNCS, vol. 5912 (Springer, Heidelberg, 2009), pp. 88–105
15. S. Goldwasser, S. Micali, Probabilistic encryption. J. Comput. Syst. Sci. **28**, 270–299 (1984)
16. V. Goppa, A new class of linear error-correcting codes (in Russian). Probl. Peredachi Informacii **6**, 24–30 (Russian Academy of Sciences) (1970)
17. S. Heyse, Low-Reiter: Niederreiter Encryption Scheme for Embedded Microcontrollers, *PQCrypto 2010*. LNCS, vol. 7071 (Springer, Heidelberg, 2011), pp. 165–181
18. R. Hu, K. Morozov, T. Takagi, Proof of Plaintext Knowledge for Code-Based Public-Key Encryption Revisited, in *ASIACCS 2013*, pp. 535–540, ACM (2013)
19. G. Kabatiansky, E. Krouk, S. Semenov, *Error Correcting Codes and Security for Data Networks* Wiley, New York (2005)
20. J. Katz, J. Shin, Parallel and Concurrent Security of the HB and HB^+ Protocols, *EUROCRYPT 2006*. LNCS, vol. 4004 (Springer, Heidelberg, 2006), pp. 73–87
21. K. Kobara, Imai, Semantically Secure McEliece Public-Key Cryptosystems—Conversions for McEliece PKC, in *PKC2001*. LNCS, vol. 1992 (Springer, Heidelberg, 2001), pp. 19–35
22. P. Lee, E. Brickell, An Observation on the Security of McEliece's Public Key Cryptosystem, *EUROCRYPT 1988*. LNCS, vol. 330 (Springer, Heidelberg, 1988), pp. 275–280
23. J. Leon, A probabilistic algorithm for computing minimum weights of large error-correcting codes. IEEE Trans. Inf. Theory **34**, 1354–1359 (1988)
24. Y. Li, R. Deng, X. Wang, The equivalence of McEliece's and Niederreiter's public-key cryptosystems. IEEE Trans. Inf. Theory **40**, 271–273 (1994)
25. P. Loidreau, N. Sendrier, Weak keys in the McEliece public-key cryptosystem. IEEE Trans. Inf. Theory **47**(3), 1207–1211 (2001)
26. F. MacWilliams, N.J.A. Sloane, *The Theory of Error-Correcting Codes* (North-Holland, Amsterdam, 1992)
27. P. Mathew, S. Vasant, S. Venkatesan, C.P. Rangan, An Efficient IND-CCA2 Secure Variant of the Niederreiter Encryption Scheme in the Standard Model, *ACISP 2012*. LNCS, vol. 7372, (Springer, Heidelberg, 2012), pp. 166–179
28. R.J. McEliece, A Public-Key Cryptosystem Based on Algebraic Coding Theory. Deep Space Network Progress Report (1978)
29. R. Misoczki, P.S.L.M. Barreto, Compact McEliece Keys from Goppa Codes, *Selected Areas in Cryptography 2009*. LNCS, vol. 5867 (Springer, Heidelberg, 2009), pp. 376–392
30. R. Niebuhr, M. Meziani, S. Bulygin, J. Buchmann, Selecting parameters for secure McEliece-based cryptosystems. Int. J. Inf. Secur. **11**(3), 137–147 (2012)
31. H. Niederreiter, Knapsack-type cryptosystems and Algebraic coding theory. Probl. Control Inf. Theory **152**, 159–166 (Russian Academy of Sciences) (1986)

32. R. Nojima, H. Imai, K. Kobara, K. Morozov, Semantic security for the McEliece cryptosystem without random oracles. Des. Codes Cryptogr. **49**(1–3), 289–305 (2008)
33. R. Overbeck, N. Sendrier, Code-based cryptography, in *Post-Quantum Cryptography*, ed. by D.J. Bernstein, J. Buchmann, E. Dahmen (Springer, Berlin, 2009), pp. 95–145
34. N.J. Patterson, The Algebraic decoding of Goppa codes. IEEE Trans. Inf. Theory **21**, 203–207 (1975)
35. R. Perlner, D. Cooper, Quantum resistant public key cryptography: a survey. IDtrust **2009**, 85–93 (2009)
36. E. Persichetti, Compact McEliece keys based on quasi-dyadic Srivastava codes. J. Math. Cryptol. **6**(2), 149–169 (Walter de Gruyter) (2012)
37. C. Peters, Information-Set Decoding for Linear Codes over F_q, *PQCrypto 2010*. LNCS, vol. 6061 (Springer, Heidelberg, 2010), pp. 81–94
38. R. Rivest, A. Shamir, L. Adleman, A method for obtaining digital signatures and public-key cryptosystems. Commun. ACM **21**(2), 120–126 (1978)
39. R. Roth, *Introduction to Coding Theory* (Cambridge University Press, Cambridge, 2006)
40. N. Sendrier, On the security of the McEliece public-key cryptosystem, in *Information, Coding and Mathematics—Proceedings of Workshop honoring Prof. Bob McEliece on his 60th birthday*, pp. 141–163, Kluwer (2002)
41. N. Sendrier, Finding the permutation between equivalent linear codes: the support splitting algorithm. IEEE Trans. Inf. Theory **46**(4), 1193–1203 (2000)
42. Sidelnikov V., Shestakov, S.: On the insecurity of cryptosystem based on generalized reed-solomon codes. Discrete Math. Appl. **2**(4), 439–444 (Walter de Gruyter) (1992)
43. J. Stern, A Method for Finding Codewords of Small Weight, *Coding Theory and Applications*. LNCS, vol. 388 (Springer, Heidelberg, 1988), pp. 106–133
44. F. Strenzke, A Smart Card Implementation of the McEliece PKC, *WISTP 2010*. LNCS, vol. 6033 (Springer, Heidelberg, 2010), pp. 47–59

Gröbner Basis and Its Applications

Takafumi Shibuta

Abstract Computer Algebra is a field of mathematics and computer science that studies algorithms for symbolic computation. A fundamental tool in computer algebra to study polynomial ideals is the theory of Geöbner basis. The notion of the Gröbner basis and the Buchberger's algorithm for computing it was proposed by Bruno Buchberger in 1965. Gröbner bases have numerous applications in commutative algebra, algebraic geometry, combinatorics, coding theory, cryptography, theorem proving, etc. The Buchberger's algorithm is implemented in many computer algebra systems, such as *Risa/Asir*, *Macaulay2*, *Singular*, *CoCoa*, *Maple*, and *Mathematica*. In this chapter, we will give a short introduction on Gröbner basis theory, and then we will present some applications of Gröbner bases.

Keywords Gröbner basis · Toric ideal

1 Affine Varieties

First, we give a preliminary on affine varieties.

Let $\mathbb{K}$ be an infinite field (e.g., the rational number field $\mathbb{Q}$, the real number field $\mathbb{R}$, the complex number field $\mathbb{C}$, and the algebraic closure of a finite field $\overline{\mathbb{F}}_p$), and $\mathbb{K}[x_1, \ldots, x_n]$ the polynomial ring over $\mathbb{K}$ with indeterminates $\mathbf{x} = (x_1, \ldots, x_n)$. For $\mathbf{a} = (a_1, \ldots, a_n) \in \mathbb{Z}^n_{\geq 0}$, we use the multi-index notation $\mathbf{x}^{\mathbf{a}} = x_1^{a_1} \ldots x_n^{a_n}$, $|\mathbf{a}| = a_1 + \cdots + a_n$.

Algebraic Geometry is the study of sets of common zeros of a family of polynomials.

T. Shibuta (✉)
Institute of Mathematics for Industry, Kyushu University, 744 Motooka, Nishi-ku, Fukuoka 819-0395, Japan
e-mail: shibuta@imi.kyushu-u.ac.jp

R. Nishii et al. (eds.), *A Mathematical Approach to Research Problems of Science and Technology*, Mathematics for Industry 5,
DOI: 10.1007/978-4-431-55060-0_5, © Springer Japan 2014

Definition 1 (*Affine Variety*) Given a set of polynomials $P \subset \mathbb{K}[x_1, \ldots, x_n]$, the set of common zeros of P

$$\mathscr{V}(P) = \mathscr{V}_{\mathbb{K}}(P) := \{(u_1, \ldots, u_n) \in \mathbb{K}^n \mid f(u_1, \ldots, u_n) = 0 \text{ for all } f \in P\}$$

is called an *affine (algebraic) variety*.

The operator $\mathscr{V}$ reverses the inclusion; $P_1 \subset P_2$ implies $\mathscr{V}(P_2) \supset \mathscr{V}(P_2)$. If P is a finite set $\{f_1, \ldots, f_r\}$, we denote $\mathscr{V}_{\mathbb{K}}(P)$ by $\mathscr{V}_{\mathbb{K}}(f_1, \ldots, f_r)$.

Example 1

- $\mathscr{V}_{\mathbb{R}}\left(x^2 + y^2 - 1\right)$ is the unit circle.
- $\mathscr{V}_{\mathbb{R}}\left(x^2 + y^2 + z^2 - 1, z - x^2 - y^2\right)$ is the intersection of the sphere $x^2 + y^2 + z^2 = 1$ and the paraboloid $z = x^2 + y^2$.
- Let $GL_n(\mathbb{K})$ be the general linear group of degree n (the set of $n \times n$ invertible matrices). Let $\mathbb{K}[t, x_{ij} \mid 1 \leq i, j \leq n]$ be the polynomial ring with indeterminates t and x_{ij}, $1 \leq i, j \leq n$. Then there exists one-to-one correspondence between the affine variety $\mathscr{V}_{\mathbb{K}}(t \cdot \det(x_{ij})_{i,j} - 1)$ and $GL_n(\mathbb{K})$:

$$\begin{aligned} \mathscr{V}_{\mathbb{K}}(t \cdot \det(x_{ij})_{i,j} - 1) &\longleftrightarrow GL_n(\mathbb{K}) \\ (d, (a_{ij})_{i,j}) &\mapsto (a_{ij})_{i,j} \\ (\det(a_{ij})_{i,j}^{-1}, (a_{ij})_{i,j}) &\leftarrow\!\shortmid (a_{ij})_{i,j}. \end{aligned}$$

  Thus one can view $GL_n(\mathbb{K})$ as an affine variety.

The expression of an affine algebraic variety as common zeros of polynomials is not unique. For example, $\mathscr{V}_{\mathbb{K}}(x, y) = \mathscr{V}_{\mathbb{K}}(x + y, x - y) = \{(0, 0)\} \subset \mathbb{K}^2$. To avoid this problem, we consider ideals.

Definition 2 (*Ideal*) (1) A subset $I \subset \mathbb{K}[x_1, \ldots, x_n]$ is called an *ideal* if

1. $f_1 + f_2 \in I$ for all $f_1, f_2 \in I$,
2. $gf \in I$ for all $f \in I$ and $g \in \mathbb{K}[x_1, \ldots, x_n]$.

(2) For $f_1, \ldots, f_r \in \mathbb{K}[x_1, ..., x_n]$, the ideal

$$\langle f_1, \ldots, f_r \rangle := \{g_1 f_1 + \cdots + g_r f_r \mid g_i \in \mathbb{K}[x_1, \ldots, x_n]\}$$

is called the ideal generated by $f_1, \ldots, f_r$.

(3) For a subset $S \subset \mathbb{K}^n$, we define

$$I(S) = \{f \in \mathbb{K}[x_1, \ldots, x_n] \mid f(u_1, \ldots, u_n) = 0 \text{ for all } (u_1, \ldots, u_n) \in S\}.$$

For an affine variety V, $V = \mathscr{V}(\mathscr{I}(V))$ holds. We call $\mathscr{I}(V)$ the *defining ideal* of V. The operator $\mathscr{I}$ reverses the inclusion; $S_1 \subset S_2$ implies $\mathscr{I}(S_1) \supset \mathscr{I}(S_2)$. By the

Hilbert's basis theorem any ideal I, there exist finitely many polynomials $f_1, \ldots, f_r$ such that $I = \langle f_1, \ldots, f_r \rangle$. Hence $\mathscr{I}(S)$ has a finite system of generators for any S. We note that $\mathscr{V}_{\mathbb{K}}(f_1, \ldots, f_r) = \mathscr{V}_{\mathbb{K}}(\langle f_1, \ldots, f_r \rangle)$. Hence $\mathscr{V}_{\mathbb{K}}(f_1, \ldots, f_r) = \mathscr{V}_{\mathbb{K}}(g_1, \ldots, g_s)$ if $\langle f_1, \ldots, f_r \rangle = \langle g_1, \ldots, g_s \rangle$.

Definition 3 (*Radical ideal*) For an ideal $I \subset \mathbb{K}[x_1, \ldots, x_n]$, we define

$$\sqrt{I} = \{f \in \mathbb{K}[x_1, \ldots, x_n] \mid f^n \in I \text{ for some } n \in \mathbb{Z}_{\geq 0}\},$$

and call it the *radical* of I. We say that I is a radical ideal if $I = \sqrt{I}$.

Example 2 (Monomial ideal) An ideal I is called a *monomial ideal* if it can be generated by set of monomials. We say that a monomial $x_1^{a_1} \ldots x_n^{a_n}$ is *square-free* if $a_1 = 0$ or 1 for all i. A monomial ideal generated by square-free monomials is called a *square-free monomial ideal*. For a monomial $\mathbf{x}^{\mathbf{a}} = x_1^{a_1} \ldots x_n^{a_n}$, $a_i \in \mathbb{Z}_{\geq 0}$, we set $\sqrt{\mathbf{x}^{\mathbf{a}}} := x_1^{b_1} \ldots x_n^{b_n}$ where $b_1 = 1$ if $a_i \neq 0$ and $b_i = 0$ otherwise. Let $I = \langle \mathbf{x}^{\mathbf{a}_1}, \ldots, \mathbf{x}^{\mathbf{a}_r} \rangle$ be a monomial ideal. Then

1. $\sqrt{I} = \left\langle \sqrt{\mathbf{x}^{\mathbf{a}_1}}, \ldots, \sqrt{\mathbf{x}^{\mathbf{a}_r}} \right\rangle$,
2. I is a radical ideal if and only if I is square-free.

For instance, $\sqrt{\langle x^3 y, y^2 z^2 \rangle} = \langle xy, yx \rangle$.

It is easy to show that $\mathscr{V}_{\mathbb{K}}(I) = \mathscr{V}_{\mathbb{K}}(\sqrt{I})$. For any subset $S \subset \mathbb{K}^n$, $I(S)$ is a radical ideal, and $\mathscr{V}_{\mathbb{K}}(I(S))$ is the minimal affine variety containing S.

Definition 4 We call $\mathscr{V}_{\mathbb{K}}(I(S))$ the *Zariski closure* of S and denote by $\overline{S}$.

The union and the intersection of affine varieties can be computed using these ideal operations. There is a map from the set of radical ideals to the set of the affine varieties $I \mapsto \mathscr{V}_{\mathbb{K}}(I)$. There is also a map in the inverse direction $V \mapsto \mathscr{I}(V)$. The Nullstellensatz states that these maps are inverse map to each other if $\mathbb{K}$ is algebraically closed (e.g. $\mathbb{K} = \mathbb{C}$).

Theorem 1 (Nullstellensatz) *Assume the $\mathbb{K}$ is algebraically closed. Then the maps $\mathscr{V}$ and $\mathscr{I}$ give an inclusion reversing one-to-one correspondence*

$$\{\textit{Radical ideal} \subset \mathbb{K}[x_1, \ldots, x_n]\} \underset{\mathscr{I}}{\overset{\mathscr{V}}{\leftrightarrows}} \{\textit{Affine variety} \subset \mathbb{K}^n\}.$$

Thus, we can view the set of radical ideals as a dictionary of affine varieties. Using this dictionary, we can translate geometric problems to algebraic ones.

For ideals I and J, the summation $I + J = \{f + g : f \in I, g \in J\}$, the intersection $I \cap J$ are again ideals.

Proposition 1 1. $\mathscr{V}(I) \cap \mathscr{V}(J) = \mathscr{V}(I + J)$.
2. $\mathscr{V}(I) \cup \mathscr{V}(J) = \mathscr{V}(I \cap J) = \mathscr{V}(IJ)$.

We note that if I and J are radical ideals, then so is $I \cap J$.

The image of projection corresponds to the restriction of the defining ideal.

Proposition 2 *For $n > i$, let $pr : \mathbb{K}^n \to \mathbb{K}^i$, $(u_1, \ldots, u_n) \mapsto (u_1, \ldots, u_i)$ be the projection map. Let $I \subset \mathbb{K}[x_1, \ldots, x_n]$ be an ideal. Then*

$$\overline{pr(\mathscr{V}(I))} = \mathscr{V}(I \cap \mathbb{K}[x_1, \ldots, x_i]) \subset \mathbb{K}^i.$$

Definition 5 We say that an affine variety is *irreducible*, and it cannot be represented as the union of two proper affine subvarieties.

Any affine variety V can be expressed as a finite union of irreducible proper affine subvarieties $W_1 \cup \cdots \cup W_r$, $W_i \subsetneq V$. This expression is called the *irreducible decomposition* of V. If $V \neq W_1 \cup \cdots W_{i-1} \cup W_{i+1} \cup \cdots W_r$ for any $1 \leq i \leq n$, this decomposition is said to be *irredundant*. The irredundant irreducible decomposition of an affine variety is unique up to reordering.

Definition 6 We say that an ideal I is *irreducible* it can not be represented as the intersection $I_1 \cap I_2$ of two ideals I_1 and I_2 such that $I \subsetneq I_i$.

Any ideal I can be expressed as a finite intersection of irreducible ideals $I_1 \cap \cdots \cap I_r$, $I \subsetneq I_i$. This expression is called the *irreducible decomposition* of I. If $I \neq I_1 \cap \cdots I_{i-1} \cap I_{i+1} \cap \cdots I_r$ for any $1 \leq i \leq n$, this decomposition is said to be *irredundant*. An irreducible ideal appearing in the irredundant irreducible decomposition of I is called an *irreducible component* of I. Clearly, irreducible decomposition of ideals corresponds to irreducible decomposition of affine varieties. Let V be an affine variety, and $\mathscr{I}(V) = I_1 \cap \cdots \cap I_r$ an irreducible decomposition of $\mathscr{I}(V)$. Then $V = \mathscr{V}(I_1) \cup \cdots \cup \mathscr{V}(I_r)$ is an irreducible decomposition of V.

A monomial ideal is irreducible if and only if it is of form $\langle x_{i_1}^{a_1+1}, \ldots, x_{i_\ell}^{a_\ell+1} \rangle$ for some $1 \leq i_1 < \cdots < i_\ell \leq n$ and $a_i \in \mathbb{Z}_{\geq 0}$. Any component of the irredundant irreducible decomposition of a monomial ideal I is also a monomial ideal. In particular, a square-free monomial ideal can be expressed as a finite intersection of ideals of form $\langle x_{i_1}, \ldots, x_{i_\ell} \rangle$ for some $1 \leq i_1 < \cdots < i_\ell \leq n$.

Example 3

$$\langle x^3y, y^2z^2 \rangle = \langle x^3, y^2 \rangle \cap \langle z^2, x^3 \rangle \cap \langle y \rangle$$

is the irredundant irreducible decomposition of $\langle x^3y, y^2z^2 \rangle$. Let $V = \mathscr{V}(x^3y, y^2z^2)$. Then, $\mathscr{I}(V) = \sqrt{\langle x^3y, y^2z^2 \rangle} = \langle xy, yz \rangle = \langle x, z \rangle \cap \langle y \rangle$. Thus,

$$V = \mathscr{V}(x, z) \cup \mathscr{V}(y)$$

is the irredundant irreducible decomposition of V.

As we have seen, the operations on affine varieties can be translated to the corresponding operations on ideals. The theory of Gröbner basis gives algorithms for computing them.

2 Gröbner Bases

Here, we give a short introduction on the theory of Gröbner bases. See [3, 4, 6, 7] for details.

Let $\mathbb{K}[x_1, \ldots, x_n]$ be a polynomial ring over a field K. A total order $\prec$ on the set of monomials $\{\mathbf{x}^{\mathbf{a}} \mid \mathbf{a} \in \mathbb{Z}^n_{\geq 0}\}$ is a *term order* on $\mathbb{K}[x_1, \ldots, x_n]$ if $\mathbf{x}^0 = 1$ is the unique minimal element, and $\mathbf{x}^{\mathbf{a}} \prec \mathbf{x}^{\mathbf{b}}$ implies $\mathbf{x}^{\mathbf{a}+\mathbf{c}} \prec \mathbf{x}^{\mathbf{b}+\mathbf{c}}$ for all $\mathbf{a}, \mathbf{b}, \mathbf{c} \in \mathbb{Z}^n_{\geq 0}$. Let $\mathbf{a} = {}^t(a_1, \ldots, a_n)$ and $\mathbf{b} = {}^t(b_1, \ldots, b_n) \in \mathbb{Z}^n_{\geq 0}$.

Definition 7 (*Lexicographic order*) The term order $\prec_{\text{lex}}$ called a *lexicographic order* with $x_1 \prec_{\text{lex}} \cdots \prec_{\text{lex}} x_n$ is defined as follows: $\mathbf{x}^{\mathbf{a}} \prec_{\text{lex}} \mathbf{x}^{\mathbf{b}}$ if $a_j < b_j$ where $j = \max\{i \mid a_i \neq b_i\}$.

Definition 8 (*Reverse lexicographic order*) The term order $\prec_{\text{rlex}}$ called a *reverse lexicographic order* with $x_1 \prec_{\text{rlex}} \cdots \prec_{\text{rlex}} x_n$ is defined as follows: $\mathbf{x}^{\mathbf{a}} \prec_{\text{rlex}} \mathbf{x}^{\mathbf{b}}$ if $|\mathbf{a}| < |\mathbf{b}|$ or $|\mathbf{a}| = |\mathbf{b}|$ and $a_j > b_j$ where $j = \min\{i \mid a_i \neq b_i\}$.

Definition 9 (*Weighted order*) For a weight vector $\omega \in \mathbb{Q}^n_{\geq 0}$ and a term order $\prec$, we define a new term order $\prec_\omega$ as follows: $\mathbf{x}^{\mathbf{a}} \prec_\omega \mathbf{x}^{\mathbf{b}}$ if $\omega \cdot \mathbf{a} < \omega \cdot \mathbf{b}$, or $\omega \cdot \mathbf{a} = \omega \cdot \mathbf{b}$ and $\mathbf{x}^{\mathbf{a}} \prec \mathbf{x}^{\mathbf{b}}$.

Definition 10 (*Product order*) Let $\prec_1$ and $\prec_2$ be term orders on $\mathbb{K}[x_1, \ldots, x_n]$ and $\mathbb{K}[y_1, \ldots, y_m]$, respectively. We define a new term order $\prec$ on $\mathbb{K}[x_1, \ldots, x_n, y_1, \ldots, y_m]$ as follows:

$$\mathbf{x}^{\mathbf{a}}\mathbf{y}^{\mathbf{c}} \prec \mathbf{x}^{\mathbf{b}}\mathbf{y}^{\mathbf{d}} \text{ if } \mathbf{x}^{\mathbf{a}} \prec_1 \mathbf{x}^{\mathbf{b}}, \text{ or } \mathbf{a} = \mathbf{b} \text{ and } \mathbf{y}^{\mathbf{c}} \prec_2 \mathbf{y}^{\mathbf{d}}.$$

This term order is called the *product order* of $\prec_1$ and $\prec_2$.

Definition 11 (*Elimination order*) Let $\prec$ be a term order on $\mathbb{K}[x_1, \ldots, x_n, y_1, \ldots, y_m]$. We say that $\prec$ is an elimination order $\{x_1, \ldots, x_n\} \gg \{y_1, \ldots, y_n\}$ if

$$\mathbf{x}^{\mathbf{a}} \prec \mathbf{x}^{\mathbf{b}} \text{ implies } \mathbf{x}^{\mathbf{a}}\mathbf{y}^{\mathbf{c}} \prec \mathbf{x}^{\mathbf{b}}\mathbf{y}^{\mathbf{d}} \text{ for any } \mathbf{a}, \mathbf{b} \in \mathbb{Z}^n, \mathbf{c}, \mathbf{d} \in \mathbb{Z}^m_{\geq 0}.$$

The product order $\prec$ of $\prec_1$ and $\prec_2$ is an elimination order $\{x_1, \ldots, x_n\} \gg \{y_1, \ldots, y_n\}$, and the lexicographic order $\prec_{\text{lex}}$ is an elimination order $\{x_{i+1}, \ldots, x_n\} \gg \{x_1, \ldots, x_i\}$ for any i.

Definition 12 Let $\prec$ be a term order on R, $f \in R$, and I an ideal of R. The *initial term* of f, denoted by $\text{in}_\prec(f)$, is the highest term of f with respect to $\prec$. We call

$$\text{in}_\prec(I) = \langle \text{in}_\prec(f) \mid f \in I \rangle$$

the *initial ideal* of I with respect to $\prec$. We say that a finite collection of polynomials $G \subset I$ is a *Gröbner basis* of I with respect to $\prec$ if $\langle \text{in}_\prec(g) \mid g \in G \rangle = \text{in}_\prec(I)$.

It is known that Gröbner basis of I is a system of generators of I. For simplicity, we assume that the coefficient of the initial term of any element of a Gröbner basis is 1.

Proposition 3 *Let $G = \{g_1, \dots, g_\ell\} \subset \mathbb{K}[x_1, \dots, x_n]$ be a finite set of polynomials such that the coefficient of the initial term of g_i if 1, and $\prec$ a term order. For a polynomial $f \in \mathbb{K}[x_1, \dots, x_n]$, there exists $h_1, \dots, h_\ell, r \in \mathbb{K}[x_1, \dots, x_n]$ such that*

$$f = h_1 g_1 + \cdots h_\ell g_\ell + r$$

and r is zero or any term of r is not reducible by $\mathrm{in}_\prec(g_i)$ for all i.

We call this r a remainder of f on division by G with respect to $\prec$. The remainder can be computed by the following so-called division algorithm.

Definition 13 (*Division algorithm*) Let the situation as in Proposition 3. If a term of f, say $c\mathbf{x}^{\mathbf{a}}$, is divisible by $\mathrm{in}_\prec(g_i)$ for some i, then we write

$$f \overset{G}{\to} f - \frac{c\mathbf{x}^{\mathbf{a}}}{\mathrm{in}_\prec(g_i)} g_i.$$

Continuing this procedure, $f \overset{G}{\to} f_1 \overset{G}{\to} f_2 \overset{G}{\to} \dots$, we eventually obtain f_N for some $N \in \mathbb{Z}_{\geq 0}$ such that $f_N = 0$ or any term of f_N is not divisible by $\mathrm{in}_\prec(g_i)$ for all i. Then, we write $f \overset{G}{\Rightarrow} f_N$.

It is known that this division algorithm terminates in finite time. If $f \overset{G}{\Rightarrow} r$, then r is a remainder of f on division by G with respect to $\prec$. A remainder is not unique in general, but if G is a Gröbner basis of the ideal generated by G, it is known that it is determined uniquely.

Theorem 2 *Let $I \subset \mathbb{K}[x_1, \dots, x_n]$ be an ideal, and G a Gröbner basis of I with respect to a term order $\prec$. Then, for any $f \in \mathbb{K}[x_1, \dots, x_n]$ there exists a unique remainder r of f on division by G. In particular, $f \in I$ if and only if $r = 0$.*

Example 4 Let $a, b, c \in \mathbb{C}$ be complex numbers, satisfying

$$a + b + c = 5, ab + bc + ca = 7, abc = 9. \tag{1}$$

Then, let us compute the value of $a^5 + b^5 + c^5$. Let

$$I = \langle a + b + c - 5, ab + bc + ca - 7, abc - 9 \rangle$$

be an ideal of the polynomial ring $\mathbb{C}[a, b, c]$ with indeterminates a, b, c. The Gröbner basis of I with respect to the lexicographic order $c \prec_{\mathrm{lex}} b \prec_{\mathrm{lex}} a$ is

$$G = \{c^3 - 5c^2 + 7c - 9, b^2 + (c - 5)b + c^2 - 5c + 7, a + b + c - 5\}.$$

By division algorithm, we have $a^5 + b^5 + c^5 \overset{G}{\to} 785$. Thus $a^5 + b^5 + c^5 = 785$ under the conditions (1).

Fig. 1 Triangle ABC

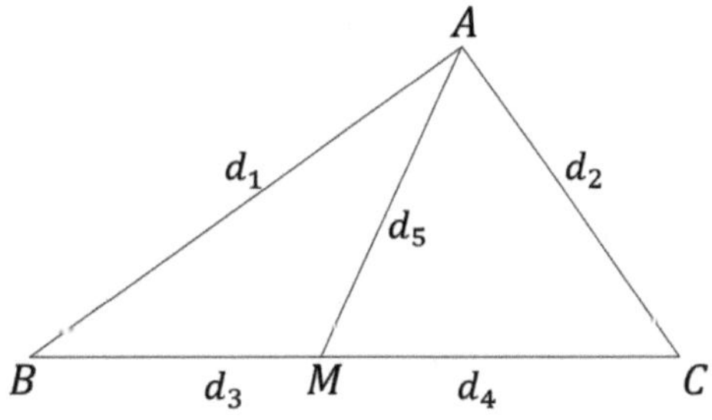

Let I be an ideal of $\mathbb{K}[x_1, \ldots, x_n, y_1, \ldots, y_m]$. Using Gröbner basis with respect to elimination order, the elimination ideal $I \cap \mathbb{K}[y_1, \ldots, y_m]$ can be computed. Recall that the elimination ideal corresponds to the image of the projection map.

Theorem 3 (Elimination theorem) *Let I be an ideal of $\mathbb{K}[x_1, \ldots, x_n, y_1, \ldots, y_m]$, and $\prec$ an elimination order $\{x_1, \ldots, x_n\} \gg \{y_1, \ldots, y_n\}$. Let G be a Gröbner basis of I with respect to $\prec$. Then $G \cap \mathbb{K}[y_1, \ldots, y_m]$ is a Gröbner basis of $I \cap \mathbb{K}[y_1, \ldots, y_m]$ with respect to the term order on $\mathbb{K}[y_1, \ldots, y_m]$ induced by $\prec$.*

Example 5 Let ABC be a triangle and M a point of the segment BC. Let $AB = d_1$, $AC = d_2$, $BM = d_3$, $MC = d_4$, and $AM = d_5$ as in Fig. 1. Using Gröbner basis, we are able to obtain the relation among the length of the segments d_i's. We place the triangle on the Euclidean plane so that $M = (0, 0)$, $B = (-d_3, 0)$, and $C = (d_4, 0)$. Let $A = (a, b)$. Then

$$(d_3 + a)^2 + b^2 = d_1^2, \quad (d_4 - a)^2 + b^2 = d_2^2, \quad a^2 + b^2 = d_5^2.$$

Let $\mathbb{K}[d_1, \ldots, d_5, a, b]$ be the polynomial ring over $\mathbb{K}$ with indeterminates $d_1, \ldots, d_5$, a, b, and $I = \langle (d_3 + a)^2 + b^2 - d_1^2, (d_4 - a)^2 + b^2 - d_2^2, a^2 + b^2 - d_5^2 \rangle \subset \mathbb{K}[d_1, \ldots, d_5, a, b]$. Let $\prec$ be the block order of the reverse lexicographic order on $\mathbb{K}[a, b]$ such that $a \succ b$, and the reverse lexicographic order on $\mathbb{K}[d_1, \ldots, d_5]$ such that $d_1 \succ \cdots \succ d_5$. Then the Gröbner basis of I with respect to $\prec$ is

$$\begin{aligned}
\{ \; & -d_4 d_3^2 + (-d_4^2 + d_2^2 - d_5^2) d_3 + d_4 d_1^2 - d_5^2 d_4, \\
& -2d_4 a + d_4^2 - d_2^2 + d_5^2, -2d_3 a - d_3^2 + d_1^2 - d_5^2, \\
& (-2d_2^2 + 2d_5^2) a + 4d_4 b^2 + d_4^3 + (-d_2^2 - 3d_5^2) d_4, \\
& (2d_1^2 - 2d_5^2) a + 4d_3 b^2 + d_3^3 + (-d_1^2 - 3d_5^2) d_3, a^2 + b^2 - d_5^2 \},
\end{aligned}$$

and thus

$$\begin{aligned}
I \cap \mathbb{K}[d_1, \ldots, d_5] &= \langle -d_4 d_3^2 + (-d_4^2 + d_2^2 - d_5^2) d_3 + d_4 d_1^2 - d_5^2 d_4 \rangle \\
&= \langle d_1^2 d_4 + d_2^2 d_3 - (d_3 + d_4)(d_5^2 + d_3 d_4) \rangle.
\end{aligned}$$

This shows that for any triangle,

$$d_1^2 d_4 + d_2^2 d_3 = (d_3 + d_4)(d_5^2 + d_3 d_4)$$

holds. This equality is known as the Stewart's theorem.

Example 6 (Intersection of ideals) Let $I, J \subset \mathbb{K}[x_1, \ldots, x_n]$ be ideals. Let t be a new indeterminate. Then

$$I \cap J = \big(tI + (1-t)J\big) \cap \mathbb{K}[x_1, \ldots, x_n]$$

where $tI + (1-t)J$ is the ideal of $\mathbb{K}[x_1, \ldots, x_n, t]$ generated by tI and $(1-t)J$. Thus, we are able to compute the intersection of ideals using the elimination theory.

Example 7 (The Zariski closure of the image of a polynomial map) Let $p_1, \ldots, p_m \in \mathbb{K}[x_1, \ldots, x_n]$ be polynomials, and $\mathbf{p} : \mathbb{K}^n \to \mathbb{K}^m$, $\mathbf{u} \mapsto (p_1(\mathbf{u}), \ldots, p_m(\mathbf{u}))$, a polynomial map. Then $\mathbf{p}$ is decomposed into the graph embedding $g : \mathbb{K}^n \to \mathbb{K}^n \times \mathbb{K}^m$ and the projection map $pr : \mathbb{K}^n \times \mathbb{K}^m \to \mathbb{K}^m$:

$$\begin{aligned} \mathbf{p} : \mathbb{K}^n &\xrightarrow{g} \mathbb{K}^n \times \mathbb{K}^m \xrightarrow{pr} \mathbb{K}^m, \\ \mathbf{u} &\mapsto (\mathbf{u}, p(\mathbf{u})) \mapsto p(\mathbf{u}). \end{aligned}$$

Thus $\text{Image}(\mathbf{p}) = pr(\text{Image}(\mathbf{g}))$. It is easy to show that $\text{Image}(\mathbf{g})$ is an affine variety $\mathscr{V}(y_1 - p_1, \ldots, y_m - p_m)$. Thus the Zariski closure of the image of a polynomial map $\mathbf{p}$ is $\mathscr{V}(\langle y_1 - p_1, \ldots, y_m - p_m\rangle \cap \mathbb{K}[y_1, \ldots, y_n])$ by Proposition 2.

3 Gröbner Bases of Toric Ideals and Its Application

We conclude this Chapter with some application of Gröbner bases of toric ideals.

The multiplicative group $(\mathbb{K}^*)^d$ where $\mathbb{K}^* = \mathbb{K}\backslash\{0\}$ is called the *algebraic torus* of dimension d.

Definition 14 (*Affine toric variety and toric ideals*) Let $A = (\mathbf{a}_1, \ldots, \mathbf{a}_n) \in \mathbb{Z}^{d\times n}$ be a $d \times n$ integral matrix. The *affine toric variety* associated to A is the Zariski closure of the image of a monomial map

$$\begin{aligned} (\mathbb{K}^*)^d &\to \mathbb{K}^n \\ \mathbf{u} = (u_1, \ldots, u_n) &\mapsto (\mathbf{u}^{\mathbf{a}_1}, \ldots, \mathbf{u}^{\mathbf{a}_1}) \end{aligned}$$

The defining ideal of this affine toric variety is called *toric ideal*of A, and is denoted by I_A.

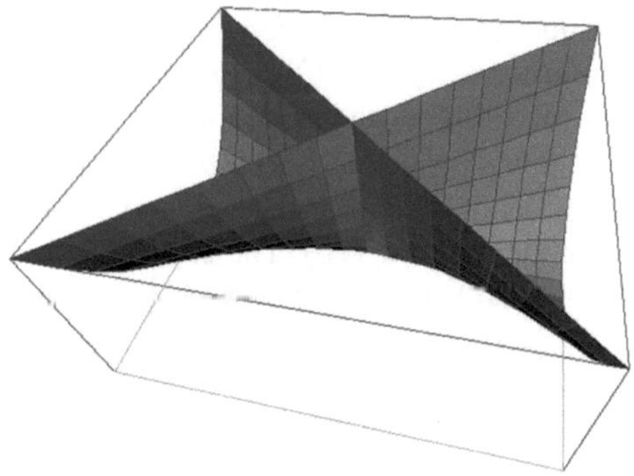

Fig. 2 Whitney umbrella

The toric ideal is generated by binomials;

$$I_A = \langle \mathbf{x}^{\mathbf{u}} - \mathbf{x}^{\mathbf{u}} \mid A\mathbf{u} = A\mathbf{v}, \mathbf{u}, \mathbf{v} \in \mathbb{Z}^n_{\geq 0} \rangle.$$

For any term order $\prec$, I_A admits a Gröbner basis consisting of binomials.

Example 8 The *Whitney umbrella* is the surface $\{(uv, u, v^2) \mid u, v \in \mathbb{R}\} \subset \mathbb{R}^3$ (Fig. 2). The Zariski closure of this surface is an affine toric variety $\mathbb{V}_{\mathbb{R}}(x^2 - y^2 z)$. We note that

$$\mathbb{V}_{\mathbb{R}}(x^2 - y^2 z) \supsetneq \{(uv, u, v^2) \mid u, v \in \mathbb{R}\}$$

since the negative z-axis is contained in the left-hand side.

Let $A = (\mathbf{a}_1, \dots, \mathbf{a}_n) \in \mathbb{Z}^{d \times n}$ be a $d \times n$ integral matrix, $\mathbf{c} = (c_1, \dots, c_n) \in \mathbb{Q}^n_{\geq 0}$ a cost vector, and $\mathbf{b} \in \mathbb{Z}^d_{\geq 0}$. We are concerned with an *integer programming*

$$\begin{aligned} \text{Minimize } & \mathbf{c} \cdot \mathbf{u} \\ \text{subject to } & A\mathbf{u} = \mathbf{b}, \\ & \mathbf{u} \in \mathbb{Z}^n_{\geq 0}. \end{aligned} \tag{2}$$

There is an algorithm by Conti and Traverso for integer programming using the Gröbner basis of the toric ideal.

Theorem 4 (Conti-Traverso algorithm [2]) *Let $\prec_{\mathbf{c}}$ be a weighted term order on $\mathbb{K}[x_1, \dots, x_n]$, and G a Gröbner basis of the toric ideal I_A. Take $\mathbf{u} \in \mathbb{Z}^n_{\geq 0}$ such that $A\mathbf{u} = \mathbf{b}$. Let $\mathbf{x}^{\mathbf{v}}$ be the remainder of $\mathbf{x}^{\mathbf{u}}$ on division by G. Then $\mathbf{v}$ is an optimal solution to the integer programming (2).*

The linear programming relaxation of the integer programming (2) is

$$\begin{aligned} \text{Minimize } & \mathbf{c} \cdot \mathbf{u} \\ \text{subject to } & A\mathbf{u} = \mathbf{b}, \\ & \mathbf{u} \in \mathbb{Q}^n_{\geq 0}. \end{aligned} \tag{3}$$

The maximum difference of the optimal values of (3) and (2) as $\mathbf{b}$ ranges in the semigroup $\mathbb{Z}_{\geq 0}A = \{\sum_{i=1}^{n} m_i\mathbf{a}_i \mid m_i \in \mathbb{Z}_{\geq 0}\}$ is called the *integer programming gap* of A and $\mathbf{c}$, and is denoted by gap$(A, \mathbf{c})$. We note that (2) is feasible if and only if $\mathbf{b} \in \mathbb{Z}_{\geq 0}A$. Hoşten and Sturmfels gave a method to compute gap$(A, \mathbf{c})$. Fix a weighted term order $\prec_{\mathbf{c}}$, and let $M = \langle x_{i_1}^{u_1+1}, \ldots, x_{i_\ell}^{u_\ell+1}\rangle$ be an irreducible component of $\mathrm{in}_{\prec_\mathbf{c}}(I_A)$. The *gap value* of M is $c_{i_1}u_1 + \cdots + c_{i_\ell}u_\ell - c^*$ where c^* is the optimal value of the liner programming

$$\begin{aligned} &\text{Minimize } \mathbf{c}\cdot\mathbf{v} \\ &\text{subject to } A\mathbf{v} = u_i\mathbf{a}_{i_1} + \cdots + u_\ell\mathbf{a}_{i_\ell}, \\ &\qquad\qquad \mathbf{v} = (v_1, \ldots, v_n) \in \mathbb{Q}^n, v_{i_1}, \ldots, v_{i_\ell} \in \mathbb{Q}_{\geq 0}. \end{aligned}$$

Theorem 5 ([8]) *The integer programming gap* gap(A, c) *equals the maximum gap value of any irreducible component of* $\mathrm{in}_{\prec_\mathbf{c}}(I_A)$.

The set of nonnegative integer vectors $\mathscr{F}_A(\mathbf{b}) := \{\mathbf{u} \in \mathbb{Z}_{\geq 0}^n \mid A\mathbf{u} = \mathbf{b}\}$ is called the *fiber space* of A over $\mathbf{b}$. This is the feasible region of the integer programming (2). As stated in [7], the enumeration of the fibers can be achieved by an algorithm based on the *reverse search* technique [1].

Let $G = \{g_1, \ldots, g_r\}$ where $g_i = \mathbf{x}^{\mathbf{u}_i} - \mathbf{x}^{\mathbf{v}_i}$ be a Gröbenr basis of the toric ideal I_A with respect to $\prec_\mathbf{c}$. We assume that $\mathrm{in}_{\prec_\mathbf{c}}(g_i) = \mathbf{x}^{\mathbf{u}_i}$. We define a directed graph $\mathscr{G} = (V_\mathscr{G}, E_\mathscr{G})$ as follows; the vertex set $V_\mathscr{G}$ is the fiber space $\mathscr{F}_A(\mathbf{b})$, and the edge set $E_\mathscr{G}$ is $\{(\mathbf{u}, \mathbf{v}) \mid \mathbf{x}^\mathbf{u} \xrightarrow{G} \mathbf{x}^\mathbf{v}\}$. Let $\mathbf{v}^*$ the optimal solution to the integer programming (2) obtained by the Coti–Traverso algorithm. Since the division algorithm terminates in finite time, $\mathscr{G}$ has no loop, and by the Conti–Traverso algorithm, $\mathscr{G}$ has the unique sink $\mathbf{v}^*$. For each $\mathbf{u} \in \mathscr{F}_A(\mathbf{b})$, let $j = \min\{i \mid \mathbf{x}^\mathbf{u}$ is divisible by $\mathrm{in}_{\prec_\mathbf{c}}(g_i)\}$, and define $f(\mathbf{u}) = \mathbf{u} + (\mathbf{v}_j - \mathbf{u}_j)$. Then, $\mathbf{x}^\mathbf{u} \xrightarrow{G} \mathbf{x}^{f(\mathbf{u})}$, and the subgraph $\mathscr{T} = \{\mathscr{F}_A(\mathbf{b}), \{(\mathbf{u}, f(\mathbf{u})) \mid \mathbf{u} \in \mathscr{F}_A(\mathbf{b}\}\}$ of $\mathscr{G}$ is a spanning tree of $\mathscr{G}$. The following algorithm with the input $\mathbf{v}^*$ outputs all vectors in $\mathscr{F}_A(\mathbf{b})$.

Algorithm $B(\mathbf{u})$
Input: A fiber $\mathbf{u} \in \mathscr{F}_A(\mathbf{b})$.
Output: All descendant nodes of $\mathbf{u}$ of the tree $\mathscr{T}$.

1: Output $\mathbf{u}$.
2: Compute the set $S = \{\mathbf{v} \mid \mathbf{x}^\mathbf{v} \xrightarrow{G} \mathbf{x}^\mathbf{u}\}$.
3: For every $\mathbf{v} \in S$, if $(f(\mathbf{v}) = \mathbf{u})\{\ B(\mathbf{v})\ \}$.

The size of the fiber space can be too large to enumerate completely. Alternatively, we are able to sample a set of fibers by performing a random walk on the connected graph $\mathscr{G}$. Using this sampling algorithm, Diaconis and Sturmfels [5] developed a Markov Chain Monte Carlo method (MCMC) for sampling contingency tables.

References

1. D. Avis, K. Fukuda, Reverse search for enumeration. Discrete Appl. Math. **65**, 21–46 (1996)
2. P. Conti, C. Traverso, Buchberger algorithm and integer programming, in *Proceedings of the AAECC-9*. LNCS, vol. 539 (Springer, New Orleans, 1991), pp. 130–139
3. D. Cox, J. Little, D. O'Shea, *Ideals, Varieties and Algorithms* (Springer, Berlin, 1992)
4. D. Cox, J. Little, D. O'Shea, *Using Algebraic Geometry* (Springer, Berlin, 1998)
5. P. Diaconis, B. Sturmfels, Algebraic algorithms for sampling from conditional distributions. Ann. Statist. **26**(1), 363–397 (1998)
6. D. Eisenbud, *Commutative Algebra with a View Toward Algebraic Geometry* (Springer, Berlin, 1995)
7. B. Sturmfels, *Gröbner Bases and Convex Polytopes*, (Lectures Series), vol. 8, (American Mathematics Society, Providence, 1996)
8. S. Hoşten, B. Sturmfels, Computing the integer programming gap. Combinatorica **27**, 367–382 (2007)

Part II
Geometry

Stability Analysis for Variational Problems for Surfaces with Constraint

Miyuki Koiso

Abstract Surfaces with constant mean curvature (CMC surfaces) are critical points of the area functional among surfaces enclosing the same volume. Therefore, they are a simple example of solutions of variational problem with constraint. A CMC surface is said to be stable if the second variation of the area is nonnegative for all volume-preserving variations satisfying the given boundary condition. The purpose of this article is to show some fundamental methods to study the stability for CMC surfaces. Especially, we give a criterion on the stability for compact CMC surfaces with prescribed boundary. Another concept that is closely related to the stability for CMC surfaces is the so-called bifurcation. We give sufficient conditions on a one-parameter family of CMC surfaces so that there exists a bifurcation. Moreover, we give a criterion for CMC surfaces in the bifurcation branch to be stable.

Keywords Bifurcation · Constant mean curvature · Pitchfork bifurcation · Stability · Symmetry breaking · Variational problem

1 Introduction

Surfaces with constant mean curvature (CMC surfaces) are critical points of the area functional among surfaces enclosing the same volume and satisfying the given boundary condition (see Sect. 2). For this reason, sometimes they serve as mathematical model of soap bubbles. If we consider physical phenomena, it is important to judge whether a CMC surface attains a local minimum of the area functional or not. A CMC surface is said to be stable if the second variation of the area is nonnegative for all volume-preserving variations satisfying the given boundary condition.

M. Koiso (✉)
Institute of Mathematics for Industry, Kyushu University, 744, Motooka, Nishi-ku, Fukuoka 819-0395, Japan
e-mail: koiso@imi.kyushu-u.ac.jp

R. Nishii et al. (eds.), *A Mathematical Approach to Research Problems of Science and Technology*, Mathematics for Industry 5,
DOI: 10.1007/978-4-431-55060-0_6,

Judging stability is an important subject both in mathematics and in applications. In this article, we give some fundamental methods to study the stability for CMC surfaces. Although we treat mainly immersed CMC surfaces in the three-dimensional euclidean space $\mathbf{R}^3$, our methods are generalized to hypersurfaces in more general Riemannian manifolds. Also, the area functional can be generalized to more general functionals for which the Jacobi operator is self-adjoint (see Sect. 3).

The contents of this article are as follows. In Sect. 2, we give the definition of mean curvature and the simplest examples of CMC surfaces: CMC surfaces of revolution. In Sect. 3, we give the first variation formulas for the area, volume, and mean curvature. We also give the second variation formula for the area and the definition of the stability. Section 4 is devoted to the uniqueness theorem on closed stable CMC hypersurfaces in the $(n+1)$-dimensional euclidean space $\mathbf{R}^{n+1}$. Actually such hypersurfaces are only round spheres [1]. We will give an outline of the proof because it is a good example to use a nice geometric comparison function. In fact, except for round spheres, for any closed CMC hypersurface there is a volume-preserving variation, which diminishes the area. In Sect. 5, we give a criterion on the stability for compact CMC surfaces with prescribed boundary and give some simple examples. In Sect. 6, we also study compact CMC surfaces with prescribed boundary. A so-called bifurcation is closely related to the stability of CMC surfaces. We give sufficient conditions on a one-parameter family of CMC surfaces so that there exists a bifurcation. Moreover, we give a criterion for CMC surfaces in this bifurcation branch to be stable. We will give sufficient conditions so that we have the so-called pitchfork bifurcations, and so that there exists an interesting phenomenon that a one-parameter family of stable solutions with high symmetry bifurcates to unstable solutions with the high symmetry and stable solutions with a lower symmetry.

For simplicity, from Sects. 2 to 5, we assume that all functions and mappings are of C^∞ if we do not state anything special.

2 Definition of Mean Curvature and Examples of Surfaces with Constant Mean Curvature

Let Σ be a connected oriented two-dimensional C^∞ manifold (with or without boundary). We denote by (u^1, u^2) the local coordinates of Σ. Let $X = (x^1, x^2, x^3) : \Sigma \to \mathbf{R}^3$ be an immersion. We denote by $\nu = (\nu^1, \nu^2, \nu^3) : \Sigma \to S^2 := \{\nu = (\nu^1, \nu^2, \nu^3) \in \mathbf{R}^3;\ |\nu| = 1\}$ the Gauss map of X. That is, ν is the unit normal vector field along X and $\{X_{u^1}, X_{u^2}, \nu\}$ gives a frame in $\mathbf{R}^3$ with positive orientation.

Remark 1 A C^1 mapping $X = (x^1, x^2, x^3) : \Sigma \to \mathbf{R}^3$ is an immersion if

$$\operatorname{rank}\begin{pmatrix} x^1_{u^1} & x^2_{u^1} & x^3_{u^1} \\ x^1_{u^2} & x^2_{u^2} & x^3_{u^2} \end{pmatrix} = 2$$

is satisfied at every point in Σ.

We denote by $\langle \mathbf{u}, \mathbf{v} \rangle$ the canonical inner product of $\mathbf{u}, \mathbf{v} \in \mathbf{R}^3$.

The first and the second fundamental forms of X are denoted by $\mathrm{d}s^2$, II, respectively, and they are given by

$$\mathrm{d}s^2 = g_{ij}\mathrm{d}u^i\mathrm{d}u^j, \quad \left(g_{ij} := \langle X_i, X_j \rangle,\ X_i := \partial X/\partial u^i\right),$$
$$\mathrm{II} = h_{ij}\mathrm{d}u^i\mathrm{d}u^j, \quad \left(h_{ij} := \langle \nu, X_{ij} \rangle = -\langle \nu_i, X_j \rangle,\ X_{ij} := \partial^2 X/\partial u^i \partial u^j\right),$$

here we used the Einstein convention, that is, for example, $h_{ij}\mathrm{d}u^i\mathrm{d}u^j$ means $\sum_{i,j=1}^{2} h_{ij}\mathrm{d}u^i\mathrm{d}u^j$. Set $(g^{ij}) := (g_{ij})^{-1}$. Then the mean curvature H and the Gauss curvature K of X are defined as

$$H = h_{ij}g^{ij}/2, \quad K = \det(h_{ij})/\det(g_{ij}).$$

$H\nu$ is called the mean curvature vector of X. The equation "H = constant" is a second-order quasi-linear elliptic partial differential equation.

Hereafter, an immersed surface is called a surface. An immersion with constant mean curvature is called a CMC surface.

Example 1 CMC surfaces of revolution are named Delaunay surfaces after a French mathematician C.-E. Delaunay of the nineteenth century. Consider a smooth curve $\Gamma : (x(s), z(s))$ $(x \geq 0)$ with arc-length s. Γ generates the surface of revolution

$$X(s, \theta) = (x(s)\cos\theta, x(s)\sin\theta, z(s)),$$

and the mean curvature of X is $H = (x''z' - x'z'' - x^{-1}z')/2$. By simple calculations, we see that Γ is represented as

$$z = \pm \int \frac{c - Hx^2}{\sqrt{x^2 - \left(c - Hx^2\right)^2}}\,\mathrm{d}x, \tag{1}$$

where c is a constant. Hence, Delaunay surfaces are two-parameter family of surfaces. They are classified into six classes: plane, catenoid, cylinder, unduloid, sphere, and nodoid (see Fig. 1).

Definition 1 A compact surface without boundary is called a closed surface. For example, spheres are closed surfaces.

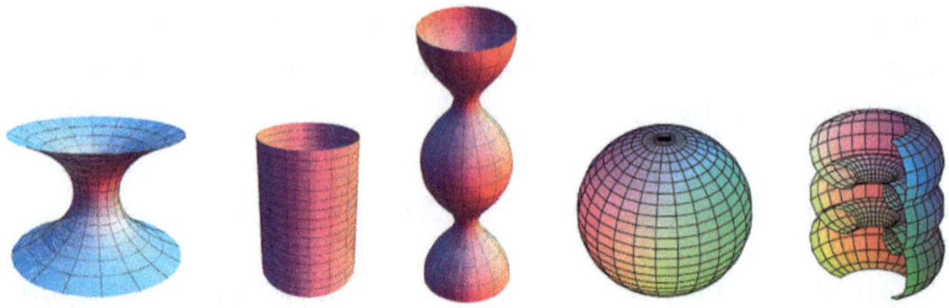

Fig. 1 Delaunay surfaces. *From the left* catenoid, cylinder, unduloid, sphere, and nodoid

3 Variation Formulas and the Definition of Stability

Assume that Σ is compact with or without boundary. For an immersion $X : \Sigma \to \mathbf{R}^3$, the area $A(X)$ and the volume $V(X)$ of X are defined as follows.

$$A(X) = \int_\Sigma \mathrm{d}\Sigma, \quad V(X) = \frac{1}{3} \int_\Sigma \langle X, \nu \rangle \, \mathrm{d}\Sigma,$$

where $\mathrm{d}\Sigma := \sqrt{\det(g_{ij})}\ \mathrm{d}u^1 \mathrm{d}u^2$ is the volume form (area element) of Σ induced by X.

Remark 2 $V(X)$ is the "algebraic volume" of the cone generated by the immersed surface $X(\Sigma)$ and the origin of $\mathbf{R}^3$. If $X(\Sigma)$ is an embedded closed surface (that is, Σ is a compact manifold without boundary and X is an injective mapping: roughly speaking, $X(\Sigma)$ is a smooth closed surface without self-intersection) and ν points outward from the domain Ω bounded by $X(\Sigma)$, then $V(X)$ coincides with the canonical volume of Ω.

Let X_ε be a variation of X that fixes the boundary with variation parameter ε. This means that $X_* : \Sigma \times (-\varepsilon_0, \varepsilon_0) \to \mathbf{R}^3$ is a C^∞ mapping and satisfies

$$X_0(w) = X(w), \quad \forall w \in \Sigma, \qquad X_\varepsilon(\zeta) = X(\zeta), \quad \forall \zeta \in \partial\Sigma.$$

In other words, X_ε can be written as $X_\varepsilon = X + (\xi + f\nu)\varepsilon + \mathcal{O}(\varepsilon^2)$, where ξ and $f\nu$ are the tangential and the normal component of the variation vector field $\delta X := (\partial X_\varepsilon / \partial \varepsilon)_{\varepsilon=0} = \xi + f\nu$, and both of them are of C^∞ on Σ and vanish on $\partial\Sigma$.

Lemma 1 *The first variation formulas of A and V are given by the following.*

$$\delta A := \frac{\mathrm{d}}{\mathrm{d}\varepsilon} A(X_\varepsilon)|_{\varepsilon=0} = -2 \int_\Sigma Hf \, \mathrm{d}\Sigma, \qquad \delta V = \int_\Sigma f \, \mathrm{d}\Sigma. \tag{2}$$

Proof The first formula is standard, and the second formula is proved in [2]. □

Remark 3 If the variation $X_\varepsilon = X + (\xi + f\nu)\varepsilon + \mathcal{O}(\varepsilon^2)$ does not fix the boundary, then

$$\delta A = -2\int_\Sigma Hf \, d\Sigma + \int_{\partial\Sigma} \langle \xi, \eta \rangle \, ds, \tag{3}$$

where η is the outward-pointing unit conormal to X along $\partial\Sigma$, and ds is the line element of $\partial\Sigma$ induced by X. The first variation of V is the same as that in (2).

Remark 4 By the first formula in (2) and the formula (3), we know that the variation $X_\varepsilon = X + \varepsilon H\nu$ of X in the direction $H\nu$ diminishes the area of the surface.

The following lemma is important to study volume-preserving variations.

Lemma 2 *If a variation* $X_\varepsilon = X + (\xi + f\nu)\varepsilon + \mathcal{O}(\varepsilon^2)$ *of* X *fixes the boundary and is volume-preserving, then* $f \in C_0^\infty(\Sigma)$ *and* $\int_\Sigma f \, d\Sigma = 0$ *hold. Conversely, for any function* $f \in C_0^\infty(\Sigma)$ *satisfying* $\int_\Sigma f \, d\Sigma = 0$*, there exists a volume-preserving variation* $X_\varepsilon = X + f\nu\varepsilon + \mathcal{O}(\varepsilon^2)$ *of* X *that fixes the boundary.*

Proof The first half is an immediate consequence of Lemma 1. The proof of the second half is given in [1], where the implicit mapping theorem is essentially used. □

Remark 5 In Lemma 2, we can weaken the assumption about the regularity of f: Even if we assume that f is only C^0 and piecewise C^∞ on Σ, a similar result holds.

In order to study variational problems with constraint, it is useful to consider the so-called Lagrange multiplier. So we define a new functional: For any $H \in \mathbf{R}$, set

$$J_H := A + 2HV.$$

The following result is proved by using Lemmas 1 and 2.

Theorem 1 ([1]) *Assume that* $X : \Sigma \to \mathbf{R}^3$ *is an immersion with mean curvature* H*. Set*

$$H_0 := (A(X))^{-1} \int_\Sigma H \, d\Sigma.$$

Then, the following (i)–(iii) *are equivalent.*

(i) *The mean curvature of* X *is constant* H_0 *on* Σ.
(ii) *For any volume-preserving variation of* X *that fixes the boundary,* $\delta A = 0$.
(iii) *For any variation of* X *that fixes the boundary,* $\delta J_{H_0} = 0$.

The first variation of the mean curvature is given by the following Proposition. The proof can be found, for example, in [12, pp. 150–151].

Proposition 1 *Let $X : \Sigma \to \mathbf{R}^3$ be an immersion. Let $X_\varepsilon = X + \left(\xi^i X_i + f\nu\right)\varepsilon + \mathcal{O}\left(\varepsilon^2\right)$ be a variation of X. Then, the first variation of the mean curvature H is given by the following.*

$$\delta H = L[f]/2 + \xi^i H_i, \tag{4}$$

where

$$L[f] := \Delta f + \|\mathrm{d}\nu\|^2 f \tag{5}$$

is the self-adjoint operator. Especially, if X is CMC, then $\delta H = L[f]/2$.

Definition 2 We call L the Jacobi operator for X.

Remark 6 $\Delta f = g^{ij} f_{ij} + \sqrt{g}^{-1}\left(\sqrt{g}g^{ij}\right)_i f_j$, $g := \det\left(g_{ij}\right)$, $\|\mathrm{d}\nu\|^2 = 4H^2 - 2K$.

Proposition 2 *Assume that X has CMC H_0. Then, for any volume-preserving variation of X that fixes the boundary, the second variation of the area A is given by the following.*

$$\delta^2 A := \frac{\mathrm{d}^2}{\mathrm{d}\varepsilon^2} A(X_\varepsilon)|_{\varepsilon=0} = -\int_\Sigma f L[f]\,\mathrm{d}\Sigma, \quad f := \langle \delta X, \nu\rangle, \tag{6}$$

where δX is the variation vector field. For any variation of X that fixes the boundary, the second variation of $J_{H_0} = A + 2H_0 V$ is given by the same form as follows.

$$\delta^2 J_{H_0} = -\int_\Sigma f L[f]\,\mathrm{d}\Sigma, \quad f := \langle \delta X, \nu\rangle. \tag{7}$$

Proof Let $X_\varepsilon = X + \varepsilon(\xi + f\nu) + \mathcal{O}\left(\varepsilon^2\right)$ be a volume-preserving variation of X that fixes the boundary. Then,

$$\delta A = \delta A + 2H_0 \delta V = -2\int_\Sigma (H - H_0)\, f\, \mathrm{d}\Sigma.$$

Therefore,

$$\delta^2 A = -2\int_\Sigma (\delta(H - H_0)) f\, \mathrm{d}\Sigma - 2\int_\Sigma (H - H_0)\delta(f\, \mathrm{d}\Sigma)$$

holds. Remark that, when $\varepsilon = 0$, $H \equiv H_0$ holds. Hence, by Proposition 1, we obtain (6). (7) is proved in a similar way. □

Set

$$I(f) := -\int_\Sigma f L[f]\,\mathrm{d}\Sigma.$$

We define the stability of CMC surfaces as follows.

Definition 3 Assume that X is a CMC immersion. X is said to be stable if $\delta^2 A \geq 0$ for all volume-preserving variations of X that fix the boundary. If X is not stable, it is said to be unstable.

From Lemma 2 and Proposition 2, we immediately obtain the following.

Lemma 3 *Assume that $X; \Sigma \to \mathbf{R}^3$ is CMC. Set*

$$F_0 := \left\{ f \in C_0^\infty(\Sigma) \,;\, \int_\Sigma f \, d\Sigma = 0 \right\}.$$

Then, X is stable if and only if $I(f) \geq 0$ holds for all $f \in F_0$.

Remark 7 It is obvious that, if $X : \Sigma \to \mathbf{R}^3$ is a stable CMC surface, then, for any subdomain $\Sigma_1 \subset \Sigma$, $X|_{\Sigma_1}$ is stable.

In order to study the stability and bifurcation for CMC surfaces, the following eigenvalue problem is useful (see Sect. 5, 6).

$$L[f] = -\lambda f, \qquad f \in C_0^\infty(\Sigma). \tag{8}$$

The following lemma is sometimes useful to estimate the eigenvalues of (8).

Lemma 4 *Assume that X has CMC H with Gauss map $\nu = \left(\nu^1, \nu^2, \nu^3\right)$. Set $E_1 :=$ $(1, 0, 0)$, $E_2 := (0, 1, 0)$, $E_3 := (0, 0, 1)$. Then, the following equalities hold.*

$$L\left[\nu^j\right] = 0, \quad L\left[\langle E_j \times X, \nu\rangle\right] = 0, \quad (j = 1, 2, 3), \tag{9}$$

$$L[\langle X, \nu\rangle] = -2H. \tag{10}$$

Proof By Proposition 1, for any variation X_ε of X with $\left\langle \delta X, \nu \right\rangle = f$, we have $2\delta H = L[f]$. Let u be any constant vector in $\mathbf{R}^3$. Since the translation $X_\varepsilon = X + \varepsilon u$ does not change the mean curvature H, $\nu_u := \langle \nu, u\rangle$ satisfies $L[\nu_u] = 0$. By applying this to $u = E_j$, we have the first equality in (9). Similarly, since the rotation $X_\varepsilon =$ $X + \varepsilon E_j \times X + \mathscr{O}\left(\varepsilon^2\right)$ does not change H, we obtain the second equality in (9). On the other hand, by the homothetic transformation $X_\varepsilon = (1 + \varepsilon)X$, H becomes $H/(1 + \varepsilon)$. This gives (10). □

Remark 8 The function $\sigma := \langle X, \nu\rangle$ appeared in (10) is called the support function of X. For each $w \in \Sigma$, $\sigma(w)$ is the $(\pm)$ distance between the origin in $\mathbf{R}^3$ and the tangent plane of the surface at $X(w)$.

4 Uniqueness for Stable Closed CMC Hypersurfaces

The discussion in Sect. 3 is generalized to immersed hypersurfaces in $\mathbf{R}^{n+1}$. In this section, we give the uniqueness result on stable closed CMC hypersurfaces in $\mathbf{R}^{n+1}$ and give an outline of the proof.

Theorem 2 (Barbosa-do Carmo [1]) *Let $\Sigma = \Sigma^n$ be a compact connected orientable C^∞ manifold, and let $X : \Sigma \to \mathbf{R}^{n+1}$ be an immersion with nonzero CMC. Then, X is stable if and only if $X(\Sigma)$ is a round sphere.*

An outline of the proof of Theorem 2 given in [1] is as follows. Denote by H, ν, the mean curvature and the Gauss map of X, respectively. Set

$$f := H\sigma + 1, \quad \sigma = \langle X, \nu \rangle. \tag{11}$$

Then, by using the constancy of H, it is proved that

$$\int_\Sigma f \, \mathrm{d}\Sigma = 0$$

holds. And it is proved that $I(f) \geq 0$ holds if and only if all principal curvatures are the same at each point of $X(\Sigma)$, which is equivalent to that $X(\Sigma)$ is a round sphere.

Wente [24] gave an excellent explanation of the variation vector field $f\nu$ given by (11) as follows. Recall that the variation $X_\varepsilon = X + \varepsilon H\nu$ decreases the area (Remark 4). However, in general, X_ε does not preserve the volume. So, consider the variation $\tilde{X}_\varepsilon = t(\varepsilon)(X + \varepsilon H\nu)$ of X. Here, $t(\varepsilon)$ is chosen so that $\tilde{X}_\varepsilon$ is volume-preserving. Then, it is proved that $\delta^2 A \geq 0$ holds if and only if the principal curvatures of X are all the same, which means that X is a round sphere. Moreover, [24] pointed out that $\tilde{f} := \langle \delta\tilde{X}_\varepsilon, \nu \rangle$ coincides with f given by (11) up to constant multiple.

Remark 9 There are many unstable closed CMC surfaces in $\mathbf{R}^3$. For example, for each $g \in \mathbf{N}$, there exist closed CMC surfaces in $\mathbf{R}^3$ with genus g. Such examples were constructed by Wente [23] for $g = 1$, and by Kapouleas (in [5] for $g \geq 3$ and in [6] for $g = 2$) for the first time.

5 A Criterion for the Stability

Let $X : \Sigma \to \mathbf{R}^3$ be an immersion with constant mean curvature H. We consider the following eigenvalue problem associated with the second variation of the area.

$$L[u] = -\lambda u, \quad u \in C_0^\infty(\Sigma). \tag{12}$$

Since L is a second order self-adjoint elliptic operator, all eigenvalues are real and they constitute a countably many nondecreasing sequence [17, Lemma 1]. We denote them by $\lambda_1 < \lambda_2 \leq \lambda_3 \leq \cdots$. The number of negative eigenvalues (with multiplicity) is called the Morse index of X, and we denote it by $\mathrm{Ind}(X)$. $\mathrm{Ind}(X)$ is the dimension of the space of variation vector fields that fix the boundary and diminish the functional $A + 2HV$. Therefore, we cannot judge the stability only by the eigenvalues of (12).

On the other hand, the eigenvalue problem associated with the second variation of the area for volume-preserving variations is given by

$$\tilde{L}[u] := L[u] + c = -\tilde{\lambda} u \text{ on } \Sigma, \quad \exists c \in \mathbf{R}, \quad u \in F_0 - \{0\}. \tag{13}$$

We can denote the eigenvalues of (13) by $\tilde{\lambda}_1 \leq \tilde{\lambda}_2 \leq \tilde{\lambda}_2 \leq \cdots$. Here, the condition that $u \in F_0$ is the condition so that $u\nu$ is the normal component of a variation of X that fixes the boundary and preserves the volume.

Lemma 5 (i) *X is stable if and only if $\tilde{\lambda}_1 \geq 0$.*
(ii) *The inequalities $\lambda_1 < \tilde{\lambda}_1 \leq \lambda_2$ hold.*

In general, it is more difficult to estimate the eigenvalues for (13) than that for (12). For example, sometimes Lemma 4 gives information about eigenvalues of (12). So, we will give a criterion for the stability by use of (12).

Hereafter, for any one-parameter family $\{X_t\}_t$ of immersions from Σ to $\mathbf{R}^3$, we will use the following notations.

$$H(t) := \text{the mean curvature of } X_t, \quad V(t) := \text{the volume enclosed by } X_t,$$
$$L_t := \text{the Jacobi operator for } X_t.$$

First we give a criterion for the stability that looks geometric.

Theorem 3 (Criterion for the stability I, [8, 15, 18]) *Assume that X is CMC.*

(I) *If $\lambda_1 \geq 0$, then X is stable.*
(II) *Assume that $\lambda_1 < 0 \leq \lambda_2$ holds. If there exists a variation X_t of X that fixes the boundary and satisfies the condition that $H'(0) = constant \neq 0$, then the following* (i) *and* (ii) *hold.*
  (i) *If $H'(0)V'(0) \geq 0$, then X is stable.*
  (ii) *If $H'(0)V'(0) < 0$, then X is unstable.*

  If there is no such variation, then X is unstable.
(III) *If $\lambda_2 < 0$, then X is unstable.*

Next we will give a criterion for the stability that looks analytic. Denote by E the eigenspace of (12) belonging to the zero eigenvalue if zero is an eigenvalue of (12). If zero is not an eigenvalue, then set $E := \{0\}$. Denote by $E^{\perp}$ the orthogonal compliment of E in $L^2(\Sigma)$. Here, $L^2(\Sigma)$ is the completion of $C^{\infty}(\Sigma)$ by the metric given by the inner product $(u, v)_{L^2} = \int_{\Sigma} uv \, d\Sigma$. Also, we denote by $H_0^1(\Sigma)$ the completion of $C_0^\infty(\Sigma)$ by the inner product $(u, v)_{H^1} = \int_\Sigma (uv + \nabla u \nabla v)\, d\Sigma$, and by $\hat{H}(\Sigma)$ the pre-Hilbert space $H_0^1(\Sigma)$ with inner product $(\,,\,)_{L^2}$.

Then, one can choose eigenfunctions $\varphi_j \in C_0^\infty(\Sigma)$ belonging to λ_j so that they form an orthonormal basis for $L^2(\Sigma)$ [17, Lemma 2]. Moreover, each λ_j has the following minimum property on the pre-Hilbert space $\hat{H}(\Sigma)$ [17, Lemma 4].

$$\lambda_1 = I(\varphi_1) = \min\left\{ I(u)\ ;\ u \in H_0^1(\Sigma) \text{ and } \int_\Sigma u^2\, d\Sigma = 1 \right\}, \tag{14}$$

$$\lambda_j = I(\varphi_j) = \min\left\{ I(u)\ ;\ u \in H_0^1(\Sigma),\ \int_\Sigma u^2\, d\Sigma = 1, \right.$$
$$\left. \int_\Sigma u\varphi_k\, d\Sigma = 0,\ \forall k \in \{1, \dots, j-1\} \right\},\quad j = 2, 3, \cdots. \tag{15}$$

This implies that the number of negative eigenvalues (counted with multiplicities) coincides with the dimensions of the space of variation vector fields that diminish the function $A + 2HV$.

The following theorem is essentially the same as Theorem 3.

Theorem 4 (Criterion of the stability II, [8, 15, 18]) *Let $X : \Sigma \to \mathbf{R}^3$ be a CMC immersion.*

(I) *If $\lambda_1 \geq 0$, then X is stable.*
(II) *If $\lambda_1 < 0 < \lambda_2$, then there exists a unique function $u \in C_0^\infty(\Sigma)$ that satisfies $L[u] = 1$, and the following* (II-1) *and* (II-2) *hold.*

(II-1) *If $\int_\Sigma u\, d\Sigma \geq 0$, then X is stable.*
(II-2) *If $\int_\Sigma u\, d\Sigma < 0$, then X is unstable.*

(III) *If $\lambda_2 = 0$, then the following* (III-A) *and* (III-B) *hold.*

(III-A) *If there exists a function $u \in E$ that satisfies $\int_\Sigma u\, d\Sigma \neq 0$, then X is unstable.*
(III-B) *If $\int_\Sigma u\, d\Sigma = 0$ for all $u \in E$, then there exists a unique function $u \in E^\perp \cap C_0^\infty(\Sigma)$ that satisfies $L[u] = 1$ and the following* (III-B1) *and* (III-B2) *hold.*
(III-B1) *If $\int_\Sigma u\, d\Sigma \geq 0$, then X is stable.*
(III-B2) *If $\int_\Sigma u\, d\Sigma < 0$, then X is unstable.*

(IV) *If $\lambda_2 < 0$, then X is unstable.*

Remark 10 (II) and (III) in Theorem 4 can be stated in the following manner: Assume $\lambda_1 < 0 \leq \lambda_2$ holds. If there exists a function $u \in C_0^\infty(\Sigma)$ that satisfies $L[u] = 1$, then, X is stable if and only if $\int_\Sigma u\, d\Sigma \geq 0$ holds. If there is no such function u, then X is unstable.

Remark 11 Theorems 3, 4 can be generalized or modified to similar results for more general variational problems or to similar results for variational problems with various (fixed, free, partially-free, or without) boundary conditions.

Remark 12 Similar results to (I), (II), (IV) in Theorem 4 were obtained by Maddocks and Vogel for more general or a different variational problems, and (III) was obtained by Koiso [8, 13–15, 18–22].

Remark 13 Assume that $X : \Sigma \to \mathbf{R}^3$ is CMC. Then, for any $w \in \Sigma$, there exists a small closed neighborhood U of w such that $\lambda_1(U) > 0$ and therefore $X|_U$ is stable.

Proof First we assume that there exists a positive function $u \in C^\infty(U)$ such that $L[u] = 0$ holds. Then, for any function $\varphi \in C_0^\infty(U)$, φ can be written as $\varphi = uf$ by using a function $f \in C_0^\infty(U)$. We then compute by using the Stokes' theorem to obtain

$$I(\varphi) = -\int_U \left(\Delta\varphi + \|\mathrm{d}\nu\|^2\varphi\right)\varphi\, \mathrm{d}\Sigma = \int_U u^2|\nabla f|^2 - uf^2 L[u]\, \mathrm{d}\Sigma$$
$$= \int_U u^2|\nabla f|^2\, \mathrm{d}\Sigma > 0,$$

which implies that $\lambda_1(U) > 0$ because of (14). Now, in a small neighborhood U of w, the surface is a graph of a function of a domain in the tangent space at $X(w)$. By rotating the surface if necessary, $\nu^3 > 0$ on U. Set $u := \nu^3$. Then, by Lemma 4, $L[u] = 0$ holds. □

Remark 14 If the mean curvature of $X : \Sigma \to \mathbf{R}^3$ vanishes on Σ, then X is called a minimal surface. A minimal surface X is said to be stable if $\delta^2 A \geq 0$ holds for all variations of X that fix the boundary. Hence, a minimal surface X is stable if and only if $\lambda_1 \geq 0$ holds.

Example 2 (i) Spheres are stable. In fact, a sphere is the minimizer of the surface area among all closed surfaces enclosing the same volume.

(ii) Consider a part C of a right circular cylinder bounded by two parallel circles, which are orthogonal to the rotation axis. Denote by r the radius of the cylinder and by h the height of C. Then, C is stable if and only if $h \leq 2\pi r$.

(iii) Let $\mathscr{U}_0$ be the one period of an unduloid $\mathscr{U}$, which is from a neck to the next neck. Then $\mathscr{U}_0$ is stable. Any part $\mathscr{U}_1 \subset \mathscr{U}_0$ is stable, and any part $\mathscr{U}_2 \subset \mathscr{U}$ that includes $\mathscr{U}_0$ properly is unstable.

(iv) If an unduloid $\mathscr{U}$ is sufficiently close to a cylinder, then the one period $\mathscr{U}_0$ from a bulge to the next bulge is stable.

(v) If an unduloid $\mathscr{U}$ is sufficiently close to spheres, then the one period $\mathscr{U}_0$ from a bulge to the next bulge is unstable.

We give an outline of the proofs of (ii)–(v). C is represented as

$$X(z, \theta) = (r\cos\theta, r\sin\theta, z), \quad -\pi r \leq z \leq \pi r.$$

The function $\varphi(z, \theta) := \sin(z/r)$ gives an eigenfunction belonging to $\lambda_2 = 0$. $\mathscr{U}_0$ is represented as

$$X(s, \theta) = (x(s)\cos\theta, x(s)\sin\theta, z(s)), \quad -s_0 \le s \le s_0.$$

Now we use Lemma 4. The function $\varphi = \nu^3$ gives an eigenfunction belonging to $\lambda_2 = 0$. For all cases, the support function $\sigma = \langle X, \nu\rangle$ is an even function with respect to z (or s) which is a solution of the equation $L[\sigma] = \text{constant} \neq 0$, and φ is an odd function which is a solution of $L[\varphi] = 0$. Let ψ be a nonzero even solution of $L = 0$. Take a suitable linear combination $u = a\sigma + b\psi$ so that u vanishes on the boundary. Note $L[u] = \text{constant} \neq 0$ and use Theorem 4. □

6 Bifurcation and the Stability

A so-called bifurcation is closely related to the stability of CMC surfaces. In this section, we consider surfaces whose boundary values are prescribed. The set of admissible surfaces is denoted by $\mathscr{S}$. Each CMC surface in $\mathscr{S}$ is called a solution. We give sufficient conditions on a one-parameter family of solutions so that there exists a bifurcation of solutions. Moreover, we give a criterion for solutions in this bifurcation branch to be stable.

First, we give the definition of bifurcation in our context. Consider a one-parameter family of variational problems (VP$_t$) with parameter t. Let I be the domain of definition of t. Denote by $\Phi(*, t) = 0$, $(* \in \mathscr{S})$, be the Euler–Lagrange equation of (VP$_t$).

Definition 4 Assume that there is a one-parameter family of solutions:

$$\Phi(X_t, t) = 0, \quad (\forall t \in I).$$

Let $t_0 \in I$. If every neighborhood of (X_{t_0}, t_0) contains zeroes of Φ not lying on the curve $\mathscr{C} := \{(X_t, t) \mid t \in I\}$, then (X_{t_0}, t_0) is called a bifurcation point of Φ with respect to $\mathscr{C}$.

If we consider CMC surfaces, we may choose H, V, the boundary condition, etc. as the parameter. In this section, we take H or V as bifurcation parameter.

For a CMC immersion $X \in C^{3+\alpha}(M, \mathbf{R}^3)$, we set

$$E := \{u \in C_0^{2+\alpha}(\Sigma);\ L[u] = 0\}.$$

The following theorem gives a sufficient condition for nonexistence of bifurcation.

Theorem 5 (Existence and uniqueness of CMC deformation. Koiso [8]) *Let* $0 < \alpha < 1$. *Let* $X \in C^{3+\alpha}(M, \mathbf{R}^3)$ *be a CMC immersion. Assume either the following* (i) *or* (ii) *holds.*

(i) $E = \{0\}$. (ii) $\dim E = 1$ *and* $\int_{\Sigma} e \, d\Sigma \neq 0$ *for all* $e \in E - \{0\}$.

Then, in a small neighborhood of X *in* $C^{2+\alpha}(\Sigma, \mathbf{R}^3)$, *there exists a unique (up to* $C^{2+\alpha}$*-diffeomorphisms of* Σ*) one-parameter family* $\{X_t\}$, $(X_t : \Sigma \to \mathbf{R}^3, X_0 = X)$, *of CMC immersions with the same boundary values as* X.

Therefore, there is no bifurcation in this case. It is well-known that the multiplicity of λ_1 is one and that any eigenfunction belonging to λ_1 does not change sign. Hence, if $\lambda_1 = 0$, then (ii) in Theorem 5 is satisfied. Therefore, bifurcation may occur only in the case where $\lambda_k = 0$ for some $k \geq 2$.

We will give two sufficient conditions for existence of bifurcations of CMC surfaces with fixed boundary condition. For a one-parameter family $X_t : \Sigma \to \mathbf{R}^3$ of CMC immersions, as in Sect. 5, we will denote the mean curvature of X_t, the volume of X_t, by $H(t)$, $V(t)$, respectively. We will denote by L_t, $\tilde{L}_t$ the self-adjoint operators associated with the second variation of the area for X_t (see (5), (13)).

Theorem 6 ([10]) *Assume we have a one-parameter family* $X_t = X + \varphi(t)\nu : \Sigma \to \mathbf{R}^3$, $(t \in I = (-\varepsilon, \varepsilon) \subset \mathbf{R})$, *of CMC* $C^{3+\alpha}$ *immersions with* $X = X_0$ *and* $X|_{\partial\Sigma} = X_t|_{\partial\Sigma}$, *which satisfy the following* (i)–(iii).

(i) X_t *is differentiable with respect to* t.
(ii) $H'(0) \neq 0$.
(iii) $E = \{ae;\ a \in \mathbf{R}\}$, $\exists e \in \left(C_0^{2+\alpha}(\Sigma) - \{0\}\right)$.
Then, $\int_{\Sigma} e \, d\Sigma = 0$. *And there exists a one-parameter family* $\lambda(t)$ $(t \in (-\varepsilon_0, \varepsilon_0) \subset I)$ *of real values such that* $\lambda(t)$ *is differentiable with respect to* t, $\lambda(0) = 0$, $\lambda(t)$ *is a simple eigenvalue of* L_t, *and there is no other eigenvalue of* L_t *near* 0.
Assume further that
(iv) $\lambda'(0) \neq 0$.
Let $E^{\perp}$ *be any complement of* E *in* $C_0^{3+\alpha}(\Sigma)$. *Then there exists an open interval* $\hat{I}$ $(0 \in \hat{I} \subset \mathbf{R})$ *and* C^1 *functions* $\zeta : \hat{I} \to E^{\perp}$ *and* $t : \hat{I} \to \mathbf{R}$, *such that* $t(0) = 0$, $\zeta(0) = 0$, *and* $Y(s) := X + (\varphi(t(s)) + se + s\zeta(s))\nu$ *is a CMC immersion with mean curvature* $\hat{H}(s) := H(t(s))$. *Moreover, in a small neighborhood of* X, *CMC immersions with the same boundary values as* X *consists of* $\{X_t;\ t \in I\}$ *and* $\left\{Y(s);\ s \in \hat{I}\right\}$. *Furthermore, surfaces* $\{X_t;\ t \in I\}$ *and* $\{Y(s);\ s \in \hat{I}\}$ *are all different except for* $X_0 = Y(0)$.

Theorem 6 is proved by applying a general result on bifurcation by Crandall-Rabinowitz [3]. The bifurcation parameter in Theorem 6 is the mean curvature H. On the other hand, Patnaik [16] was trying to obtain a similar result to Theorem 6, where he used the volume instead of the mean curvature as bifurcation parameter. The following result is proved by a modification of the proof of Theorem 4.6 in [16], which is proved essentially by using Theorem 1.7 in [3].

Theorem 7 ([10]) *Assume we have a one-parameter family* $X_t = X + \varphi(t)\nu : \Sigma \to \mathbf{R}^3$, $(t \in I = (-\varepsilon, \varepsilon) \subset \mathbf{R})$, *of CMC* $C^{3+\alpha}$ *immersions with* $X = X_0$ *and* $X|_{\partial\Sigma} = X_t|_{\partial\Sigma}$, *which satisfy the following* (i)–(iii).

(i) X_t *is differentiable with respect to* t.
(ii) $V'(0) \neq 0$, $H'(0) \neq 0$.
(iii) $E = \{ae;\ a \in \mathbf{R}\}$, $\exists e \in (C_0^{2+\alpha}(\Sigma) - \{0\})$.
Then, $\int_\Sigma e\, d\Sigma = 0$, *and* $\lambda_j = \tilde{\lambda}_k = 0$ *for* $\exists j \geq 2$ *and* $\exists k \geq 1$. *There exists a one-parameter family* $\tilde{\lambda}(t)$ $(t \in (-\varepsilon_0, \varepsilon_0) \subset I)$ *of real values such that* $\tilde{\lambda}(t)$ *is differentiable with respect to* t, $\tilde{\lambda}(0) = 0$, $\tilde{\lambda}(t)$ *is a simple eigenvalue of* $\tilde{L}_t$, *and there is no other eigenvalue of* $\tilde{L}_t$ *near* 0. *Assume further that*
(iv) $\tilde{\lambda}'(0) \neq 0$.
Let $E^\perp$ *be any complement of* E *in* $C_0^{3+\alpha}(\Sigma)$. *Then there exists an open interval* $\hat{I}$ $(0 \in \hat{I} \subset \mathbf{R})$ *and* C^1 *mappings* $\eta : \hat{I} \to C_0^{3+\alpha}(\Sigma)$ *and* $\tau : \hat{I} \to \mathbf{R}$, *such that* $\tau(0) = 0$, $\eta(0) = 0$, *and* $Y(s) := X + (\varphi(\tau(s)) + se + s\eta(s))\nu$ *is a CMC immersion with volume* $\hat{V}(s) := V(\tau(s))$. $\eta(s)$ *can be written as* $\eta(s) = c(s)\varphi'(0) + \xi(s)$, *where* $c : \hat{I} \to \mathbf{R}$ *and* $\xi : \hat{I} \to \left\{u \in C_0^{3+\alpha}(\Sigma) \mid \int_\Sigma u\, d\Sigma = 0\right\} \cap E^\perp$ *are* C^1 *mappings such that* $c(0) = 0$, $\xi(0) = 0$. *Moreover, in a small neighborhood of* X, *CMC immersions with the same boundary values as* X *consists of* $\{X_t;\ t \in I\}$ *and* $\{Y(s);\ s \in \hat{I}\}$. *Furthermore, surfaces* $\{X_t;\ t \in I\}$ *and* $\left\{Y(s);\ s \in \hat{I}\right\}$ *are all different except for* $X_0 = Y(0)$.

Remark 15 Let us denote by "˙" the derivative with respect to t. In Theorem 6, the variation vector field of X_t at $t = 0$ is $(\dot{\varphi}(0))\nu$, that of $Y(s)$ is $(t'(0)\dot{\varphi}(0) + e)\nu$, and $\int_\Sigma e\, d\Sigma = 0$. It seems that this implies that, when X_t has a certain symmetry, $Y(s)$ does not have the same symmetry as X_t, that is, the so-called "symmetry breaking" seems to occur. In Theorem 7, the same formula is valid by exchanging t for τ. Bifurcation and symmetry breaking from nodoids are studied in [9].

In view of Theorems 3–5, in order to study the stability of CMC surfaces in a bifurcation branch, we need to study only the case where $\lambda_2 = 0$ holds.
From Theorem 3, we obtain the following lemma.

Lemma 6 *We assume* (i)–(iii) *in Theorem* 7. *We use the same notations as those in Theorem* 7. *Also, we assume that* $\lambda_2 = 0$ *holds. Then, the following* (A) *and* (B) *hold.*

(A) *If* $H'(0)V'(0) \geq 0$, *then* X *is stable and* $\tilde{\lambda}_1 = 0$ *holds.*
(B) *If* $H'(0)V'(0) < 0$, *then* X *is unstable and* $\tilde{\lambda}_2 = 0$ *holds.*

By using Theorem 7, Lemma 6, and generalizations of results in Crandall-Rabinowitz [4], we obtain the following result about stability for surfaces in the bifurcation branch.

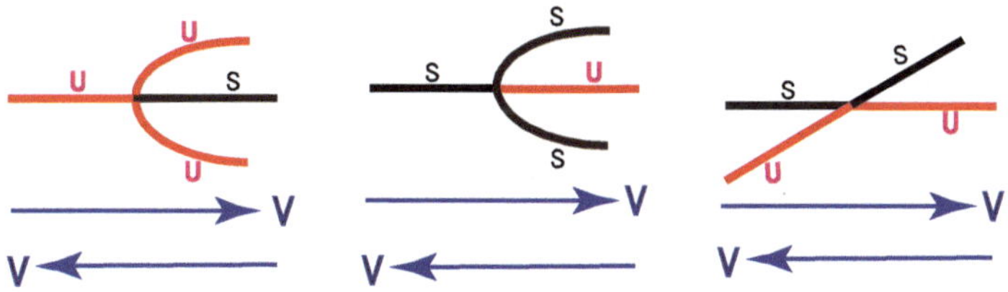

Fig. 2 The *pictures* represent, *from the left*, a subcritical pitchfork bifurcation, a supercritical pitchfork bifurcation, and a transcritical bifurcation. The bifurcation parameter is the volume V

Theorem 8 (Stability of bifurcation branch) *We assume* (i)–(iv) *in Theorem* 7. *We use the same notations as those in Theorem* 7. *Moreover, we assume that* $\lambda_2 = 0$ *holds. Then, the following* (A1), (A2) *and* (B) *hold near* $s = 0$.

(A1) *Assume* $H'(0) > 0$ *and* $V'(0) > 0$ *holds. Then,* X *is stable. If* $\tilde{\lambda}'(0) < 0$ *(resp.* $\tilde{\lambda}'(0) > 0$*), then, for the CMC-bifurcation* $Y(s)$ *with volume* $\hat{V}(s)$ *obtained in Theorem* 7*, the following result about stability holds. If* $\hat{V}'(s) \equiv 0$*, then* $Y(s)$ *is stable. Assume that* $\hat{V}'(s) \neq 0$ *holds. Then, for* $s > 0$*,* $Y(s)$ *is stable if and only if* $\hat{V}'(s) > 0$ *(resp.* $\hat{V}'(s) < 0$*) holds. And for* $s < 0$*,* $Y(s)$ *is stable if and only if* $\hat{V}'(s) < 0$ *(resp.* $\hat{V}'(s) > 0$*) holds.*

(A2) *Assume* $H'(0) < 0$ *and* $V'(0) < 0$ *holds. Then,* X *is stable. If* $\tilde{\lambda}'(0) < 0$ *(resp.* $\tilde{\lambda}'(0) > 0$*), then, for the CMC-bifurcation* $Y(s)$ *with volume* $\hat{V}(s)$ *obtained in Theorem* 7*, the following result about stability holds. If* $\hat{V}'(s) \equiv 0$*, then* $Y(s)$ *is stable. Assume that* $\hat{V}'(s) \neq 0$ *holds. Then, for* $s > 0$*,* $Y(s)$ *is stable if and only if* $\hat{V}'(s) < 0$ *(resp.* $\hat{V}'(s) > 0$*) holds. And for* $s < 0$*,* $Y(s)$ *is stable if and only if* $\hat{V}'(s) > 0$ *(resp.* $\hat{V}'(s) < 0$*) holds.*

(B) *If* $H'(0)V'(0) < 0$*, Then,* X *is unstable, and* $Y(s)$ *is unstable for small* $|s|$.

Remark 16 Theorem 8 implies that if $H'(0)V'(0) > 0$ (that is, the original surface X is stable), then, only the following three types of bifurcations can occur: a supercritical pitchfork bifurcation, a subcritical pitchfork bifurcation, and a transcritical bifurcation (see Fig. 2). If, for the surfaces $Y(s)$ in Theorem 8, $Y(-s) = \Phi \circ Y(s) \circ \Psi$ holds for an isometry Φ of $\mathbf{R}^3$ and a diffeomorphism Ψ of Σ, then if $H'(0)V'(0) > 0$, only pitchfork bifurcations can occur.

Remark 17 As we saw in Remark 15, If X_t has a certain symmetry, it seems that $Y(s)$ does not have the same symmetry as X_t. This implies that Theorem 8 may give a sufficient condition for existence of the interesting phenomenon that a one parameter family of stable solutions with a certain symmetry bifurcates to unstable solutions with the same symmetry and stable solutions with lower symmetry.

Remark 18 In the above theorems on bifurcation, we assumed that zero-eigenspace of the bifurcation point is one-dimensional, which means that the Morse index jumps at that point and the jump is one. It seems that if the jump of Morse index is an odd number, a bifurcation may occur. In fact, recently [11] gives a sufficient condition for existence of bifurcation on triply-periodic minimal surfaces in $\mathbf{R}^3$, which gives new examples of triply-periodic minimal surfaces. There they only assume that the jump of Morse index is an odd number. However, they obtain only discrete bifurcation and they have not yet obtained a continuous bifurcation as the above theorems. They use a general bifurcation result given in [7].

Acknowledgments The author is partially supported by Grant-in-Aid for Scientific Research (B) No. 25287012 of the Japan Society for the Promotion of Science, and the Kyushu University Interdisciplinary Programs in Education and Projects in Research Development.

References

1. J.L. Barbosa, M. do Carmo, Stability of hypersurfaces of constant mean curvature. Math. Zeit. **185**, 339–353 (1984)
2. J.L. Barbosa, M. do Carmo, J. Eschenburg, Stability of hypersurfaces of constant mean curvature in Riemannian manifolds. Math. Zeit. **197**, 123–138 (1988)
3. M.G. Crandall, P.H. Rabinowitz, Bifurcation from simple eigenvalues. J. Func. Anal. **54**, 321–340 (1971)
4. M.G. Crandall, P.H. Rabinowitz, Bifurcation, perturbation of simple eigenvalues, and linearized stability. Arch. Rat. Mech. Anal. **52**, 161–180 (1973)
5. N. Kapouleas, Complete constant mean curvature surfaces in euclidean three-space. Ann. Math. **131**(2), 239–330 (1990)
6. N. Kapouleas, Constant mean curvature surfaces constructed by fusing Wente tori. Invent. Math. **119**, 443–518 (1995)
7. H. Kielhöfer, in *Bifurcation Theory: An Introduction with Applications to Partial Differential Equations*. Applied Mathematical Sciences, 2nd edn., vol. 156. (Springer, New York, 2012)
8. M. Koiso, Deformation and stability of surfaces with constant mean curvature. Tohoku Math. J. **54**(2), 145–159 (2002)
9. M. Koiso, B. Palmer, P. Piccione, in, *Bifurcation and Symmetry Breaking of Nodoids with Fixed Boundary*. Advances in Calculus of Variation (to appear)
10. M. Koiso, B. Palmer, P. Piccione, Stability and bifurcation for surfaces with constant mean curvature (in preparation)
11. M. Koiso, P. Piccione, T. Shoda, On bifurcation and local rigidity of triply periodic minimal surfaces in R^3 (preprint)
12. N. Koiso, *Variational Problem* (Kyoritsu, Tokyo, Japan, 1998) (In Japanese)
13. J.H. Maddocks, Stability of nonlinearly elastic rods. Arch. Rat. Mech. Anal. **85**, 311–354 (1984)
14. J.H. Maddocks, Restricted quadratic forms and their application to bifurcation and stability in constrained variational principles. SIAM J. Math. Anal. **16**, 47–68 (1985)
15. J.H. Maddocks, Stability and folds. Arch. Rat. Mech. Anal. **99**, 301–328 (1987)
16. U. Patnaik, Volume constrained Douglas problem and the stability of liquid bridges between two coaxial tubes. Dissertation, University of Toledo, USA, 1994
17. S. Smale, On the Morse index theorem. J. Math. Mech. **14**, 1049–1055 (1965)
18. T.I. Vogel, Stability of a liquid drop trapped between two parallel planes. SIAM J. Appl. Math. **47**, 516–525 (1987)
19. T.I. Vogel, Stability of a liquid drop trapped between two parallel planes II. General contact angles. SIAM J. Appl. Math. **49**, 1009–1028 (1989)
20. T.I. Vogel, On constrained extrema. Pac. J. Math. **176**, 557–561 (1996)
21. T.I. Vogel, Sufficient conditions for multiply constrained extrema. Pac. J. Math. **180**, 377–383 (1997)

22. T. I. Vogel, Non-linear stability of a certain capillary surfaces. Dynam. Contin. Discrete Impuls. Syst. **5**, 1–15 (1999)
23. H.C. Wente, Counterexample to a conjecture of H. Hopf. Pac. J. Math. **121**, 193–243 (1986)
24. H.C. Wente, A note on the stability theorem of J. L. Barbosa and M. do Carmo for closed surfaces of constant mean curvature. Pac. J. Math. **147**, 375–379 (1991)

Discrete Models of Isoperimetric Deformation of Plane Curves

Jun-ichi Inoguchi, Kenji Kajiwara, Nozomu Matsuura and Yasuhiro Ohta

Abstract We consider the isoperimetric deformation of smooth curves on the Euclidean plane. It naturally gives rise to a nonlinear partial differential equation called the modified KdV(mKdV) equation as a deformation equation of the curvature, which is known as one of the most typical example of the soliton equations or the integrable systems. The Frenet equation and the deformation equation of the Frenet frame of the curve are the auxiliary linear problem or the Lax pair of the mKdV equation. Based on this formulation, we present two discrete models of isoperimetric deformation of plane curves preserving underlying integrable structure: the discrete deformation described by the discrete mKdV equation and the continuous deformation described by the semi-discrete mKdV equation.

Keywords Plane curve · Discrete plane curve · Isoperimetric deformation · mKdV equation · Semi-discrete mKdV equation · Discrete mKdV equation

---

J. Inoguchi
Department of Mathematical Sciences, Yamagata University, Yamagata 990-8560, Japan
e-mail: inoguchi@sci.kj.yamagata-u.ac.jp

K. Kajiwara (✉)
Institute of Mathematics for Industry, 744 Motooka, Nishi-ku, Fukuoka 819-0395, Japan
e-mail: kaji@imi.kyushu-u.ac.jp

N. Matsuura
Department of Applied Mathematics, Fukuoka University, Nanakuma 8-19-1, Fukuoka 814-0180, Japan
e-mail: nozomu@fukuoka-u.ac.jp

Y. Ohta
Department of Mathematics, Kobe University, Rokko, Kobe 657-8501, Japan
e-mail: ohta@math.sci.kobe-u.ac.jp

R. Nishii et al. (eds.), *A Mathematical Approach to Research Problems of Science and Technology*, Mathematics for Industry 5,
DOI: 10.1007/978-4-431-55060-0_7,

1 Isoperimetric Deformation of Plane Curves

1.1 Frenet Formula

Let $\gamma(x) = \begin{bmatrix} X(x) \\ Y(x) \end{bmatrix} \in \mathbb{R}^2$ be an arc-length parameterized plane curve and x be the arc-length. This implies from the definition $\mathrm{d}x = \sqrt{(\mathrm{d}X)^2 + (\mathrm{d}Y)^2}$ that

$$|\gamma'| = \sqrt{\left(\frac{\mathrm{d}X}{\mathrm{d}x}\right)^2 + \left(\frac{\mathrm{d}Y}{\mathrm{d}x}\right)^2} = 1, \tag{1}$$

where $'$ is the derivative with respect to x. We define the *tangent vector* T of γ by $T = \gamma'$. Then it follows from (1) that T admits a parameterization as

$$T = \gamma' = \begin{bmatrix} \cos\theta \\ \sin\theta \end{bmatrix}. \tag{2}$$

Note that the geometric meaning of $\theta = \theta(x)$ is the angle between T and the horizontal axis, and it is called the *turning angle*. We next define the normal vector N by (see Fig. 1)

$$N = R\left(\frac{\pi}{2}\right) T = \begin{bmatrix} 0 & -1 \\ 1 & 0 \end{bmatrix} T, \quad R(\phi) = \begin{bmatrix} \cos\phi & -\sin\phi \\ \sin\phi & \cos\phi \end{bmatrix}. \tag{3}$$

Let $\langle \cdot, \cdot \rangle$ be the standard Euclidean scalar product. Since we have $\langle T, T \rangle = 1$ from (1), differentiating this by x yields $\langle T', T \rangle = 0$, which implies that T and T' are orthogonal. Therefore, there exists a function $\kappa = \kappa(x)$ such that

$$T' = \kappa N = \begin{bmatrix} 0 & -\kappa \\ \kappa & 0 \end{bmatrix} T, \tag{4}$$

which we call the *curvature*. Differentiating the both sides of (2) and comparing with (4), we see that $\kappa = \theta'$. From this fact, θ is also called the *potential function*. We define the *Frenet frame* $\Phi \in \mathrm{SO}(2)$ by $\Phi = [T, N]$, which is an orthonormal frame of $\mathbb{R}^2$ defined along the curve. Then we see from (3) and (4) that

$$\Phi' = \Phi L, \quad L = \begin{bmatrix} 0 & -\kappa \\ \kappa & 0 \end{bmatrix}. \tag{5}$$

Equation (5) is called the *Frenet formula* of γ.

Fig. 1 Smooth plane curve

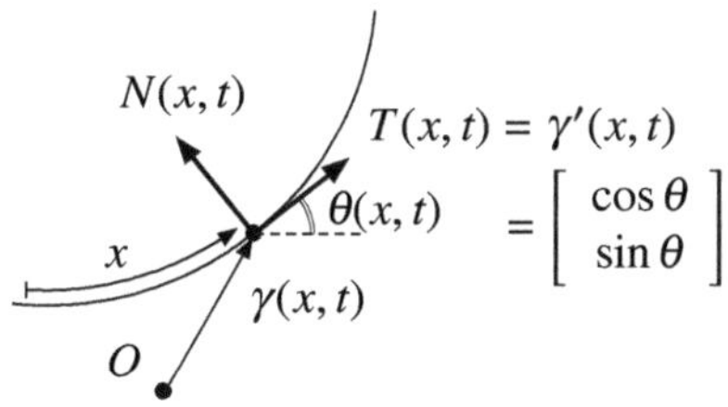

1.2 *Isoperimetric Deformation of Plane Curves and mKdV Equation*

Suppose that the curve $\gamma(x)$ and associated quantities, such as the turning angle $\theta(x)$ and the curvature $\kappa(x)$, depend also on the deformation parameter t. We require the condition

$$|\gamma'(x,t)| = 1 \tag{6}$$

for all t, namely, we consider the deformation such that the arc-length is preserved. We call such deformation the *isoperimetric deformation*, and (6) is referred to as the *isoperimetric condition*. Noticing that Φ is the orthonormal frame of $\mathbb{R}^2$, let us express the deformation of γ as

$$\frac{\partial \gamma}{\partial t} = f(x,t)T + g(x,t)N. \tag{7}$$

A simple calculation by using (5) and (7) yields the equation describing the deformation of the Frenet frame Φ:

$$\frac{\partial \Phi}{\partial t} = \Phi M, \quad M = \begin{bmatrix} 0 & -(g' + \kappa f) \\ g' + \kappa f & 0 \end{bmatrix}, \quad f' = \kappa g. \tag{8}$$

Note that we have used the fact that $\partial T/\partial t$ is orthogonal to T which is verified by differentiating the both sides of $\langle T, T \rangle = 1$ by t. The compatibility condition $\Phi_{xt} = \Phi_{tx}$ of the system of linear partial differential Eqs. (5) and (8) for Φ implies that L and M satisfy

$$\frac{\partial L}{\partial t} - \frac{\partial M}{\partial x} - LM + ML = 0. \tag{9}$$

Writing down the equations for the entries of (9), we see that κ satisfies

$$\kappa_t = g_{xx} + \kappa_x f + \kappa^2 g, \quad f_x = \kappa g. \tag{10}$$

If we choose f and g as

$$f = -\frac{\kappa^2}{2}, \quad g = -\kappa_x, \tag{11}$$

then the deformation Eq. (7) reads

$$\frac{\partial \gamma}{\partial t} = -\frac{\kappa^2}{2}T - \kappa_x N, \tag{12}$$

and (10) yields

$$\kappa_t + \frac{3}{2}\kappa^2\kappa_x + \kappa_{xxx} = 0, \tag{13}$$

or in terms of θ,

$$\theta_t + \frac{1}{2}(\theta_x)^3 + \theta_{xxx} = 0. \tag{14}$$

Equations (13) and (14) are called the *modified Korteweg-de Vries (mKdV) equation* and the *potential mKdV equation*, respectively, which are typical integrable systems. Namely, the (potential) mKdV equation describes an isoperimetric deformation of the curves on the Euclidean plane[4].

Remark 1 We note that (10) can be rewritten as

$$\kappa_t = \Omega g, \quad \Omega = \partial^2 + \kappa^2 + \kappa_x \partial^{-1}\kappa, \tag{15}$$

where ∂^{-1} denotes the formal integration operator. The operator Ω is well-known as the *recursion operator of the mKdV hierarchy*, which yields a family of infinitely many integrable differential equations called the *mKdV hierarchy*

$$\kappa_t = -\Omega^{n-1}\kappa_x, \quad n = 1, 2, \ldots. \tag{16}$$

The system of linear Eqs. (5) and (8) for Φ is called the *auxiliary linear problem* or the *Lax pair* of the mKdV hierarchy in the theory of the integrable systems(see, for example, [1]). In this way, the mKdV hierarchy naturally arises in the framework of the isoperimetric deformation of the plane curves [4].

2 Isoperimetric Deformation of Discrete Plane Curves

2.1 Discrete Plane Curves and Discrete Frenet Formula

Let us consider the deformation of the discrete plane curves. For $\gamma_n \in \mathbb{R}^2 (n \in \mathbb{Z})$, if any consecutive three points $\gamma_{n-1}, \gamma_n, \gamma_{n+1}$ are not colinear, we call γ_n a *discrete plane curve*. We put

$$a_n = |\gamma_{n+1} - \gamma_n| \tag{17}$$

and define the *tangent vector* T_n and the *normal vector* N_n by

Fig. 2 Discrete plane curve

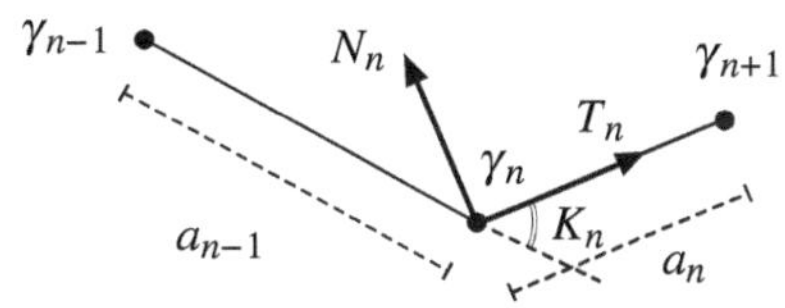

$$T_n = \frac{\gamma_{n+1} - \gamma_n}{a_n}, \quad N_n = R\left(\frac{\pi}{2}\right) T_n, \tag{18}$$

respectively. From

$$\left| \frac{\gamma_{n+1} - \gamma_n}{a_n} \right| = 1, \tag{19}$$

T_n is parameterized as

$$T_n = \frac{\gamma_{n+1} - \gamma_n}{a_n} = \begin{bmatrix} \cos \Psi_n \\ \sin \Psi_n \end{bmatrix}, \tag{20}$$

where Ψ_n is the angle between T_n and the horizontal axis. We call Ψ_n the *turning angle* similarly to the case of the smooth curve. Let K_n be the angle between the tangent vectors T_{n-1} and T_n. Then it follows immediately from the definition that (see Fig. 2)

$$\frac{\gamma_{n+1} - \gamma_n}{a_n} = R(K_n) \frac{\gamma_n - \gamma_{n-1}}{a_{n-1}}, \quad K_n = \Psi_n - \Psi_{n-1}. \tag{21}$$

Introducing the *discrete Frenet frame* $\Phi_n \in \mathrm{SO}(2)$ by

$$\Phi_n = [T_n, N_n], \tag{22}$$

it follows from (18) and (21) that

$$\Phi_{n+1} = \Phi_n L_n, \quad L_n = R(K_{n+1}). \tag{23}$$

Equation (21) or (23) is called the *discrete Frenet formula*.

2.2 Discrete Isoperimetric Deformation of Discrete Plane Curves

We consider a discrete isoperimetric deformation of the discrete curve, which reduces to the isoperimetric deformation of the smooth curves described by the mKdV equation in Sect. 1 in the continuous limit [8, 11]. Let $m \in \mathbb{Z}$ be the discrete time and γ_n^m be the isoperimetric deformation of $\gamma_n = \gamma_n^0$. Namely, γ_n^m satisfies

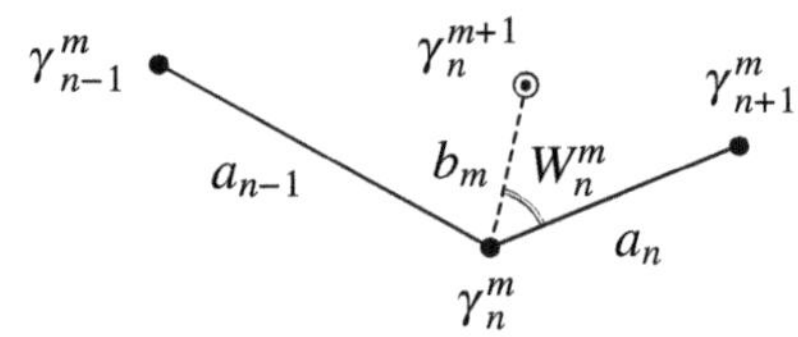

Fig. 3 Discrete isoperimetric deformation of discrete curves: b_m and W_n^m

$$\left|\frac{\gamma_{n+1}^{m} - \gamma_n^m}{a_n}\right| = 1 \tag{24}$$

for all m so that the length of the edge $\gamma_{n+1}^{m} - \gamma_n^m$ is constant with respect to m. We also require the *equidistant condition*

$$\left|\frac{\gamma_{n}^{m+1} - \gamma_n^m}{b_m}\right| = 1 \tag{25}$$

for all n, where b_m is an arbitrary function of m. Then, as shown in Fig. 3, putting the angle between the vectors $\gamma_{n+1}^{m} - \gamma_n^m$ and $\gamma_n^{m+1} - \gamma_n^m$ as W_n^m, we have

$$\frac{\gamma_{n}^{m+1} - \gamma_n^m}{b_m} = \cos W_n^m\ T_n^m + \sin W_n^m\ N_n^m. \tag{26}$$

Substituting (26) into the isoperimetric condition

$$|\gamma_{n+1}^{m+1} - \gamma_n^{m+1}| = a_n \tag{27}$$

gives

$$\sin\left(\frac{W_{n+1}^m + K_{n+1}^m + W_n^m}{2}\right) = \frac{b_m}{a_n}\sin\left(\frac{W_{n+1}^m + K_{n+1}^m - W_n^m}{2}\right), \tag{28}$$

or equivalently

$$W_{n+1}^m = -K_{n+1}^m + 2\arctan\left(\frac{b_m + a_n}{b_m - a_n}\tan\frac{W_n^m}{2}\right). \tag{29}$$

The deformation of the discrete curve is determined by (26) and (29) as follows. Let γ_n^0 be a given initial curve, and accordingly K_n^0 and a_n are given. Then:

1. Choose a point, e.g, γ_0^0, and move it to an arbitrary point γ_0^1 on the plane. Put $b_0 = |\gamma_0^1 - \gamma_0^0|$ and $W_0^0 = \angle(\gamma_0^1 - \gamma_0^0, \gamma_1^0 - \gamma_0^0)$.
2. Compute W_n^0 from W_0^0 and K_n^0 by using (29) with $m = 0$ successively.
3. Compute γ_n^1 by using (26).

Applying this procedure successively, we obtain γ_n^m ($m = 1, 2, 3\ldots$) from the initial curve γ_n^0.

Now we see that the discrete Frenet frame Φ_n^m satisfies

$$\begin{aligned}
\Phi_{n+1}^m &= \Phi_n^m L_n^m, \quad L_n^m = R(K_{n+1}^m), \\
\Phi_n^{m+1} &= \Phi_n^m M_n^m, \quad M_n^m = R(W_{n+1}^m + K_{n+1}^m + W_n^m),
\end{aligned} \tag{30}$$

by using (26) and (29). Note that the first equation in (30) is nothing but the discrete Frenet formula. The compatibility condition $L_n^m M_{n+1}^m = M_n^m L_n^{m+1}$ of the system of linear difference Eq. (30) gives

$$K_n^{m+1} - K_{n+1}^m = W_{n+1}^m - W_{n-1}^m. \tag{31}$$

Eliminating K_n^m from (29) and (31), we find that W_n^m satisfies the *discrete mKdV equation* [11]

$$\begin{aligned}
\frac{W_{n+1}^{m+1}}{2} - \frac{W_n^m}{2} &= \arctan\left(\frac{b_{m+1} + a_n}{b_{m+1} - a_n} \tan \frac{W_n^{m+1}}{2}\right) \\
&\quad - \arctan\left(\frac{b_m + a_{n+1}}{b_m - a_{n+1}} \tan \frac{W_{n+1}^m}{2}\right).
\end{aligned} \tag{32}$$

By virtue of (31), the potential function θ_n^m is introduced as

$$K_n^m = \frac{\theta_{n+1}^m - \theta_{n-1}^m}{2}, \quad W_n^m = \frac{\theta_n^{m+1} - \theta_{n+1}^m}{2}, \quad \Psi_n^m = \frac{\theta_{n+1}^m + \theta_n^m}{2}. \tag{33}$$

Then θ_n^m satisfies the *discrete potential mKdV equation* [6]:

$$\tan \frac{\theta_{n+1}^{m+1} - \theta_n^m}{2} = \frac{b_m + a_n}{b_m - a_n} \tan \frac{\theta_n^{m+1} - \theta_{n+1}^m}{2}. \tag{34}$$

Therefore we have formulated the isoperimetric deformation of discrete curves by (26) and (29) which is described by the discrete mKdV Eq. (32) or the discrete potential mKdV equation (34).

Remark 2 The isoperimetric condition (27) is also satisfied if

$$\sin\left(\frac{W_{n+1}^m + K_{n+1}^m - W_n^m}{2}\right) = 0 \tag{35}$$

holds. This implies that T_n^m and T_n^{m+1} are parallel, and in a sense, trivial as shown in Fig. 4. At each point γ_n^m, one can choose the deformation according to either (29) or (35), but we do not consider the latter deformation here.

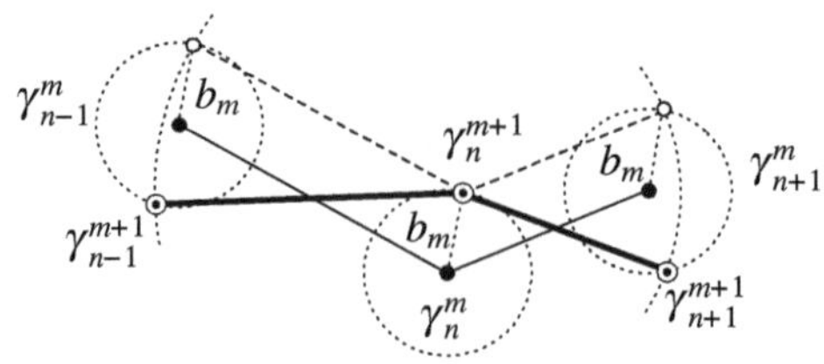

Fig. 4 Discrete isoperimetric deformation of discrete curves. *Thick line* deformation according to (29). *Broken line* deformation according to (35)

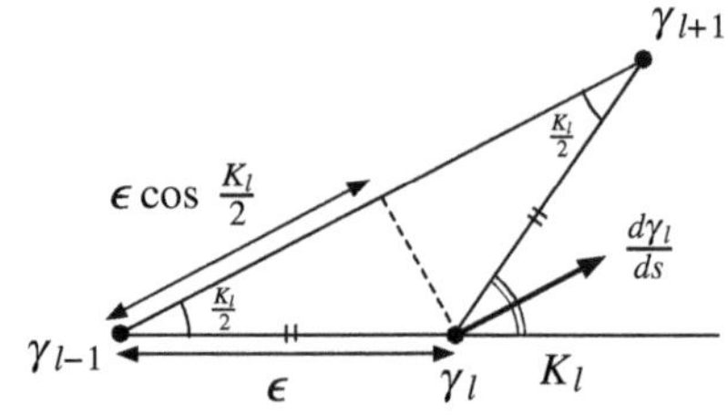

Fig. 5 Tangential flow of discrete curves

2.3 Continuous Isoperimetric Deformation of Discrete Plane Curves

Let $\gamma_l \in \mathbb{R}^2$ ($l \in \mathbb{Z}$) be the discrete curve with the length of tangent vector $a_l = \varepsilon$ (constant) and the turning angle Ψ_l. Namely, γ_l satisfies

$$\left|\frac{\gamma_{l+1}-\gamma_l}{\varepsilon}\right| = 1, \quad \frac{\gamma_{l+1}-\gamma_l}{\varepsilon} = \begin{bmatrix} \cos\Psi_l \\ \sin\Psi_l \end{bmatrix}, \tag{36}$$

$$\frac{\gamma_{l+1}-\gamma_l}{\varepsilon} = R(K_l)\frac{\gamma_l-\gamma_{l-1}}{\varepsilon}. \tag{37}$$

Let s be a continuous deformation parameter. As shown in Fig. 5, we consider the deformation in the direction of the *vertex tangential vector* $\gamma_{l+1} - \gamma_{l-1}$ [2, 7, 9]

$$\frac{d}{ds}\gamma_l = 2\varepsilon\alpha\frac{\gamma_{l+1}-\gamma_{l-1}}{|\gamma_{l+1}-\gamma_{l-1}|^2}, \tag{38}$$

where α is a constant, or equivalently

$$\frac{d}{ds}\gamma_l = \frac{\alpha}{\cos\frac{K_l}{2}}R\left(-\frac{K_l}{2}\right)\frac{\gamma_{l+1}-\gamma_l}{\varepsilon}, \tag{39}$$

which is called the *tangential flow*. It is also expressed in terms of the tangent vector T_l and the normal vector N_l as

$$\frac{d}{ds}\gamma_l = \alpha\left(T_l - \tan\frac{K_l}{2}N_l\right). \tag{40}$$

Fig. 6 Osculating circle and discrete curvature

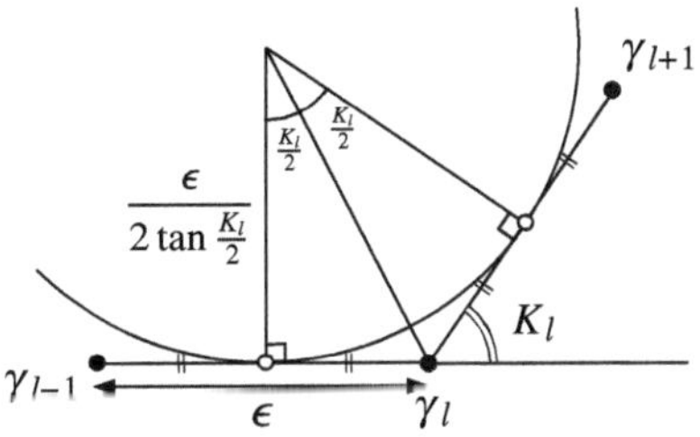

We may put $\alpha = 1$ without loss of generality. We can verify that this deformation is isoperimetric, namely, $\frac{d\varepsilon}{ds} = 0$ by direct calculation. Then it follows that the Frenet frame $\Phi_l = [T_l, N_l]$ satisfies

$$\begin{aligned} &\Phi_{l+1} = \Phi_l L_l, \quad L_l = R(K_{l+1}),\\ &\frac{d}{ds}\Phi_l = \Phi_l M_l, \quad M_l = \frac{1}{\varepsilon}\begin{bmatrix} 0 & -\left(\tan\frac{K_{l+1}}{2} + \tan\frac{K_l}{2}\right) \\ \tan\frac{K_{l+1}}{2} + \tan\frac{K_l}{2} & 0 \end{bmatrix}. \end{aligned} \tag{41}$$

The compatibility condition $\frac{d}{ds}L_l = L_l M_{l+1} - M_l L_l$ of the linear system (41) yields the *semi-discrete mKdV equation*

$$\frac{dK_l}{ds} = \frac{1}{\varepsilon}\left(\tan\frac{K_{l+1}}{2} - \tan\frac{K_{l-1}}{2}\right). \tag{42}$$

Note that (42) is also derived by differentiating $\cos K_l = \langle T_{l-1}, T_l\rangle$ and by using (41). Introducing the "discrete curvature"[7] κ_l by $\kappa_l = \frac{2}{\varepsilon}\tan\frac{K_l}{2}$, which is the inverse of the radius of the circle osculating at the mid-points of two adjacent edges of γ_l (see Fig. 6), (42) is rewritten as

$$\frac{d\kappa_l}{ds} = \frac{1}{2\varepsilon}\left(1 + \frac{\varepsilon^2\kappa_l^2}{4}\right)(\kappa_{l+1} - \kappa_{l-1}). \tag{43}$$

The potential function θ_l introduced by

$$\Psi_l = \frac{\theta_{l+1} + \theta_l}{2}, \quad K_l = \frac{\theta_{l+1} - \theta_{l-1}}{2} \tag{44}$$

satisfies the *potential semi-discrete mKdV equation*[5]:

$$\frac{d\theta_l}{ds} = \frac{2}{\varepsilon}\tan\frac{\theta_{l+1} - \theta_{l-1}}{4}. \tag{45}$$

2.4 Continuous Limit

The discrete potential mKdV Eq. (34) yields the semi-discrete potential mKdV Eq. (45) and the potential mKdV Eq. (14) by the following limiting procedures [5, 6, 9]:
(34)⟶(45)

$$a_n = a \text{ (const.)}, \quad b_m = b \text{ (const.)}, \quad \delta = \frac{a+b}{2}, \quad \varepsilon = \frac{a-b}{2}, \tag{46}$$
$$\frac{s}{\delta} = n+m, \quad l = n-m, \quad \delta \to 0$$

(45)⟶(14)

$$x = \varepsilon l + s, \quad t = -\frac{\varepsilon^2}{6}s, \quad \varepsilon \to 0 \tag{47}$$

The semi-discrete mKdV Eq. (45) and the mKdV Eq. (13) are derived from the discrete mKdV Eq. (34) by the same limiting procedure. The same procedures also apply to the turning angle, the Frenet frame, and the curve itself. These continuous limits can be verified simply by applying the Taylor expansion with respect to the limiting parameters after the designated variable transformations.

3 Concluding Remarks

It is possible to construct various exact solutions to the deformation of the curves, which are expressed explicitly in terms of the τ functions by using the theory of integrable systems, such as the solution describing the interaction of loop solitons [8, 9]. It is known that the Wadati-Konno-Ichikawa (WKI) elastic beam equation admitting the loop soliton solutions is related to the mKdV equation by a certain variable transformation called the *reciprocal transformation* (or sometimes referred to as the *hodograph transformation*). This transformation includes the independent variable transformation in which the dependent variable is incorporated. From the geometric formulation of the mKdV equation presented in this article, the reciprocal transformation can be regarded as a variable change from the Lagrangian description to the Eulerian description of the curves. Based on this observation, one can construct the discrete analog of the reciprocal transformation, which yields the integrable discretization of the WKI elastic beam equation. It is possible to apply the similar procedure to obtain the integrable (semi-)discretization of various soliton equations admitting the loop solitons, such as the short pulse equation which describes the interaction of the ultrashort pulses in the optical fiber [3]. We also remark that the deformations of space curves described by the semi-discrete and discrete mKdV equations are presented in [10].

References

1. M.J. Ablowitz, H. Segur, Solitons and the Inverse Scattering Transform. SIAM Studies in Applied Mathematics **4** (SIAM, Philadelphia, 1981)
2. A. Doliwa, P.M. Santini, Integrable dynamics of a discrete curve and the Ablowitz-Ladik hierarchy. J. Math. Phys. **36**, 1259–1273 (1995)
3. B.F. Feng, J. Inoguchi, K. Kajiwara, K. Maruno, Y. Ohta, Discrete integrable systems and hodograph transformations arising from motions of discrete plane curves. J. Phys. A Math. Theor. **44**, 395201 (2011)
4. R.E. Goldstein, D.M. Petrich, The Korteweg-de Vries hierarchy as dynamics of closed curves in the plane. Phys. Rev. Lett. **67**, 3203–3206 (1991)
5. R. Hirota, Exact N-soliton solution of nonlinear lumped self-dual network equation. J. Phys. Soc. Jpn. **35**, 289–294 (1973)
6. R. Hirota, Discretization of the potential modified KdV equation. J. Phys. Soc. Jpn. **67**, 2234–2236 (1998)
7. T. Hoffmann, *Discrete differential geometry of curves and surfaces, COE Lecture Notes*, vol. 18 (Kyushu University, Fukuoka, 2009)
8. J. Inoguchi, K. Kajiwara, N. Matsuura, Y. Ohta, Motion and Bäcklund transformations of discrete plane curves. Kyushu J. Math. **66**, 303–324 (2012)
9. J. Inoguchi, K. Kajiwara, N. Matsuura, Y. Ohta, Explicit solutions to the semi-discrete modified KdV equation and motion of discrete plane curves. J. Phys. A: Math. Theor. **45**, 045206 (2012)
10. J. Inoguchi, K. Kajiwara, N. Matsuura, Y. Ohta, Discrete mKdV and discrete sine-Gordon flows on discrete space curves. J. Phys. A: Math. Theor. **47**, 235202 (2014)
11. N. Matsuura, Discrete KdV and discrete modified KdV equations arising from motions of discrete planar curves. Int. Math. Res. Notices **2012**, 1681–1698 (2012)

Computing Optimal Cycles of Homology Groups

Emerson G. Escolar and Yasuaki Hiraoka

Abstract This is a brief survey concerning the problem of computing optimal cycles of homology groups through linear optimization. While homology groups encode information about the presence of topological features such as holes and voids of some geometrical structure, optimal cycles tighten the representatives of the homology classes. This allows us to infer additional information concerning the location of those topological features. Moreover, by a slight modification of the original problem, we extend it to the case where we have multiple nonhomologous cycles. By considering a more general class of combinatorial structures called complexes, we recast this multiple nonhomologous cycles problem as a single cycle optimization problem in a modified complex. Finally, as a numerical example, we apply the optimal cycles problem to the 3D structure of human deoxyhemoglobin.

Keywords Computational topology · Homology groups · Optimal cycles

1 Introduction

Recent advances in applied computational topology, particularly persistent homology [15], have brought into light the usefulness of topological methods in applied settings. One motivation for this work is provided by the paper [9]. In that paper, protein compressibility is characterized with high correlation by a

E. G. Escolar
Graduate School of Mathematics, Kyushu University, 744, Motooka, Nishi-ku, Fukuoka 819-0395, Japan
e-mail: eescolar@math.kyushu-u.ac.jp

Y. Hiraoka (✉)
Institute of Mathematics for Industry, Kyushu University, 744, Motooka, Nishi-ku, Fukuoka 819-0395, Japan
e-mail: hiraoka@imi.kyushu-u.ac.jp

R. Nishii et al. (eds.), *A Mathematical Approach to Research Problems of Science and Technology*, Mathematics for Industry 5,
DOI: 10.1007/978-4-431-55060-0_8, © Springer Japan 2014

topological quantity obtained through persistent homology. One insight is that compressibility is related to the sizes of holes in the structure. As a proxy for studying size, the paper [9] uses the idea of persistence (robustness) throughout different scales to measure holes.

In this work, we focus directly on the sizes of holes by computing optimal cycles in homology groups. In abstract, homology groups are quotient groups of kernels by images of so-called boundary maps. When applied to concrete geometrical structures, the homology classes encode information about connected components, holes, voids, and higher dimensional analogs.

As a quotient, a homology group is a group of homology classes, each composed of cycles that are topologically equivalent, or to be precise, homologous to each other. For a homology class, we choose a representative cycle to stand for it. Given appropriate inputs, a set of representative cycles for the homology classes of a homology group can be computed [11].

However, representative cycles are determined only up to homology class. There is room for a representative cycle to be deformed among its topologically equivalent cycles. The idea behind finding optimal cycles is to find good representatives for the homology classes so that we gain additional information about the underlying geometric structure. The criterion for "good" varies according to what information is desired. Typical criteria include the number of cells of the representative cycle or the total length/surface area/volume. Particularly helpful for intuition is that in dimension 1, the problem of finding optimal cycles can be thought of as tightening loops around holes. See Fig. 2 for an example.

We also mention that the problem of finding optimal cycles of homology groups is closely related to the localization of homology classes, as studied in [3, 4, 16]. There, the focus is on the optimization of homology classes by some definition of locality on homology classes. Here, we optimize directly on the representatives of homology classes.

We follow Dey et al. [6] in casting the problem of finding an optimal cycle as an integer linear optimization problem. While integer linear optimization is NP-hard in general, Dey et al. derive several conditions where the problem can be converted to linear optimization without losing integrality in the solution. Of course, one can instead solve the optimization problem with real coefficients to avoid the theoretical NP-hardness. However, cycles with a fractional/irrational number of cells tend to be awkward to interpret geometrically.

We also tackle the case where there is more than one hole in the structure. A given cycle may go around two or more holes, so that even after optimizing the cycle, it still winds around multiple holes. We modify the optimization problem to address this, further showing the flexibility in using an optimization-based approach.

Moreover, by considering a more general class of combinatorial objects—complexes, we show that this modification in the optimization problem can be realized as simply pasting extra cells. That is, minimization with multiple nonhomologous cycles is simply minimization of a single cycle in a modified complex.

2 Background

In this section, we review basic ideas from both algebraic topology and integer linear optimization. For a more thorough reference, we refer the reader, for example, to [12, 13].

2.1 Constructions

A *complex* K over $\mathbb{Z}$ is a graded set $K = \sqcup_{p \geq 0} K_p$ with elements called *cells*, together with an *incidence map* $\kappa : K \times K \to \mathbb{Z}$ satisfying

1. $\kappa(\sigma, \tau) \neq 0$ implies $\dim \sigma = \dim \tau + 1$,
2. $\sum_{\sigma \in K} \kappa(\rho, \sigma)\kappa(\sigma, \tau) = 0$.

The dimension of a cell $\sigma \in K$ is given by $\dim \sigma = p$ if and only if $\sigma \in K_p$. Throughout this work, we assume that our complexes only have a finite number of cells.

The definition above is very abstract. To motivate our minimization problem, we consider certain complexes called simplicial complexes. A *simplicial complex* is a set of vertices V together with a set of *simplices* $S \subset 2^V$ satisfying the property that every singleton $\{v\}$ is in S for $v \in V$, and if $A \in S$, $B \subset A$, then $B \in S$. For $A \in S$, any subset of A is called a *face* of A. In this work, we only consider simplicial complexes with V finite, called finite simplicial complexes.

One can think of $A = \{v_0, \ldots, v_p\} \in S$, for $p = 0, 1, 2, 3, \ldots$, as a point, line, triangle, tetrahedron, and so on, respectively. For example, $\{v_0, v_1, v_2\}$ in Fig. 1 is the triangle with vertices v_0, v_1, and v_2. For every simplex in S, we consistently choose an orientation by giving the vertices some order $[v_0, v_1, \ldots, v_p]$. Consistency means that the faces of a simplex inherit from that order. Now, let

$$K_p = \{\sigma = [v_0, \ldots, v_p] \mid \{v_0, \ldots, v_p\} \in S\}$$

and define

$$\kappa([v_0, \ldots, v_p], \tau) = \begin{cases} (-1)^i, & \text{if } \tau = [v_0, \ldots, v_{i-1}, \hat{v_i}, v_{i+1}, \ldots, v_p], \\ 0, & \text{otherwise,} \end{cases}$$

where the hat $\hat{}$ denotes removal of the vertex v_i. For simplicity, we call a p-dimensional simplex $\sigma \in K_p$ as a *p-simplex*. It is easy to check that this construction gives us a complex from a simplicial complex.

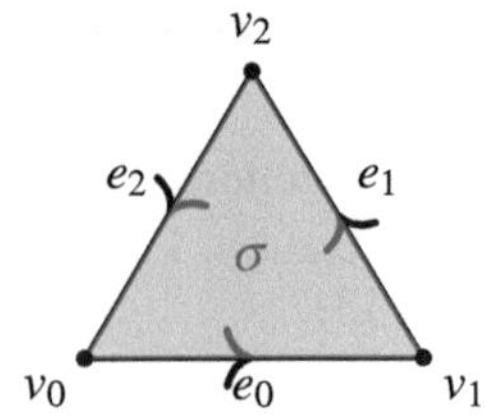

Fig. 1 A 2-simplex is a triangle. The cell $\sigma = [v_0, v_1, v_2]$ is oriented in a counter-clockwise direction

2.2 Homology Groups

Given a complex, we construct its homology groups by first defining its *p-th chain groups*

$$C_p(K) = \left\{ \sum_{\sigma \in K_p} n_\sigma \sigma \;\middle|\; n_\sigma \in \mathbb{Z} \right\}$$

for $p \geq 0$. The elements of $C_p(K)$ are simply formal sums of p-dimensional cells with integral coefficients and are called *p-chains* of K. The *boundary maps* $\partial_p : C_p(K) \to C_{p-1}(K)$, for $p \geq 1$, are defined by linearly extending

$$\partial_p \sigma = \sum_{\tau \in K} \kappa(\sigma, \tau)\tau = \sum_{\tau \in K_{p-1}} \kappa(\sigma, \tau)\tau$$

for σ with dimension p. We set $\partial_0 : C_0(K) \to \{0\}$ as the zero map. By property 2 of the incidence map, it can be shown that $\partial_p \partial_{p+1} = 0$ for $p \geq 0$. By choosing the set of all p-dimensional cells as the basis for $C_p(K)$, for each $p \geq 0$, we write down each ∂_p in matrix form, and call these the *boundary matrices* of $C_p(K)$.

If the complex K is obtained from a simplicial complex, we regain the classical formula

$$\partial_p[v_0, \ldots, v_p] = \sum_{i=0}^{p} (-1)^i [v_0 \ldots, \hat{v_i}, \ldots, v_p]$$

for $\sigma = [v_0, \ldots, v_p] \in K_p$. Here, for any permutation $g \in S_{p+1}$ of the $p+1$ vertices of σ, we identify

$$[v_0, \ldots, v_p] = \mathrm{sgn}(g)[v_{g(0)}, \ldots, v_{g(p)}].$$

As an example, consider the triangle $\sigma = [v_0, v_1, v_2]$ in Fig. 1. We have

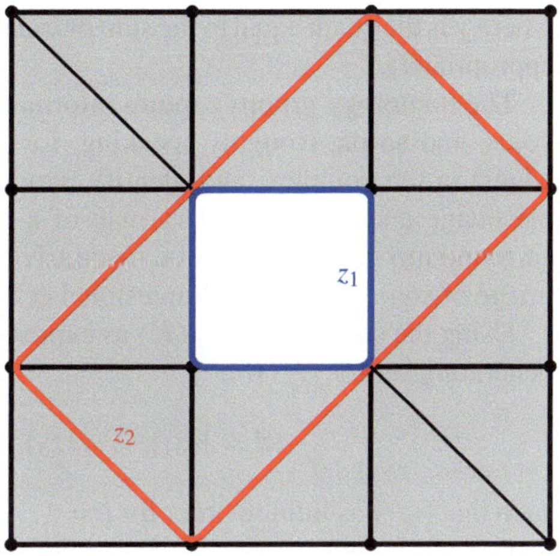

Fig. 2 Representative cycles do not always precisely describe the hole. Here, z_1 and z_2 loop around the same hole and are homologous

$$\begin{aligned}\partial_2\sigma &= 1[v_1, v_2] + (-1)[v_0, v_2] + 1[v_0, v_1]\\ &= [v_1, v_2] + [v_2, v_0] + [v_0, v_1]\\ &= e_1 + e_2 + e_0,\end{aligned}$$

which is a chain composed of the edges on the geometrical boundary of the triangle.

Define the group of *p-cycles of K* and the group of *p-boundaries of K* by $Z_p(K) = \ker \partial_p$ and $B_p(K) = \operatorname{im} \partial_{p+1}$, respectively. By the fact that $\partial_p \partial_{p+1} = 0$, we have $B_p(K) \subset Z_p(K)$ and we obtain the *p-th homology group of K* as

$$H_p(K) = \frac{\ker \partial_p}{\operatorname{im} \partial_{p+1}} = \frac{Z_p(K)}{B_p(K)}.$$

Since we are assuming that K is finite, $C_p(K)$, $Z_p(K)$, and $B_p(K)$ are all finitely generated, free $\mathbb{Z}$-modules. In general, however, $H_p(K)$ is not free, but has structure given by

$$H_p(K) \cong \mathbb{Z}^b \oplus \mathbb{Z}_{n_1} \oplus \ldots \oplus \mathbb{Z}_{n_t}, \tag{1}$$

with b called the *p-th Betti number of K*. The summand $\mathbb{Z}_{n_1} \oplus \ldots \oplus \mathbb{Z}_{n_t}$ is called the *torsion part* of $H_p(K)$.

Elements of $H_p(K)$ are called *homology classes*, denoted by $[a] = a + B_p(K)$ for some $a \in Z_p(K)$. Two elements $a, b \in Z_p(K)$ are said to be *homologous* if their projections $[a]$, $[b]$ into $H_p(K)$ are equal. This occurs if and only if

$$a = b + \partial_{p+1} y$$

for some $y \in C_{p+1}(K)$. In Fig. 2 for example, the cycles z_1 and z_2 are homologous since their difference $z_1 - z_2$ is the image under the boundary map of a $y \in C_2(K)$, where y is the chain equal to the sum of triangles enclosed between z_1 and z_2, oriented appropriately.

The homology groups capture information about connected components, holes, voids, and so on. Roughly speaking, for $H_1(K)$, we take all the cycles (loops of edges) in the complex, and identify two loops as being the same if they differ by the image under the boundary map of a chain of 2-dimensional cells. Loops that surround any holes will not be made trivial, since they cannot be expressed as the image of some chain of 2-dimensional cells under the boundary map.

Using the structure of $H_p(K)$ as expressed in Eq. (1), assume that we are given a generating set for $H_p(K)$,

$$\mathscr{B} = \{[z_1], \ldots, [z_b], [z_{b+1}], \ldots, [z_{b+t}]\},$$

such that $[z_i]$ has infinite order for $i = 1, \ldots, b$ and order n_i for $i = b+1, \ldots b+t$. From each homology class $[z_i]$ we choose a representative cycle z_i. If $H_p(K)$ has no torsion, $\mathscr{B}$ will be a basis for $H_p(K)$. If $H_p(K)$ has torsion, linear independence fails since we have $n_{b+t}[z_{b+t}] = 0$ even though $n_{b+t} \neq 0$.

Popular implementations of homology group calculations such as CHomP [5] can compute a set of representative cycles for $H_p(K)$. These are typically computed by Smith normal form operations on the boundary matrices. Since these computations do not control what representatives are chosen, the resulting cycles do not directly tell us anything about the locations of the topological features. We illustrate this in Fig. 2. Consider the simplicial complex K, with one hole as counted by $H_1(K) \cong \mathbb{Z}^1$ with Betti number 1. In homological terms, there is no difference between z_1 and z_2 since they generate the same homology class $[z_1] = [z_2]$. In practical applications, we would like to be able to identify z_1.

2.3 Integer Linear Optimization

An *integer linear optimization problem* [13] is the following. Given a rational matrix A, and rational vectors b, c of appropriate dimensions, determine

$$\min\{c^T x \mid Ax \leq b,\ x \text{ is integral}\}, \tag{2}$$

where ^{T} denotes vector transposition and $Ax \leq b$ means component-wise inequality. A vector x is said to be integral if each coordinate entry is an integer. We write such a problem in the form

$$\begin{array}{ll} \text{minimize} & c^T x \\ \text{subject to} & \begin{cases} Ax \leq b, \\ x \text{ is integral.} \end{cases} \end{array}$$

Each row in the inequality $Ax \leq b$, together with any other conditions imposed on x, is called a *constraint*. The set of vectors x satisfying the constraints of an optimization problem is called its *feasible region* F. In problem (2), the feasible region is

$$F = \{x \mid Ax \leq b,\ x \text{ is integral}\},$$

and thus the integer linear optimization problem is simply to determine

$$\min_{x \in F} c^T x.$$

This form of the integer linear optimization problem is equivalent to many other forms of integer linear optimization. For example, a maximization problem can be expressed by setting $\max_{x \in F} c^T x = -\min_{x \in F} -c^T x$. Instead of inequalities in the constraints, we can have equality $Ax = b$, by setting

$$\begin{bmatrix} A & -A \end{bmatrix} x \leq \begin{bmatrix} b & -b \end{bmatrix}.$$

Certain coordinates in the vector x may be also be restricted to be nonnegative by including those inequalities to the constraints.

In general, integer linear optimization is known to be NP-hard. We consider certain conditions where it can be solved in polynomial time. Given problem (2), we consider a related problem, called its *linear relaxation*, as follows:

$$\min\{c^T x \mid Ax \leq b,\ x \text{ is a real vector}\}. \qquad (3)$$

Of course, the solutions to problems (2) and (3) may be different. Of particular interest in integer linear optimization problems are totally unimodular matrices. A matrix A with integral entries is said to be *totally unimodular* if and only if all of its submatrices have determinant 0, -1, or 1.

The following is also used in the paper [6] to show that certain optimal cycle problems can be solved in polynomial time.

Proposition 1 *Let A be a totally unimodular matrix, b an integral vector. Then, problem* (2) *can be solved in time polynomial in the dimensions of A.*

We recall some definitions and give a sketch of the proof. The feasible region $F = \{x \mid Ax \leq b,\ x \text{ is a real vector}\}$ of the linear relaxation is a polyhedron. A polyhedron is said to be *integral* if and only if it is the convex hull of integral vectors. In other words, its vertices are integral.

It is known [14] that if A is a totally unimodular matrix and b is an integral vector, then F is integral. Since the optimum solution to the linear relaxation problem, if any, will be on a vertex, this shows that solving the linear relaxation suffices to obtain a solution to problem (2). Finally, it is known [13] that a linear optimization problem can be solved in polynomial time.

3 Optimal Cycles Through Integer Linear Optimization

3.1 Optimization of a Single Cycle

Following Dey et al. [6], we consider the problem of optimizing cycles using integer linear optimization. By fixing a standard basis for $C_p(K)$ consisting of all the p-dimensional cells of K, we identify

$$C_p(K) \cong \mathbb{Z}^m.$$

Let $\{\tau_k\}_{k=1}^m$ be the set of all p-dimensional cells in K. We can thus write the elements $x = \sum_{k=1}^m c_k \tau_k \in C_p(K)$ as vectors

$$x = \begin{bmatrix} c_1 \dots c_m \end{bmatrix}^T$$

in $\mathbb{Z}^m$. Using this, we define a 1-norm on $C_p(K)$ by

$$||x||_1 = \sum_{k=1}^{m} |c_k|.$$

Even though we do not use it in this work, we mention that a different norm may be used depending on what criterion for good representative cycles is needed. For example, let W be a diagonal matrix with diagonal entries w_{kk}. Depending on dimension p, the entries w_{kk} can be chosen to be the length, surface area, volume, and so on, of the p-dimensional cells τ_k. Then, the weighted norm $||x||_w = ||Wx||_1 = \sum_{k=1}^m |c_k w_{kk}|$ incorporates additional geometric information about the p-dimensional cells τ_k in computing the norm of x.

With bases fixed for $C_p(K)$ for every $p \geq 0$, we write down the boundary maps ∂_p in matrix form relative to these bases. Henceforth, we use the same symbol for the map ∂_p and its matrix representation.

The optimal homologous cycle problem is the following optimization problem. Given $z \in Z_p(K)$, we solve the problem

$$\begin{array}{ll} \text{minimize} & ||x||_1 \\ \text{subject to} & \begin{cases} x = z + \partial_{p+1} y, \\ x \in C_p(K),\ y \in C_{p+1}(K). \end{cases} \end{array} \tag{4}$$

In other words, we are solving for the smallest 1-norm of the chains homologous to z. Even though we did not require it explicitly, any minimizer $\hat{z}$ for problem (4) is a cycle. Since $\hat{z}$ must satisfy $\hat{z} = z + \partial_{p+1} y$, we compute $\partial_p \hat{z} = \partial_p z + \partial_p \partial_{p+1} y = 0$, showing that $\hat{z} \in Z_p(K)$.

Using a standard trick in linear optimization, problem (4) is equivalent to solving the integer linear optimization problem

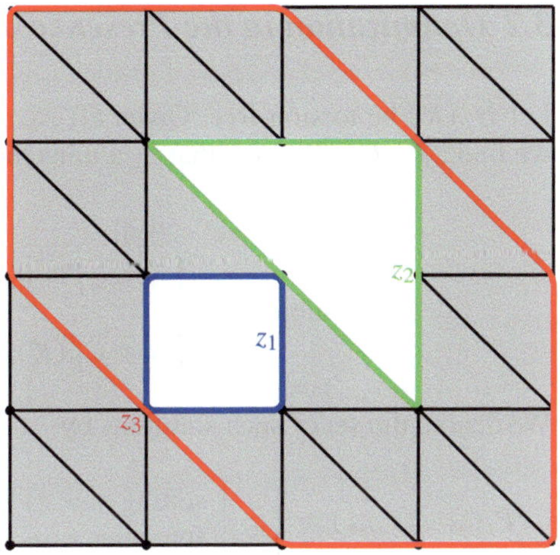

Fig. 3 The loop z_3 is wound around two holes

$$\begin{aligned} &\text{minimize} \quad \sum_{k=1}^{m}(x_k^+ + x_k^-) \\ &\text{subject to} \quad \begin{cases} x^+ - x^- = z + \partial_{p+1} y, \\ x^+, x^- \in \mathbb{Z}^m, x^+, x^- \geq 0, y \in \mathbb{Z}^n, \end{cases} \end{aligned} \tag{5}$$

and taking $\hat{z} = \hat{x}^+ - \hat{x}^-$ as a minimizer for problem (4), where $\hat{x}^+, \hat{x}^-$ is a minimizer for problem (5).

If ∂_{p+1} is totally unimodular, so is the constraint matrix in problem (5). For a fixed $p \geq 0$, Dey et al. [6] show that in the following cases the matrix ∂_{p+1} will be totally unimodular.

1. K is a finite simplicial complex triangulating a compact, $(p + 1)$-dimensional orientable manifold.
2. K is a finite simplicial complex embedded in $\mathbb{R}^{p+1}$.
3. Only for $p \leq 1$, the simplicial complex K contains no so-called Mobius subcomplexes of dimension $p + 1$.

In these cases then, problem (5) can be solved in polynomial time by Proposition 1.

In certain cases however, solving problem (5) may not be enough. Consider the simplicial complex in Fig. 3 with two holes. Suppose that we are trying to find an optimal cycle homologous to z_3 by solving problem (5) with $z = z_3$. With some computation, the minimal 1-norm is 8, attained by the cycle $\hat{z}$ that loops around both z_1 and z_2 by going through the square of side length 2. Note that neither z_3 nor $\hat{z}$ are homologous to z_1 or z_2.

3.2 Optimization in the Presence of Multiple Cycles

Let $H_p(K)$ be torsion-free. Given $z_1, \ldots, z_r$ in $Z_p(K)$ and for a fixed $1 \le j \le r$, we find $\hat{z}_j \in C_p(K)$ such that $\hat{z}_j$ attains the minimum in the following problem:

$$\begin{aligned} &\text{minimize} \quad ||x||_1 \\ &\text{subject to} \begin{cases} x = z_j + \partial_{p+1} y + \sum_{i \neq j} a_i z_i, \\ x \in C_p(K), y \in C_{p+1}(K), a_i \in \mathbb{Z}. \end{cases} \end{aligned} \tag{6}$$

We denote the set of such solutions by

$$P_j(z_1, \ldots, z_r) = \left\{ \hat{z}_j \;\middle|\; \begin{array}{l} \text{setting } x = \hat{z}_j \text{ attains the minimum in problem (6)} \\ \text{for some } y, a_i \text{ satisfying its constraints} \end{array} \right\}.$$

To be precise, the feasible region of problem (6) is

$$F = \left\{ \begin{bmatrix} x \\ y \\ \mathbf{a}_j \end{bmatrix} \;\middle|\; \begin{array}{l} x = z_j + \partial_{p+1} y + \sum_{i \neq j} a_i z_i, \\ x \in C_p(K), y \in C_{p+1}(K), \mathbf{a}_j \in \mathbb{Z}^{r-1} \end{array} \right\}$$

where $\mathbf{a}_j = [a_1 \ldots a_{j-1}\, a_{j+1} \ldots a_r]^T$. We are defining $P_j(z_1, \ldots, z_r)$ only by looking at the x components of the vectors of F. In other words, $P_j(z_1, \ldots, z_r)$ is the set of vectors x that attain the minimum in $\min_{x \in F_x} ||x||_1$, where F_x is the projection of F to the x component.

As before, we can convert problem (6) to an integer linear problem:

$$\begin{aligned} &\text{minimize} \quad \sum_{k=1}^{m} (x_k^+ + x_k^-) \\ &\text{subject to} \begin{cases} x^+ - x^- = z_j + \partial_{p+1} y + \sum_{i \neq j} a_i z_i, \\ x^+, x^- \in \mathbb{Z}^m, x^+, x^- \ge 0, y \in \mathbb{Z}^n, a_i \in \mathbb{Z}. \end{cases} \end{aligned} \tag{7}$$

The $\sum_{i \neq j} a_i z_i$ term allows us to drag z_j across $\{z_i\}_{i \neq j}$ in minimizing z_j. This is similar to the "sealing technique" used by Chen and Freedman [4] except that we are sealing all the other cycles at the same time. In effect, it adds additional cells with boundaries equal to z_i for $i \neq j$ to the $(p+1)$-boundary matrix. This can be easily seen by writing out the constraint in matrix form:

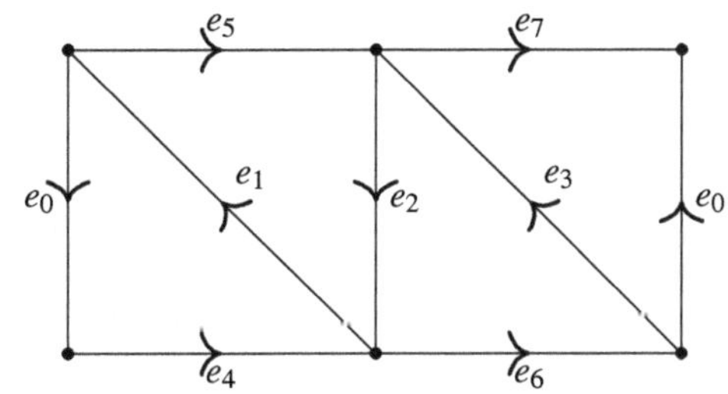

Fig. 4 We cut out the triangles from a triangulation of the Mobius band

$$
\begin{aligned}
&\text{minimize} \quad \sum_{k=1}^{m}(x_k^+ + x_k^-) \\
&\text{subject to} \quad \begin{cases} \begin{bmatrix} I & -I & -\partial_{p+1} & -z_1 \ldots - z_{j-1} & -z_{j+1} \ldots - z_r \end{bmatrix} \begin{bmatrix} x^+ \\ x^- \\ y \\ \mathbf{a}_j \end{bmatrix} = z_j, \\ x^+, x^- \in \mathbb{Z}^m, x^+, x^- \geq 0, y \in \mathbb{Z}^n, \mathbf{a}_j \in \mathbb{Z}^{r-1} \end{cases}
\end{aligned} \tag{8}
$$

where

$$\mathbf{a}_j = \begin{bmatrix} a_1 \ldots a_{j-1} \; a_{j+1} \ldots a_r \end{bmatrix}^T.$$

The main insight here is that optimization in the presence of the other cycles $\{z_i\}_{i \neq j}$ can be recast as the optimization of a single cycle by modifying the complex K. For every $z_i = \sum_{k=1}^{m} c_{ik}\tau_k$, $i \neq j$, we include a new $(p+1)$-dimensional cell σ_i. The new complex is

$$K' = K \cup \{\sigma_i\}_{i \neq j}$$

with $\kappa' : K' \times K' \to \mathbb{Z}$ such that κ' is equal to κ on $K \times K$, and $\kappa'(\sigma_i, \tau_k) = c_{ik}$ for $\tau_k \in K_p$ and 0 otherwise.

After some computation, we obtain the $(p+1)$th boundary map for K' as

$$\partial'_{p+1} = \begin{bmatrix} \partial_{p+1} \; z_1 \ldots z_{j-1} \; z_{j+1} \ldots z_r \end{bmatrix}.$$

Thus, the problem (6) is simply

$$
\begin{aligned}
&\text{minimize} \quad ||x||_1 \\
&\text{subject to} \quad \begin{cases} x = z_j + \partial'_{p+1} y', \\ x \in C_p(K') \cong C_p(K), \; y' \in C_{p+1}(K'). \end{cases}
\end{aligned}
$$

Note that $C_p(K)$ is left unchanged. This problem is essentially the optimal homologous cycle problem (4) for K'.

In general, integer linear optimization is known to be NP-hard. We might hope that even in this modified problem, we may use total unimodularity to convert it to a linear optimization problem. However, even if ∂_{p+1} were totally unimodular, the concatenation of the z_i columns may destroy total unimodularity.

For example, consider the complex in Fig. 4, the 1-skeleton $M^{(1)}$ of a certain triangulation of the Mobius band. It has $H_1(M^{(1)}) \cong \mathbb{Z}^5$, but we need only consider four cycles to see how total unimodularity could be destroyed. Consider $z_1, \ldots, z_4 \in Z_1(M^{(1)})$ given by

$$z_1 = e_0 + e_1 + e_4, \quad z_2 = e_1 + e_2 + e_5, \quad z_3 = e_2 + e_3 + e_6, \quad z_4 = -e_0 + e_3 + e_7.$$

In $e_0, e_1, e_2, e_3, e_4, e_5, e_6, e_7$ basis, we have

$$\begin{bmatrix} z_1 \; z_2 \; z_3 \; z_4 \end{bmatrix} = \begin{bmatrix} 1 & 0 & 0 & -1 \\ 1 & 1 & 0 & 0 \\ 0 & 1 & 1 & 0 \\ 0 & 0 & 1 & 1 \\ 1 & 0 & 0 & 0 \\ 0 & 1 & 0 & 0 \\ 0 & 0 & 1 & 0 \\ 0 & 0 & 0 & 1 \end{bmatrix},$$

containing a (4×4)-submatrix with determinant

$$\begin{vmatrix} 1 & 0 & 0 & -1 \\ 1 & 1 & 0 & 0 \\ 0 & 1 & 1 & 0 \\ 0 & 0 & 1 & 1 \end{vmatrix} = 2.$$

Suppose that we have some complex K containing $M^{(1)}$ as a subcomplex, such that the homology classes of z_1, z_2, z_3, z_4 form a subset of some basis for $H_1(K)$. Then, appending $[z_1 \; z_2 \; z_3 \; z_4]$ will ensure that the resulting $\partial_2' = [\partial_2 \; z_1 \; z_2 \; z_3 \; z_4]$ cannot be totally unimodular.

Nevertheless, let us state some properties of P_j.

Lemma 1 *Let* $z_1, \ldots, z_r \in Z_p(K)$.

1. $P_j(z_1, \ldots, z_r) \neq \emptyset$
2. $\hat{z}_j \in P_j(z_1, \ldots, z_r)$ *implies* $\hat{z}_j \in Z_p(K)$.
3. *Suppose that* $[z_i] = [s_i]$ *for* $i = 1, \ldots, r$. *Then* $P_j(z_1, \ldots, z_r) = P_j(s_1, \ldots, s_r)$.
4. *Let* $H_p(K)$ *be free and* $\mathscr{B} = \{[z_1], \ldots, [z_r]\}$ *be a basis for* $H_p(K)$. *Suppose that* $\hat{z}_j \in P_j(z_1, \ldots, z_r)$. *Then,* $\mathscr{B}' = \{[z_1], \ldots, [\hat{z}_j], \ldots, [z_r]\}$ *is also a basis for* $H_p(K)$.

Proof 1. It is obvious that $P_j(z_1, \ldots, z_r)$ is nonempty.

2. One of the constraints that $\hat{z}_j \in P_j(z_1, \ldots, z_r)$ must satisfy is

$$\hat{z}_j = z_j + \partial_{p+1} y + \sum_{i \neq j} a_i z_i$$

for some $y \in C_{p+1}(K)$. Thus $\partial_p \hat{z}_j = \partial_p z_j + \partial_p \partial_{p+1} y + \sum_{i \neq j} a_i \partial_p z_i = 0$.

3. Without loss of generality, we show that if $[z_1] = [s_1]$ in $H_p(K)$, then $P_j(z_1, z_2, \ldots, z_r) = P_j(s_1, z_2, \ldots, z_r)$. Let F_x^1 and F_x^2 be the feasible regions, projected to the x component, of the optimization problems associated to $P_j(z_1, \ldots, z_r)$ and $P_j(s_1, z_2, \ldots, z_r)$ respectively. We have $z_1 = s_1 + \partial_{p+1} b$ for some $b \in C_{p+1}(K)$. Suppose that $j \neq 1$. Then, for any $x \in F_x^1$, we have

$$\begin{aligned} x &= z_j + \partial_{p+1} y + \sum_{i \neq j} a_i z_i \\ &= z_j + \partial_{p+1} y + \sum_{i \neq 1, j} a_i z_i + a_1 (s_1 + \partial_{p+1} b) \\ &= z_j + \partial_{p+1} (y + a_1 b) + a_1 s_1 + a_2 z_2 + \ldots + a_{j-1} z_{j-1} + a_{j+1} z_{j+1} \\ &\quad + \ldots + a_r z_r \in F_x^2. \end{aligned}$$

A similar argument shows the opposite inclusion.

The case $j = 1$ is similar. For any $x \in F_x^1$,

$$\begin{aligned} x &= z_1 + \partial_{p+1} y + \sum_{i \neq 1} a_i z_i \\ &= s_1 + \partial_{p+1} b + \partial_{p+1} y + \sum_{i \neq 1} a_i z_i \\ &= s_1 + \partial_{p+1} (y + b) + \sum_{i \neq 1} a_i z_i \\ &\in F_x^2. \end{aligned}$$

Showing that $F_x^2 \subset F_x^1$ can be done in an analogous manner.

In both cases we have $F_x^1 = F_x^2$. Thus the elements of both $P_j(z_1, \ldots, z_r)$ and $P_j(s_1, z_2, \ldots, z_r)$ are the minimizers of the same optimization problem. Consequently, $P_j(z_1, z_2, \ldots, z_r) = P_j(s_1, z_2, \ldots, z_r)$.

4. For $x \in H_p(K)$,

$$\begin{aligned} x &= \sum_{i=1}^{r} c_i [z_i] \\ &= \sum_{i \neq j} (c_i - c_j a_i)[z_i] + \sum_{i \neq j} c_j a_i [z_i] + c_j [z_j] \\ &= \sum_{i \neq j} (c_i - c_j a_i)[z_i] + c_j [\hat{z}_j] \end{aligned}$$

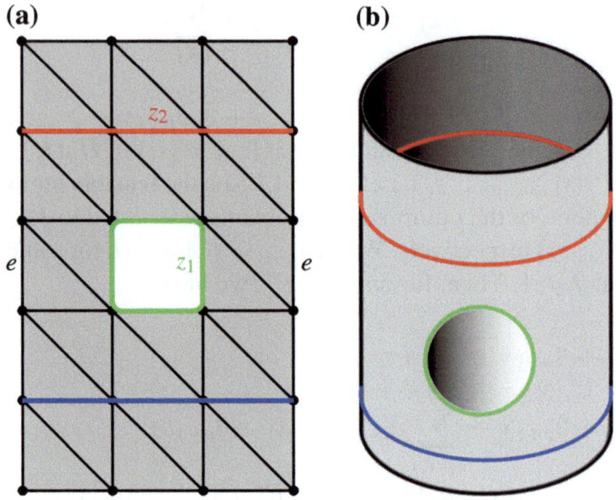

Fig. 5 $P_j(z_1, \ldots, z_r)$ may contain more than one $\hat{z}_j$

shows that $\mathscr{B}'$ generates $H_p(K)$. To show linear independence, suppose that we have some integers c_i such that

$$\sum_{i \neq j} c_i[z_i] + c_j[\hat{z}_j] = 0.$$

Then,

$$\sum_{i \neq j} c_i[z_i] + \sum_{i \neq j} a_i c_j[z_i] + c_j[z_j] = 0$$

or

$$\sum_{i \neq j} (c_i + c_j a_i)[z_i] + c_j[z_j] = 0.$$

Since $\mathscr{B}$ is linearly independent, $c_i + c_j a_i = 0$ and $c_j = 0$, implying $c_i = 0$ for all i. □

We note that $P_j(z_1, \ldots, z_r)$ may contain more than one cycle. In Fig. 5, we have a cylinder with a hole on its surface. In its triangulation K, the left and right edges are pasted together. The set $P_2(z_1, z_2)$ is the set of minimizers to

$$\begin{array}{ll} \text{minimize} & ||x||_1 \\ \text{subject to} & \begin{cases} x = z_2 + \partial_2' y', \\ x \in C_1(K'), \, y' \in C_2(K'), \end{cases} \end{array}$$

where K' is K with an additional cell σ_1 such that $\partial_2' \sigma_1 = z_1$, as defined above. Clearly, z_2 is in $P_2(z_1, z_2)$, with $||z_2||_1 = 3$. However, so are any of the chains of edges parallel to z_2 in the triangulation.

4 Algorithm and Numerical Example

Let $H_p(K)$ be free. Given a basis $\mathscr{B} = \{[z_1], \ldots, [z_r]\}$ for $H_p(K)$, we choose representative cycles $z_1, \ldots, z_r$ for each of the basis elements. Through Algorithm 1, we go through all the cycles z_j, optimizing each of them. We implement computation of $\hat{z}_j \in P_j(\hat{z}_1, \ldots, \hat{z}_{j-1}, z_j, \ldots, z_r)$ using IBM ILOG CPLEX Optimization Studio [10]. Note that we do not have to compute the entire set

$$P_j(\hat{z}_1, \ldots, \hat{z}_{j-1}, z_j, \ldots, z_r)$$

since we only need one $\hat{z}_j$ from it.

Recall problem (7) in Sect. 3.2, which we use to define $P_j(z_1, \ldots, z_r)$. In solving

$$\begin{aligned} &\text{minimize} \quad \sum_{k=1}^{m} (x_k^+ + x_k^-) \\ &\text{subject to} \quad \begin{cases} x^+ - x^- = z_j + \partial_{p+1} y + \sum_{i \neq j} a_i z_i, \\ x^+, x^- \in \mathbb{Z}^m, x^+, x^- \geq 0, \, y \in \mathbb{Z}^n, a_i \in \mathbb{Z} \end{cases} \end{aligned}$$

to find $\hat{z}_j$, we first try to solve its linear relaxation

$$\begin{aligned} &\text{minimize} \quad \sum_{k=1}^{m} (x_k^+ + x_k^-) \\ &\text{subject to} \quad \begin{cases} x^+ - x^- = z_j + \partial_{p+1} y + \sum_{i \neq j} a_i z_i, \\ x^+, x^- \in \mathbb{R}^m, x^+, x^- \geq 0, \, y \in \mathbb{R}^n, a_i \in \mathbb{R}. \end{cases} \end{aligned}$$

If the minimizer for the linear relaxation has x^+, x^-, y, a_i all integral, we do not need to solve problem (7), and just set $\hat{z}_j = x^+ - x^-$. Otherwise, we have to solve the integer linear optimization problem.

After step j, we update our set of representative cycles by replacing z_j with $\hat{z}_j$. By Lemma 1, the output of algorithm MinimizeGenerators forms a basis $\mathscr{B}' = \{[\hat{z}_1], \ldots, [\hat{z}_r]\}$ for $H_p(K)$. In essence, what we are doing in each step is to modify the homology basis one homology class at a time.

Unfortunately, the choice of $\hat{z}_j$ in the loop implies that the output may depend on the choices made. Suppose that all the cycles in $P_j(\hat{z}_1, \ldots, \hat{z}_{j-1}, z_j, \ldots, z_r)$ are homologous. Then, by Lemma 1, we are guaranteed that $P_{j+1}(\hat{z_1}, \ldots, \hat{z}_{j-1}, \hat{z}_j,$

Algorithm 1 MinimizeGenerators Algorithm

Require: $\{z_1, \ldots, z_r\}$ such that $\{[z_1], \ldots, [z_r]\}$ is a basis for $H_p(K)$
  function MINIMIZEGENERATORS($\{z_1, \ldots, z_r\}$)
    for $j = 1, \ldots, r$ **do**
      Choose a $\hat{z}_j$ from $P_j(\hat{z}_1, \ldots, \hat{z}_{j-1}, z_j, \ldots, z_r)$
    end for
    return $\{\hat{z}_1, \ldots, \hat{z}_r\}$
  end function

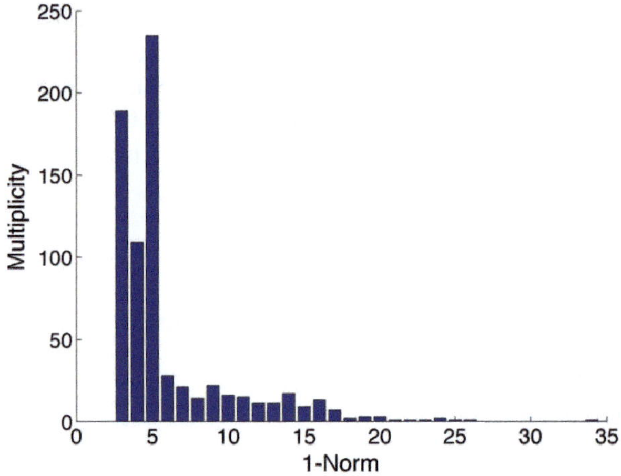

Fig. 6 Multiset of 1-norms of $H_1(2\text{HHB}_0)$

$z_{j+1}, \ldots, z_r)$ does not depend on the choice of $\hat{z}_j$. In general however, this is not the case.

Given a set of optimized cycles $\{\hat{z}_1, \ldots, \hat{z}_r\}$ from Algorithm 1, we record its 1-norms as the multiset

$$L = \{||\hat{z}_1||_1, \ldots, ||\hat{z}_r||_1\}.$$

As noted above, the $\hat{z}_j$ may vary depending on the choices made in the algorithm. Thus, L is not guaranteed to be invariant.

Given a point cloud (set of points in Euclidean space) together with weights for every point, the weighted α-shape [7] is a sequence of simplicial complexes constructed on the weighted point cloud that generalizes α-shapes. Without going into details, α-shapes are used to define what we mean by the "shape" of a point cloud. The α value controls the level of detail or scale. Since the weighted α-shape is dual to the union of balls space-filling model, it is appropriate for modeling molecular structures. We refer the reader to [7] for more details.

For a numerical example, we take point cloud data of the atomic structure of the protein human deoxyhemoglobin (PDB ID: 2HHB [8]) from the Protein Data Bank [1] at http://www.rcsb.org. Using CGAL [2], we construct its weighted alpha shape at $\alpha = 0$, where each atom is given a weight equal to the square of its van

Fig. 7 We detect an optimal cycle of length 34, going around the *central* hole

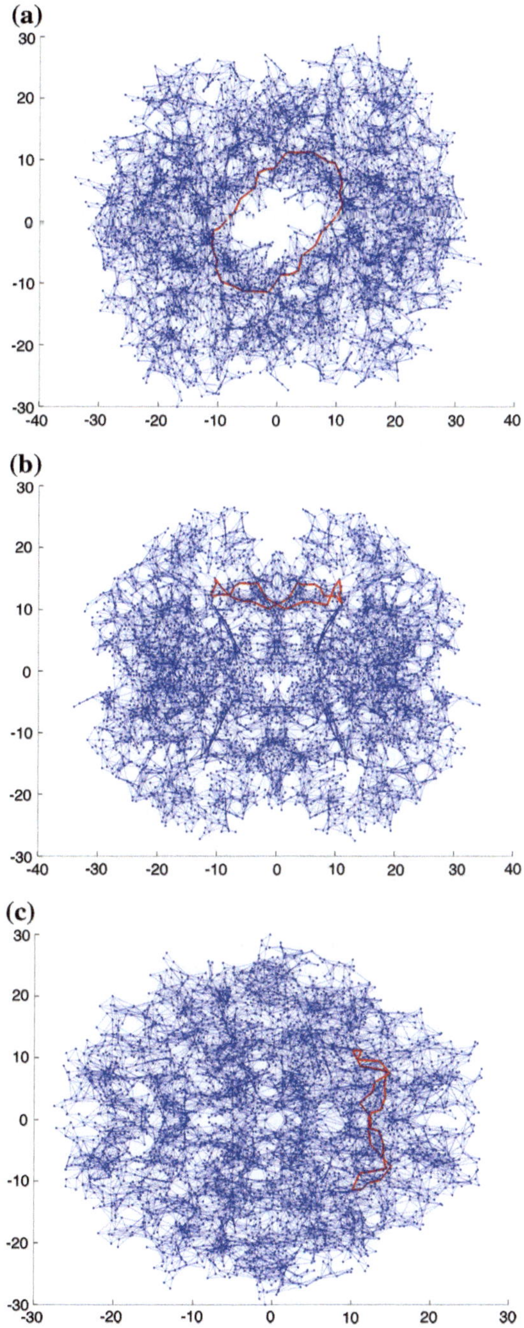

der Waals radius. We denote this simplicial complex by $2\mathrm{HHB}_0$. We compute a set of representative cycles for a homology basis of $H_1(2\mathrm{HHB}_0)$ using CHomP [5], and apply the MinimizeGenerators algorithm to it.

In Fig. 6, we plot the multiset of 1-norms of the set of optimized cycles as a histogram. Of interest is the optimal cycle with length 34. We plot it together with the point cloud in the Fig. 7, projecting to the xz, xy, and yz planes respectively. In particular, we observe that the optimal cycle of length 34 is wrapped around the central hole of the structure.

Acknowledgments The authors would like to thank Hayato Waki for valuable comments and discussions and for introducing optimization software.

References

1. H.M. Berman, J. Westbrook, Z. Feng, G. Gilliland, T.N. Bhat, H. Weissig, I.N. Shindyalov, P.E. Bourne, The protein data bank. Nucleic Acids Res. **28**(1), 235–242 (2000)
2. Cgal, Computational geometry algorithms library. http://www.cgal.org
3. C. Chen, D. Freedman, Quantifying homology classes, in *Symposium on Theoretical Aspects of Computer, Science*, 169–180 (2008)
4. C. Chen, D. Freedman, Measuring and computing natural generators for homology groups. Comput. Geom. **43**(2), 169–181 (2010)
5. CHomP homology software. http://chomp.rutgers.edu/
6. T.K. Dey, A.N. Hirani, B. Krishnamoorthy, Optimal homologous cycles, total unimodularity, and linear programming. SIAM J. Comput. **40**(4), 1026–1044 (2011)
7. H. Edelsbrunner, Weighted alpha shapes. Technical report, Department of Computer Science, University of Illinois at Urbana-Champaign, 1992
8. G. Fermi, M.F. Perutz, B. Shaanan, R. Fourme, The crystal structure of human deoxy-haemoglobin at 1.74 a resolution. J. Mol. Biol. **175**(2), 159–174 (1984)
9. M. Gameiro, Y. Hiraoka, S. Izumi, M. Kramar, K. Mischaikow, V. Nanda, Topological Measurement of Protein Compressibility via Persistence Diagrams, MI Preprint Series, 6 (2012)
10. International Business Machines Corp, IBM ILOG CPLEX optimization studio. http://www-03.ibm.com/software/products/en/ibmilogcpleoptistud/
11. T. Kaczynski, K. Mischaikow, M. Mrozek, *Computational Homology* (Springer, New York, 2004)
12. J. Munkres, *Elements of Algebraic Topology* (The Benjamin/Cummings Publishing Company Inc, Menlo Park, 1984)
13. A. Schrijver, in *Theory of Linear and Integer Programming*. Wiley-Interscience Series in Discrete Mathematics and Optimzation (Wiley, New York, 2000)
14. A.F. Veinott Jr, G.B. Dantzig, Integral extreme points. SIAM Rev. **10**(3), 371–372 (1968)
15. A. Zomorodian, G. Carlsson, Computing persistent homology. Discrete Comput. Geom. **33**, 249–274 (2004)
16. A. Zomorodian, G. Carlsson, Localized Homology. Comput. Geom. **41**(3), 126–148 (2008)

Singularity Theory of Differentiable Maps and Data Visualization

Osamu Saeki

Abstract In many scientific situations, a given set of large data, obtained through simulation or experiment, can be considered to be a discrete set of sample values of a differentiable map between Euclidean spaces or between manifolds. From such a viewpoint, this article explores how the singularity theory of differentiable maps is useful in the visualization of such data. Special emphasis is put on Reeb graphs for scalar functions and on singular fibers of multi-variate functions.

Keywords Data visualization · Multi-variate function · Differential topology · Singularity theory · Reeb graph · Singular fiber

1 Introduction

In general, data obtained through scientific simulation or experiment can often be formulated as a set of discrete sample points of a differentiable map $f : \mathbb{R}^n \to \mathbb{R}^p$ between Euclidean spaces. In this article, in order to explore mathematical technologies, based on differential topology, for analyzing and visualizing such big data, we would like to present some fundamental materials from the theory of singularities of differentiable maps.

In the following, M will be a C^∞ manifold of dimension n, N a C^∞ manifold of dimension p, and we assume $n \geq p \geq 1$. If the reader is not familiar with the theory of differentiable manifolds, then M and N can be safely assumed to be open subsets of $\mathbb{R}^n$ and $\mathbb{R}^p$, respectively. Furthermore, $f : M \to N$ will be a differentiable map, or more precisely, a map of class C^∞.

O. Saeki (✉)
Institute of Mathematics for Industry, Kyushu University, 744, Motooka, Nishiku, Fukuoka 819-0395, Japan
e-mail: saeki@imi.kyushu-u.ac.jp

R. Nishii et al. (eds.), *A Mathematical Approach to Research Problems of Science and Technology*, Mathematics for Industry 5,
DOI: 10.1007/978-4-431-55060-0_9, © Springer Japan 2014

2 Scalar Functions and Their Level Sets

As to data analysis of maps as described in Sect. 1, the case of scalar functions has been extensively studied. In fact, this case does not require theoretically complicated technologies and is very useful in many real-world situations. In this section, we describe the case of such scalar functions $f : M \to \mathbb{R}$.

For the feature analysis of scalar functions, the following notions of a level set play an important role.

Definition 1 For a value $c \in \mathbb{R}$, the set

$$f^{-1}(c) = \{x \in M \mid f(x) = c\}$$

is called a *level set* (or an *iso-value set*, *iso-line*, *iso-surface*, etc.).

In general, a level set is of dimension $n - 1$, where $n = \dim M$, although it may not be a manifold. For example, for elevation data, we have $n = 2$ and the level sets are nothing but the contours.

Given such a set of elevation data, for example, in order to read-off the characteristic features of the data, what is important? It is clear that the contours play important roles, and the reader may notice that there are characteristic contours among them. They, in fact, correspond to the peaks, passes, and pits (see Fig. 1).

If we slightly change the values corresponding to these feature contours, then we have the birth–death (around a peak or a pit) or split–merge (around a pass) of contours. Therefore, the values around which the corresponding contours change their topology are essential for grasping the characteristic features of the given data.

In order to integrate this type of information, the following notion is often extremely useful.

Definition 2 [12] For a scalar function $f : M \to \mathbb{R}$, the graph obtained by contracting each connected component of a level set to a point is called the *Reeb graph*. (Depending on the situation, this is also called a *contour tree*, *volume skeleton tree*, *topological volume skeleton*, *level-set graph*, *Stein factorization*, etc.) See Fig. 2. In mathematical terms, the Reeb graph is the topological space endowed with the quotient topology induced by the quotient map from the domain M. It is known that for a generic scalar function f, its Reeb graph is actually a graph consisting of vertices and edges.

These graphs have been extensively studied as a tool for describing the topological change in the contours for elevation data [14] and have been applied to various situations.

The algorithm for obtaining the Reeb graph from a given set of data has been established when the domain dimension n is equal to 2, 3, and 4 (for example, see [2, 5, 11]).

What is important here is the fact that the vertices of a Reeb graph correspond to the values where a topological change in the level sets occurs. For obtaining a global

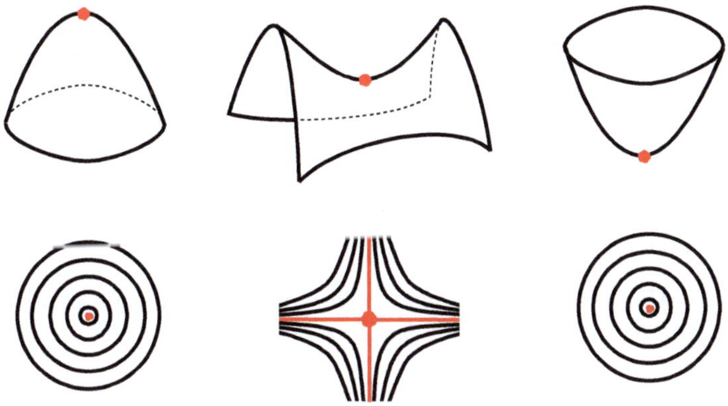

Fig. 1 Feature contours

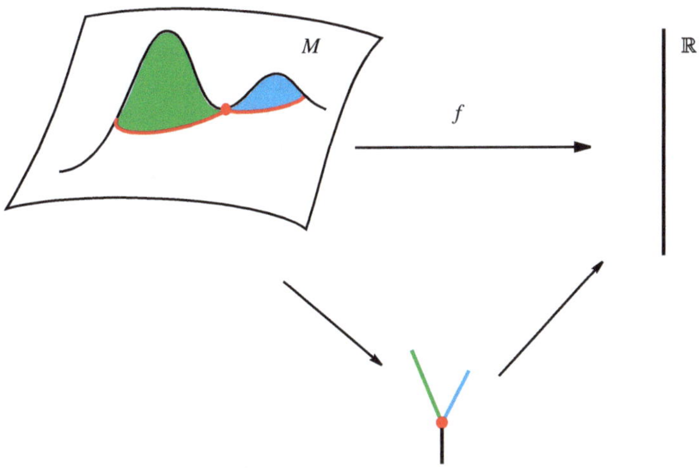

Fig. 2 Example of Reeb graph

feature of the given data, these vertices play important roles. These correspond to the points as described in Fig. 1. They are formulated as follows.

Definition 3 (1) Let us consider a differentiable function $f : M \to \mathbb{R}$. A point $x \in M$ such that all the partial derivatives of f with respect to local coordinates vanish at x, is called a *critical point* (or a *singular point*) of f. Furthermore, its corresponding value $f(x)$ is called a *critical value*.

(2) Suppose that the $n \times n$ symmetric matrix

$$\left(\frac{\partial^2 f}{\partial x_i \partial x_j}(x)\right)_{i,j}$$

consisting of the second-order partial derivatives of f at a critical point x is regular, where $(x_1, x_2, \ldots, x_n)$ are the local coordinates around x. Then, the critical point x is said to be *nondegenerate*.

It is known that every differentiable function can be perturbed arbitrarily slightly in such a way that all of its critical points are nondegenerate (for example, see [9, 10]).

In differential topology, the following is fundamental.

Theorem 1 (Morse Lemma) *Around each nondegenerate critical point, we can choose a set of local coordinates so that f can be written as*

$$f = \pm x_1^2 \pm x_2^2 \pm \cdots \pm x_n^2 + c$$

for a constant c, which is equal to the corresponding critical value.

In the above quadratic form, the number of minus signs is called the *index* of the critical point. The topology of a given function near a nondegenerate critical point is determined by the index. In fact, the topological change of level sets occurs only near critical points, and such a change can be described by the above quadratic form. (For example, when $n = 2$, the index is equal either to 0, 1 or 2, and the topology of the level sets near the critical point is exhausted in Fig. 1.) In this sense, the Morse Lemma is fundamental for chasing the topological changes of level sets.

For example, when $n = 3$ and M is an open subset of the 3-dimensional Euclidean space, the global change in the level sets (iso-surfaces) can be classified mathematically, including the information on whether the function value increases or decreases into the outer region of the surface; in other words, whether the function value corresponding to the region occluded by the iso-surface has smaller or larger values (see [15, 16]). This type of classification is essential for visualizing volume data.

As an explicit example, [4] applied this technique to analyze the simulation data of the electron density function for a proton and hydrogen atom collision. These are spatio-temporal data, corresponding to the case $n = 4$; however, for each fixed time T, they analyzed the corresponding 3-dimensional space data. By varying the time T, they extracted Reeb graph changes and sought the characteristic time for the data. With this method, they showed that the electron distribution change and its characteristic features at the collision can be clearly understood; much better than a straightforward visualization with a simple video.

3 Singular Points of Differentiable Maps and Their Singular Fibers

Now, let us consider a general differentiable map $f : M \to N$ with $n = \dim M \geq p = \dim N \geq 1$, where p may not necessarily be equal to 1. When $N = \mathbb{R}^p$, we have

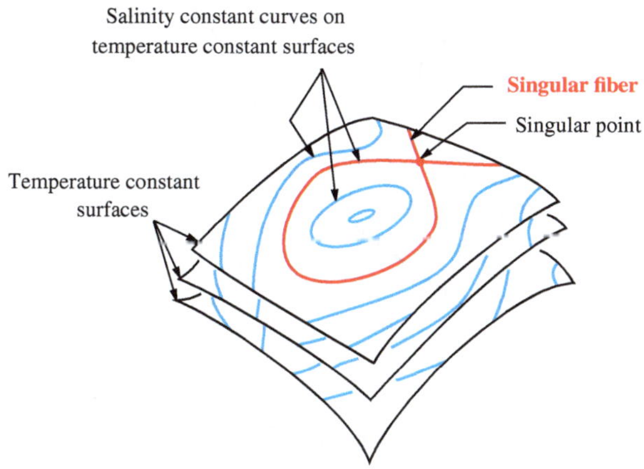

Fig. 3 Example of fibers for 2-variate function

$$f = (f_1, f_2, \ldots, f_p)$$

for a set of p scalar functions. Thus, such an f can also be called a *p-variate function* (or a *multi-variate function*). In this case, we can naturally analyze each coordinate function f_i, $i = 1, 2, \ldots, p$, independently, as described in the previous section. However, the relation among the coordinate functions often cannot be seen, or in a worse case, one may not grasp the global features of the given data.

In this article, the idea of analyzing coordinate functions individually and studying their relations is discarded: instead, the given set of data is analyzed based on the principle with which f is considered as a single map.

Definition 4 For a point $c \in N$, the set

$$f^{-1}(c) = \{x \in M \mid f(x) = c\}$$

is called a *fiber*. Sometimes, it is also called a *level set*; however, the former is preferred to emphasize that a multi-variate function is considered.

For example, consider the case where $n = 3$, M is a 3-dimensional domain filled with sea water, and $f{:}M \to \mathbb{R}^2$ is given by $f =$ (temperature, salt density). This situation is shown in Fig. 3.

A fiber containing a singular point is called a singular fiber, which will be explained in detail later. It can be naturally expected, as an analogy of scalar functions, that these singular fibers play important roles in extracting characteristic features of a given set of multi-variate data.

Definition 5 Consider a differentiable map $f : M \to N$. For a point $x \in M$, we choose local coordinates around x and $f(x)$. Then, let the map $df_x : \mathbb{R}^n \to \mathbb{R}^p$ be defined as the linear map associated with the *Jacobian matrix* of f (the real $p \times n$

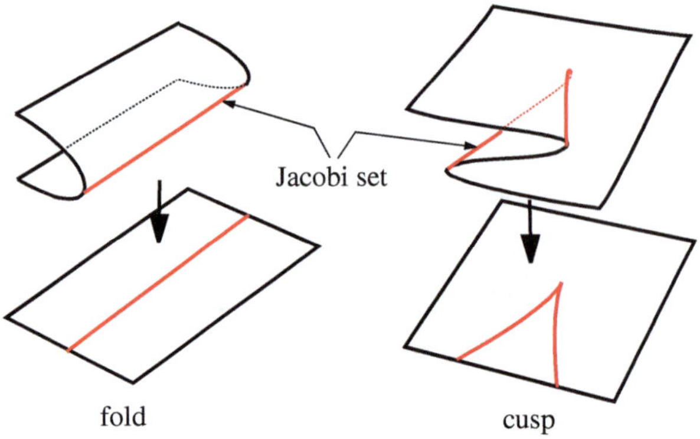

Fig. 4 Generic singularities for maps from dimension 2 to dimension 2

matrix consisting of the first derivatives of the coordinate functions of f at x). The linear map df_x is called the *differential* of f at x. If rank $df_x < p$, then x is called a *singular point* of f. It is easy to show that this definition does not depend on a particular choice of local coordinates. The set

$$J(f) = \{x \in M \mid \text{rank } df_x < p\}$$

of all singular points is called the *Jacobi set* (or *singular point set*) of f. Furthermore, the image of a singular point by f is called a *singular value*, and a fiber containing a singular point is called a *singular fiber*.

In general, a Jacobi set $J(f)$ is of dimension $p - 1$.

When $p = 1$, i.e. for a scalar function, we have the Morse Lemma, which enables us to clearly understand the level set changes. However, in the general case where $p \geq 2$, no such lemma is known, except for a few special cases: or rather, it has been even mathematically proved that such a lemma is impossible in general (for details, see [6–8], etc.).

In this article, let us explain the case where such a result similar to the Morse Lemma is available. More precisely, the case in which $p = 2, 3$ is the focus.

Let us begin by the case $p = 2$. For simplicity, we assume $n = 2$. Then, it is classically known that a "generic map" has only the following two types of singularities, a *fold* and *cusp* (see [1, 17], etc.). They are described, with respect to local coordinates, as

$$(x, y) \mapsto (x, y^2), \qquad (x, y) \mapsto (x, -xy + y^3)$$

(see Fig. 4).

The above classification is possible even when $n \geq 3$, where n is the dimension of the domain. However, as in the Morse Lemma, we have several different types depending on the indices.

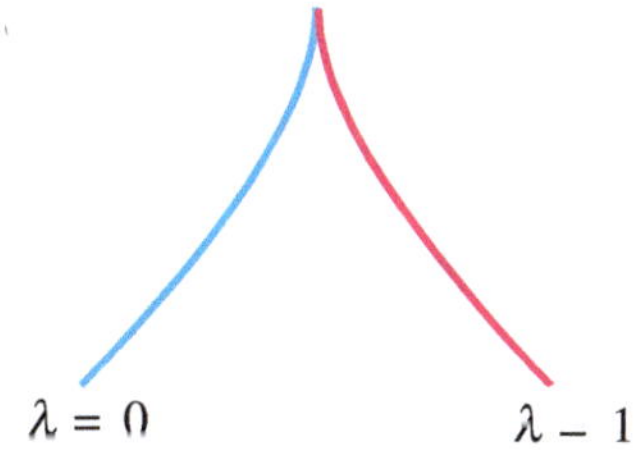

Fig. 5 Fold indices near cusp point (in the case $n = 3$)

For example, let us consider a differentiable map $f : M \to N$, where $p = \dim N = 2$ and $\dim M = n \geq 3$ is odd. Then, each fold point has its own *index* λ, where $\lambda = 0, 1, \ldots, (n-1)/2$. Inside the Jacobi set, a cusp is theoretically identifiable; however, in the case of a set of real data, we need an additional scheme in order to identify a cusp point. For example, a cusp is characterized as a singular point where the above-mentioned fold index changes (see Fig. 5). In fact, it is not algorithmically difficult to compute the index of a given fold point.

Edelsbrunner and Harer (2004) [3] proposed an algorithm for obtaining the Jacobi set of a differentiable map by approximating it with a piecewise linear map. Theoretically, it gives us the Jacobi set; however, in practice, if we directly apply the algorithm for real datasets, then a Jacobi set curve may appear as an irregular zig-zag curve and may not necessarily be clearly visualized. Figure 6 shows examples of Jacobi sets that have been computed with the help of this extracting algorithm. The blue and green curves on the (X, Y)-plane on the left represent the Jacobi set in the domain, and the curves on the right represent their images in the range.

Figure 6 gives examples of singular fibers: the relevant fibers are the intersections of the level set surfaces of the two scalar functions. In the figures, the domain $\mathbb{R}^3$ and range $\mathbb{R}^2$ correspond to each other, and the singular fiber corresponding to the black dot in $\mathbb{R}^2$ on the right is depicted in $\mathbb{R}^3$ as a red curve on the left. The upper pair corresponds to the birth-death of a component, while the lower one corresponds to the split-merge of two components. If we move the black dot in the range, then the corresponding fiber changes accordingly. With the help of this interface, we can observe the topological change in the fibers as the corresponding value changes. This method for extracting singular fibers turns out to be useful for explicit analytic multivariate functions and provides us with appropriate differential topological feedbacks. However, if we apply it to a real volume dataset, then we often encounter problems: sometimes, most of the domain is covered by singular fibers because of the noise in the data or the sparse discrete data.

Furthermore, even if we know where the singular points are and where the singular fibers are, their types may be unknown. If we use the theory of differentiable maps and their singularities in differential topology, then we can more efficiently identify the types of such singular points and fibers, which would significantly contribute to the analysis of a given large dataset and its visualization.

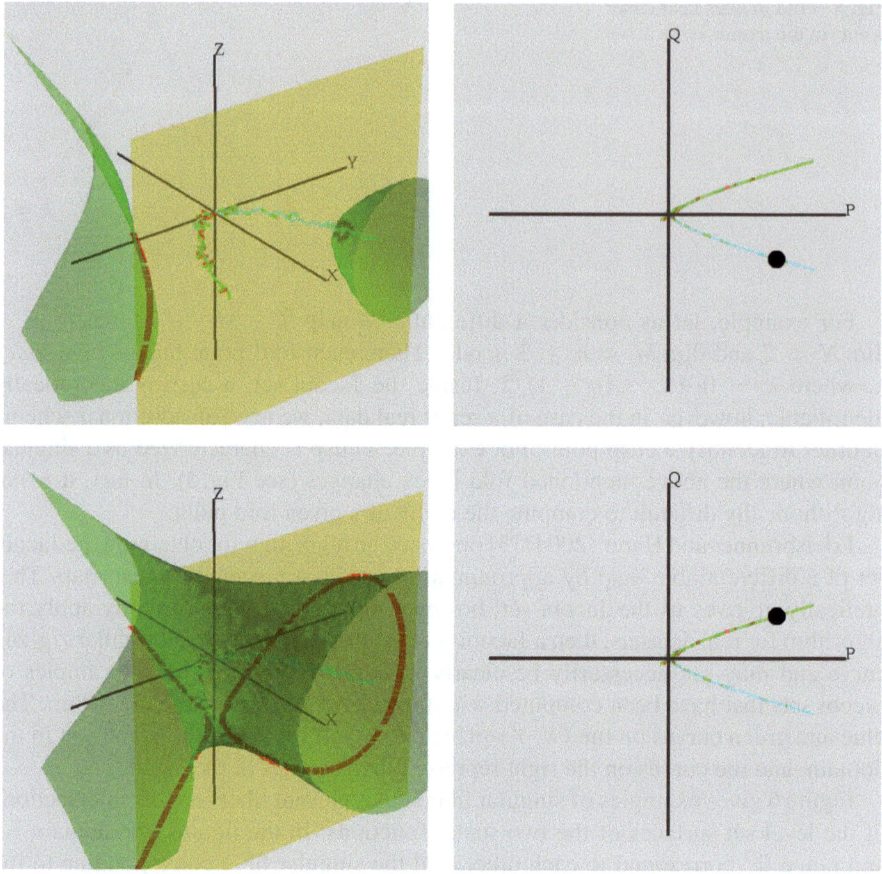

Fig. 6 Example of singular fibers for maps from dimension 3 to dimension 2

4 For Visualization of Multi-Variate Data

In order to visualize multi-variate data, we need to identify the following items.

(1) The Jacobi set.
(2) The type of each singular point.
(3) The Jacobi set image.
(4) The singular fiber type over each region of the range divided by the Jacobi set image.

In particular, for item (4), it is indispensable to identify the fiber change near the singular fiber. An example of a fiber change is shown in Fig. 7 for $f : M \to N$ with $\dim M = n = 3$ and $N = \mathbb{R}^2$, where the red curves represent the Jacobi set image, the black curves represent the fibers over the regions in $\mathbb{R}^2$ divided by the Jacobi set image, the blue curves represent the fibers over the points in the Jacobi set image

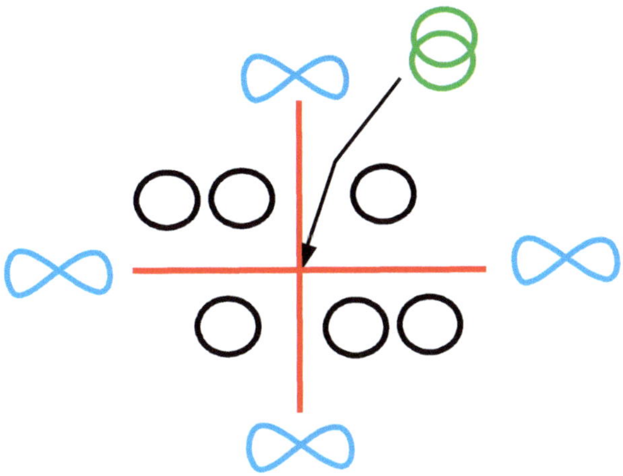

Fig. 7 Example of fiber change for map from dimension 3 to dimension 2

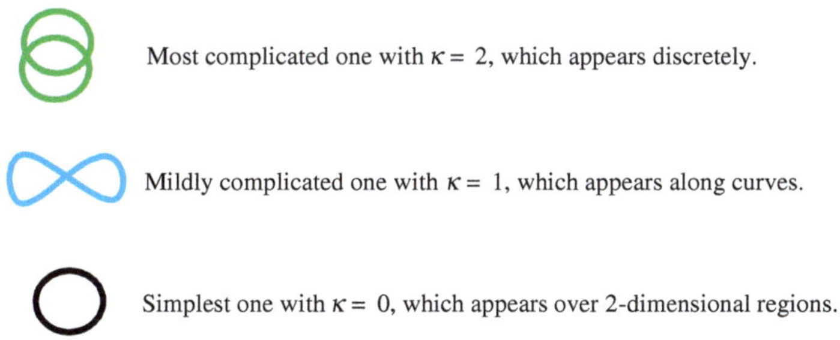

Fig. 8 Example of fibers and their codimensions for maps from dimension 3 to dimension 2

other than the crossing point, and the fiber over that crossing point is represented in green.

Looking at Fig. 7, we notice that the fibers have hierarchies depending on their complexities. More specifically, for each fiber type $\mathcal{F}$, we consider the subset

$$\mathcal{F}(f) = \{y \in N \mid \text{the fiber } f^{-1}(y) \text{ is of type } \mathcal{F}\}$$

of the range N, and

$$\kappa = \dim N - \dim \mathcal{F}(f)$$

is called the *codimension* of fiber type $\mathcal{F}$. An example of the codimensions of fibers for the case of $n = 3$ and $p = 2$ is shown in Fig. 8.

Let us now consider the case of $n = 4$ and $p = 3$ (for details, see [13]). This case corresponds, for example, to analyzing a set of given triplets of spatio-temporal data. First, the Jacobi set image of a "generic map" $f: M \to N$ with $\dim M = 4$

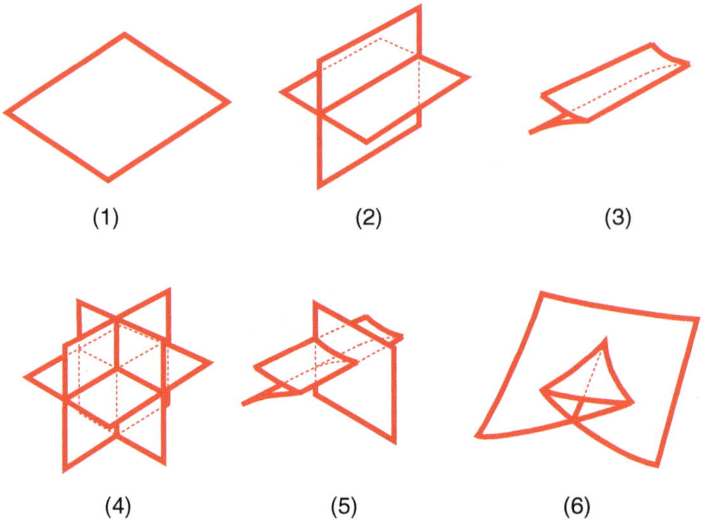

Fig. 9 Local configurations of Jacobi set images of maps from dimension 4 to dimension 3: (1), (2) and (4) correspond to one, two or three fold points, respectively; (3) corresponds to a single cusp point; (5) corresponds to a fold point and a cusp point; and (6) corresponds to a single swallowtail

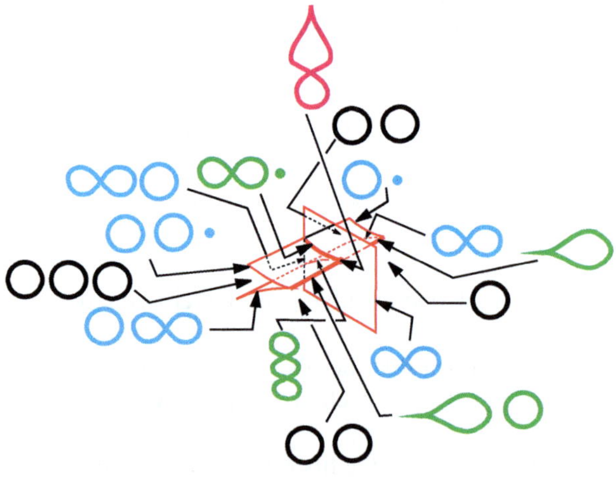

Fig. 10 Example of fiber change for map from dimension 4 to dimension 3

and dim $N = 3$ appears as a singular surface in N. The neighborhood of each of its points is classified in Fig. 9.

For example, the fiber change in a neighborhood of the singular fiber over a point, as in Fig. 9 (5), has several types, one of which is depicted in Fig. 10.

The list of singular fibers for the case of $n = 4$ and $p = 3$ is shown in Fig. 11.

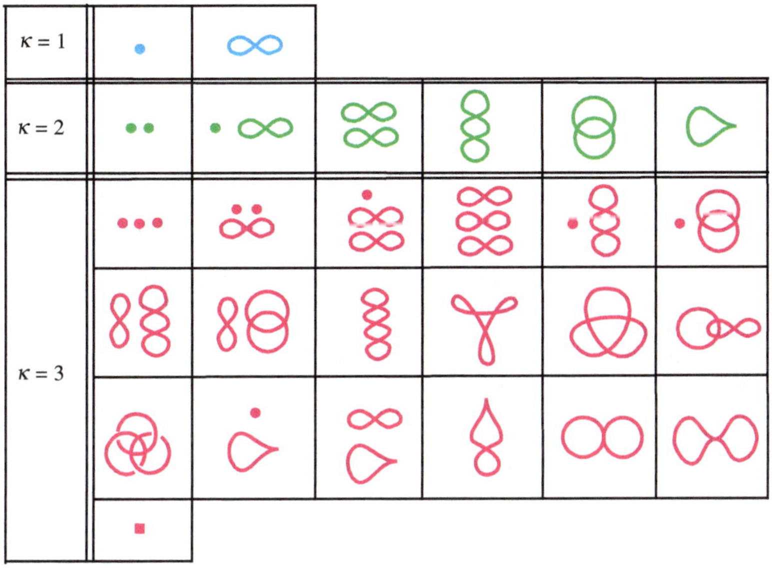

Fig. 11 List of singular fibers for maps from dimension 4 to dimension 3

With the help of such a list, we can extract characteristic fibers and their types when analyzing a given set of data, and it is expected that this will contribute to the visualization of the given data.

We have CT scan data as an example of a possible application. A CT scan measures volume data by piling up the sliced images with respect to a fixed direction. In this process, if we can take the sliced image datasets with respect to several directions, then it is expected that the accuracy of the reconstruction of the real object data is improved. If we just consider one direction, then it corresponds to constructing a Reeb graph. Therefore, if we consider two or more directions and consider a unified representation of the Reeb graphs of the corresponding directions, or if we construct a graph containing the adjacency information of the singular fibers, then we would be able to extract that information, which would have never been obtained by considering only one direction.

The differential topological method that we have explored has the advantage in helping to efficiently represent, with a fairly small amount of data, important intrinsic features of a given large dataset, which could not be handled using current methods. It is also clearly a significant benefit that the visual information band is quite broad compared with character information such as numerical data. It is therefore largely expected that this kind of method will play an essential role in data analysis.

Acknowledgments Some of the contents of this article are the results of joint work with Professor Shigeo Takahashi (Graduate School of Frontier Sciences, the University of Tokyo), who kindly provided important materials for this article. The author would like to take this opportunity to express his sincere gratitude to him.

References

1. J. Callahan, Singularities and plane maps. Amer. Math. Mon. **81**, 211–240 (1974)
2. H. Carr, J. Snoeyink, U. Axen, Computing contour trees in all dimensions. Comput. Geom. Theor. Appl. **24**, 75–94 (2003)
3. H. Edelsbrunner, J. Harer, Jacobi sets of multiple Morse functions. Foundations of Computational Mathematics, Minneapolis, 2002, pp. 37–57, London Mathematical Society Lecture Note Series, vol. 312 (Cambridge University Press, Cambridge, 2004)
4. I. Fujishiro, R. Otsuka, S. Takahashi, Y. Takeshima, T-Map: A topological approach to visual exploration of time-varying volume data, *High-Performance Computing*, ed. by J. Labarta, K. Joe, T. Sato, pp. 176–190, Lecture Notes in Computer Science, vol. 4759 (Springer, Berlin, Heidelberg, 2008)
5. X. Ge, I. Safa, M. Belkin, Y. Wang, Data skeletonization via Reeb graphs, in *Twenty-Fifth Annual Conference on Neural Information Processing Systems* 2011, pp. 837–845
6. R. Gilmore, *Catastrophe theory for scientists and engineers* (Wiley, New York, 1981)
7. M. Golubitsky, V. Guillemin, *Stable Mappings and Their Singularities*. Graduate Texts in Mathematics, vol. 14 (Springer, New York, Heidelberg, 1973)
8. J.N. Mather, Stability of C^∞ mappings. VI: The nice dimensions, in *Proceedings of the Liverpool Singularities-Symposium, I (1969/70)*, pp. 207–253, Lecture Notes in Mathematics, vol. 192 (Springer, Berlin, 1971)
9. Y. Matsumoto, *An introduction to Morse theory*, Translated from the 1997 Japanese original by Kiki Hudson and Masahico Saito, Translations of Mathematical Monographs, vol. 208, American Mathematical Society, Providence, RI, Iwanami Series in Modern Mathematics (2002)
10. J. Milnor, *Morse theory*, based on lecture notes by M. Spivak and R. Wells, Ann. of Math. Stud. **51** (Princeton University Press, Princeton, NJ, 1963)
11. V. Pascucci, G. Scorzelli, P.-T. Bremer, A. Mascarenhas, Robust on-line computation of Reeb graphs: simplicity and speed. ACM Trans. Graph. **26**(3), Article 58, 58.1–58.9 (2007)
12. G. Reeb, Sur les points singuliers d'une forme de Pfaff complètement intégrable ou d'une fonction numérique. C. R. Acad. Sci. Paris **222**, 847–849 (1946)
13. O. Saeki, *Topology of singular fibers of differentiable maps*. vol. 1854, Lecture Notes in Mathematics (Springer-Verlag, Berlin, 2004)
14. S. Takahashi, T. Ikeda, Y. Shinagawa, T.L. Kunii, M. Ueda, Algorithms for extracting correct critical points and constructing topological graphs from discrete geographical elevation data. Comput. Graph. Forum **14**, 181–192 (1995)
15. S. Takahashi, Y. Takeshima, I. Fujishiro, Topological volume skeletonization and its application to transfer function design. Graph. Models **66**, 24–49 (2004)
16. Y. Takeshima, S. Takahashi, I. Fujishiro, G.M. Nielson, Introducing topological attributes for objective-based visualization of simulated datasets, in *Proceedings of the Volume Graphics, 2005*, pp. 137–145 (2005)
17. H. Whitney, On singularities of mappings of euclidean spaces: mappings of the plane into the plane. Ann. of Math. (2) **62**, 374–410 (1955)

Part III
Analysis

Mathematical Analysis for Pattern Formation Problems

Shin-ichiro Ei

Abstract We explain our theoretical treatment of various kinds of patterns appearing in nature in this paper. We introduce one of our typical approaches to focus on the pattern boundaries and to derive a curvature flow equation for the motion of these boundaries. This approach is based on the idea that patterns are defined by their boundaries.

Keywords Interface · Curvature flow · Reaction-diffusion model

1 Patterns Appearing in Nature

Let us consider how we can theoretically explain the various patterns in nature. There are many different kinds of defining scales, such as the cosmos is a huge scale and atoms or electrons are small scale, in which we have corresponding theories, such as in astronomy and quantum theory. In this report, we consider the phenomena appearing in our everyday life. Typical examples are snow crystals, fire shapes, and animal coat patterns (Figs. 1, 2, 3, and 4). Of course, there are a lot of examples in our life.

We would like to clarify the mechanism for theoretically creating these patterns in this report. We are going to introduce one approach for doing this. In addition, there are many kinds of objects with different mechanisms. For example, the problem with differing snow crystal patterns may belong to material sciences and for the animal coat patterns it may in biology. Thus, to discuss all of them in one story would be impossible, so we concentrate on one simple explicit example.

We should make this one "pattern" clear before starting with the main ideas. As in the examples mentioned above, patterns are represented by the boundaries between two different regions. For example, the pattern of a snow crystal is expressed by

S. Ei (✉)
Institute of Math-for-Industry, Kyushu University, Fukuoka 819-0395, Japan
e-mail: ichiro@imi.kyushu-u.ac.jp

R. Nishii et al. (eds.), *A Mathematical Approach to Research Problems of Science and Technology*, Mathematics for Industry 5,
DOI: 10.1007/978-4-431-55060-0_10, © Springer Japan 2014

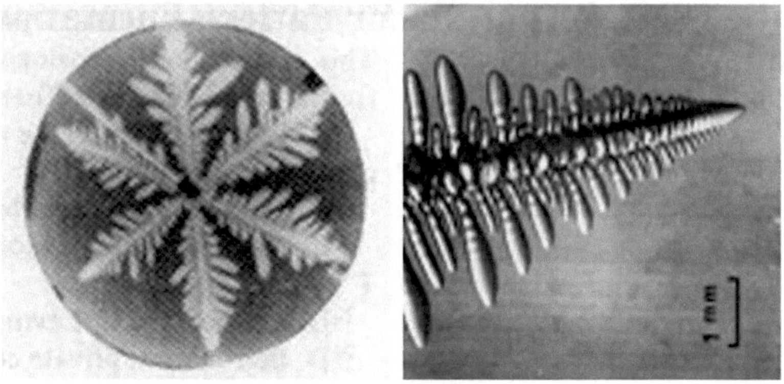

Fig. 1 Snow crystal (*left*) and dendritic crystal (*right*) ([1])

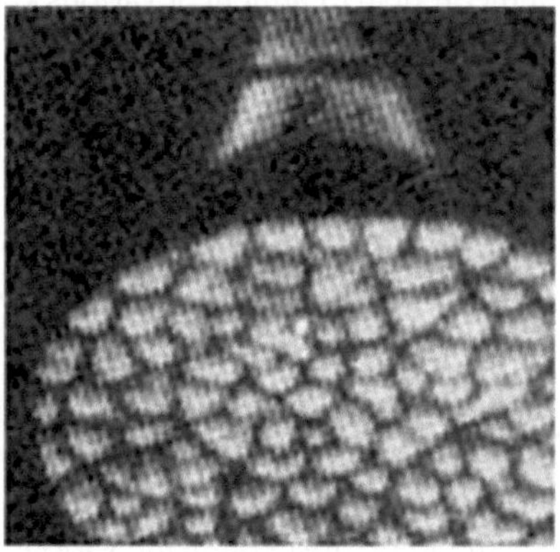

Fig. 2 Fire shapes ([2])

Fig. 3 Animal coat patterns (Tama Zoo)

the boundaries between the solid and liquid regions. The coat patterns on animal skins are expressed by the boundaries between the regions of high and low pigment concentrations on the skins. Let us focus on the neighborhoods of the boundaries, in which two different states adjoin. The state will change from one to the other

Fig. 4 Spiral patterns appearing in oxidation-reduction reaction ([1])

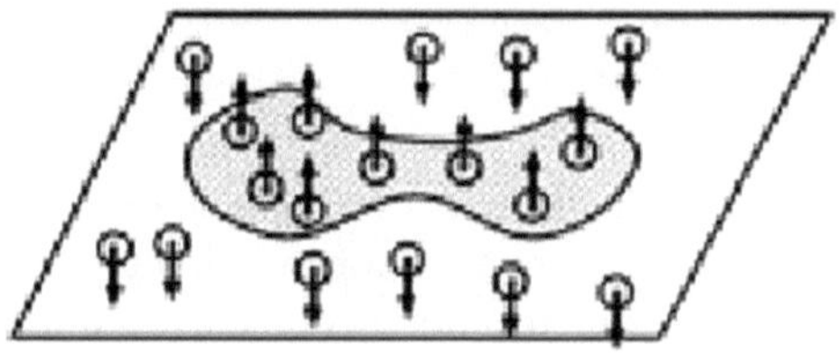

Fig. 5 Magnetic-like units spread over a plane

when we cross the boundaries together with intermediate states. If the region of an intermediate state is large, the boundaries become ambiguous and the patterns are then not clearly recognizable. Thus, the intermediate regions need to be small in order to clearly differentiate between the patterns. Therefore, we can say that the patterns are clearly recognized when the intermediate regions are sufficiently small.

2 Theoretical Treatment of Shapes

We explained in the previous section that patterns are expressed by their boundaries. We regard them as hyperplanes (possibly curves in two-dimensional spaces) with zero width in order to deal with the boundaries because the intermediate regions are assumed to be very small. We analyze the patterns discussed in the following paragraphs by adopting one of the simplest examples, which has almost been completely analyzed.

We assume that some microscopic magnetic-like units are spread all over a plane without any space and that each unit is stable in the right upper or under directions. On the other hand, if they come into contact with each other, they are assumed to be inclined in the same direction see Fig. 5.

For the convenience in understanding, the upper- and under-directed units are colored black and white, respectively. Then, we consider the movements of the colored regions over time.

First, let us consider a case in which the strengths toward the upper and under directions of each unit are not equal, in particular, when the upper direction is much

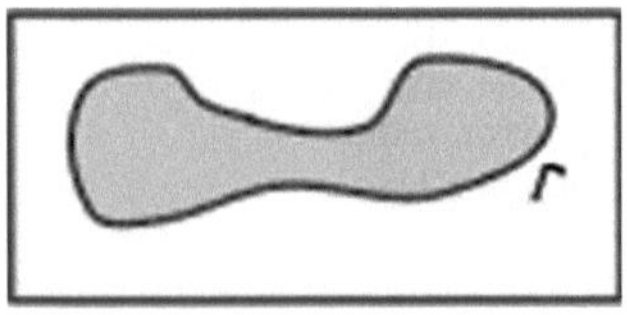

Fig. 6 Initial distribution

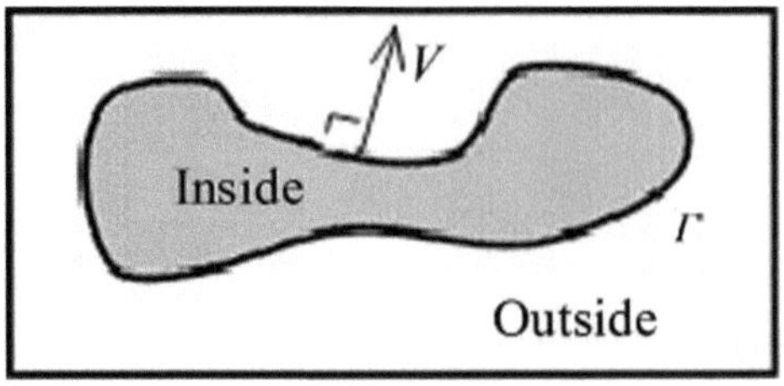

Fig. 7 Movement of Γ

stronger than the under one. We can easily expect that all of the units will eventually point in the upper direction even if there are many units initially pointing in the under direction, which is not theoretically attractive.

Therefore, we assume as a final assumption that the strengths toward the upper and under directions of each unit are completely equal, by which our intuition does not work and theoretical consideration is necessary.

Finally, we assume that the units are sufficiently small and regard the system as a continuous system with an initial distribution, such as that in Fig. 6.

Then, the problem is to analyze the movement of the interfaces (say Γ) between the black and white regions. The following have been known since the 1980's under the above-mentioned assumptions together with several additional conditions (see Fig. 7):

Theorem 1 *The curve Γ moves according to*

$$V = -\kappa. \tag{1}$$

Here, V and κ denote the outward normal velocity and the curvature of Γ, respectively.

The movement according to (1) is called the curvature flow.

Let us further explain Eq. (1). First, V is the outward normal velocity, which means we have to determine the inside and outside of Γ in advance. We call the region corresponding to the black region surrounded by Γ the "inside of Γ." Thus, V denotes the velocity from the black region to the white region.

Next, we explain the curvature κ. Fixing a point, say P on Γ, we draw a maximal tangential circle at P, which is uniquely determined. If the tangential circle is drawn inside of Γ, then we define the curvature by $\kappa := 1/r$ by using the radius of the circle while we define $\kappa := -1/r$ if the circle is drawn outside Γ (see Fig. 8).

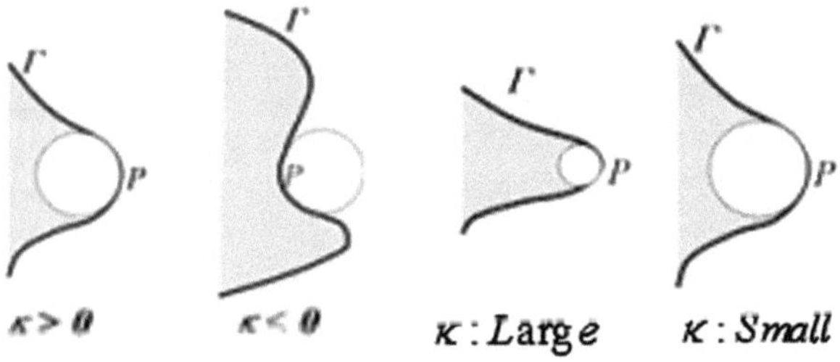

Fig. 8 Sign of curvature κ (*left*) and magnitude of curvature κ (*right*)

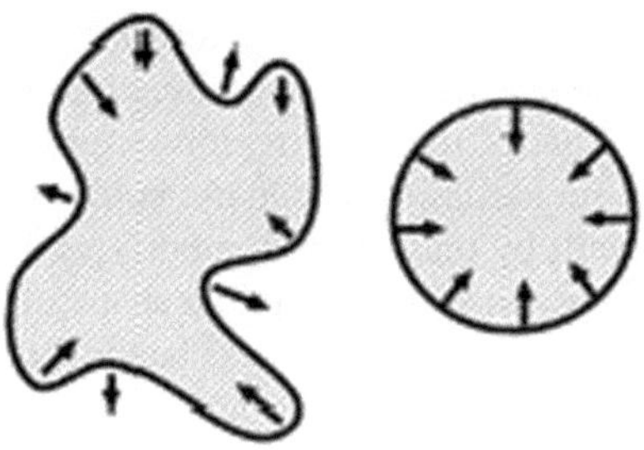

Fig. 9 Movement of Γ

In particular, if Γ is a line, then we can draw an infinitely large tangential circle at P, and thus, the radius is $R = \infty$, which means $\kappa = 0$. In general, the curvature denotes how Γ curves toward the inside, that is, the more curved toward the inside of Γ is, the smaller r and κ is positively larger. On the other hand, if Γ is more curved outside, κ is negatively larger. These properties of κ and (1) lead to the movement of Γ as follows: at convex parts toward the outside of Γ, $\kappa > 0$ and the outward normal velocity $V = -\kappa < 0$, which means the movement of Γ is toward the inside while the movement toward the outside is induced in the concave parts. This means Γ moves toward an unrelieved shape and approaches a circle. Since a shape close to a circle has a positive curvature everywhere, V is negative everywhere. Thus, once Γ is close to a circle, it eventually shrinks to one point and disappears at a finite time. This means the final occupying color is determined by the initial distribution (see Fig. 9).

In practice, the movement of Γ according to (1) is more well known than the above style. For example, the length of Γ is monotonically decreasing over time. This property implies that Γ approaches a minimal line and remains in the dumbbell-shaped region, as shown in Fig. 10.

We previously considered that the strengths toward the upper and under directions of each unit were completely equal but this is unnatural and unrealistic. Therefore, let us consider a case in which the strengths are slightly different. Then, the corresponding equation to (1) is

$$V = -\kappa + c. \tag{2}$$

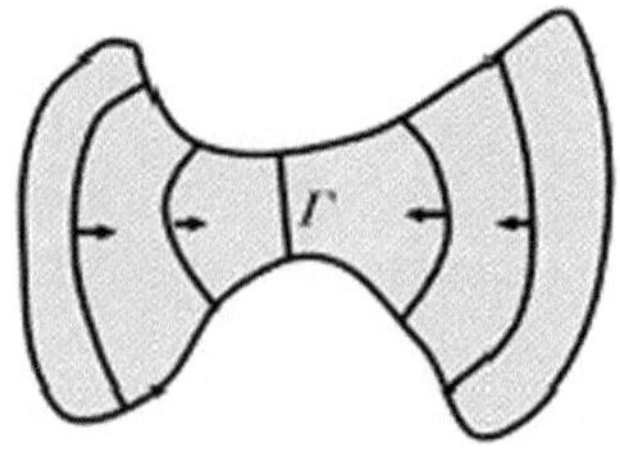

Fig. 10 Movement of Γ in dumbbell-shaped region

This is also called the curvature flow. The constant c is determined by the difference in the strengths and $c = 0$ means there are completely equal strengths. Hereafter, we assume $c > 0$, which means that the upper direction is slightly easier to tend to than the under direction. That is, the black regions are easier to extend toward than the white regions. All the properties for (1) do not hold and the theoretical treatment becomes drastically difficult by only slightly modifying Eqs. (1) to (2). Therefore, we consider the simplest case in which the initial shape of Γ is a circle. Then, Γ is known to always be a circle at any time. Let $r(t)$ be the radius of the circle at time t. Then, (2) is represented by

$$\frac{dr}{dt} = -\frac{1}{r} + c. \tag{3}$$

We hope that readers can obtain (3) from (2) by themselves as an exercise. Equation (3) is an ordinary differential equation of $r(t)$ and can be explicitly solved. Here, we analyze (3) in another way.

First, we define $r^* := 1/c$. If the initial radius $r(0) = r^*$, then the right-hand side of (3) is just zero and $\frac{dr}{dt} = 0$ holds, which means $r(t) = r^*$ for any $t > 0$. This special solution $r(t) \equiv r^*$ is called "equilibrium."

Next, consider the case if $r(0) < r^*$. Then, $\frac{1}{r(0)} > c$, and therefore, $\frac{dr}{dt} < 0$ holds. That is, the radius $r(t)$ decreases over time $t > 0$ and decreases even more for a smaller r (i.e., larger $\frac{1}{r}$). Thus, $r(t) = 0$ over a finite time, that is, Γ disappears.

Conversely, if $r(0) > r^*$, then by a similar discussion to the above, $r(t)$ increases over time $t > 0$. Thus, whether Γ can grow or shrink over time is determined by whether or not the initial radius $r(0)$ is larger than r^*. r^* is called the "critical radius," which is known to be effective for explaining the growth of initially small ice blocks in solidification phenomena.

Curvature flows like (1) and (2) are frequently observed in other fields. For example, in two almost balanced competing species, the movement of the boundaries between the occupied regions based on the species is known to be essentially governed by (2).

Various patterns described by the boundaries between two different regions, such as the solid and liquid regions, regions with high and low concentrations of pigments, and regions occupied by two competing species, are all known to be related to the curvature flow dynamics and can be reestablished by conducting numerical simulations although they are in quite different scientific fields from each other.

The approaches for patterns research by focusing on the boundaries and deriving the equations for the motions of the boundaries have just begun to be created, by which many unsolved problems based on patterns have been related to the geometrical properties and are currently being solved [3, 4].

Referenccs

1. P. Pelce, *Dynamics of Curved Fronts: Perspectives in Physics* (Academic Press, New York, 1988)
2. B. Lewis, G. von Elbe, *Combustion, Flames, and Explosions of Gases*. (Academic Press, New York, 1961)
3. J.D. Murray, *Mathematical Biology, Biomathematics*, vol. 19. (Springer, Berlin, 1989)
4. A.R. Winfree, *When Time Breaks Down*. (Princeton University Press, Princeton, 1986)

Models and Applications of Organism Transportation

Atsushi Tero

Abstract Organism makes various transportation networks. These many networks have adaptive character, in which the link grows with high-use and degenerates with low-use. In this chapter the mathematical model of adaptive network is introduced. Next, this chapter shows the simulation results by this mathematical model with various parameter. As a result, this chapter shows that how the organism can gain the global function only with the local growth law.

Keywords Adaptive network · Mathematical model · Optimal network · Shortest path · True slime mold

1 Introduction

Transportation networks, such as roads and railroads are accompanied with particularly high costs, but they are essential to our daily lives and as such, road networks and railroad networks continue to be serviced. In railroad networks, the number railroad lines and trains are increased for those routes with a high number of users using the funds collected from passenger fares, while those with a low number of users are reduced or discontinued. Since the optimum configurations of transport networks in urban areas change in response to the population distribution and the like, networks that are optimum when they are planned and when construction work starts do not necessarily remain optimum when their operations begin. In addition, transportation networks may be disconnected owing to accidents or disasters, and since they are also required to transport medical aid and materials when such accidents and disasters occur, they must be able to respond to situations that change

A. Tero (✉)
Institute of Mathematics for Industry, Kyushu University, 744, Motooka, Nishiku, Fukuoka 819-0395, Japan
e-mail: tero@imi.kyushu-u.ac.jp

R. Nishii et al. (eds.), *A Mathematical Approach to Research Problems of Science and Technology*, Mathematics for Industry 5,
DOI: 10.1007/978-4-431-55060-0_11,

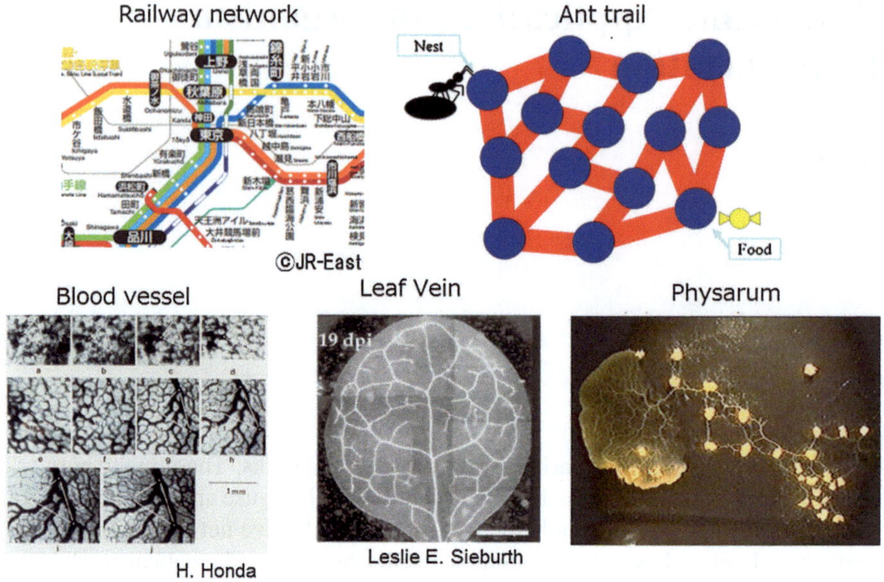

Fig. 1 Adaptive transport networks created by organisms

in ways that cannot be predicted in either the short or long term. The most suitable urban planning methods must be explored by implementing a variety of methods and through trial and error. Such efforts, however, are not realistic, because they require a large amount of cost and time (Fig. 1).

However, there are occasions where groups of organisms other than humans create transportation networks. Ants for instance, disperse pheromones when bringing food back to their nests to make it easier for other ants to follow the path. As a result, ants form a queue along an efficient route. Similarly, transportation networks also exist inside individual organisms. Many organisms including humans have a blood vessel network inside their bodies to distribute nutrients and oxygen properly within the body. Blood vessels grow from blood flows and shearing stress on the blood vessels to form such blood vessel networks. Additionally, leaf veins in plants have the function of transporting moisture and nutrients throughout the leaves, and a chemical substance known as auxin is believed to affect the formation of leaf veins significantly [1].

Such networks in all cases are built in an adaptive manner, with development occurring along routes that are in use, while those routes that are not in use decay and become extinct. The description of a theory that can be applied to all adaptive networks is provided in this chapter. Furthermore, a comparison is made between biological networks that enabled organisms to survive through the process of natural selection occurring over many years and railroad networks built by humans. The result obtained from considerations of these optimized networks is explained.

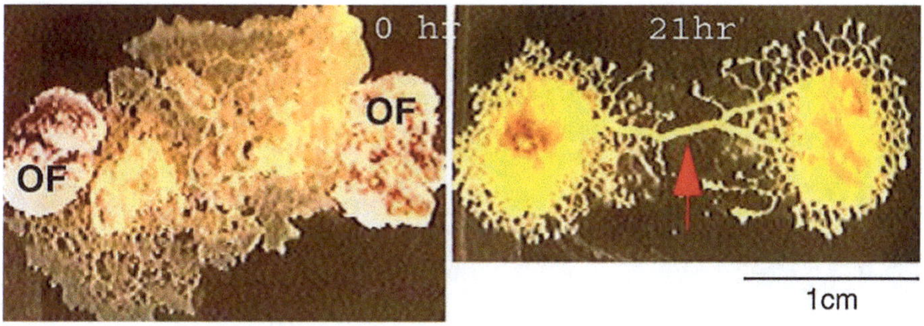

Fig. 2 Slime molds gathering around food

2 Tubular Network of a Slime Mold

True slime mold is an organism that has such an adaptive network (hereinafter referred to as "slime mold"). When a slime mold comes into contact with food, it engulfs it to feed and absorb its nutrients. When a slime mold comes into contact with multiple food sources at the same time, it engulfs all the food sources and connects them with a tubular structure (Fig. 2). Furthermore, slime molds have multiple nuclei within a plasmodium, but no cell membranes or cell walls separating them, even though they are considered unicellular organisms. Since a slime mold has multiple nuclei within a single plasmodium, it has the characteristics of both a singular individual and a group. When a slime mold is physically cut in two using a knife, the separated pieces both start to behave as individual entities, and when two slime molds come into contact with each other, they combine to form a single organism. Since this property makes it possible for a slime mold to control its shape, it has become a good model for understanding adaptive networks.

3 Slime Mold Solving a Maze

First of all, the maze-solving phenomenon of the slime mold, which is an important boundary for determining a network topology, is introduced here [2]. This experiment was performed by Professor Toshiyuki Nakagaki. First, the initial state is prepared as a maze on an overhead projection (OHP) film, which is placed over a nutrient agar (Fig. 3a). Since agar contains moisture, it becomes a penetrable pathway for a slime mold to enter, while the OHP film is dry and becomes an impenetrable barrier of the maze for the slime mold. Next, food is placed at two locations, the start and the goal. This prompts the slime molds to gather around the food and form a network with a tubular structure over the pathway of the maze (hereinafter referred to as the tube). Those tubes that end up in a dead end cease to exist (Fig. 3b), while only the route with the shortest distance through the maze remains, subject to conditions (Fig. 3c).

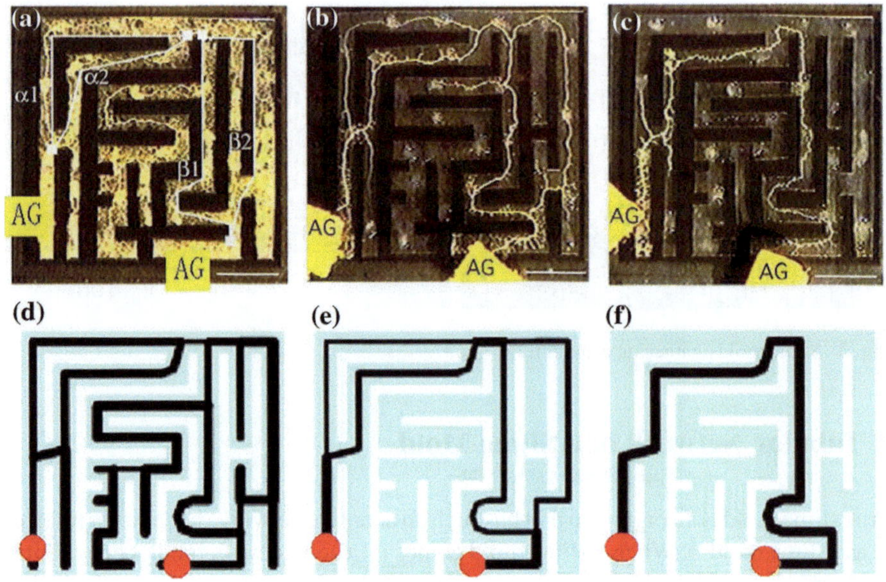

Fig. 3 **a–c** Maze solving by a slime mold. **d–f** Numerical calculation results with a mathematical model

4 Mathematical Model

What is the mechanism involved for a slime mold, which has no brain, to solve the maze? This mechanism is explained by using a mathematical model, with the maze expressed in terms of a discrete graph. Intersections and dead ends of the maze are expressed by nodes N_i, and the pathways of the maze that connect them are expressed as links E_{ij}. N_1, N_2 is the location where food is placed for our purpose. Each node N_i is assigned a variable p_i, which express the force with which the tubes of the slime molds press onto the internal protoplast at each given instant in time. Furthermore, each link E_{ij} is assigned variables $L_{ij}, a_{ij}(t), Q_{ij}(t)$. L_{ij} represents the physical length of the link, while $a_{ij}(t)$ represents the radius of the tube, and $Q_{ij}(t)$ represents the flow rate.

Poiseuille's flow is assumed here, based on the assumption that the flow of protoplasts inside the tube can be considered as an incompressible Newtonian fluid flowing through a circular tube.

$$Q_{ij} = \frac{\pi a_{ij}^4}{8\kappa} \frac{p_i - p_j}{L_{ij}}, \tag{1}$$

where κ, which is a constant, is the viscosity of the protoplast. Next, by defining $D_{ij}(t) = \frac{\pi a_{ij}^4}{8\kappa}$ the following can be formulated:

$$Q_{ij} = \frac{D_{ij}}{L_{ij}}(p_i - p_j), \tag{2}$$

where $D_{ij}(t)$ is a monotonically increasing function with respect to the tube radius a_{ij} and represents the weight of the link.

Furthermore, the following equation can be formulated, since the amount of protoplast that flows into each node N_i is equal to the amount of protoplast that flows out of it.

$$\sum_j Q_{ij} + I_i = 0. \tag{3}$$

$I_i(t)$ expresses the amount of protoplast that flows into the tube network from the slime mold that gathers around the food. At a node that is not connected to the food, $N_i(i \neq 1, 2)$ it is $I_i = 0$.

Furthermore, by defining $R_{ij} = \frac{L_{ij}}{D_{ij}}$ as resistance for our purpose here, the Eqs. (2) and (3) become equivalent of the Ohm's Law and Kirchoff's Law, respectively, for electric circuits.

The growth of the route is considered next. The thickness a tube often grows when the amount of protoplast flowing inside the tube is large, or it decays or become extinct when the amount is small. Because of $D_{ij}(t) = \frac{\pi a_{ij}^4}{8\kappa}$, the weight of the link $D_{ij}(t)$ changes in response to the flow rate $Q_{ij}(t)$.

$$\frac{\mathrm{d}}{\mathrm{d}t} D_{ij} = f(|Q_{ij}|) - r D_{ij}. \tag{4}$$

where $f(q)$ satisfies $f(q) = 0$, and is a monotonously increasing function for q. The shape of the final network varies, depending on the function of $f(q)$.

5 Numerical Calculation Result of $f(q) = |q|^{\mu}$

Numerical calculation results for a case where the typical numerical calculation result for the mathematical model described in the previous section $f(q) = |q|^{\mu}$ (Fig. 4). While almost all of the routes within the entire network remain in case of $\mu < 1$ (Fig. 4a), only one route is selected for the initial values or network shape in the case of $\mu > 1$. In this manner the network topology changes significantly, using the parameters of $\mu = 1$ as the boundary [3]. In the case of $\mu = 1$, only the shortest route remains, and the experiment on the slime mold solving the baze can be reproduced (Fig. 3d–f) [4]. The fact that a tube remains the shortest route is also mathematically proven [5, 6].

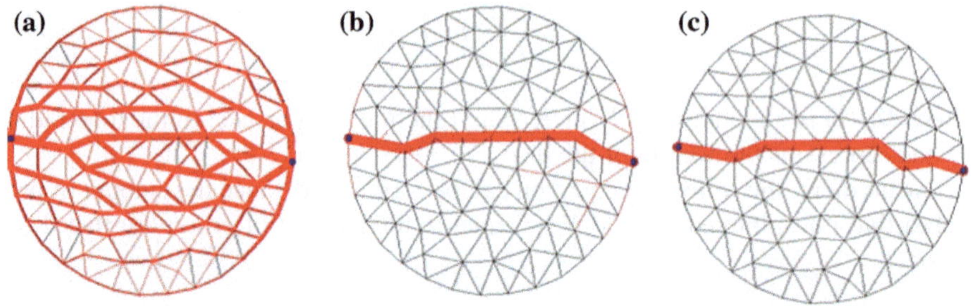

Fig. 4 Numerical calculation results for $f(q) = |q|^{\mu}$. **a** $\mu = 0.8$. **b** $\mu = 1.0$. **c** $\mu = 1.2$

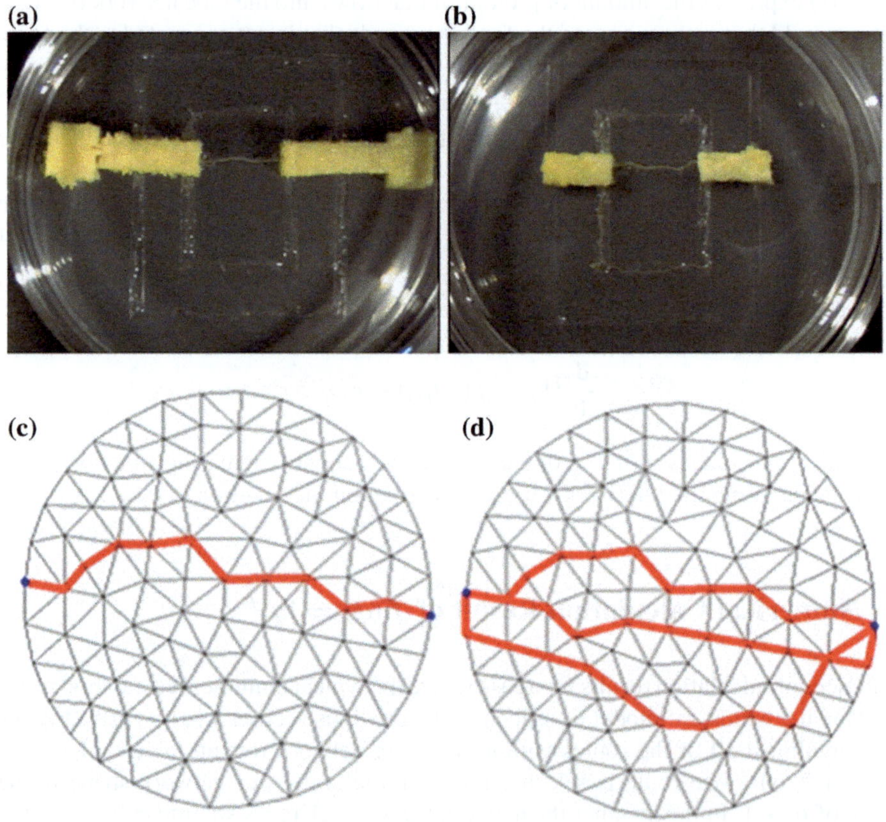

Fig. 5 An experiment on the risk distribution in a slime mold network. Both (**a**) and (**b**) have same amount of slime mold, but (**a**) has more food. When the slime molds gather around the food, the flow rate between the groups increases for (**b**), and the number of tubes changes. **c** and **d** are results of numerical calculations. Increasing the flow rate from the vicinity of the food, I_i, results in an increased number of tubes

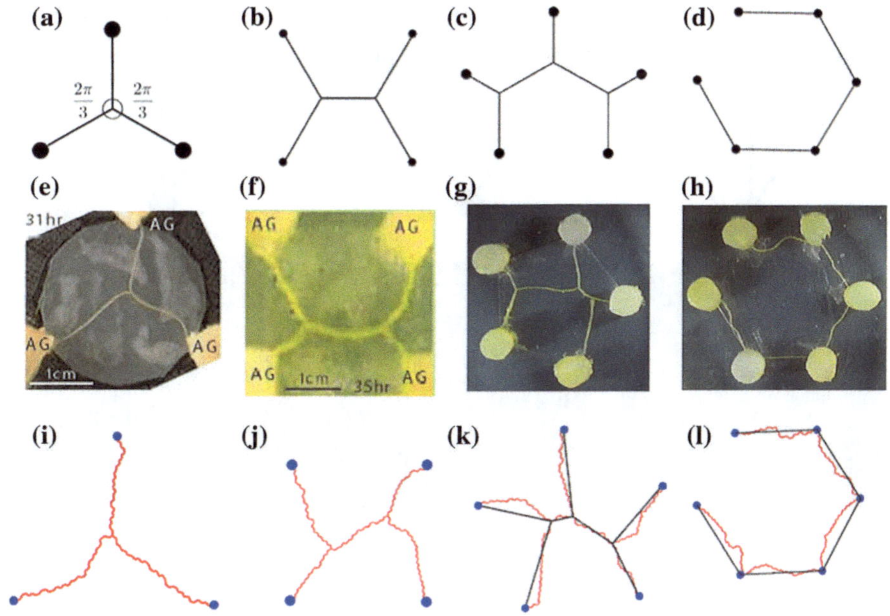

Fig. 6 The shortest network that connects the peaks of a regular polygon (**a–d**), the solution of the slime mold (**e–h**), and the result of numerical calculation (**i–l**). Both experimental and numerical calculation results can potentially offer a network topology that is different from the correct solution, but a solution can be obtained quickly with little error

6 Risk Distribution of Network

Slime molds are also known to use a number of tubes instead of creating a tube with larger flow rate capacity when the flow rate increases (Fig. 5a, b), [7]. However, if a thick tube with an excessively large flow rate capacity is created, and if such a tube is cut off due to an accident, the slime mold will sustain significant damage. This leads us to believe that the slime molds incorporates a mechanism that voluntarily distributes such risks. This phenomenon can be reproduced by applying the S-shape function to $f(q)$, which is the law of growth for tubes (for our purpose here it is defined as: $f(q) = \frac{|q|^3}{1+|q|^3}$).

7 Shortest Network with Multiple Points

Next, a description is provided for an optimized network that connects three or more points. Slime molds create a straight-forward network when the flow rate is small, as described thus far. This property was used to derive the shortest possible network between multiple points, and the result is shown in Fig. 6. This algorithm does not

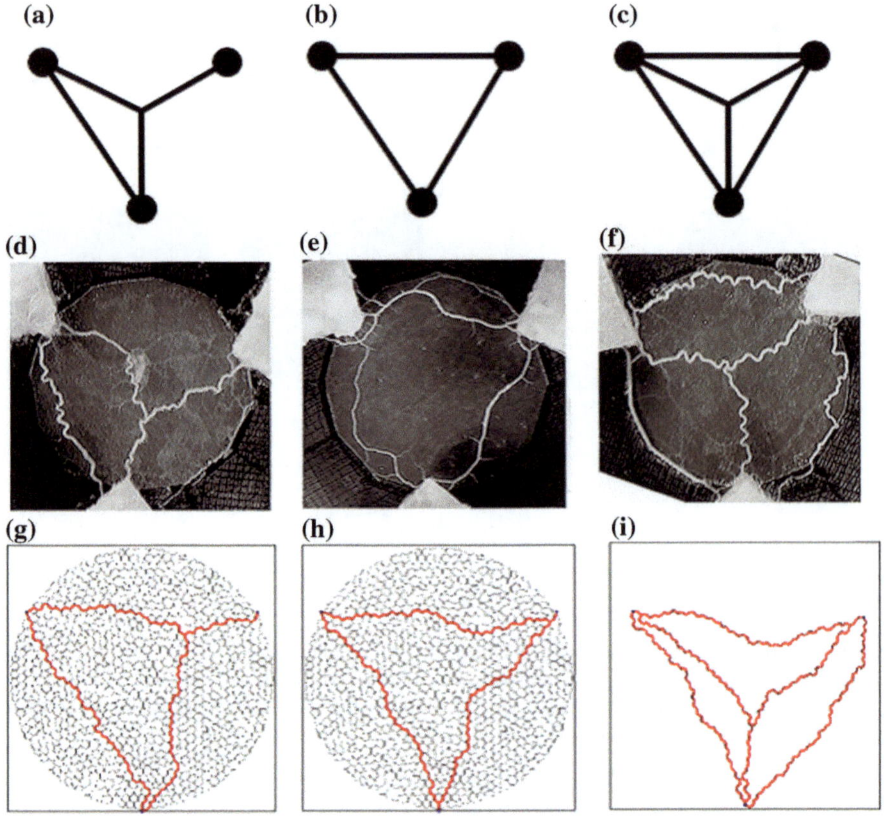

Fig. 7 An optimized network with three points. The total distance of the network increases as risk distribution is performed for an increasing flow rate

necessarily provide a correct solution, but it does provide a solution with a certain level of accuracy in a short time [8].

8 Optimized Network with Multiple Points

A network having three or more points with considerations for risk distribution is examined here. The slime mold configures a complex network as the flow rate increases (Fig. 7d–f). A complex network is also derived with this mathematical model by significantly increasing the flow rate $I_j(t)$ (Fig. 7g–i) .

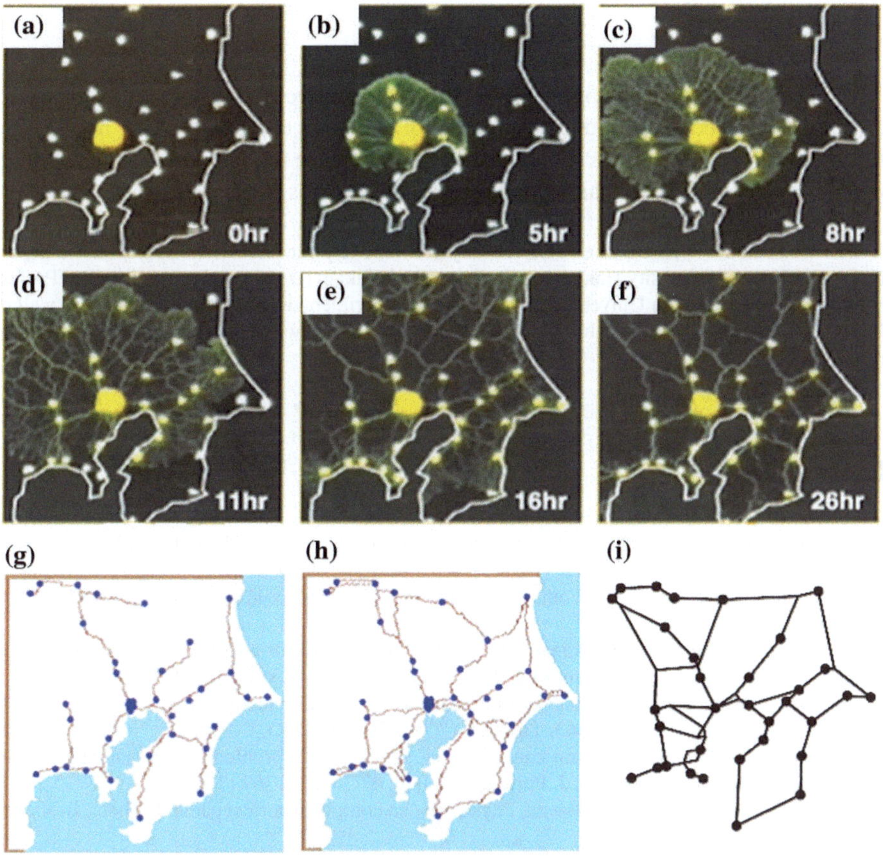

Fig. 8 **a–f** The railroad network as proposed by the slime molds. **g**, **h** The results of numerical calculations. **i** Actual railroad network in the Kanto region

9 Comparison of Optimized Networks Made by Organisms

Finally, the result of a comparison between the network of slime molds with an actual man-made railroad network is introduced [9]. Figure 8g is an actual railroad network in the Kanto region in Japan. Figure 8a–f are photographs taken to indicate how the slime molds spread on a culture prepared in the shape of there. Figure 8g, h are the results of simulations. Similar to an actual railroad network, the shape of the final network becomes condensed as the number of users increases.

10 Conclusion

A network theory that is applicable to various adaptive networks was proposed in this chapter, based on our numerical model of networks created by slime molds. The fact that networks in various fields and organisms can be understood in such a unified manner can be considered an advantage of mathematics.

Acknowledgments I would like to express my gratitude to joint researchers Professor Toshiyuki Nakagaki (Future University Hakodate), Professor Seiji Takagi (Hokkaido University), and Professor Toru Saigusa (Kyushu University) who provided experimental data for this research, as well as Professor Ryo Kobayashi (Hiroshima University) for assisting us with consultation on mathematical models.

References

1. L.E. Sieburth, Auxin is required for leaf vein pattern in *arabidopsis*. Plant Phys. **121**, 1179–1190 (1999)
2. T. Nakagaki, H. Yamada, A. Tóth, Maze-solving by an amoeboid organism. Nature 407, 470 (2000)
3. A. Tero, R. Kobayashi, T. Nakagaki, A mathematical model for adaptive transport network in path finding by the true slime mold. J. Theor. Biol. **244**, 553–564 (2007) (ELSEVIER)
4. A. Tero, R. Kobayashi, T. Nakagaki, Physarum solver: a biologically inspired method of road-network navigation. Phys. A **363**, 115–119 (2006) (ELSEVIER)
5. T. Miyaji, I. Ohnishi, Physarum can solve the shortest path problem on riemannian surface mathematically rigorously. Int. J. Pure Appl. Math. **47**(3), 353–369 (2008)
6. V. Bonifaci, K. Mehlhorn, G. Varma, Physarum can compute shortest paths. J. Theor. Biol. **309**, 121–133 (2012)
7. T. Nakagaki, T. Saigusa, A. Tero, R. Kobayashi, *Effects of Amount of Food on Path Selection in the Transport Network of an Amoeboid Organism. Topological Aspects of Critical Systems and Networks*, 2007/07 pp. 94–100
8. A. Tero, K. Toyabe, K. Yumiki, R. Kobayashi, T. Nakagaki, A method inspired by Physarum for solving the Steiner problem. Int. J. Unconventional Comput. **6**, 109–123 (2010)
9. A. Tero, S. Takagi, T. Saigusa, K. Ito, D.P. Bebber, M.D. Fricker, K. Yumiki, R. Kobayashi, T. Nakagaki, Rules for biologically inspired adaptive network design. Science **327**(5964), 439–442 (2010/1/22)

The Renormalization Group Method for Ordinary Differential Equations

Hayato Chiba

Abstract The renormalization group (RG) method is one of the singular perturbation methods which provides asymptotic behavior of solutions of differential equations. In this article, how to construct approximate solutions by the RG method is shown with several examples and basic theorems on the RG method, such as an error estimate and the existence of invariant manifolds are given.

Keywords Renormalization group · Dynamical systems · Perturbation method

1 Introduction

Differential equations form a fundamental topic in mathematics and its application to natural sciences. In particular, perturbation methods occupy an important place in the theory of differential equations. Although most of the differential equations can not be solved exactly, some of them are close to solvable problems in some sense, so that perturbation methods, which provide techniques to handle such class of problems, have been long studied.

In this article, we investigate a system of ordinary differential equations on $\mathbf{R}^n$ of the form

$$\frac{\mathrm{d}x}{\mathrm{d}t} = f(x) + \varepsilon g(t, x, \varepsilon), \quad x \in \mathbf{R}^n, \tag{1}$$

with appropriate assumptions, where $\varepsilon \in \mathbf{R}$ is a small parameter.

Since ε is small, it is natural to try to construct a solution of this system as a power series in ε of the form

H. Chiba (✉)
Institute of Mathematics for Industry, Kyushu University, 744, Motooka, Nishiku, Fukuoka 819-0395, Japan
e-mail: chiba@imi.kyushu-u.ac.jp

R. Nishii et al. (eds.), *A Mathematical Approach to Research Problems of Science and Technology*, Mathematics for Industry 5,
DOI: 10.1007/978-4-431-55060-0_12,

$$x = \hat{x}(t) = x_0(t) + \varepsilon x_1(t) + \varepsilon^2 x_2(t) + \cdots . \tag{2}$$

Substituting Eq. (2) into Eq. (1) yields a system of equations of $x_0, x_1, x_2, \ldots$ to obtain $\hat{x}(t)$. The method to construct $\hat{x}(t)$ in this manner is called the *regular perturbation method*.

It is known that if the function $g(t, x, \varepsilon)$ is analytic in ε, the series (2) converges to an exact solution of (1) while if it is not analytic, (2) diverges and no longer provides an exact solution. However, the problem is that one can not calculate infinite series as (2) in general whether it converges or not because it involves infinitely many equations of $x_0, x_1, x_2, \ldots$. If the series is truncated at a finite-order term in ε, another problem arises. For example, suppose that Eq. (1) admits an exact solution $x(t) = \sin(\varepsilon t)$, and that we do not know the exact solution. In this case, the regular perturbation method provides a series of the form

$$\hat{x}(t) = \varepsilon t - \frac{1}{3!}(\varepsilon t)^3 + \frac{1}{5!}(\varepsilon t)^5 + \cdots . \tag{3}$$

If truncated, the series becomes a polynomial in t, which diverges as $t \to \infty$ although the exact solution is periodic in t. Thus, the perturbation method fails to predict qualitative properties of the exact solution. Methods which handle such a difficulty and provide acceptable approximate solutions are called *singular perturbation methods*. Many singular perturbation methods had been proposed so far and many authors reported that some of them produced the same results though procedures to construct approximate solutions were different from each other.

The renormalization group (RG) method is the relatively new method proposed by Chen et al. [3], which reduces a problem to a more simple equation called the *RG equation*. In their paper, it is shown (without a proof) that the RG method unifies conventional singular perturbation methods such as the multiple time scale method, the boundary layer technique, the WKB analysis and so on. Their results are mathematically justified by Chiba [1]. Indeed, it is proved that the RG method extends and unifies other traditional singular perturbation methods, such as the averaging method, the multiple time scale method, the (hyper-) normal forms theory, the center manifold reduction, the geometric singular perturbation method and the phase reduction. Furthermore, the RG method is also applicable to some class of partial differential equations [2]. The purpose of this article is to show how to construct approximate solutions by the RG method and give basic theorems on the RG method with several examples.

2 Example

We explain how to construct approximate solutions by using a simple example.

Example 1 Let us consider the following second order differential equation.

$$\ddot{x} + x + \varepsilon x^3 = 0, \quad x \in \mathbf{R}, \tag{4}$$

where $(\dot{\ })$ denotes the derivative with respect to a time t. At first, we apply the regular perturbation method and obtain a secular term explicitly. For this purpose, put $x(t) = x_0(t) + \varepsilon x_1(t) + O(\varepsilon^2)$ and substitute it to the equation.

$$\ddot{x}_0 + \varepsilon\ddot{x}_1 + x_0 + \varepsilon x_1 + \varepsilon(x_0 + \varepsilon x_1)^3 + O(\varepsilon^2) = 0.$$

Comparing the coefficients of ε^0 and ε^1 in the both sides of the equation, we obtain

$$\ddot{x}_0 + x_0 = 0, \tag{5}$$

$$\ddot{x}_1 + x_1 = -x_0^3, \tag{6}$$

respectively. The equation of x_0 is just the unperturbed system obtained by putting $\varepsilon = 0$ in Eq. (4). Equation (5) is solved as

$$x_0(t) = Ae^{it} + \overline{A}e^{-it}, \tag{7}$$

where $A \in \mathbf{C}$ is an arbitrary constant. Substituting the x_0 into Eq. (6) provides

$$\ddot{x}_1 + x_1 = -(A^3e^{3it} + 3|A|^2Ae^{it} + 3|A|^2\overline{A}e^{-it} + \overline{A}^3e^{-3it}).$$

Since this equation is an inhomogeneous linear equation, we can obtain a solution explicitly as

$$x_1(t) = \frac{A^3}{8}e^{3it} + \frac{3i}{2}|A|^2Ate^{it} + \text{c.c.}, \tag{8}$$

where c.c. denotes the complex conjugate of the preceding term. The first term is periodic in t, while the second term diverges as $t \to \infty$, which is called the secular term. Because of the secular term, the regular perturbation solution

$$x(t) = x_0(t) + \varepsilon x_1(t) = Ae^{it} + \varepsilon\left(\frac{A^3}{8}e^{3it} + \frac{3i}{2}|A|^2Ate^{it}\right) + \text{c.c.} \tag{9}$$

does not give a nice approximate solution.

Now we introduce the RG approach to remove the secular term and to obtain an effective approximate solution. The first step is to introduce a dummy parameter τ and rewrite Eq. (9) as

$$x(t;\tau) = Ae^{it} + \varepsilon\left(\frac{A^3}{8}e^{3it} + \frac{3i}{2}|A|^2A(t-\tau)e^{it}\right) + \varepsilon\frac{3i}{2}|A|^2A\tau e^{it} + \text{c.c.} \tag{10}$$

The secular term t was replaced with $t - \tau$, and the last term including τ was added for the consistency. Next, we renormalize the last term into the constant A. This means that we assume $A = A(\tau)$ to be an unknown function of τ, and rewrite Eq. (10) as

$$x(t;\tau) = A(\tau)e^{it} + \varepsilon\left(\frac{A(\tau)^3}{8}e^{3it} + \frac{3i}{2}|A(\tau)|^2 A(\tau)(t-\tau)e^{it}\right) + \text{c.c.} \tag{11}$$

We shall determine the function $A(\tau)$. Since the exact solution $x(t)$ is independent of the dummy parameter τ, we require that the derivative of $x(t;\tau)$ with respect to τ becomes zero. Hence, we assume the equality

$$\left.\frac{\mathrm{d}x}{\mathrm{d}\tau}\right|_{\tau=t}(t;\tau) = 0. \tag{12}$$

For Eq. (11), this condition yields the differential equation of A of the form

$$\frac{\mathrm{d}A}{\mathrm{d}t} = \varepsilon\frac{3i}{2}|A|^2 A + O(\varepsilon^2),$$

which is called the *RG equation*. The equation

$$\frac{\mathrm{d}A}{\mathrm{d}t} = \varepsilon\frac{3i}{2}|A|^2 A, \tag{13}$$

obtained by truncating the higher order with respect to ε is called the *first order RG equation*.

This RG procedure is motivated by the following idea. The secular term including a factor t diverges as $t \to \infty$, which makes the accuracy of an approximate solution bad. Thus, we introduce the dummy parameter τ, replace the factor t by $t-\tau$ and assume that τ is also large when t is large so that $t-\tau$ is bounded. Then, the constant A is regarded as some function of τ for the consistency.

Equation (13) is explicitly solved as

$$A(t) = \frac{1}{2}a\exp i\left(\frac{3\varepsilon}{8}a^2 t\right), \tag{14}$$

with an arbitrary constant a. Substituting $A(t)$ into Eq. (11) and putting $\tau = t$, we obtain the desired nice approximate solution

$$x(t) = \frac{1}{2}a\exp i\left(t + \frac{3\varepsilon}{8}a^2 t\right) + \frac{\varepsilon}{8}\cdot\frac{1}{8}a^3\exp i\left(3t + \frac{9\varepsilon}{8}a^2 t\right) + \text{c.c.} \tag{15}$$

See Fig. 1 for the graphs of an exact solution and the approximate solution.

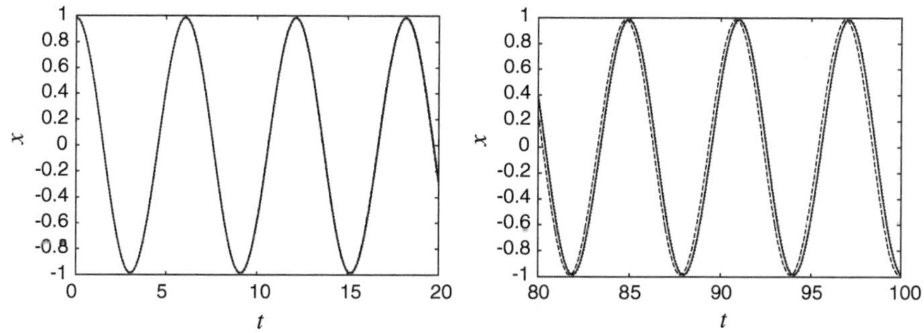

Fig. 1 The *solid line* and the *dashed line* denote an exact solution of Eq. (4) and the approximate solution (15), respectively. The *dotted line*, which is almost overlapped with the *solid line*, denotes an approximate solution obtained by employing the second order RG equation, which is not explained in this example

3 Settings

We give a general settings of the problem. Let us consider the perturbative problem on $\mathbf{R}^n$

$$\dot{x} = f(x) + \varepsilon g(t, x), \quad x \in \mathbf{R}^n. \tag{16}$$

We suppose that
(A1) the flow $\varphi_t(y)$ of the unperturbed system $\dot{x} = f(x)$ is periodic in t, where the flow $\varphi_t(y)$ is a solution of the equation $\dot{x} = f(x)$ satisfying the initial condition $x(0) = y$.
(A2) the function g is periodic in t.

In many applications, Eq. (16) is a perturbation of a linear system like as the previous example. In this case, $f(x)$ is expressed as $f(x) = Fx$ with a matrix F. Then, $\varphi_t(y) = \mathrm{e}^{Ft}y$ and the assumption (A1) implies that all eigenvalues of the matrix F lie on the imaginary axis. These assumptions can be weakened in several ways, see [1] for more general results.

Let us derive the RG equation for Eq. (16). For this purpose, put $x = x_0 + \varepsilon x_1 + \cdots$ and substitute it into the equation. From the zero-th order and the first order of ε, we obtain the equations

$$\dot{x}_0 = f(x_0), \quad \dot{x}_1 = Df(x_0)x_1 + g(t, x_0),$$

where Df is the Jacobi matrix of f. A solution of the former equation is written as $x_0 = \varphi_t(y)$. Thus, the equation for x_1 takes

$$\dot{x}_1 = Df(\varphi_t(y))x_1 + g(t, \varphi_t(y)).$$

When $f(x) = Fx$, then $Df(\varphi_t(y))x_1 = Fx_1$. In any case, this is an inhomogeneous linear equation solved explicitly as

$$x_1 = D\varphi_t(y)\int (D\varphi_t(y))^{-1} g(t, \varphi_t(y))\mathrm{d}t.$$

Due to the assumption on periodicity, the integrand consists of a constant term and a periodic term with respect to t;

$$(D\varphi_t(y))^{-1} g(t, \varphi_t(y)) = (\text{constant}) + (\text{periodic}).$$

Integrating it, it turns out that the integral of the constant term yields the polynomial of t, while the integral of the periodic term is still periodic;

$$\int (D\varphi_t(y))^{-1} g(t, \varphi_t(y))dt = (\text{constant}) \cdot t + (\text{periodic}).$$

Hence, the secular term arises from the constant term of the integrand. This constant term may be explicitly expressed as

$$R_1(y) := \lim_{t\to\infty} \frac{1}{t}\int (D\varphi_t(y))^{-1} g(t, \varphi_t(y))\mathrm{d}t.$$

Note that this is a function of y.

We define a function $h_t^{(1)}(y)$ by removing the secular term from x_1 as

$$h_t^{(1)}(y) = D\varphi_t(y)\int ((D\varphi_t(y))^{-1} g(t, \varphi_t(y)) - R_1(y))\mathrm{d}t.$$

Therefore, the regular perturbation solution up to the first order is given by

$$\begin{aligned} x(t) &= x_0 + \varepsilon x_1 \\ &= \varphi_t(y) + \varepsilon D\varphi_t(y)\int (D\varphi_t(y))^{-1} g(t, \varphi_t(y))\mathrm{d}t \\ &= \varphi_t(y) + \varepsilon D\varphi_t(y)\int \Big((D\varphi_t(y))^{-1} g(t, \varphi_t(y)) - R_1(y) + R_1(y)\Big)\,\mathrm{d}t \\ &= \varphi_t(y) + \varepsilon D\varphi_t(y) R_1(y)t + \varepsilon h_t^{(1)}(y). \end{aligned}$$

In particular, we have obtained the secular term $D\varphi_t(y)R_1(y)t$.

Now we apply the RG method to remove the secular term. The dummy parameter τ is introduced and the factor t is replaced by $t - \tau$. After that, the constant y is assumed to be an unknown function $y = y(\tau)$:

$$x(t;\tau) = \varphi_t(y(\tau)) + \varepsilon D\varphi_t(y(\tau))R_1(y(\tau)) \cdot (t - \tau) + \varepsilon h_t^{(1)}(y(\tau)). \tag{17}$$

Since this expression is independent of the dummy parameter τ, we suppose

$$\left.\frac{dx}{d\tau}\right|_{\tau=t} = 0.$$

Then, Eq. (17) yields

$$0 = \left.\frac{dx}{d\tau}\right|_{\tau=t} = D\varphi_t(y(t))\frac{dy}{dt} - \varepsilon D\varphi_t(y(t))R_1(y(t)) + \varepsilon Dh_t^{(1)}(y(t))\frac{dy}{dt}.$$

This is rearranged as

$$\frac{dy}{dt} = \varepsilon R_1(y(t)) + O(\varepsilon^2). \tag{18}$$

Truncating the higher order terms $O(\varepsilon^2)$, we obtain the first order RG equation

$$\frac{dy}{dt} = \varepsilon R_1(y). \tag{19}$$

Let $y(t)$ be a solution of the RG equation. Substituting $y(y)$ into Eq. (17) and putting $\tau = t$, we obtain the first order approximate solution as

$$\hat{x}(t) = \varphi_t(y(t)) + \varepsilon h_t^{(1)}(y(t)) + O(\varepsilon^2). \tag{20}$$

This procedure can be continued to obtain a more higher order approximation, see [1] for the detail.

4 Main results

The above RG procedure to obtain an approximate solution should be mathematically justified. What we want to show is
(i) quantitative nature: how good is the accuracy of an approximate solution and how long is it valid in a time t?
(ii) qualitative nature: does it provide qualitative properties of solutions such as the existence of a periodic solution?
(iii) is it easier to solve the RG equation than the original equation?

Regarding these questions, the following theorems hold. See [1] for the proof.

Theorem 1 (error estimate) *Let $\hat{x}(t)$ be an approximate solution obtained by the RG method given as Eq. (20) and $x(t)$ an exact solution of Eq. (16) satisfying the initial condition $x(0) = \hat{x}(0)$. There exist positive constants $C, T > 0$ such that the inequality*

$$||x(t) - \hat{x}(t)|| < C\varepsilon$$

holds for the time interval $0 \le t \le T/\varepsilon$.

In this sense, the RG method actually provides a good approximate solution. If we use the m-th order RG equation, then the inequality is refined as $||x(t)-\hat{x}(t)|| < C\varepsilon^m$. Since the time interval $0 \leq t \leq T/\varepsilon$ is finite, however, this theorem does not describe the asymptotic behavior of solutions as $t \to \infty$. For the asymptotic behavior, we can prove the next theorem, in which we suppose that a given Eq. (16) is autonomous (that is, g is independent of t) for simplicity.

Theorem 2 (existence of invariant manifolds) *Suppose that the RG equation $\dot{y} = \varepsilon R_1(y)$ has a normally hyperbolic invariant manifold M. Then the given Eq.* (16) *also has an invariant manifold M_ε within an ε-neighborhood of M. The manifold M_ε is diffeomorphic to M and the stability of M_ε is the same as that of M.*

Roughly speaking, a normally hyperbolic invariant manifold is an invariant manifold which is exponentially attracting or repelling, see a textbook of dynamical systems theory for the precise definition. In applications, it is important to find a periodic solution. When M is a stable periodic orbit, the above theorem is restated as follows.

Theorem 3 (existence of a periodic orbit) *If the RG equation $\dot{y} = \varepsilon R_1(y)$ has an asymptotically stable periodic orbit, then the given Eq.* (16) *also has an asymptotically stable periodic orbit.*

These theorems imply that near a stable invariant manifold, we can detect the asymptotic behavior of solutions as $t \to \infty$. Finally, we show that the RG equation is simpler than the original equation.

Theorem 4 (symmetry)

(i) *If the given Eq.* (16) *is invariant under the action of a Lie group H, the RG equation is also invariant under the action of H.*
(ii) *The RG equation is invariant under the action of the flow φ_t of f; that is, the equality*

$$R_1(\varphi_t(y)) = D\varphi_t(y)R_1(y)$$

holds.

This theorem means that the RG equation has a symmetry larger than that for the original equation. In this sense, the RG equation is easier to solve than the original equation.

Example 2 Consider the following system

$$\begin{cases} \dot{x} = y + \varepsilon(x - x^3) \\ \dot{y} = -x. \end{cases} \tag{21}$$

This is a perturbed harmonic oscillator. The RG equation is given by

$$\dot{A} = \frac{\varepsilon}{2}(A - 3A|A|^2).$$

It is easy to show that this equation has a stable periodic orbit given by $|A| = 1/\sqrt{3}$. Hence, Theorem 3 proves that the system (21) also has a stable periodic orbit whose radius is approximately estimated as $1/\sqrt{3} + O(\varepsilon)$.

References

1. H. Chiba, Extension and unification of singular perturbation methods for ODEs based on the renormalization group method. SIAM J. Appl. Dyn. Syst. **8**, 1066–1115 (2009)
2. H. Chiba, Reduction of weakly nonlinear parabolic partial differential equations. J. Math. Phys. **54**, 101501 (2013)
3. L.Y. Chen, N. Goldenfeld, Y. Oono, Renormalization group and singular perturbations: multiple scales, boundary layers, and reductive perturbation theory. Phys. Rev. E **54**, 376–394 (1996)

A Phase Field Approach to Mathematical Modeling of Crack Propagation

Masato Kimura and Takeshi Takaishi

Abstract We consider a phase field model for crack propagation in an elastic body. The model is derived as an irreversible gradient flow of the Francfort-Marigo energy with the Ambrosio-Tortorelli regularization and is consistent to the classical Griffith theory. Some numerical examples computed by adaptive mesh finite element method are presented.

Keywords Adaptive mesh finite element method · Crack propagation · Fracture mechanics · Irreversible system · Phase field model

1 Introduction

Crack propagation phenomenon appears in various situations from tiny size to huge scale and often causes serious problems, for example, in tiny precision machine and its parts, the body of a car or a ship, a building, a large structure, the ground, or the crust of the earth. Since the propagation of a relatively small crack in such material may cause the collapse of the whole structure, to understand the behavior of the crack propagation is very important.

Among various crack propagation models in fracture mechanics, the phase field approximation [20] of the crack seems to be one of very interesting ideas. A number of engineering-oriented discrete models, such as extended finite element method

---

M. Kimura (✉)
Institute of Science and Engineering, Kanazawa University,
Kakuma, Kanazawa 920-1192, Japan
e-mail: mkimura@se.kanazawa-u.ac.jp

T. Takaishi
Faculty of Information Design, Hiroshima Kokusai Gakuin University,
6-20-1 Nakano, Aki-ku, Hiroshima 739-0321, Japan
e-mail: t.takaishi@hkg.ac.jp

R. Nishii et al. (eds.), *A Mathematical Approach to Research Problems of Science and Technology*, Mathematics for Industry 5,
DOI: 10.1007/978-4-431-55060-0_13,

(XFEM) [3, 21], rigid body spring model (RBSM) [16], discrete element method (DEM) [9, 22], or particle discretization scheme-FEM (PDS-FEM) [13, 14], etc., are widely used for fracture analysis in engineering simulations. On the other hand, such discrete models are dependent on the FEM mesh and other numerical parameters and algorithms. From the mathematical point of view, a mathematically closed continuous model is also preferable.

In [24, 25], the authors proposed a phase filed model for mode III crack propagation on a two dimensional plate and showed several numerical examples. In this paper, we consider some generalizations of our phase field model and discuss their properties based on the idea of [19].

In Sect. 2, we derive our phase field model with two-dimensional linear elasticity. The model is derived as the gradient flow of the Francfort-Marigo type energy [7, 10] with the Ambrosio-Tortorelli regularization [2]. We also introduce a non-repair condition of the crack without destructing the gradient flow structure.

In Sect. 3, we give some numerical examples of crack propagation for mode III case and modes I and II case. We also show how it works in the case that the fracture toughness is spatially variable. For the simulation, we use P2 adaptive mesh finite element method with FreeFEM++ software [12].

2 Derivation of Crack Propagation Models

We suppose that $\Omega \subset \mathbb{R}^2$ is a bounded elastic body without crack. Let $u(x) \in \mathbb{R}^2$ be an in-plane displacement field at $x \in \Omega$. The strain tensor is denoted by $e[u] = (e_{ij}[u](x))$, where

$$e_{ij}[u](x) := \frac{1}{2}\left(\frac{\partial u_i}{\partial x_j}(x) + \frac{\partial u_j}{\partial x_i}(x)\right) \quad (i, j = 1, 2).$$

We use the Einstein summation convention for spatial indices $i, j, k, l \in \{1, 2\}$. We suppose that the elasticity tensor $c_{ijkl}(x)$ satisfies the symmetries $c_{ijkl}(x) = c_{klij}(x) = c_{jikl}(x)$ and the positivity condition $c_{ijkl}(x)\,\xi_{ij}\,\xi_{kl} \ge c_* \,|\xi|^2$ for $x \in \Omega,\ \xi \in \mathbb{R}^{n\times n}_{\text{sym}}$, where $|\xi| := \sqrt{\xi_{ij}\,\xi_{ij}}$. The stress tensor is denoted by $\sigma[u] = (\sigma_{ij}[u](x))$ and is defined as

$$\sigma_{ij}[u](x) = c_{ijkl}(x)e_{kl}[u](x).$$

Then the equilibrium equation is given by

$$-\mathrm{div}\sigma[u] = f(x) \quad x \in \Omega, \tag{1}$$

where $f(x) \in \mathbb{R}^2$ is a given body force at x. It is known that the solution u is obtained as the minimizer of the following elastic energy including the body force under a suitable boundary condition:

$$E_0(u) = \frac{1}{2}\int_\Omega \sigma[u] : e[u]\,dx - \int_\Omega f(x)\cdot u(x)\,dx,$$

where $\sigma : e = \sigma_{ij}e_{ij}$.

Let us assume that there is a crack Σ in Ω, where Σ is a smooth curve in Ω with finite length and that $\Omega \setminus \Sigma$ is open and connected. We have crack opening modes I and II with two-dimensional displacement. We derive a crack propagation model of modes I and II.

We introduce the following smooth function $z(x)$ for $x \in \Omega$ to represent the approximate profile of the crack. We assume that $0 \le z(x) \le 1$ and $z(x) \approx 1$ for around the crack Σ and $z(x) \approx 0$ for the other region. We call $z(x)$ the phase field for the crack shape and derive a time evolution model of z.

We suppose that the damaged stress tensor is defined as

$$\tilde{\sigma}[u] := (1-z)^2\sigma[u]. \tag{2}$$

The function z also can be considered as a damage variable which represents the damage ratio of the material in the sense of (2).

Then we have the modified elastic energy:

$$E_1(u,z) = \frac{1}{2}\int_\Omega (1-z)^2\sigma[u] : e[u]\,\mathrm{d}x - \int_\Omega f(x)\cdot u(x)\,\mathrm{d}x.$$

The surface energy on the crack is approximated by

$$E_2(z) = \frac{1}{2}\int_\Omega \gamma(x)\left(\varepsilon|\nabla z|^2 + \frac{1}{\varepsilon}z^2\right)\mathrm{d}x,$$

where the fracture toughness $\gamma(x) > 0$ is a given function and $\varepsilon > 0$ is a small regularization parameter. This is called the Ambrosio-Tortorelli regularization and they proved that the energy $E_2(z)$ approximates the surface energy $\int_\Sigma \gamma(x)\mathrm{d}s$ in the sense of Γ-convergence for some special cases [2].

Following the derivation of the phase field model in [25], we consider a regularized Francfort-Marigo type energy [10]:

$$E(u,z) := E_1(u,z) + E_2(z),$$

and derive our model as a gradient flow of E. It is shown that this energy approach is compatible to the classical Griffith theory [7, 10, 11].

Let the boundary $\Gamma = \partial\Omega$ be Lipschitz and piece-wise smooth, and the unit outward normal vector on Γ is denoted by n. We suppose that Γ_D is a nonempty open piece-wise smooth portion of Γ, and define $\Gamma_N := \Gamma \setminus \Gamma_D$. The displacement on Γ_D is given as $u = g(x)$ and the traction free condition is assumed on Γ_N. We consider the following boundary condition:

$$u = g \quad \text{on} \quad \Gamma_D, \qquad \sigma[u]n = 0 \quad \text{on} \quad \Gamma_N, \qquad \frac{\partial z}{\partial n} = 0 \quad \text{on} \quad \Gamma.$$

If f and g do not depend on t, the first variations of the energy $E(u, z)$ with respect to u and z are formally derived as follows. For arbitrary $v(x)$ with $v = 0$ on Γ_D, we have

$$\frac{\mathrm{d}}{\mathrm{d}\rho} E(u + \rho v, z)\bigg|_{\rho=0} = -\int_\Omega \left\{ \mathrm{div}\left((1 - z)^2 \sigma[u] \right) + f \right\} v \, \mathrm{d}x$$

Hence, the gradient flow equation of the displacement $u(x, t)$ becomes

$$\alpha_1 \frac{\partial u}{\partial t} = \mathrm{div}\left((1 - z)^2 \sigma[u] \right) + f, \tag{3}$$

where $\alpha_1 \geq 0$ is a small time constant. It must be remarked that the case $\alpha_1 = 0$ corresponds to the equilibrium state of forces (1), however, for numerical simulation, we can set $0 < \alpha_1 << 1$ to stabilize the numerical solution even in the case of $z = 1$, where the ellipticity is broken.

For any $\zeta(x)$, we derive the first variation of the energy $E(u, z)$ with respect to z as

$$\frac{\mathrm{d}}{\mathrm{d}\rho} E(u, z + \rho\zeta)\bigg|_{\rho=0} = -\int_\Omega \left\{ \varepsilon \, \mathrm{div}\, (\gamma(x) \nabla z) - \frac{\gamma(x)}{\varepsilon} z + \sigma[u] : e[u](1 - z) \right\} \zeta \, \mathrm{d}x. \tag{4}$$

The gradient flow equation of the damage variable $z(x, t)$ becomes

$$\alpha_2 \frac{\partial z}{\partial t} = \varepsilon \mathrm{div}\, (\gamma(x) \nabla z) - \frac{\gamma(x)}{\varepsilon} z + \sigma[u] : e[u](1 - z),$$

where $\alpha_2 > 0$ is a time constant.

The resulted phase field model is as follows:

$$\begin{cases} \alpha_1 \dfrac{\partial u}{\partial t} = \mathrm{div}\left((1 - z)^2 \sigma[u] \right) + f(x, t) & x \in \Omega, \ t > 0 \\ \alpha_2 \dfrac{\partial z}{\partial t} = \left(\varepsilon \, \mathrm{div}\, (\gamma(x) \nabla z) - \dfrac{\gamma(x)}{\varepsilon} z + \sigma[u] : e[u](1 - z) \right)_+ & x \in \Omega, \ t > 0 \\ u = g(x, t) & x \in \Gamma_D, \ t > 0 \\ \sigma[u]n = 0 & x \in \Gamma_N, \ t > 0 \\ \dfrac{\partial z}{\partial n} = 0 & x \in \Gamma, \ t > 0 \\ u(x, 0) = u_0(x) & x \in \Omega \\ & \text{(omit if } \alpha_1 = 0) \\ z(x, 0) = z_0(x) \in [0, 1] & x \in \Omega \end{cases} \tag{5}$$

Since a crack once generated can be no longer repaired, we put $(\ )_+$ to the right-hand side of the second equation, where $(a)_+ = \max(a, 0)$. It guarantees the non-repair condition for the crack: $\frac{\partial z}{\partial t} \geq 0$. The second equation expresses the crack evolution due to the magnitude of the elastic energy density $\sigma : e$. The fracture toughness $\gamma(x) > 0$ prescribes the critical value of the energy release rate in the Griffith criterion. It is harder for the crack to grow, if the value of γ is larger.

We remark that the second equation is a fully nonlinear parabolic equation and it is called an irreversible system in the mathematical field of evolution equation. One of the authors recently established a global existence of a unique strong solution for the irreversible diffusion equation $u_t = (\Delta u + f(x,t))_+$ in [1].

If f and g do not depend on t, under suitable regularity assumptions, formally we have the following energy decay property:

$$
\begin{aligned}
\frac{\mathrm{d}}{\mathrm{d}t} E(u(\cdot,t), z(\cdot,t)) &= -\int_\Omega \left\{ \operatorname{div}((1-z)^2\sigma[u]) + f \right\} \frac{\partial u}{\partial t}\,\mathrm{d}x \\
&\quad - \int_\Omega \left\{ \varepsilon \operatorname{div}(\gamma \nabla z) - \frac{\gamma}{\varepsilon} z + \sigma[u] : e[u](1-z) \right\} \frac{\partial z}{\partial t}\,\mathrm{d}x \\
&= -\alpha_1 \int_\Omega \left| \frac{\partial u}{\partial t} \right|^2 \mathrm{d}x - \alpha_2 \int_\Omega \left| \frac{\partial z}{\partial t} \right|^2 \mathrm{d}x \leq 0.
\end{aligned}
$$

This stands for the gradient flow structure of our phase field model (5) even with the non-repair condition.

We remark that the phase field model (5) can be also considered in the three-dimensional case, $u(x,t) \in \mathbb{R}^3$, $x \in \Omega \subset \mathbb{R}^3$.

In [25], the authors studied our phase field model in two dimension with scalar anti-plane displacement $u(x,t) \in \mathbb{R}$, $x \in \Omega \subset \mathbb{R}^2$:

$$
\begin{cases}
\alpha_1 \dfrac{\partial u}{\partial t} = \mu \operatorname{div}\left((1-z)^2 \nabla u\right) + f(x,t) & x \in \Omega,\ t > 0 \\
\alpha_2 \dfrac{\partial z}{\partial t} = \left(\varepsilon \operatorname{div}(\gamma(x)\nabla z) - \dfrac{\gamma(x)}{\varepsilon} z + \mu |\nabla u|^2 (1-z)\right)_+ & x \in \Omega,\ t > 0 \\
u = g(x,t) & x \in \Gamma_D,\ t > 0 \\
\dfrac{\partial u}{\partial n} = 0 & x \in \Gamma_N,\ t > 0 \\
\dfrac{\partial z}{\partial n} = 0 & x \in \Gamma,\ t > 0 \\
u(x,0) = u_0(x) & x \in \Omega \\
 & (\text{omit if } \alpha_1 = 0) \\
z(x,0) = z_0(x) \in [0,1] & x \in \Omega,
\end{cases}
\tag{6}
$$

where $\mu > 0$ denotes the rigidity, one of the Lamé constant. This is a mode III crack propagation model. See [25] for detail of the derivation of our phase field models.

We cannot find any difficulty to compute crack growth numerically. This model equation makes possible to analyze the crack growth phenomena by well-known numerical method, only taking notice about the mesh size that you must choose carefully. In the next section, we show some numerical results of crack growth from this model equation.

Similar mathematical models for simulation based on the Fracfort-Marigo type energy are also found in [4–8, 15].

3 Numerical Results

We show some numerical results of our phase field models of crack growth derived in the previous section. We use a free software FreeFem++ [12] for our computation. It is a useful tool of finite element method for our purpose. Due to the small regularization parameter ε introduced in our model, the profiles of u and z have small spatial patterns of ε-scale. We, however, use an adaptive mesh finite element method with the help of FreeFEM++.

We fix $\tau > 0$ as a constant time step, $u^k(x)$ and $z^k(x)$ are the approximated solution of u and z, respectively, at $t = k\tau$ $(k = 0, 1, 2, \ldots)$.

First, we make two types of numerical simulation of the single-line crack growth of the mode III type (6). We set minimum mesh size more than 2×10^{-3} and maximum number of the vertices of triangular mesh less than 5,000. The numerical solution (u^k, z^k) for mode III model (6) are computed from (u^{k-1}, z^{k-1}) with the semi-implicit scheme:

$$\begin{cases} \alpha_1 \dfrac{u^k - u^{k-1}}{\tau} = \mu \operatorname{div}\left((1 - z^{k-1})^2 \nabla u^k\right) \\ \alpha_2 \dfrac{\tilde{z}^k - z^{k-1}}{\tau} = \varepsilon \operatorname{div}\left(\gamma(x) \nabla \tilde{z}^k\right) - \dfrac{\gamma(x)}{\varepsilon} \tilde{z}^k + \mu |\nabla u^{k-1}|^2 (1 - \tilde{z}^k) \\ z^k = \min(1, \ \max(\tilde{z}^k, \ z^{k-1})) \end{cases} \tag{7}$$

We remark that $0 \leq z^k \leq 1$ is guaranteed when $z^0 \in [0, 1]$ at $t = 0$. For spatial discretization, we use adaptive mesh P2 finite element method. The model of modes I and II (5) is similarly discretized. It is necessary to set the mesh of small size near crack region ($z \sim 1$), because the values of u and z change drastically around there. We solve (7) by FreeFem++ with adaptive P2-element, where z is evaluated for remeshing at each time step.

Initial crack is set as Fig. 1a, single-line from left hand side. Figure 2 shows that a crack propagates to the another side when fracture toughness γ is homogeneous. Adaptive mesh is effective to follow the crack path. As the crack grows, FreeFem++ adapts to set small size mesh near the crack, and the number of vertices becomes larger till it breaks down (Fig. 3).

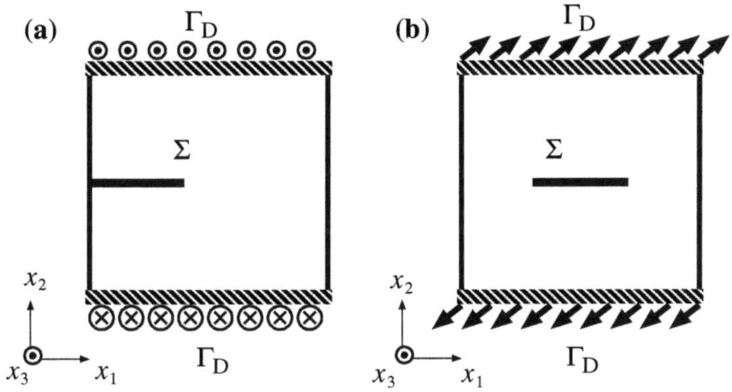

Fig. 1 Domain of numerical simulation of mode III crack growth (**a**) and mode I+II crack growth (**b**)

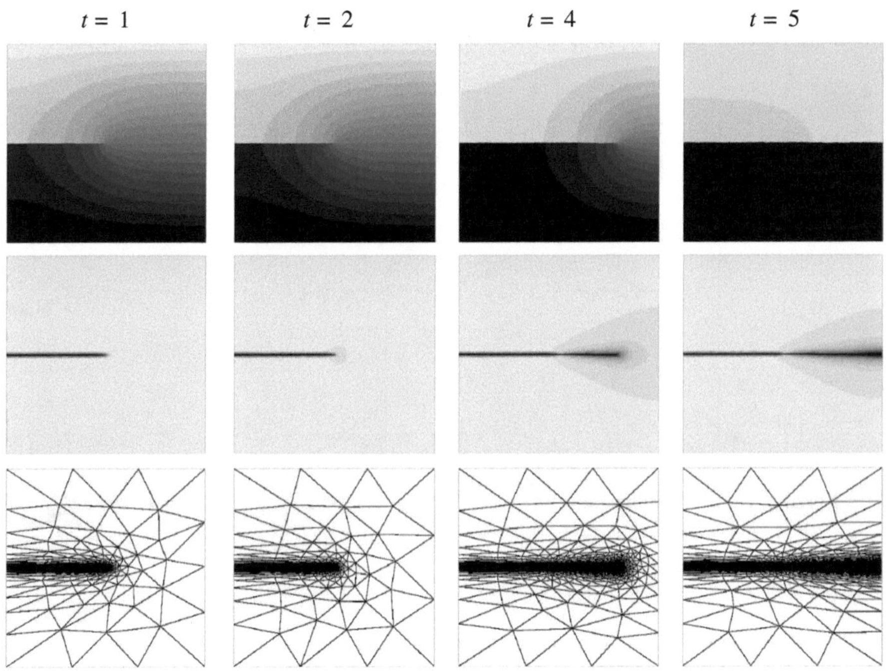

Fig. 2 An initial crack grows in isotropic media with given boundary displacement $g = (0, 0, 10) \times t$ (u (*upper*), z (*middle*), mesh (*lower*))

When γ is inhomogeneous, the crack intends to follow the weak region. Figure 4 shows the crack growth when γ has the profile as stripe ($\gamma(x, y) = 0.5(1 + 0.2 \sin 20(x + y))$). First crack grows following weak position, after that, sub-branch emerges repeatedly.

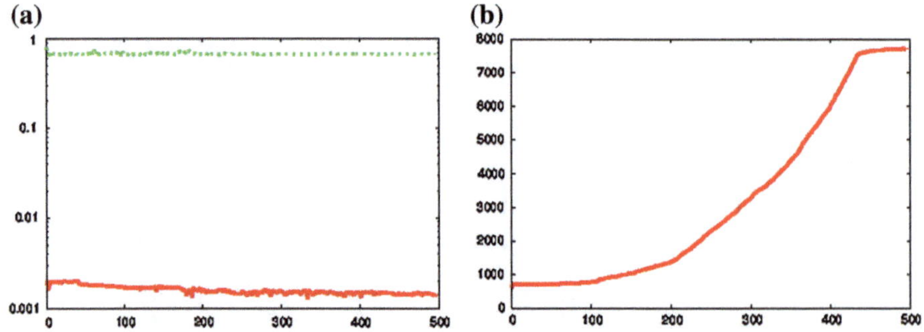

Fig. 3 Temporal evolution of **a** the minimum and maximum mesh size, (**b**) number of vertices

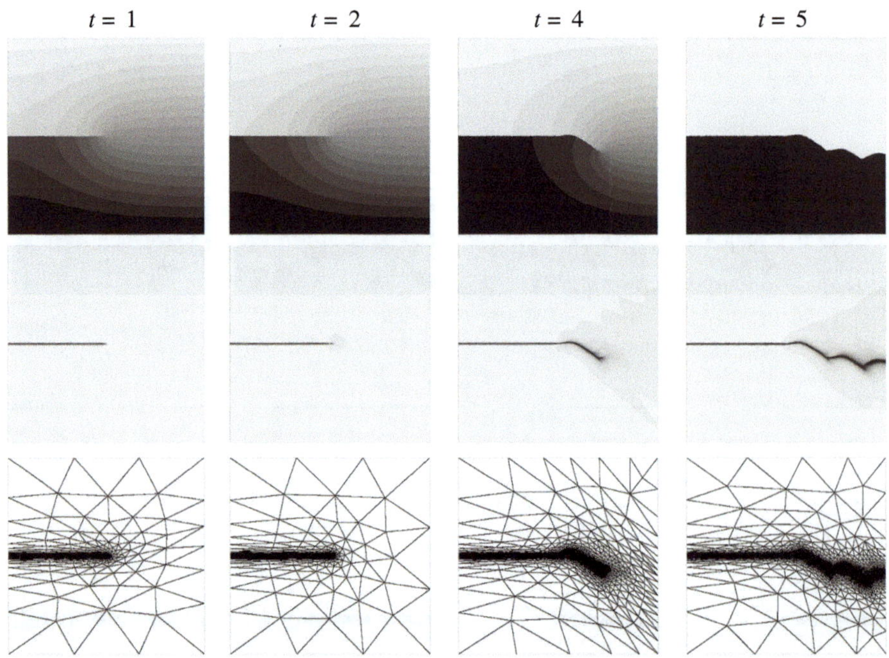

Fig. 4 An initial crack grows in an anisotropic media ($\gamma(x, y) = 0.5*(1+0.2*sin(20*(x+y)))$) with given boundary displacement $g = (0, 0, 10) \times t$ (u (*upper*), z (*middle*), mesh (*lower*))

Finally, we show the numerical results of mode I+II crack growth. We use the phase field model (5) with isotropic elasticity tensor. Lamé constants are set as $\lambda = 26.76, \mu = 19.38$, where Young's modulus and Poisson's ratio are set as $E = 50, \sigma = 0.29$, respectively. Initial crack which is shaped as slit changes its direction to perpendicular to the displacement on Γ_D (Fig. 1b). It shows that crack kinks to annihilate mode II at the front (Fig. 5).

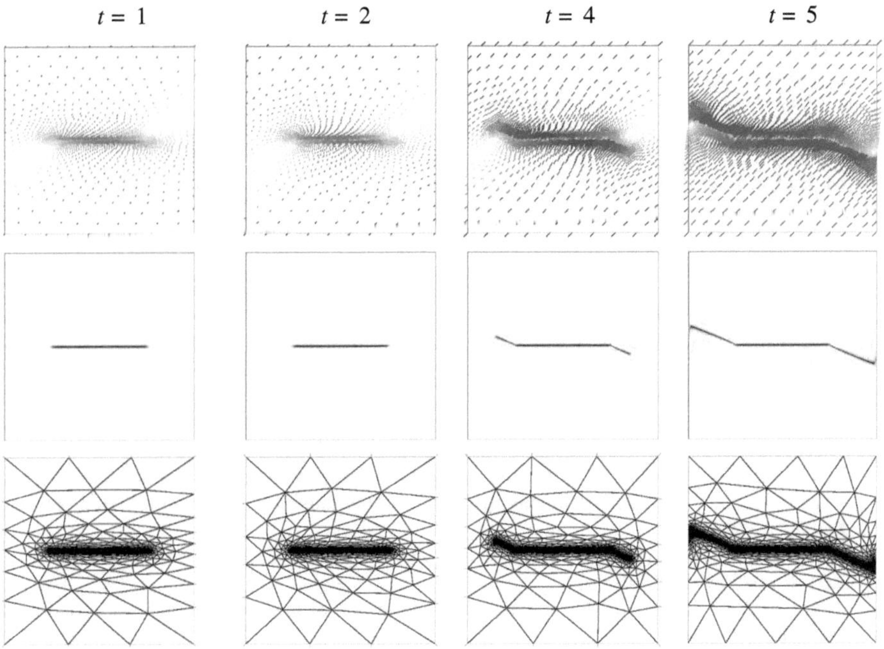

Fig. 5 An initial crack grows in an isotropic media with given boundary displacement $g = (1, 1, 0) \times t$ (u (*upper*), z (*middle*), mesh (*lower*))

From these results, using adaptive mesh method is effective to calculate the crack path. For the similar purpose, we use ALBERTA toolbox [23] in [24, 25]. Adaptive mesh method is also useful in other free boundary or pattern formation problem, such as reaction-diffusion model [17, 18].

Acknowledgments This work was partially supported by JSPS KAKENHI Grant Number 00268666.

References

1. G. Akagi, M. Kimura, Well-posedness and long time behavior for an irreversible diffusion equation (in preparation)
2. L. Ambrosio, V.M. Tortorelli, On the approximation of free discontinuity problems. Boll. Un. Mat. Ital. **6-B**(7), 105–123 (1992)
3. T. Belytschko, T. Black, Elastic crack growth in finite elements with minimal remeshing. Int. J. Numer. Methods Eng. **45**(5), 601–620 (1999)
4. B. Bourdin, The variational formulation of brittle fracture: numerical implementation and extensions, in *IUTAM Symposium on discretization methods for evolving discontinuities*, ed. by T. Belytschko, A. Combescure, R. de Borst. (Springer, 2007), pp. 381–393

5. B. Bourdin, Numerical implementation of the variational formulation of brittle fracture. Interfaces Free Bound. **9**, 411–430 (2007)
6. B. Bourdin, G.A. Francfort, J.-J. Marigo, Numerical experiments in revisited brittle fracture. J. Mech. Phys. Solids **48**, 797–826 (2000)
7. B. Bourdin, G.A. Francfort, J.-J. Marigo, *The Variational Approach to Fracture*. (Springer, 2008)
8. M. Buliga, Energy minimizing brittle crack propagation. J. Elast. **52**, 201–238 (1998/99)
9. P.A. Cundall, A computer model for simulating progressive large scale movements in blocky rock systems, in *Proceedings of the Symposium of the International Society for Rock Mechanics, Nancy*, vol. 2, pp. 129–136 (1971)
10. G.A. Francfort, J.-J. Marigo, Revisiting brittle fracture as an energy minimization problem. J. Mech. Phys. Solids **46**, 1319–1342 (1998)
11. A.A. Griffith, The phenomenon of rupture and flow in solids. Phil. Trans. Royal Soc. London **A221**, 163–198 (1921)
12. F. Hecht, New development in freefem++. J. Numer. Math. **20**, 251–265 (2012)
13. M. Hori, K. Oguni, H. Sakaguchi, Proposal of FEM implemented with particle discretization for analysis of failure phenomena. J. Mech. Phys. Solids **53**, 681–703 (2005)
14. T. Ichimura, M. Hori, M.L.L. Wijerathne, Linear finite elements with orthogonal discontinuous basis functions for explicit earthquake ground motion modeling. Int. J. Numer. Methods Eng. **86**, 286–300 (2011)
15. A. Karma, H. Levine, D. Kessler, Phase-field model of mode-III dynamic fracture. Phys. Rev. Lett. **87**, 045501 (2001)
16. T. Kawai, New discrete models and their application to seismic response analysis. Nucl. Eng. Des. **48**, 207–229 (1978)
17. M. Kimura, H. Komura, M. Mimura, H. Miyoshi, T. Takaishi, D. Ueyama, Adaptive mesh finite element method for pattern dynamics in reaction-diffusion systems, in *Proceedings of the Czech-Japanese Seminar in Applied Mathematics 2005, COE Lecture Note*, vol. 3, Faculty of Mathematics, Kyushu University ISSN 1881–4042 (2006), pp. 56–68
18. M. Kimura, H. Komura, M. Mimura, H. Miyoshi, T. Takaishi, D. Ueyama, Quantitative study of adaptive mesh FEM with localization index of pattern. in *Proceedings of the Czech-Japanese Seminar in Applied Mathematics 2006, COE Lecture Note*, vol. 6, Faculty of Mathematics, Kyushu University ISSN 1881–4042 (2007), pp. 114–136
19. M. Kimura, T. Takaishi, Phase field models for crack propagation. Theor. Appl. Mech. Jpn. **59**, 85–90 (2011)
20. R. Kobayashi, Modeling and numerical simulations of dendritic crystal growth. Phys. D **63**, 410–423 (1993)
21. N. Moës, J. Dolbow, T. Belytschko, A finite element method for crack growth without remeshing. Int. J. Numer. Methods Eng. **46**, 131–150 (1999)
22. A. Munjiza, *The Combined Finite-Discrete Element Method*. (Wiley, New York, 2004)
23. A. Schmidt, K.G. Siebert, Design of adaptive finite element software. the finite element toolbox ALBERTA, in *Lecture Notes in Computational Science and Engineering*, vol. 42, (Springer-Verlag, Berlin, 2005)
24. T. Takaishi, Numerical simulations of a phase field model for mode III crack growth. Trans. Jpn. Soc. Ind. Appl. Math. **19**, 351–369 (2009) (in Japanese)
25. T. Takaishi, M. Kimura, Phase field model for mode III crack growth. Kybernetika **45**, 605–614 (2009)

Variational Methods in Differential Equations

Michiaki Onodera

Abstract This chapter concerns classical variational methods in boundary value problems and a free boundary problem, with a special emphasis on how to view a differential equation as a variational problem. Variational methods are simple, but very powerful analytical tools for differential equations. In particular, the unique solvability of a differential equation reduces to a minimization problem, for which a minimizer is shown to be a solution to the original equation. As a model problem, the Poisson equation with different types of boundary conditions is considered. We begin with the derivation of the equation in the context of potential theory, and then show successful applications of variational methods to these boundary value problems. Finally, we study a free boundary problem by developing the idea to a minimization problem with a constraint.

Keywords Boundary value problem · Free boundary problem · Variational method

1 Introduction

The variational method, or calculus of variations, is an infinite-dimensional version of calculus: minimize a (real-valued) "function" defined in an infinite-dimensional phase space. Such a phase space is often taken to be a function space; thus, the variational method is, so to speak, calculus for a "function" of functions. Such a generalized "function" is called a functional.

In nature we often encounter a situation where an observable state of a phenomenon minimizes a certain "energy" functional. A well-known example is a spherical soap film, which minimizes its surface area (energy) under the constraint that the volume of the enclosed region is prescribed. Thus, a physically reasonable energy functional serves as an explanation of a natural phenomenon.

M. Onodera (✉)
Institute of Mathematics for Industry, Kyushu University,
744 Motooka, Nishi-ku, Fukuoka 819-0395, Japan
e-mail: onodera@imi.kyushu-u.ac.jp

R. Nishii et al. (eds.), *A Mathematical Approach to Research Problems of Science and Technology*, Mathematics for Industry 5,
DOI: 10.1007/978-4-431-55060-0_14,

Furthermore, the variational method is used as an analytical tool to reveal some properties of a solution to a given mathematical problem. Specifically, for a differential equation, we can virtually construct an energy functional in such a way that a minimizer of that energy becomes a solution to the original problem.[1] In this way, the original problem can be reduced to a minimization problem. To illustrate this idea, let us consider the equation

$$x = g(x) \tag{1}$$

where g is a real-valued function defined in $\mathbb{R}$. To prove the existence of a solution x to (1), let us define the energy

$$E(x) := \frac{1}{2}x^2 - G(x), \tag{2}$$

where

$$G(x) := \int_0^x g(s)\,\mathrm{d}s,$$

and find a minimum point $x \in \mathbb{R}$ of E. In fact, x is a solution to (1) if and only if x is a critical point of E, i.e., $E'(x) = 0$. Note that a minimum point x necessarily becomes a critical point.

The existence of a minimum point x will be proved, for example, by taking a minimizing sequence $\{x_k\}_{k=1}^{\infty}$ such that $E(x_k) \to \inf_{y\in\mathbb{R}} E(y)$ and showing that it contains a convergent subsequence (denoted again by $\{x_k\}_{k=1}^{\infty}$) with the limit $x \in \mathbb{R}$. Then,

$$E(x) = \lim_{k\to\infty} E(x_k) = \inf_{y\in\mathbb{R}} E(y).$$

In particular, when g is a bounded function, i.e., $|g(x)| \leq M$ for some constant $M > 0$, we easily see that E satisfies the coerciveness condition[2]

$$\lim_{|x|\to\infty} E(x) = \infty.$$

Hence, every minimizing sequence is bounded, so the Bolzano–Weierstrass theorem guarantees the existence of a convergent subsequence. We can thus obtain a solution x to (1). We also note that, if E is (strictly) convex, i.e.,

$$E(tx_1 + (1-t)x_2) < tE(x_1) + (1-t)E(x_2) \quad (x_1 \neq x_2 \text{ and } 0 < t < 1), \tag{3}$$

then its critical point is unique. Indeed, at any critical point x, (3) implies that

[1] The energy thus constructed also has a physical meaning.

[2] This terminology is not standard (see Kinderlehrer and Stampacchia [3, Definition 4.4]).

$$0 = E'(x)(y-x) = \lim_{t\to 0} \frac{E(x+t(y-x)) - E(x)}{t} \le E(y) - E(x)$$

for all $y \in \mathbb{R}$. Thus, every critical point is a minimum point. Furthermore, it also follows from (3) that there is at most one minimum point.

In the following sections, we will show how variational methods can be applied to several problems in differential equations.

2 Problems in Potential Theory

One of the basic problems in potential theory is to find the gravitational potential u induced by a prescribed mass distribution f, with an additional condition. Here, the mass f can be regarded as a function (or measure) defined in the n-dimensional Euclidean space $\mathbb{R}^n$ ($n \ge 2$). Then, the Newtonian potential u of f is given by

$$u(x) := \int_{\mathbb{R}^n} \Gamma(x-y) f(y)\, \mathrm{d}y, \tag{4}$$

where Γ is defined by

$$\Gamma(x) := \begin{cases} -\dfrac{1}{2\pi} \log|x| & (n=2), \\ \dfrac{1}{n(n-2)\omega_n |x|^{n-2}} & (n \ge 3), \end{cases} \tag{5}$$

with ω_n being the volume of the unit ball in $\mathbb{R}^n$. The function Γ is the Newtonian potential induced by a unit point mass located at the origin $x = 0$; hence, its gradient

$$\nabla u(x) = -\frac{1}{n\omega_n |x|^{n-1}} \frac{x}{|x|}$$

represents the gravitational field induced by the point mass. When $n = 3$, this is nothing but Newton's law of universal gravitation: the induced force is inversely proportional to the square of the distance from the origin. Thus, formula (4) is its generalization, since each factor $\Gamma(x-y) f(y)\, \mathrm{d}y$ is the potential induced by a small portion $f(y)\, \mathrm{d}y$ of mass.

Formula (4) is expressed by $u(x) = (\Gamma * f)(x)$, and it is important to note that the operation $f \mapsto \Gamma * f$ is the inverse operator to $-\Delta := -\sum_{j=1}^{n} (\partial/\partial x_j)^2$, i.e., $-\Delta u = f$. This fact can be checked mathematically under some regularity condition on f, but for our purposes it suffices to note that $-\Delta\Gamma = \delta$ (in the distribution sense), where δ is the Dirac measure (i.e., point mass) supported at the origin $x = 0$. Then, formally,

$$-\Delta u(x) = \int_{\mathbb{R}^n} -\Delta\Gamma(x-y)f(y)\,\mathrm{d}y = \int_{\mathbb{R}^n} \delta(x-y)f(y)\,\mathrm{d}y = f(x).$$

One may wish to control the potential u in a bounded domain $\Omega \subset \mathbb{R}^n$ by locating an additional mass f_0 outside Ω (i.e., $f_0 = 0$ in Ω). A standard question is whether or not we are able to construct the potential u in such a way that $u = 0$ on the boundary $\partial\Omega$. More precisely, for given f and Ω with $f(x) = 0$ $(x \notin \Omega)$, we ask if there is f_0 such that $f_0(x) = 0$ $(x \in \Omega)$ and that the potential

$$u(x) := \int_{\mathbb{R}^n} \Gamma(x-y)f(y)\,\mathrm{d}y + \int_{\mathbb{R}^n} \Gamma(x-y)f_0(y)\,\mathrm{d}y \tag{6}$$

satisfies $u = 0$ on $\partial\Omega$. This problem is actually equivalent to the following Poisson equation

$$\begin{cases} -\Delta u = f & (x \in \Omega), \\ u = 0 & (x \in \partial\Omega). \end{cases} \tag{7}$$

Indeed, (7) can be derived from (6) by applying $-\Delta$ to (6). Conversely, if u is a solution to (7), then by extending u (which is defined only in $\overline{\Omega}$) smoothly to a function $\overline{u}$ defined in the whole space $\mathbb{R}^n$ (more precisely, $\overline{u}$ should be extended such that $\overline{u} = 0$ near $|x| = \infty$) and setting $f_0 := -\Delta\overline{u} - f$, we see that (6) holds for $\overline{u}$.

The Poisson equation also appears in different contexts in physics, biology, and even in mathematics itself. One of the advantages of mathematical abstraction lies in the fact that one mathematical result can lead to several consequences in other fields of science.

3 Variational Methods in Boundary Value Problems

Our purpose here is to show how variational methods can be successfully applied to boundary value problems, including the Dirichlet problem (7). The primary step is to view a given equation as a variational equation, where by a variational equation we mean an equation of the form $E'(u) = 0$ with a functional E. As we will see later, the concept of solutions will be weakened so that a given problem is reformulated to a variational setting, and such weak (or generalized) solutions are not necessarily twice differentiable. The variational equation thus obtained will be treated in a similar way to that presented for (1).

We remark that the regularity theory of elliptic equations tells us that, under some regularity condition on given data (e.g., f and $\partial\Omega$), generalized solutions are eventually shown to be smooth enough, and hence they are actual solutions as expected. This regularity theory is beyond our scope, so we do not go into the details here. The interested reader may consult, for example, a book by Gilbarg and Trudinger [1]. Throughout this section, we assume that Ω is a bounded domain in $\mathbb{R}^n$.

3.1 How to View a Differential Equation as a Variational Problem

Let us reformulate (7) into a variational setting by introducing an appropriate energy functional

$$E : X \to \mathbb{R}$$

together with some linear function space X equipped with the norm $\|\cdot\|_X$ (e.g., see Example 1 below). For problem (7), the boundary condition shall be incorporated into the space X so that each function $u \in X$ satisfies $u = 0$ on $\partial\Omega$ in a certain sense. Thus, what we need to do is to identify

$$-\Delta u = f \quad \text{and} \quad E'(u) = 0.$$

To clarify the meaning of the derivative $E'(u)$ in a function space X, we recall that a function

$$F : \mathbb{R} \to \mathbb{R}$$

is differentiable at $x \in \mathbb{R}$ if and only if there is a constant $A \in \mathbb{R}$ such that

$$F(x+h) = F(x) + Ah + o(h), \quad \text{where } \lim_{h\to 0} \frac{o(h)}{|h|} = 0.$$

Then, $F'(x) = A$. This fact leads to the notion of the Fréchet derivative of a functional on X: E is said to be differentiable at $u \in X$ if there is a continuous linear functional $A : X \to \mathbb{R}$, such that

$$E(u+v) = E(u) + A[v] + o(v), \quad \text{where } \lim_{v\to 0} \frac{o(v)}{\|v\|_X} = 0.$$

Then, we write $E'(u) = A$. Here, we use the notation $A[v]$ instead of $A(v)$ merely to emphasize that A is linear in v. Note that $v \mapsto E(u) + E'(u)[v]$ is an affine approximation of $v \mapsto E(u+v)$.

Example 1 Let $X = C_0^1(\Omega)$, that is, the space of all continuously differentiable functions u satisfying $u = 0$ on $\partial\Omega$. The space $C_0^1(\Omega)$ is equipped with the norm

$$\|u\|_{C_0^1(\Omega)} := \sup_{x\in\Omega} |u(x)| + \sup_{x\in\Omega} |\nabla u(x)|.$$

We consider the functional

$$E(u) := \int_\Omega |\nabla u|^2 \, dx \quad (u \in C_0^1(\Omega)).$$

Then, the Fréchet derivative $E'(u)$ at $u \in C_0^1(\Omega)$ is given by

$$E'(u)[v] = 2\int_\Omega \nabla u \cdot \nabla v \,\mathrm{d}x,$$

since

$$E(u+v) = E(u) + 2\int_\Omega \nabla u \cdot \nabla v \,\mathrm{d}x + \int_\Omega |\nabla v|^2 \,\mathrm{d}x$$

and the functional $v \mapsto 2\int_\Omega \nabla u \cdot \nabla v \,\mathrm{d}x$ is linear and continuous, and

$$\int_\Omega |\nabla v|^2 \,\mathrm{d}x \le \|v\|^2_{C^1_0(\Omega)} |\Omega| = o(v), \quad \text{i.e.,} \quad \lim_{v\to 0} \frac{\int_\Omega |\nabla v|^2 \,\mathrm{d}x}{\|v\|_{C^1_0(\Omega)}} = 0,$$

where $|\Omega|$ denotes the volume of Ω.

The dual space X^* of X consists of all continuous linear functionals on X. Thus, since $E'(u) \in X^*$, Eq. (7) should be interpreted as an equation in X^*. One of the ways to view a function f defined in a domain $\Omega \subset \mathbb{R}^n$ as a functional in X^* for a function space X is to consider the correspondence

$$X \ni \varphi \mapsto \int_\Omega f\varphi \,\mathrm{d}x \in \mathbb{R} \tag{8}$$

and identify the above correspondence (8) with f. Indeed, with an appropriate choice of X, (8) becomes a continuous linear functional on X. For example, the choice of $X = C_c^\infty(\Omega)$ (the space of infinitely differentiable functions φ with compact support in Ω, equipped with a certain topology) gives the identification of a function f as, what we call, a distribution. Another choice is to take X to be the Lebesgue space $L^2(\Omega)$: the space of all square-integrable functions on Ω, in which the integral in (8) is finite for $f, \varphi \in L^2(\Omega)$ (see Hölder's inequality in the Appendix).

Remark 1 The Lebesgue space $L^2(\Omega)$ has the inner product

$$(\varphi, \Psi)_{L^2(\Omega)} := \int_\Omega \varphi\Psi \,\mathrm{d}x \qquad (\varphi, \Psi \in L^2(\Omega)), \tag{9}$$

and the norm is defined by $\|\varphi\|_{L^2(\Omega)} := (\varphi, \varphi)^{1/2}_{L^2(\Omega)}$. Note that $L^2(\Omega)$ is a natural extension of the Euclidean space $\mathbb{R}^n$, since (9) is the sum of all $\varphi(x)\Psi(x)$: the multiple of each component of φ and each component of Ψ, and this is similar to the inner product in $\mathbb{R}^n$: $x \cdot y = \sum_{j=1}^n x_j y_j$. In fact, $L^2(\Omega)$ becomes a Hilbert space with this inner product.

On the other hand, there are several natural ways to identify $-\Delta u$ as a functional. When u and $\varphi \in C_c^\infty(\Omega)$ are smooth, by integration by parts, we have

$$-\int_{\Omega} \Delta u \varphi \, dx = \int_{\Omega} \nabla u \cdot \nabla \varphi \, dx = -\int_{\Omega} u \Delta \varphi \, dx. \tag{10}$$

Thus, we ca use each of them in (10) as an identification of $-\Delta u$ as a functional. In variational methods, (i) it is natural to impose condition $u \in X$, i.e., both u, φ live in the same space X. Moreover, (ii) it is preferable to take the space X as large as possible, but keeping the identification to make sense. These conditions lead to the Sobolev space $H^1(\Omega)$: the space of differentiable (in the distribution sense) functions φ with $\varphi, \nabla\varphi \in L^2(\Omega)$, and we choose the second identification in (10). Note that, of course, (8) still makes sense for $X = H^1(\Omega)$ and $f \in L^2(\Omega)$.

Remark 2 The Sobolev space $H^1(\Omega) = H^{1,2}(\Omega)$ has the inner product

$$(\varphi, \Psi)_{H^1(\Omega)} := \int_{\Omega} \varphi \Psi \, dx + \int_{\Omega} \nabla \varphi \cdot \nabla \Psi \, dx$$

and the norm is defined by $\|\varphi\|_{H^1(\Omega)} := (\varphi, \varphi)_{H^1(\Omega)}^{1/2}$. Note that $H^1(\Omega)$ becomes a Hilbert space with the inner product.

We are now in a position to introduce the variational formulation of the Poisson equation (7) with $f \in L^2(\Omega)$. We choose $X = H_0^1(\Omega)$, where $H_0^1(\Omega)$, a subspace of $H^1(\Omega)$, consists of functions $\varphi \in H^1(\Omega)$ with $\varphi = 0$ on $\partial\Omega$ (more precisely, the completion of $C_c^\infty(\Omega)$ with the topology of $H^1(\Omega)$). Thus, the boundary condition $u = 0$ on $\partial\Omega$ is now incorporated into the space itself.

Definition 1 We call a function $u \in H_0^1(\Omega)$ a weak (or generalized) solution to (7) if

$$\int_{\Omega} \nabla u \cdot \nabla \varphi \, dx = \int_{\Omega} f \varphi \, dx \tag{11}$$

holds for all $\varphi \in H_0^1(\Omega)$.

This is an equation in the dual space $H^{-1}(\Omega) := (H_0^1(\Omega))^*$. Observe that a smooth solution u to (11) satisfies the original Eq. (7). Indeed, when u, φ are smooth, (11) becomes

$$\int_{\Omega} (-\Delta u - f)\varphi \, dx = 0.$$

Hence, if $-\Delta u - f > 0$ (or < 0) in a subset $\Omega_0 \subset \Omega$, then the choice of $\varphi = (-\Delta u - f)\eta$ with an appropriate cut-off function $\eta \in C_0^\infty(\Omega_0)$ leads to a contradiction. Thus, $-\Delta u = f$ everywhere in Ω.

3.2 The Dirichlet Problem

We derived (11) in order to view the original problem (7) as a variational equation $E'(u) = 0$ in $H^{-1}(\Omega)$. The remaining question is what energy E corresponds to (11). Recalling the derivation of (2) for problem (1), we define

$$E(u) := \frac{1}{2}\int_\Omega |\nabla u|^2 \,\mathrm{d}x - \int_\Omega f u \,\mathrm{d}x \quad (u \in H_0^1(\Omega)).$$

Then, we see that

$$E'(u)[\varphi] = \int_\Omega \nabla u \cdot \nabla\varphi \,\mathrm{d}x - \int_\Omega f\varphi \,\mathrm{d}x.$$

Therefore, $u \in H_0^1(\Omega)$ is a solution to (11) if and only if $E'(u) = 0$ in $H^{-1}(\Omega)$.

When $f \in L^2(\Omega)$, Hölder's inequality followed by the Poincaré inequality (see the Appendix) yields that

$$\begin{aligned}\left|\int_\Omega f u \,\mathrm{d}x\right| &\le \|f\|_{L^2(\Omega)}\|u\|_{L^2(\Omega)} \\ &\le C\,\|f\|_{L^2(\Omega)}\|\nabla u\|_{L^2(\Omega)} \\ &\le C\left(\varepsilon\|\nabla u\|^2_{L^2(\Omega)} + \frac{\|f\|^2_{L^2(\Omega)}}{4\varepsilon}\right).\end{aligned}$$

Hence, by taking $\varepsilon > 0$ small enough such that $C\varepsilon \le 1/2$, we have

$$E(u) \ge \frac{1}{4}\int_\Omega |\nabla u|^2 \,\mathrm{d}x - C\|f\|^2_{L^2(\Omega)}$$

with some constant $C > 0$. Applying the Poincaré inequality again, we see that E satisfies the coerciveness condition

$$E(u) \ge \delta\|u\|^2_{H_0^1(\Omega)} - C \to \infty \quad \text{as } \|u\|_{H_0^1(\Omega)} \to \infty,$$

where $\delta, C > 0$ are some constants.

With the virtue of the coerciveness, every minimizing sequence $\{u_k\}_{k=1}^\infty$ (i.e., $\lim_{k\to\infty} E(u_k) = \inf_{v\in H_0^1(\Omega)} E(v)$) is bounded in $H_0^1(\Omega)$. However, it should be emphasized that $H_0^1(\Omega)$ is infinite-dimensional, and we cannot conclude the existence of a convergent subsequence from the boundedness of the minimizing sequence. Hence, we need to use some facts in functional analysis: (i) every bounded sequence in $H_0^1(\Omega)$ contains a weakly convergent subsequence (denoted again by $\{u_k\}_{k=1}^\infty$)

with the limit $u \in H_0^1(\Omega)$; (ii) E is weakly lower semicontinuous, i.e.,

$$E(u) \leq \liminf_{k\to\infty} E(u_k) = \inf_{v\in H_0^1(\Omega)} E(v).$$

Thus, we can conclude the existence of a minimizer u of E, and hence the existence of a solution to (11). Moreover, in this case, E is convex; hence, u is a unique solution. Here, the convexity of E follows from the direct computation

$$\begin{aligned}|\nabla (tu_1 + (1-t)u_2)|^2 &= t^2 |\nabla u_1|^2 + 2t(1-t)\nabla u_1 \cdot \nabla u_2 + (1-t)^2 |\nabla u_2|^2 \\ &\leq t^2 |\nabla u_1|^2 + t(1-t)\left(|\nabla u_1|^2 + |\nabla u_2|^2\right) + (1-t)^2 |\nabla u_2|^2 \\ &= t\,|\nabla u_1|^2 + (1-t)\,|\nabla u_2|^2\end{aligned}$$

for $0 < t < 1$, where the equality holds only when $\nabla u_1 = \nabla u_2$ (see the Appendix).

Remark 3 The reason for the choice $X = H_0^1(\Omega)$ lies in the facts that

- X is compatible with E in the sense that the coerciveness holds with its norm;
- X is a Hilbert space, which guarantees fact (i) in the argument above.

Check that both conditions are not satisfied by $C_0^1(\Omega)$.

3.3 The Neumann Problem

The variational method developed above can also be applied to the Neumann problem

$$\begin{cases} -\Delta u + u = f & (x \in \Omega), \\ \dfrac{\partial u}{\partial \nu} = 0 & (x \in \partial\Omega), \end{cases} \tag{12}$$

where ν is the unit outer normal vector to $\partial\Omega$. Thus, in the Neumann problem, we control the normal derivative $\partial u/\partial \nu$, i.e., the gravitational force itself. The reason that we put the term u into the left-hand side of the equation will be made clear later, when we show the coerciveness of energy.

As in the Dirichlet problem, let us derive a variational formulation of (12). In this case, we work with H^1 and do not incorporate the boundary condition into the space. But, the boundary condition is implicitly included in the formulation. Indeed, as we will see later, the boundary condition $\partial u/\partial \nu = 0$ will be automatically satisfied by minimizers.

Definition 2 We call a function $u \in H^1(\Omega)$ a weak (or generalized) solution to (12) if

$$\int_\Omega \nabla u \cdot \nabla \varphi \, \mathrm{d}x + \int_\Omega u\varphi \, \mathrm{d}x = \int_\Omega f\varphi \, \mathrm{d}x \tag{13}$$

holds for all $\varphi \in H^1(\Omega)$.

We should observe that (10) does not hold in general for $\varphi \in C^\infty(\Omega)$ unless φ vanishes on $\partial\Omega$. However, the first equality in (10) still holds, since

$$-\int_\Omega \Delta u \varphi \,\mathrm{d}x = \int_\Omega \nabla u \cdot \nabla \varphi \,\mathrm{d}x - \int_{\partial\Omega} \frac{\partial u}{\partial \nu} \varphi \,\mathrm{d}\sigma$$
$$= \int_\Omega \nabla u \cdot \nabla \varphi \,\mathrm{d}x$$

follows from the boundary condition $\partial u/\partial \nu = 0$. Hence, even for a wider class of functions $\varphi \in H^1(\Omega)$, the identification of $-\Delta u$ with

$$H^1(\Omega) \ni \varphi \mapsto \int_\Omega \nabla u \cdot \nabla \varphi \,\mathrm{d}x \in \mathbb{R}$$

is adequate, or, what amounts to the same thing, smooth solutions u to (12) satisfy (13).

To see that the boundary condition $\partial u/\partial \nu = 0$ is incorporated into the formulation (13), let us check the converse statement: smooth solutions u to (13) satisfy (12). For this purpose, we note that (13) is reduced to a Dirichlet-type problem by restricting $\varphi \in H^1(\Omega)$ to $\varphi \in H_0^1(\Omega)$. Then, the same reasoning as in the Dirichlet problem (11) applies (see the argument immediately after Definition 1), and we deduce that $-\Delta u + u = f$ holds everywhere in Ω. Furthermore, using (13) now with $\varphi \in H^1(\Omega)$, we see that

$$\int_{\partial\Omega} \frac{\partial u}{\partial \nu} \varphi \,\mathrm{d}\sigma = \int_\Omega \nabla u \cdot \nabla \varphi \,\mathrm{d}x + \int_\Omega \Delta u \varphi \,\mathrm{d}x$$
$$= \int_\Omega (f - u + \Delta u) \varphi \,\mathrm{d}x$$
$$= 0$$

holds for smooth $\varphi \in H^1(\Omega)$. By choosing φ such that $\varphi = \partial u/\partial \nu$ on $\partial\Omega$, we find that the boundary condition $\partial u/\partial \nu = 0$ in (12) is now recovered.

Let us define the energy functional

$$E(u) := \frac{1}{2} \int_\Omega |\nabla u|^2 \,\mathrm{d}x + \frac{1}{2} \int_\Omega u^2 \,\mathrm{d}x - \int_\Omega f u \,\mathrm{d}x \quad (u \in H^1(\Omega))$$

which is similar to the one for the Dirichlet problem, but now the domain of E is taken to be $H^1(\Omega)$. As before, we can check that critical points u correspond to solutions to (13) and that E satisfies the coerciveness condition

$$\lim_{\|u\|_{H^1(\Omega)}\to\infty} E(u) \to \infty.$$

Note that the presence of the second term $\int_\Omega u^2\,\mathrm{d}x$ in $E(u)$ is necessary to have the coerciveness of E because the Poincaré inequality is not available for $u \in H^1(\Omega)$. The minimization argument with the help of functional analysis deduces the existence of a solution u to (13). The uniqueness also follows from the convexity of E.

3.4 The Robin Problem

To conclude this section, we study a boundary value problem of the third type

$$\begin{cases} -\Delta u = f & (x \in \Omega), \\ \dfrac{\partial u}{\partial \nu} + gu = 0 & (x \in \partial\Omega), \end{cases} \tag{14}$$

where $g = g(x)$ is a positive function defined on $\partial\Omega$. As in the Neumann problem (12), the boundary condition is incorporated into the variational formulation itself.

Definition 3 We call a function $u \in H^1(\Omega)$ a weak (or generalized) solution to (14) if

$$\int_\Omega \nabla u \cdot \nabla\varphi\,\mathrm{d}x + \int_{\partial\Omega} gu\varphi\,\mathrm{d}\sigma = \int_\Omega f\varphi\,\mathrm{d}x \tag{15}$$

holds for all $\varphi \in H^1(\Omega)$.

Here, $-\Delta u$ is identified with

$$H^1(\Omega) \ni \varphi \mapsto \int_\Omega \nabla u \cdot \nabla\varphi\,\mathrm{d}x + \int_{\partial\Omega} gu\varphi\,\mathrm{d}\sigma \in \mathbb{R}.$$

The validity of this can be checked by integration by parts.

If $u \in H^1(\Omega)$ is a smooth solution to (15), then, as in the Neumann problem, we can deduce that $-\Delta u = f$ in Ω by taking $\varphi \in H_0^1(\Omega)$ in (15). Then, for smooth $\varphi \in H^1(\Omega)$, we see that

$$\begin{aligned}\int_{\partial\Omega}\left(\frac{\partial u}{\partial \nu} + gu\right)\varphi\,\mathrm{d}\sigma &= \int_\Omega \nabla u \cdot \nabla\varphi\,\mathrm{d}x + \int_\Omega \Delta u\varphi\,\mathrm{d}x + \int_{\partial\Omega} gu\varphi\,\mathrm{d}\sigma \\ &= \int_\Omega (f + \Delta u)\varphi\,\mathrm{d}x \\ &= 0.\end{aligned}$$

This yields the boundary condition $\partial u/\partial \nu + gu = 0$ on $\partial\Omega$.

We define the energy functional E for problem (15) by

$$E(u) := \frac{1}{2}\int_{\Omega} |\nabla u|^2 \,\mathrm{d}x + \frac{1}{2}\int_{\partial\Omega} gu^2 - \int_{\Omega} fu \,\mathrm{d}x \quad (u \in H^1(\Omega)).$$

Again, critical points of E correspond to solutions to (15). Note that the coerciveness of E in $H^1(\Omega)$ follows when $g > 0$. In fact, in this case we have

$$\int_{\Omega} u^2 \,\mathrm{d}x \le C\left(\int_{\Omega} |\nabla u|^2 \,\mathrm{d}x + \int_{\partial\Omega} gu^2 \,\mathrm{d}\sigma\right)$$

for $u \in H^1(\Omega)$ as a substitute for the Poincaré inequality. The existence of a unique solution u follows from the same argument as presented before.

4 Free Boundary Problem

We saw in the previous section how boundary value problems can be formulated as variational equations. In this section, the variational point of view is extended to a free boundary problem, for which one needs to determine the domain Ω. We will see that our free boundary problem also has a variational formulation, known as a variational inequality, which can be treated again by minimizing an energy functional.

Let the mass f be concentrated in the sense that $f > 1$ in a domain Ω_0 and $f = 0$ outside Ω_0. Then, one of the inverse problems in potential theory asks about the existence of a uniform mass distribution χ_Ω, in a domain $\Omega \supset \Omega_0$, which is "graviequivalent" to the prescribed mass f. Namely, we ask whether there exists $\Omega \supset \Omega_0$ such that

$$\int_{\mathbb{R}^n} \Gamma(x-y) f(y) \,\mathrm{d}y = \int_{\Omega} \Gamma(x-y) \,\mathrm{d}y \tag{16}$$

holds for all $x \in \mathbb{R}^n \setminus \Omega$. Here, χ_Ω denotes the characteristic function of Ω, i.e., $\chi_\Omega = 1$ in Ω and $\chi_\Omega = 0$ outside Ω.

Example 2 Let $f = M\chi_{B(0,r)}$ with a constant $M > 1$, where $B(0, r)$ denotes the ball of radius $r > 0$ with center at the origin $x = 0$. Then, we can show that $\Omega = B(0, R)$ satisfies (16), where $R > 0$ is chosen to satisfy $|B(0, R)| = M|B(0, r)|$. Indeed, for $x \in \mathbb{R}^n \setminus \Omega$, since $\Delta_y \Gamma(x-y) = 0$ for $y \in \Omega$, the mean value formula of harmonic functions implies that

$$\begin{aligned}\int_{\mathbb{R}^n} \Gamma(x-y)f(y)\,\mathrm{d}y &= M \int_{B(0,r)} \Gamma(x-y)\,\mathrm{d}y \\ &= M|B(0,r)|\Gamma(x) \\ &= \int_{B(0,R)} \Gamma(x-y)\,\mathrm{d}y.\end{aligned}$$

It is not at all obvious that there is such a domain Ω for a general f. However, the variational point of view will give us a new insight to the problem, and we are able to solve the free boundary problem by minimizing a functional as (fixed) boundary value problems.

Our starting point is the derivation of a differential equation for (16). For this purpose, let us consider the potential difference

$$u(x) := \int_{\mathbb{R}^n} \Gamma(x-y)f(y)\,\mathrm{d}y - \int_{\Omega} \Gamma(x-y)\,\mathrm{d}y. \tag{17}$$

Then, the free boundary problem is rewritten, in terms of u, as the problem of finding Ω such that the following boundary problem has a solution u:

$$\begin{cases} -\Delta u = f - \chi_\Omega & (x \in \mathbb{R}^n), \\ u = 0 & (x \in \mathbb{R}^n \setminus \Omega). \end{cases} \tag{18}$$

It is easy to check that, if (16) holds for $x \in \mathbb{R}^n \setminus \Omega$, then u defined by (17) is a solution to (18). Conversely, if (18) possesses a solution u, then (16) holds for $x \in \mathbb{R}^n \setminus \Omega$.

Let us impose the additional requirement $u \geq 0$ $(x \in \mathbb{R}^n)$ on (18) and try to construct Ω under this stronger condition in such a way that

$$\Omega = \{x \in \mathbb{R}^n \mid u(x) > 0\}. \tag{19}$$

It will be revealed that this makes the problem proper for variational methods. To make it clear, let us first rewrite (18) combined with the side condition $u \geq 0$ as

$$\begin{cases} -\Delta u \geq f - 1 & (x \in \mathbb{R}^n), \\ u \geq 0 & (x \in \mathbb{R}^n), \\ \displaystyle\int_{\mathbb{R}^n} (-\Delta u - f + 1)u\,\mathrm{d}x = 0. \end{cases} \tag{20}$$

Indeed, the last condition in (20) asserts that at least one of the two inequalities in (20) must be equality at each point $x \in \mathbb{R}^n$. However, (20) merely implies that $-\Delta u = f - 1$ in Ω, so (18) does not seem to directly follow from (20). Nevertheless, by virtue of the regularity theory (see Kinderlehrer and Stampacchia [3]), (18) is eventually shown to be satisfied by a solution u to (20) with Ω defined by (19). This fact can be formally verified, since $u = 0$ in $\mathbb{R}^n \setminus \Omega$ implies $-\Delta u = 0$ there. Moreover, in view of the first inequality in (20), $\Omega_0 \subset \Omega$ follows from the fact that $f > 1$ on Ω_0 and $-\Delta u = 0$ on $\mathbb{R}^n \setminus \Omega$. We emphasize that formulation (20) does not contain Ω explicitly; thus, our problem is just to find a solution u to (20).

The variational structure of (20) will be further clarified by noting that $-\Delta u - f + 1 \geq 0$ if and only if

$$\int_{\mathbb{R}^n} (-\Delta u - f + 1)\varphi \, dx \geq 0$$

holds for all $\varphi \in C_c^\infty(\mathbb{R}^n)$ with $\varphi \geq 0$. Then, (20) is equivalent to finding $u \geq 0$ such that

$$\int_{\mathbb{R}^n} (-\Delta u - f + 1)(\varphi - u) \, dx \geq 0 \tag{21}$$

for all $\varphi \in C_c^\infty(\mathbb{R}^n)$ with $\varphi \geq 0$. Furthermore, by identifying (21) with

$$\int_{\mathbb{R}^n} \{\nabla u \cdot \nabla(\varphi - u) + (1 - f)(\varphi - u)\} \, dx \geq 0 \tag{22}$$

and by setting

$$K := \left\{ u \in H^1(\mathbb{R}^n) \mid u \geq 0 \right\},$$

we finally arrive at the following variational inequality.

Definition 4 We call a function $u \in K$ a weak (or generalized) solution to (21) if

$$\int_{\mathbb{R}^n} \nabla u \cdot \nabla(\varphi - u) \, dx + \int_{\mathbb{R}^n} (1 - f)(\varphi - u) \, dx \geq 0 \tag{23}$$

holds for all $\varphi \in K$.

Remark 4 The validity of (22) for $\varphi \in H^1(\mathbb{R}^n)$ follows from the fact that $H_0^1(\mathbb{R}^n) = H^1(\mathbb{R}^n)$; namely every function $u \in H^1(\mathbb{R}^n)$ can be approximated by $\varphi \in C_c^\infty(\mathbb{R}^n)$ arbitrarily closely in $H^1(\mathbb{R}^n)$.

Remark 5 Any smooth domain Ω defined by (19), where u is a smooth solution to (23), satisfies (18) and hence (16) as desired.

Now we define the energy functional

$$E(u) := \frac{1}{2} \int_{\mathbb{R}^n} |\nabla u|^2 \, dx + \int_{\mathbb{R}^n} (1 - f) u \, dx \quad (u \in K)$$

and observe that $u \in K$ is a solution to (23) if and only if $E'(u)[\varphi - u] \geq 0$ for $\varphi \in K$. The latter condition implies that u is a minimum point of E in K, since E is a convex functional on the closed convex set $K \subset H^1(\mathbb{R}^n)$, where K is said to be convex if

$$tu_1 + (1-t)u_2 \in K \quad (u_1, u_2 \in K \text{ and } 0 \leq t \leq 1).$$

This fact is illustrated by the following example.

Example 3 Let $K \subset \mathbb{R}^m$ be a closed convex set and let $E : K \to \mathbb{R}$ be a convex function. Then, for any minimum point $x \in K$,

$$E'(x)[y-x] = \nabla E(x) \cdot (y-x) = \lim_{t \to 0} \frac{E(x+t(y-x)) - E(x)}{t} \geq 0 \qquad (24)$$

holds for $y \in K$ (note that $x + t(y-x) \in K$ for $0 \leq t \leq 1$). Conversely, by the convexity of E, we have

$$E(y) \geq E(x) + \nabla E(x) \cdot (y-x) \quad (y \neq x);$$

thus, (24) implies that x is a minimum point of E.

The existence of a solution u to (23) will be proved by minimizing E in K. Indeed, the argument we have presented for the whole space H^1 in the case of boundary value problems still works even for the convex closed subset $K \subset H^1$, once we show the coerciveness condition, since the limit of every weakly convergent sequence in K still lies in K. The uniqueness of a minimizer also follows from the convexity of K. However, because of the lack of the Poincaré inequality, the coerciveness does not hold and one needs to modify the argument as follows to complete the proof:

(i) Take a large ball B and consider the minimization problem in $K_B := \{u \in H_0^1(B) \mid u \geq 0\}$ instead of K.
(ii) Check that the minimizer u_B and $\Omega_B := \{x \in B \mid u_B(x) > 0\}$ satisfy (23) if $\overline{\Omega_B} \subset B$.
(iii) Show an a priori bound of Ω_B, which is independent of B, so that B can be chosen to satisfy $\overline{\Omega_B} \subset B$.

For details of the above argument in a slightly different manner, see Sakai [4] and Gustafsson [2].

Appendix

Here, the basic inequalities in analysis known as Hölder's inequality and the Poincaré inequality are supplied for the sake of completeness.

Hölder's inequality (or the Cauchy-Schwarz inequality) states that

$$\left| \int_{\Omega} uv \, \mathrm{d}x \right| \leq \|u\|_{L^2(\Omega)} \|v\|_{L^2(\Omega)} \tag{25}$$

holds for $u, v \in L^2(\Omega)$. This is a natural generalization of the inequality $|x \cdot y| \leq |x||y|$ for $x, y \in \mathbb{R}^n$, since the left-hand side is the inner product $(u, v)_{L^2(\Omega)}$. In particular, equality holds in (25) if and only if $u = \alpha v$ for some scalar $\alpha \in \mathbb{R}$.

The Poincaré inequality states that, for a bounded domain Ω, there is a constant $C > 0$ such that

$$\|u\|_{L^2(\Omega)} \leq C \|\nabla u\|_{L^2(\Omega)}$$

holds for $u \in H_0^1(\Omega)$. The boundedness of Ω can be relaxed to some extent; however, the inequality does not hold, in general, for unbounded domains. Moreover, $H_0^1(\Omega)$ cannot be replaced by $H^1(\Omega)$ for the inequality to hold. Indeed, the constant function $u \equiv 1$ violates the inequality.

References

1. D. Gilbarg, N.S. Trudinger, in *Elliptic Partial Differential Equations of Second Order*. Reprint of the 1998 edition. Classics in Mathematics (Springer, Berlin, 2001)
2. B. Gustafsson, Applications of variational inequalities to a moving boundary problem for Hele-Shaw flows. SIAM J. Math. Anal. **16**(2), 279–300 (1985)
3. D. Kinderlehrer, G. Stampacchia, *An Introduction to Variational Inequalities and Their Applications* (Academic Press, New York, 1980)
4. M. Sakai, Application of variational inequalities to the existence theorem on quadrature domains. Trans. Amer. Math. Soc. **276**, 267–279 (1983)

Part IV
Probability and Statistics

Finite Markov Chains and Markov Decision Processes

Tomoyuki Shirai

Abstract Markov chains are important tools used for stochastic modeling in various areas of mathematical sciences. The first section of this article presents a survey of the basic notions of discrete-time Markov chains on finite state spaces together with several illustrative examples. Markov decision processes (MDPs), which are also known as stochastic dynamic programming or discrete-time stochastic control, are useful for decision making under uncertainty. The second section will provide a simple formulation of MDPs with finite state spaces and actions, and give two important algorithms for solving MDPs, value iteration and policy iteration, with an example on iPod shuffle.

Keywords Markov chain · Markov decision process · Mixing time · Coupling · Cutoff phenomenon

1 Markov Chains

Throughout this article, we assume that S is a finite set of states and we denote the set $\{0, 1, 2, \dots\}$ by $\mathbf{T}$.

A discrete-time stochastic process on S is a sequence of S-valued random variables $\{X_t\}_{t\in\mathbf{T}}$ defined on a probability space $(\Omega, \mathscr{F}, \mathbb{P})$. A *Markov chain* on S is a stochastic process having the following Markov property: for $0 \le t_0 < t_1 < \cdots < t_n < t$ and $x_0, x_1, \dots, x_n, x \in S$,

$$\mathbb{P}(X_t = x | X_{t_0} = x_0, X_{t_1} = x_1, \dots, X_{t_n} = x_n) = \mathbb{P}(X_t = x | X_{t_n} = x_n).$$

T. Shirai (✉)
Institute of Mathematics for Industry, Kyushu University, 744 Motooka, Nishi-ku, Fukuoka 819-0395, Japan
e-mail: shirai@imi.kyushu-u.ac.jp

R. Nishii et al. (eds.), *A Mathematical Approach to Research Problems of Science and Technology*, Mathematics for Industry 5,
DOI: 10.1007/978-4-431-55060-0_15,

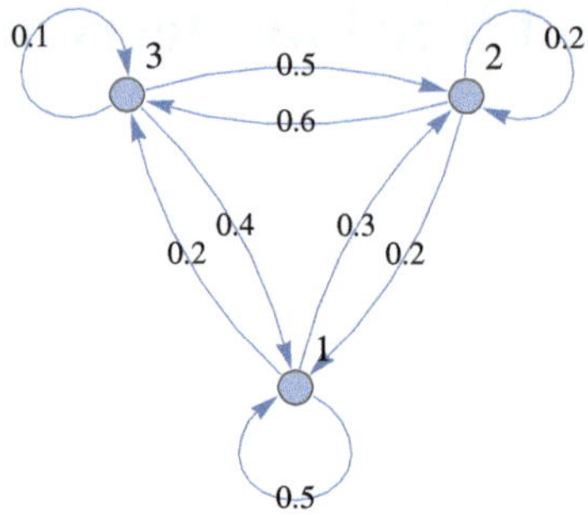

$$P = \begin{pmatrix} 0.5 & 0.3 & 0.2 \\ 0.2 & 0.2 & 0.6 \\ 0.4 & 0.5 & 0.1 \end{pmatrix}$$

Fig. 1 State transition diagram and corresponding transition matrix

In particular, a Markov chain $X = \{X_t\}_{t \in \mathbf{T}}$ is said to be *time homogeneous* if $\mathbb{P}(X_{t+1} = y | X_t = x), x, y \in S$ does not depend on t. When a Markov chain X is time homogeneous, the $|S| \times |S|$ matrix $P = (p(x, y))_{x,y \in S}$ given by the one-step transition probability $p(x, y) := \mathbb{P}(X_{t+1} = y | X_t = x)$ is called a *transition matrix*. A time homogeneous Markov chain is completely determined by a transition matrix.

Lemma 1 *Let $X = \{X_t\}_{t \in \mathbf{T}}$ be a time homogeneous Markov chain. Then, for every $s, t \in \mathbf{T}$ and $x, y \in S$, $\mathbb{P}(X_{t+s} = y | X_s = x) = P^t(x, y)$.*

Throughout this section, we treat only time homogeneous Markov chains.

1.1 Examples of Markov Chains

Example 1 Let $S = \{1, 2, \dots, n\}$. An n by n matrix $P = (p_{ij})_{i,j=1}^n$ is said to be a *stochastic matrix* if $p_{ij} \geq 0$ for all $i, j = 1, 2, \dots, n$ and $\sum_{j=1}^n p_{ij} = 1$ for all $i = 1, 2, \dots, n$. Every stochastic matrix P defines a Markov chain. If n is small, it is well described by using a diagram (Fig. 1).

Example 2 (Simple random walk (SRW) on a finite graph) Let $G = (V, E)$ be a finite connected graph and set $S = V$ with $|V| \geq 2$. An SRW on a finite graph G is a Markov chain on the vertex set S with the transition probability being $\deg(x)^{-1}$ at each vertex $x \in S$, where $\deg(x)$ is the degree of a vertex x in G. For example, in Fig. 2, $\deg(1) = \deg(2) = \deg(5) = 2$, and $\deg(3) = \deg(4) = 3$.

Example 3 (Ehrenfest's urn) In two urns, say U_1 and U_2, there are n balls in total. A ball is taken out uniformly at random and put into the other urn. Looking at the number of balls in U_1, we can regard it as a Markov chain on $S = \{0, 1, 2, \dots, n\}$ with transition probability

$$p(k, k-1) = \frac{k}{n}, \ p(k, k+1) = \frac{n-k}{n} \quad (k = 0, 1, 2, \dots, n).$$

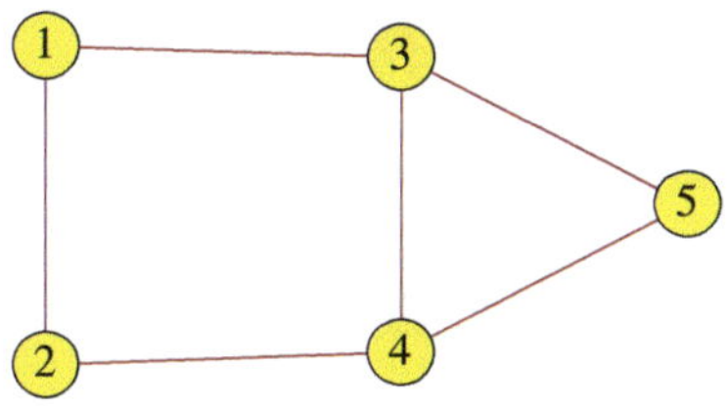

$$P = \begin{pmatrix} 0 & 1/2 & 1/2 & 0 & 0 \\ 1/2 & 0 & 0 & 1/2 & 0 \\ 1/3 & 0 & 0 & 1/3 & 1/3 \\ 0 & 1/3 & 1/3 & 0 & 1/3 \\ 0 & 0 & 1/2 & 1/2 & 0 \end{pmatrix}$$

Fig. 2 Finite graph and transition matrix of SRW on it

Example 4 (SRW on a hypercube) Let $S = \{0, 1\}^n$. We can identify S with the vertices of a square when $n = 2$ and those of a cube when $n = 3$. Since the size of the transition matrix is 2^n, it is not practical to write it down. In this case, it is more convenient to give a transition rule algorithmically. The transition rule from a point $x = (x_1, \dots, x_n) \in S$ is defined as follows:

1. Choose a coordinate i from $\{1, 2, \dots, n\}$ uniformly at random.
2. Update x_i to be 1 if $x_i = 0$ and 0 if $x_i = 1$. That is, $x_i \mapsto 1 - x_i$.

This rule defines the SRW $X = \{X_t\}_{t \in \mathbf{T}}$ on S. For example, when $n = 5$, transition proceeds like

$$(1, 0, 1, 1, 0) \xrightarrow{3} (1, 0, 0, 1, 0) \xrightarrow{5} (1, 0, 0, 1, 1) \xrightarrow{2} (1, 1, 0, 1, 1) \xrightarrow{3} (1, 1, 1, 1, 1) \xrightarrow{1} \cdots$$

The number above each arrow indicates the coordinate chosen in step 1. If we use up-spin and down-spin instead of 1 and 0, we see that

$$(\uparrow, \downarrow, \uparrow, \uparrow, \downarrow) \xrightarrow{3} (\uparrow, \downarrow, \downarrow, \uparrow, \downarrow) \xrightarrow{5} (\uparrow, \downarrow, \downarrow, \uparrow, \uparrow) \xrightarrow{2} (\uparrow, \uparrow, \downarrow, \uparrow, \uparrow) \xrightarrow{3} (\uparrow, \uparrow, \uparrow, \uparrow, \uparrow) \xrightarrow{1} \cdots$$

It seems like a transition for the stochastic Ising model (a model for magnetism). One can easily see that $N_t = \sum_{i=1}^{n} (X_t)_i$, the number of 1's in X_t, is the same Markov chain as was given in Example 3.

Example 5 (Markov chain on the set of q-colorings) Let $G = (V, E)$ be a finite connected graph. For a fixed integer $q > \max_{x \in V} \deg(x)$, we consider a map $c : V \to \{1, 2, \dots, q\}$. It can be regarded as a coloring of V by q-colors. We call a map c a q-coloring and denote the set of all q-colorings by S. If c satisfies $c(v) \neq c(w)$ whenever $vw \in E$, i.e., v and w are adjacent in G, we call it a proper q-coloring and denote the totality of proper q-colorings by S_{proper}. Even when it is difficult to identify the structure of S for a general graph G, we can define a natural Markov chain $\{c_t\}_{t \in \mathbf{T}}$ on S algorithmically:

1. A vertex in V is chosen uniformly at random.
2. If $v \in V$ is chosen at step 1, we set $A_v(c_t) = \{1, 2, \dots, q\} \setminus \{c_t(w) : vw \in E\}$, which is the set of colors admissible for the vertex v. A color is chosen from $A_v(c_t)$ uniformly at random and $c_{t+1}(v)$ is updated to that color, leaving all the other vertices unchanged (Fig. 3).

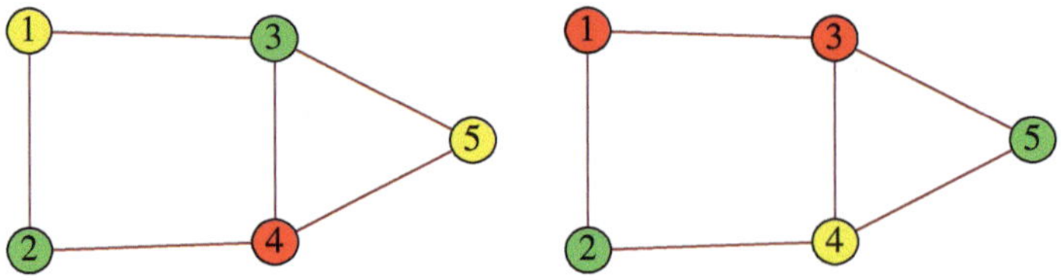

Fig. 3 *Left diagram* shows a proper 3-coloring, and *right diagram* shows an improper one

1.2 Irreducibility and Periodicity

It is important to know whether or not the Markov chain under consideration can traverse its state space.

Definition 1 We say that a Markov chain $X = \{X_t\}_{t \in \mathbf{T}}$ on S is *irreducible* if for any $x, y \in S$ there exists $t = t_{x,y} \in \mathbb{N}$ such that $\mathbb{P}(X_t = y | X_0 = x) > 0$.

Definition 2 Let $\mathrm{Per}(x) := \{t \in \mathbb{N} : \mathbb{P}(X_t = x \,| X_0 = x) > 0\}$. We call the greatest common divisor of $\mathrm{Per}(x)$ the period of a state $x \in S$. It is known that the period is constant on S when X is irreducible. In this case, the period can be considered as that of Markov chain X. If the period is 1, X is said to be *aperiodic*.

Example 6 (Random bishop/knight moves) The possible moves for a bishop and a knight from a particular square on a chessboard are shown in Fig. 4. The state space S comprises the 64 squares. The square to move to is chosen uniformly at random from the possible moves. In the example shown, the bishop chooses one of the squares with probability $1/13$ and moves to it, and the knight does the same with probability $1/8$. These transition rules define Markov chains on S. We call these chains "random bishop move" and "random knight move," respectively.

- (Irreducibility). The random bishop move is not irreducible. Indeed, by the transition rule, the bishop can only move on the squares of the same color as that of the initial place. Then, it is impossible for the bishop to jump to any square of the other color. By induction on the size of the chessboard, it can be shown that the random knight move is irreducible.
- (Periodicity). The period of the random knight move is two. Indeed, the random knight can move only to a square of the opposite color so that an even number of moves is required to return to the initial square. On the other hand, the random bishop move is aperiodic since it is clear that $\{2, 3\} \subset \mathrm{Per}(x)$ for every $x \in S$.

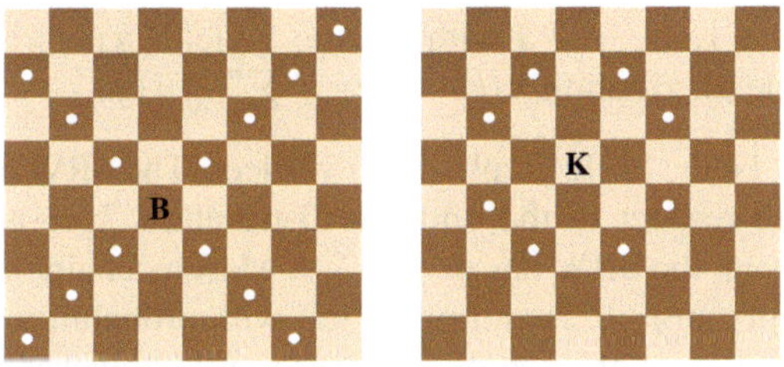

Fig. 4 Bishop (B) and knight (K) can move to a *square* with a *white dot*

1.3 Stationarity and Reversibility

It is important to study the behavior of a Markov chain $X = \{X_t\}_{t\in\mathbf{T}}$ as $t \to \infty$. By the Markov property, the distribution of X_t converges to a stationary distribution (under mild conditions) as $t \to \infty$ regardless of its initial distribution.

Definition 3 We say that π is a *stationary distribution* of a Markov chain X on S if it is a probability distribution and satisfies

$$\sum_{x\in S} \pi(x)p(x, y) = \pi(y), \ \forall y \in S.$$

We say that a Markov chain X or its transition matrix P is *reversible* with respect to π if the detailed balance condition

$$\pi(x)p(x, y) = \pi(y)p(y, x), \ \forall x, y \in S$$

holds. We call π a *reversible distribution* or a *reversible probability measure*.

It is easy to see the following.

Proposition 1 *If π is a reversible distribution, then it is also a stationary distribution.*

Remark 1 Suppose that P is irreducible. There exists a reversible distribution if and only if for any closed path $(x_1, x_2, \dots, x_n, x_1)$, it holds that

$$p(x_1, x_2)p(x_2, x_3)\cdots p(x_n, x_1) = p(x_1, x_n)p(x_n, x_{n-1})\cdots p(x_2, x_1). \quad (1)$$

For fixed $a \in S$, we define $\tilde{\pi}(x) = \frac{p(a,x_1)p(x_1,x_2)\cdots p(x_n,x)}{p(x_1,a)p(x_2,x_1)\cdots p(x,x_n)}$ by taking a path $(a, x_1, \dots, x_n, x)$. It does not depend on the choice of a path joining a and x under the condition (1), and it is a constant multiple of the reversible distribution.

Example 7 It is easy to show that the Markov chain defined in Example 3 is reversible with respect to $\pi(k) = \binom{n}{k}2^{-n}$. Indeed, the detailed balance condition $\tilde{\pi}(k)\frac{n-k}{n} =$

$\tilde{\pi}(k+1)\frac{k+1}{n}$, $k = 0, 1, \ldots, n-1$ with $\tilde{\pi}(0) = 1$ yields $\tilde{\pi}(k) = \binom{n}{k}$. Therefore, we obtain the reversible distribution $\pi(k) = \tilde{\pi}(k)/\sum_{j=0}^{n} \tilde{\pi}(j)$.

Example 8 Let C_n be the cycle graph with n vertices. The SRW on C_n is irreducible and reversible with respect to the uniform distribution. If n is odd, the SRW is aperiodic; if n is even, the SRW has period 2. A Markov chain on C_n moving to the right with probability $p(\neq 1/2)$ and to the left with probability $1 - p(\neq 1/2)$ has the uniform distribution as the stationary distribution; however, it is not reversible since the condition (1) in Remark 1 fails.

The following two propositions are useful for identifying reversible distributions:

Proposition 2 *The SRW on a finite graph $G = (V, E)$ in Example 2 has the reversible distribution $\pi(x) = \frac{\deg(x)}{2|E|}$, where $2|E| = \sum_{x \in V} \deg(x)$ by the hand-shaking lemma.*

Proposition 3 *Suppose that the transition probability of an irreducible Markov chain on S is symmetric in the sense that $p(x, y) = p(y, x)$ for every $x, y \in S$. Then, the uniform distribution $\pi(x) = \frac{1}{|S|}$, $\forall x \in S$, is the reversible distribution.*

The next theorem is one of the most important facts in Markov chain theory.

Theorem 1 *Let $X = \{X_t\}_{t \in \mathbf{T}}$ be an irreducible Markov chain on a finite state space S.*

(1) There exists a unique stationary distribution π.
(2) If X is aperiodic, then the distribution $\mathbb{P}(X_t = \cdot | X_0 = x) = P^t(x, \cdot)$ of X_t starting at x converges to the stationary distribution π as $t \to \infty$ for any $x \in S$. In other words, P^t converges to the matrix Π whose row vectors are all $\Pi(x, \cdot) = \pi$ $(x \in S)$.
(3) For each $x \in S$, $\pi(x) = \frac{1}{\mathbb{E}_x[\tau_x^+]}$, where $\tau_x^+ = \inf\{t \geq 1 : X_t = x\}$.

Example 9 The Markov chain given in Example 2 has the stationary distribution $\pi = (\frac{1}{6}, \frac{1}{6}, \frac{1}{4}, \frac{1}{4}, \frac{1}{6})$ from Proposition 2. Since P is irreducible and aperiodic, (2) of Theorem 1 implies

$$P^t = \begin{pmatrix} 0 & 1/2 & 1/2 & 0 & 0 \\ 1/2 & 0 & 0 & 1/2 & 0 \\ 1/3 & 0 & 0 & 1/3 & 1/3 \\ 0 & 1/3 & 1/3 & 0 & 1/3 \\ 0 & 0 & 1/2 & 1/2 & 0 \end{pmatrix}^t \to \begin{pmatrix} 1/6 & 1/6 & 1/4 & 1/4 & 1/6 \\ 1/6 & 1/6 & 1/4 & 1/4 & 1/6 \\ 1/6 & 1/6 & 1/4 & 1/4 & 1/6 \\ 1/6 & 1/6 & 1/4 & 1/4 & 1/6 \\ 1/6 & 1/6 & 1/4 & 1/4 & 1/6 \end{pmatrix} = \Pi \quad (t \to \infty)$$

By (3) of Theorem 1, we have $\mathbb{E}_x[\tau_x^+] = 6$ for $x = 1, 2, 5$ and $\mathbb{E}_x[\tau_x^+] = 4$ for $x = 3, 4$.

Example 10 For a random knight move starting from one of the corners on the chessboard, say c, it is easy to show that $\mathbb{E}_c[\tau_c^+] = 168$ by Proposition 2 and (3) of Theorem 1. Indeed, it is easy to check that $\deg(c) = 2$ and that

$$2|E| = \sum_{x \in S} \deg(x) = 2 \times 4 + 3 \times 8 + 4 \times 20 + 6 \times 16 + 8 \times 16 = 336.$$

Remark 2 We note that an irreducible Markov chain on S is aperiodic if there exists a state $x \in S$ such that $p(x, x) > 0$. To apply (2) of Theorem 1, we define the lazy version of a Markov chain with $P = (p(x, y))$ as the Markov chain with transition matrix $Q = (q(x, y))$ with

$$q(x, y) = \begin{cases} \frac{1}{2} p(x, y) & \text{if } y \neq x, \\ \frac{1}{2} + \frac{1}{2} p(x, x) & \text{if } y = x. \end{cases}$$

It is clear that $Q = \frac{1}{2}(I + P)$. If a fair coin is flipped and it comes up heads, then the Markov chain moves according to the original probability law P; if it comes up tails, then it stays at the present position. The stationary distribution of Q is the same as that of P. Even if P is periodic, Q becomes aperiodic.

Example 11 Let $G = (V, E)$ be a finite connected graph and suppose that $q > \max_{x \in S} \deg(x)$. In Example 5, a Markov chain was defined on the set of all q-colorings. By the transition rule, a vertex chosen in step 1 is colored differently from the vertices in its neighborhood. Through repeated transitions, at least after all the vertices are chosen in step 1, the state becomes a q-proper coloring even if it was originally a non-q-proper coloring. Moreover, once the state becomes q-proper, it will remain q-proper. This means that S_{proper} is closed with respect to this Markov chain. Although the Markov chain on S is not irreducible, that on S_{proper} is irreducible. Such a subset of a state space as S_{proper} is sometimes called an irreducible component. By Proposition 3, the stationary distribution is the uniform distribution on S_{proper}.

1.4 Coupon Collector's Problem

Coupon collector's problem is a classic problem in probability theory and has been extended in several ways. Here we consider the most basic one.

Problem 1 Suppose that there are n different kinds of coupons. One coupon is obtained with equal probability $\frac{1}{n}$ in each trial. How many trials does it take to collect a complete set of coupons?

The number of different coupons is considered to be a Markov chain $X = \{X_t\}_{t \in \mathbf{T}}$ on $S = \{0, 1, 2, \ldots, n\}$ with $X_0 = 0$. Since the probability of getting a new kind of coupon is $\frac{n-k}{n}$ if one has k different kinds already, the transition probability is given by

$$p(k, k) = \frac{k}{n}, \quad p(k, k+1) = \frac{n-k}{n} \quad (k \in S).$$

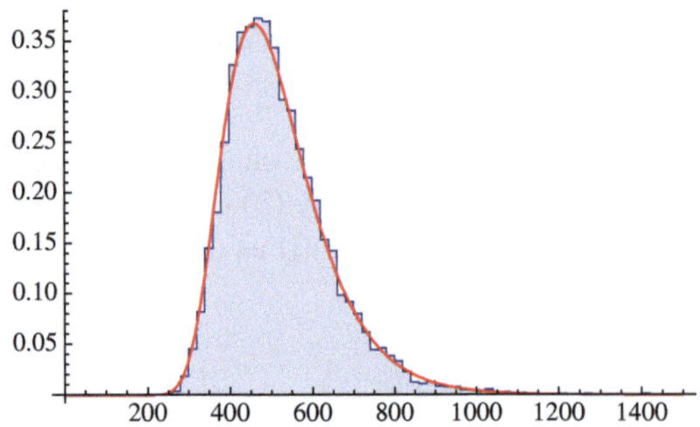

Fig. 5 Histogram of τ_{100} (simulation) and the limiting distribution $e^{-e^{-c}}$

By definition, this Markov chain only goes upwards. Let τ_n be a random variable taking values in $\mathbb{N}$ defined by

$$\tau_n = \inf\{t \in \mathbb{N} : X_t = n\},$$

which is the first time that a complete set of coupons has been collected; if the set $\{t \in \mathbb{N} : X_t = n\}$ is empty, τ_n is understood to be ∞. Problem 1 can thus be rephrased as the problem of studying the random variable τ_n.

Proposition 4 *(1)* $E[\tau_n] = n \sum_{k=1}^{n} \frac{1}{k} \sim n \log n$.[1] *(2)* $\lim_{n\to\infty} P(\tau_n \leq n \log n + cn) = e^{-e^{-c}}$ $(c \in \mathbb{R})$.

This proposition implies that the expected time to collect a complete set of coupons is about $n \log n$ and the probability that all kinds are not yet collected after $n \log n$ is exponentially small. For example, if $n = 100$, then $E[\tau_{100}] = 518.738\ldots$ (Fig. 5).

1.5 Mixing Time

The distribution at time t of an irreducible and aperiodic Markov chain on a finite state space S converges to the stationary distribution as $t \to \infty$ by Theorem 1. Here we consider the speed of convergence. For that we introduce a distance on $\mathscr{P}(S)$, the set of all probability measures on S.

Definition 4 For $\mu, \nu \in \mathscr{P}(S)$, we define the *total variation distance* by

$$\|\mu - \nu\|_{TV} = \max_{A \subset S} |\mu(A) - \nu(A)|.$$

This distance has several different expressions.

---

[1] $a_n \sim b_n$ means that $a_n/b_n \to 1$ as $n \to \infty$.

Proposition 5 *For* $\mu, \nu \in \mathscr{P}(S)$, $0 \le \|\mu - \nu\|_{TV} \le 1$ *and*

$$\|\mu - \nu\|_{TV} = \frac{1}{2}\sum_{x \in S} |\mu(x) - \nu(x)| = \sum_{\substack{x \in S \\ \mu(x) \ge \nu(x)}} |\mu(x) - \nu(x)|$$
$$= \inf\{P(X \ne Y) : (X, Y) \text{ is a coupling of}(\mu, \nu)\},$$

where a two-dimensional random variable (X, Y) *is said to be a coupling of* (μ, ν) *if the marginal distributions of* X *and* Y *are equal to* μ *and* ν*, respectively. Here we simply write* $\mu(x)$ *for* $\mu(\{x\})$.

Remark 3 When a Markov chain is irreducible and aperiodic, since S is finite, Theorem 1 implies that $d(t) := \max_{x \in S} \|P^t(x, \cdot) - \pi\|_{TV} \to 0$. Moreover, it is known that $d(t)$ is monotone decreasing.

Definition 5 From the remark above, we can define the *mixing time* by

$$t_{\text{mix}}(\varepsilon) := \inf\{t \in \mathbb{N} : d(t) \le \varepsilon\}$$

for given $\varepsilon \in (0, 1/2)$. In particular, we write $t_{\text{mix}} := t_{\text{mix}}(1/4)$. Here $1/4$ can be replaced with any $\varepsilon \in (0, 1/2)$.

Mixing time is the time when a Markov chain approaches the stationarity "sufficiently." Several results have been obtained for the following problem.

Problem 2 Given an increasing sequence of state spaces $\{S_n : n \in \mathbb{N}\}$ and Markov chains $X^{(n)} = \{X_t^{(n)}\}_{t \in \mathbf{T}}$ on S_n, one can define $t_{\text{mix}}^{(n)}$ for each $X^{(n)}$ on S_n. Analyze the asymptotic behavior of the mixing time $t_{\text{mix}}^{(n)}$ as $n \to \infty$.

1.6 Coupling of Markov Chains

The coupling method is often used for comparisons with probability distributions. In the example below, we use a coupling of Markov chains to derive an inequality.

Definition 6 (1) Let $X = \{X_t\}_{t \in \mathbf{T}}$ and $Y = \{Y_t\}_{t \in \mathbf{T}}$ be Markov chains on S starting at different initial states x and y, respectively. A Markov chain $\{(\tilde{X}_t, \tilde{Y}_t)\}_{t \in \mathbf{T}}$ on $S \times S$ is said to be a Markov *coupling* of X and Y if the probability law of $\{\tilde{X}_t\}_{t \in \mathbf{T}}$ (resp. $\{\tilde{Y}_t\}_{t \in \mathbf{T}}$) is equal to that of the given Markov chain X (resp. Y). We denote the probability law of this coupling $\{(\tilde{X}_t, \tilde{Y}_t)\}_{t \in \mathbf{T}}$ by $\mathbb{P}_{x,y}$.

(2) We define a coupling time by $\tau_{couple} = \inf\{t \ge 0 : \tilde{X}_t = \tilde{Y}_t\}$.

Example 12 Consider a Markov chain on $S = \{0, 1, 2, \ldots, n\}$. This chain jumps to one of its two neighbors with equal probability $1/2$ at $\{1, 2, \ldots, n-1\}$, to 0 or 1 with equal probability at 0, and to $n-1$ or n with equal probability at n. We construct a coupling as follows: Toss a fair coin. Both $\tilde{X}_t$ and $\tilde{Y}_t$ move upwards if it comes up heads and both move downwards if it comes up tails. The important feature of this coupling is the fact that if $x \le y$, then $\tilde{X}_t \le \tilde{Y}_t$ for any $t \ge 0$. Therefore, since $\{\tilde{X}_t = n\} \subset \{\tilde{Y}_t = n\}$, we can see that if $x \le y$ then

$$P^t(x, n) = \mathbb{P}_{x,y}(\tilde{X}_t = n) \le \mathbb{P}_{x,y}(\tilde{Y}_t = n) = P^t(y, n).$$

In other words, $P^t(x, n)$ is an increasing function of x for each t. This fact is not so easy to prove by simply using matrix computations.

1.7 Upper Estimate of Mixing Time via Coupling of Markov Chains

The expected coupling time is used as an upper bound of t_{mix}.

Proposition 6 *Let $\mathbb{P}_{x,y}$ be a coupling of two Markov chains starting at x and y. Then, $t_{\mathrm{mix}} \le 4 \max_{x,y \in S} \mathbb{E}_{x,y}[\tau_{couple}]$.*

From Proposition 6, it is important to construct a "nice" coupling with small coupling time. Here we give two examples.

1.7.1 Mixing Time of LSRW on Cycle Graph C_n

First we estimate the mixing time for the lazy version of the SRW on C_n given in Example 8. We construct a coupling $\{(\tilde{X}_t, \tilde{Y}_t)\}_{t \in \mathbf{T}}$ of two LSRWs starting at x and y respectively as follows:

1. Toss a fair coin. If it comes up heads, $\tilde{X}_t$ moves according to the transition rule; if it comes up tails, $\tilde{Y}_t$ does.
2. After the two chains meet, they move together as a single LSRW, keeping $\tilde{X}_t = \tilde{Y}_t$ for $t \ge \tau_{\mathrm{couple}}$.

Looking at either $\tilde{X}_t$ or $\tilde{Y}_t$ reveals that each chain is obviously an LSRW on C_n. Let us consider the coupling time of this chain $\{(\tilde{X}_t, \tilde{Y}_t)\}_{t \in \mathbf{T}}$. Let Z_t be the shortest path distance between $\tilde{X}_t$ and $\tilde{Y}_t$. It is thus a Markov chain on $\{0, 1, \ldots, \lfloor n/2 \rfloor\}$. (The transition rule at $\lfloor n/2 \rfloor$ is a little different depending on whether n is even or odd.) Then, the coupling time of $\tilde{X}_t$ and $\tilde{Y}_t$ is equal to the first hitting time of Z_t at 0. It is known to be of $O(n^2)$. Therefore, by Proposition 6, we can conclude that $t_{\mathrm{mix}}^{(n)} = O(n^2)$.

1.7.2 Mixing Time of LSRW on Hypercube

Consider the lazy version of the SRW on hypercube $S = \{0, 1\}^n$ given in Example 4. A coupling $\{(\tilde{X}_t, \tilde{Y}_t)\}_{t \in \mathbf{T}}$ of two LSRWs starting at different initial states is constructed as follows:

1. A coordinate i is chosen from $\{1, 2, \ldots, n\}$ uniformly at random.
2. In accordance with the heads or tails of a fair coin flip, set $\tilde{X}_t(i) = \tilde{Y}_t(i) = 1$ or $\tilde{X}_t(i) = \tilde{Y}_t(i) = 0$.

   For example, a transition when $n = 5$ proceeds like

$$\begin{pmatrix} 1\,0\,1\,1\,0 \\ 0\,0\,0\,0\,0 \end{pmatrix} \overset{3,\text{heads}}{\rightarrow} \begin{pmatrix} 1\,0\,1\,1\,0 \\ 0\,0\,1\,0\,0 \end{pmatrix} \overset{5,\text{heads}}{\rightarrow} \begin{pmatrix} 1\,0\,1\,1\,1 \\ 0\,0\,1\,0\,1 \end{pmatrix} \overset{3,\text{tails}}{\rightarrow} \begin{pmatrix} 1\,0\,0\,1\,1 \\ 0\,0\,0\,0\,1 \end{pmatrix} \overset{1,\text{tails}}{\rightarrow} \cdots$$

Suppose i is chosen at step 1. No matter what the value of the i-th coordinate is, at step 2, it keeps that value with probability $1/2$ and is updated to the other value with probability $1/2$. Therefore, if we look at either $\tilde{X}_t$ or $\tilde{Y}_t$ only, we see nothing but an LSRW. Under this coupling, once the i-th coordinate is chosen at step 1, the values of the i-th coordinate of $\tilde{X}_t$ and $\tilde{Y}_t$ will remain the same. Therefore, the coupling time of the coupled chain is the first time when all the coordinates at which the values are different at $t = 0$ (e.g., $\{1, 3, 4\}$ in the example above) are chosen. If we regard $\{1, 3, 4\}$ as the coupons yet to be collected, the coupling time is smaller than τ_n defined in coupon collector's problem in Sect. 1.4. Therefore, $\mathbb{E}_{x,y}[\tau_{\text{couple}}] \le \mathbb{E}[\tau_n] \le n \log n + n$. By Proposition 6, we see that $t_{\text{mix}}^{(n)} \le 4(n \log n + n)$. It is known that $t_{\text{mix}}^{(n)} \sim \frac{1}{2} n \log n$.

1.8 *Cutoff Phenomenon*

The cutoff phenomenon is said to occur when the total variation distance $d(t)$ keeps nearly 1 before the mixing time t_{mix} and abruptly drops to near 0 around the mixing time t_{mix}. This implies that the distribution of X_t is far from the stationarity before time t_{mix} and close to stationarity after time t_{mix}. This phenomenon is formulated as follows.

Definition 7 A sequence of Markov chains has a *cutoff* if

$$t_{\text{mix}}^{(n)}(\varepsilon) \sim t_{\text{mix}}^{(n)}(1 - \varepsilon) \quad \text{for every } \varepsilon \in (0, 1/2)$$

as $n \to \infty$, which is equivalent to

$$\lim_{n \to \infty} d_n(c t_{\text{mix}}^{(n)}) = \begin{cases} 1 & \text{if } c < 1, \\ 0 & \text{if } c > 1. \end{cases}$$

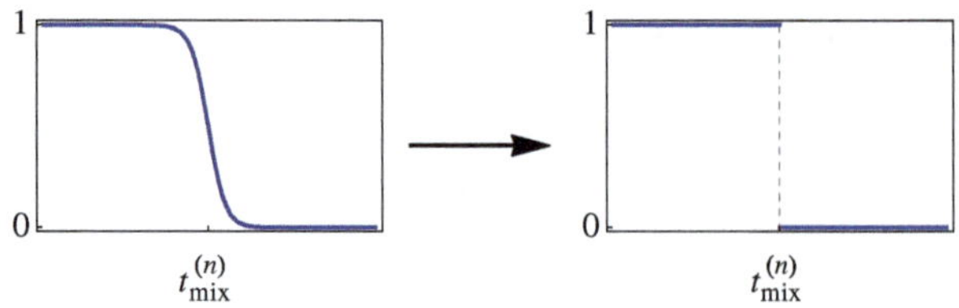

Fig. 6 Cutoff phenomenon. Graph of $d_n(t)$ rescaled by $t_{\rm mix}^{(n)}$ as $n \to \infty$

The total variation distance $d_n(t)$ converges to a step function as $n \to \infty$ by rescaling time t by $t_{\rm mix}^{(n)}$ (Fig. 6).

The following is a more precise version of the above.

Definition 8 A sequence of Markov chains has a *cutoff with a window of size* w_n if $w_n = o(t_{\rm mix}^{(n)})$ and for every $\varepsilon \in (0, 1/2)$ there exists $c_\varepsilon > 0$ such that

$$t_{\rm mix}^{(n)}(\varepsilon) - t_{\rm mix}^{(n)}(1-\varepsilon) \le c_\varepsilon w_n \quad (\forall n \in \mathbb{N}),$$

which is equivalent to

$$\lim_{c\to-\infty} \liminf_{n\to\infty} d_n(t_{\rm mix}^{(n)} + cw_n) = 1, \qquad \lim_{c\to+\infty} \limsup_{n\to\infty} d_n(t_{\rm mix}^{(n)} + cw_n) = 0.$$

Example 13 (1) The LSRW on the hypercube $\{0, 1\}^n$ has a cutoff at $t_{\rm mix}^{(n)} \sim \frac{1}{2} n \log n$ with a window of size n.

(2) The SRW on cycle graph C_n does not have a cutoff.

(3) A biased random walk on $\{0, 1, \ldots, n\}$ moves upwards with probability $p > 1/2$ and downwards with probability $1 - p$. Then its lazy version has a cutoff around $t_{\rm mix}^{(n)} \sim (p - 1/2)^{-1} n$ with a window of size $\sqrt{n}$.

2 Markov Decision Processes

On many occasions one has to make a decision to minimize a cost or maximize a reward. Markov decision processes (MDPs) provide a model for use in such situations.

Here we give a formulation of MDPs. Let S be a finite state space and A a finite set of actions. For each $a \in A$ a transition matrix $P(a) = (p_{xy}(a))_{x,y\in S}$ is given. A function $c : S \times A \to [0, \infty)$ is called a *cost function*. A *policy* is a sequence $u = \{u_t\}_{t\in\mathbf{T}}$ of functions $u_t : S^{t+1} \to A$. For given $\{P(a)\}_{a\in A}$, we define a stochastic process $\{X_t\}_{t\in\mathbf{T}}$ on S associated with policy u and initial state $\mu = (\mu_x)_{x\in S}$ by the following properties: for $x_0, x_1, \ldots, x_{t+1} \in S$,

1. $\mathbb{P}^u(X_0 = x_0) = \mu_{x_0}$.
2. $\mathbb{P}^u(X_{t+1} = x_{t+1}|X_0 = x_0, \dots, X_t = x_t) = p_{x_t x_{t+1}}(u_t(x_0, \dots, x_t))$.

When the initial state μ is the delta measure at x, we denote the probability law of $\{X_t\}_{t\in\mathbf{T}}$ by $\mathbb{P}^u_x$. This process is not, in general, a Markov chain since the conditional probability depends on the past not just on the present state. A policy u is said to be a *stationary policy* if there exists a map $u : S \to A$ such that $u_t(x_0, \dots, x_t) = u(x_t)$ for any $t = 0, 1, \dots$. Here we abuse the notation of u. If a policy u is stationary, then the corresponding stochastic process is a Markov chain.

In what follows, for simplicity, we assume the following:

(A1) There exists an absorbing state $z \in S$ in the sense that $p_{zy}(a) = \delta_{zy}$ and $c(z, a) = 0$ for any $a \in A$. We denote the set of all absorbing states by S_{abs}.
(A2) For $x \in S \setminus S_{abs}$, $c(x, a) > 0$ for every $a \in A$.
(A3) There exists a stationary policy u such that for every $x \in S \setminus S_{abs}$ there exists $t = t_x \in \mathbb{N}$ so that $\mathbb{P}^u_x(X_t \in S_{abs}) > 0$.

Let τ be the first hitting time to S_{abs}, i.e., $\tau = \inf\{t \geq 0 : X_t \in S_{abs}\}$. We define the expected total cost associated with a policy u by

$$V^u(x) = \mathbb{E}^u_x\left[\sum_{t=0}^{\tau-1} c(X_t, u_t(X_0, X_1, \dots, X_t))\right] \quad (x \in S)$$

and the optimal total cost by

$$V^*(x) = \inf_u V^u(x) \quad (x \in S).$$

It is clear that

$$c_{\min}\mathbb{E}^u_x[\tau] \leq V^u(x) \leq c_{\max}\mathbb{E}^u_x[\tau], \tag{2}$$

where $c_{\min} = \min_{x\in S\setminus S_{abs}, a\in A} c(x, a)$ and $c_{\max} = \max_{x\in S, a\in A} c(x, a)$. This implies, under (A2), that $\max_{x\in S} V^u(x) < \infty$ is equivalent to $\max_{x\in S} \mathbb{E}^u_x[\tau] < \infty$.

Lemma 2 *Let u be a stationary policy as in (A3). Then,* $\max_{x\in S} \mathbb{E}^u_x[\tau] < \infty$. *In particular,* $\max_{x\in S} V^*(x) < \infty$.

We note that if a policy u is a stationary policy associated with $u : S \to A$, then $V^u(x)$ satisfies

$$V^u(x) = c(x, u(x)) + \sum_{y\in S} p_{xy}(u(x))V^u(y). \tag{3}$$

Example 14 For $n \geq 2$, let $S = \{0, 1, 2, \dots, n\}$ be a state space with 0 being the absorbing state. There are two actions $A = \{a_1, a_2\}$. If one chooses a_1, then one goes downward by 1 in every state; if one chooses a_2, then one jumps to 0 or $n - 1$ with equal probability 1/2 at n and goes downward by 1 otherwise. Suppose that the cost of action a_1 (resp. a_2) is 1 (resp. C), i.e., $c(x, a_1) = 1$ (resp. $c(x, a_2) = C$)

for $x = 1, 2, \ldots, n$. Suppose $C > 1$ for simplicity. It is clear that $V^*(x) = x$ for $x \in \{0, 1, \ldots, n-1\}$ and

$$V^*(n) = \begin{cases} C + \frac{n-1}{2} & \text{if } 1 < C \le \frac{n+1}{2}, \\ n & \text{if } C \ge \frac{n+1}{2} \end{cases}$$

for $x = n$. One should choose action a_2 at n for the former and action a_1 for the latter.

In Example 14, we can compute the optimal cost V^* explicitly. However, it is not easy to determine the optimal cost in general. So the question is how to estimate the optimal cost V^*. Here we give upper and lower estimates for V^*.

2.1 Lower Bound: Value Iteration

For a lower bound, we define the minimum expected cost incurred before time t inductively by

$$V_t(x) = \min_{a \in A} \left\{ c(x, a) + \sum_{y \in S} p_{xy}(a) V_{t-1}(y) \right\}, \quad V_0(x) = 0 \; (\forall x \in S), \tag{4}$$

which is often called the *Bellman equation with finite horizon*. By induction, it is easy to see that $V_t(x)$ is increasing in t. Hence, there exists an increasing limit $\lim_{t\to\infty} V_t(x) \in [0, \infty]$. We can show that the limit is equal to the optimal value $V^*(x)$.

Proposition 7 *For each $x \in S$, $V_t(x)$ is increasing in t and converges to $V^*(x)$ as $t \to \infty$. In particular, $V_t(x) \le V^*(x)$ for any t.*

We apply Proposition 7 to Example 14. Since $V_t(0) \equiv 0$ and $C > 1$, we see that

$$V_t(x) = \begin{cases} 1 + V_{t-1}(x-1) & \text{for } x = 1, 2, \ldots, n-1, \\ \min\{1 + V_{t-1}(n-1), \; C + \frac{1}{2} V_{t-1}(n-1)\} & \text{for } x = n. \end{cases}$$

This implies that $V_t(x) = \min\{x, t\}$ and hence $V^*(x) = x$ for $x = 1, 2, \ldots, n-1$. When $t \ge n$, as $V_{t-1}(n-1) = n-1$, we have

$$V^*(n) = V_t(n) = \begin{cases} C + \frac{n-1}{2} & \text{if } 1 < C \le \frac{n+1}{2}, \\ n & \text{if } C \ge \frac{n+1}{2}. \end{cases}$$

2.2 Upper Bound: Policy Iteration

Next we consider an upper bound for V^*. For a given stationary policy u_0 such that $\max_{x\in S} V^{u_0}(x) < \infty$, one can choose a stationary policy u_1 such that for each $x \in S$ action $a = u_1(x)$ minimizes the function $a \mapsto c(x,a) + \sum_{y\in S} p_{xy}(a)V^{u_0}(y)$. For such a stationary policy u_0, we inductively define a sequence of stationary policies $\{u_t\}_{t\in \mathbf{T}}$ by

$$u_t(x) \in \arg\min_{a\in A} \left\{ c(x,a) + \sum_{y\in S} p_{xy}(a)V^{u_{t-1}}(y) \right\} \quad (x \in S), \tag{5}$$

where $\arg\min_{a\in A} f(a)$ is the set of arguments for which $f(a)$ attains its minimum and $u_t(x)$ is arbitrarily chosen from the right-hand side.

Proposition 8 *For a stationary policy u_0 such that $\max_{x\in S} V^{u_0}(x) < \infty$, we define $\{u_t\}_{t\in \mathbf{T}}$ as described above. Then, $V^{u_t}(x)$ is decreasing in t and converges to $V^*(x)$ as $t \to \infty$ for each $x \in S$. In particular, $V^*(x) \le V^{u_t}(x)$ for any t.*

We apply Proposition 8 to Example 14. For simplicity, we assume that $n \ge 3$. The $n = 2$ case is left to the reader as an exercise. First, we suppose a policy $u_0(x) = a_2$ for every x. Then,

$$V^{u_0}(x) = \begin{cases} Cx & \text{for } x = 0, 1, \dots, n-1, \\ \frac{C}{2}(1+n) & \text{for } x = n. \end{cases}$$

It is clear that $u_1(x) = a_1$ for $x = 0, 1, \dots, n-1$ since $C > 1$. For $x = n$,

$$c(n, a_i) + \sum_{y\in S} p_{ny}(a_i)V^{u_0}(y) = \begin{cases} 1 + C(n-1) & \text{for } i = 1, \\ C + \frac{1}{2}C(n-1) & \text{for } i = 2. \end{cases}$$

Then, it is easy to see that $u_1(n) = a_2$ when $n \ge 3$ since $C > 1$ and that

$$V^{u_1}(x) = \begin{cases} x & \text{for } x = 0, 1, \dots, n-1, \\ c + \frac{1}{2}(n-1) & \text{for } x = n. \end{cases}$$

Similarly, it is clear that $u_2(x) = a_1$ for $x = 1, \dots, n-1$ and that

$$c(n, a_i) + \sum_{y\in S} p_{ny}(a_i)V^{u_1}(y) = \begin{cases} 1 + (n-1) = n & \text{for } i = 1, \\ C + \frac{1}{2}(n-1) & \text{for } i = 2 \end{cases}$$

for $x = n$. Therefore, we have

$$u_2(x) = a_1 \ (x = 1, \ldots, n-1), \quad u_2(n) = \begin{cases} a_2 & \text{if } 1 < C \leq \frac{n+1}{2}, \\ a_1 & \text{if } C \geq \frac{n+1}{2} \end{cases} \tag{6}$$

and

$$V^{u_2}(x) = \begin{cases} x & \text{for } x = 0, 1, \ldots, n-1, \text{ or for } x = n \text{ and } C \geq \frac{n+1}{2}, \\ C + \frac{n-1}{2} & \text{for } x = n \text{ and } 1 < C \leq \frac{n+1}{2}. \end{cases}$$

It can be easily seen that $u_t(x) = u_2(x)$ for $t \geq 2$. Therefore, $u_2(x)$ given in (6) is an optimal policy.

2.3 An Example: iPod Shuffle

An iPod shuffle is an MP3 music player with a clickable control pad and an external button for switching between two different modes; one is sequential play mode and the other is random shuffle mode. Suppose that the playlist for your iPod is sorted by song title, say, $S = \{1, 2, \ldots, n\}$, and n is assumed to be the song you want to listen to. If you click the control pad in the sequential play mode, x goes to $x+1$, and if you click the control pad in the random shuffle mode, the next song is chosen uniformly at random from $\{1, 2, \ldots, n\}$. Intuitively, if you start at a song close enough to n, it might be better to stay in the sequential play mode, and if you start at a song far from n, it might be better to switch to the random shuffle mode until a song close to n is reached. The question is, what is the threshold for switching between two modes? This problem is well-modeled by a Markov decision process.

Example 15 (iPod shuffle) Let $S = \{1, 2, \ldots, n\}$ be a state space with n being the absorbing state. There are two actions $A = \{a_1, a_2\}$. If action a_1 is chosen, one moves upward by 1; if action a_2 is chosen, one jumps to a state uniformly at random. The costs of action a_1 and a_2 are 1 and T, respectively. We assume that $1 < T \ll n$. We apply Proposition 7 to this example. It follows from (4) that

$$V_t(x) = \min \left\{ 1 + V_{t-1}(x+1), \ T + \frac{1}{n} \sum_{y=1}^{n} V_{t-1}(y) \right\}, \quad x = 1, \ldots, n-1, \tag{7}$$

and $V_t(n) = 0 (\forall t = 0, 1, \ldots)$. From this expression, by induction, it is easy to see that $V_t(x)$ is decreasing in x for each t and that there exist $\nu_t > 0$ and $K_t \in \{1, \ldots, n\}$ such that

$$V_t(x) = \begin{cases} \nu_t & \text{for } x = 1, 2, \ldots, n - K_t, \\ n - x & \text{for } x = n - K_t + 1, n - K_t + 2, \ldots, n. \end{cases}$$

By Proposition 7, $V_t(x) \nearrow V^*(x)$ as $t \to \infty$, and hence we obtain $\nu > 0$ and $K \in \{1, \dots, n\}$ such that

$$V^*(x) = \begin{cases} \nu & \text{for } x = 1, 2, \dots, n-K, \\ n - x & \text{for } x = n-K+1, n-K+2, \dots, n. \end{cases}$$

On the other hand, from (7),

$$V^*(x) = \min\left\{1 + V^*(x+1),\ T + \frac{1}{n}\sum_{y=1}^{n} V^*(y)\right\}, \quad \text{for } x = 1, 2, \dots, n-1. \tag{8}$$

The second argument on the right-hand side does not depend on x and is equal to

$$C_n(\nu, K, T) = T + \frac{1}{n}\left\{(n-K)\nu + \frac{1}{2}K(K-1)\right\}.$$

Setting $x = 1$ in (8) yields

$$\begin{cases} \nu = \min\{1 + \nu,\ C_n(\nu, K, T)\} & K = 0, 1, \dots, n-2, \\ \nu = \min\{n-1,\ C_n(\nu, n-1, T)\} & K = n-1, \\ n - 1 = \min\{n-1,\ C_n(\nu, n, T)\} & K = n. \end{cases}$$

Since we assumed that $T \ll n$, we have that $C_n(\nu, n, T) < n-1$, and so $K \neq n$. It is also easy to see that $\nu = C_n(\nu, K, T)$ for $K \leq n-1$, which implies that

$$\upsilon = \frac{1}{2}(K-1) + \frac{nT}{K}. \tag{9}$$

Setting $x = n - K$ and $x = n - K + 1$ in (8) yields $\nu = \min\{K, C_n(\nu, K, T)\}$ and $K - 1 = \min\{K - 1, C_n(\nu, K, T)\}$. Hence, we have

$$K - 1 \leq \nu = C_n(\nu, K, T) \leq K.$$

By solving these inequalities together with (9), we have

$$\frac{\sqrt{1+8nT} - 1}{2} \leq K \leq \frac{\sqrt{1+8nT} + 1}{2}.$$

Therefore, we can see that $\nu \sim K \sim \sqrt{2nT}$ as $n \to \infty$. □

Remark 4 We refer the reader to Levin et al. [1] for a comprehensive account of the topics covered in Sect. 1, especially mixing time and cutoff phenomenon. Norris [2] provides additional details for the explanations in Sect. 2. The iPod example in Sect. 2.3 is taken from Norvig [3].

References

1. D.A. Levin, Y. Peres, E.L. Wilmer, *Markov Chains and Mixing Times* (American Mathematical Society, Providence, 2009)
2. J.R. Norris, *Markov Chains* (Cambridge University Press, Cambridge, 1997)
3. P. Norvig, Doing the Martin Shuffle (with your iPod). Available at http://norvig.com/ipod.html

Introduction to the Premium Principle Based on the Wang Transform

Shingo Saito

Abstract This is a self-contained introductory survey article on the premium principle based on the Wang transform. We give the definition and examples of the Wang transform and prove that the induced premium principle is a coherent risk measure.

Keywords Premium principle · Wang transform · Risk measure · Coherent risk measure

1 Introduction

Young begins his survey article [7] on premium principles in the *Encyclopedia of Actuarial Science* with the sentence,

> Loosely speaking, a *premium principle* is a rule for assigning a premium to an insurance risk.

Mathematically speaking, a premium principle is a map π that assigns to each random variable X (possibly satisfying certain conditions such as integrability or nonnegativity) a real number $\pi(X)$, viewed as the premium of the insurance risk modeled by the random variable X. The simplest examples include the *expectation premium principle* $\pi(X) = (1 + h)E[X]$ defined for integrable random variables X and the *standard deviation premium principle* $\pi(X) = E[X] + h\sigma(X)$ defined for square-integrable random variables X, where h is a non-negative constant, and $E[X]$ and $\sigma(X)$ are, respectively, the expectation and standard deviation of X.

Wang et al. [6] showed that if a premium principle π defined for nonnegative random variables satisfies certain desirable conditions, then π must be of the form

S. Saito (✉)
Faculty of Arts and Science, Kyushu University, 744, Motooka, Nishi-ku, Fukuoka 819-0395, Japan
e-mail: ssaito@artsci.kyushu-u.ac.jp

R. Nishii et al. (eds.), *A Mathematical Approach to Research Problems of Science and Technology*, Mathematics for Industry 5,
DOI: 10.1007/978-4-431-55060-0_16,

$\pi(X) = \int_0^\infty g\big(P(X > x)\big)\,dx$, where the function $g\colon [0,1] \to [0,1]$, called a *distortion*, is an increasing and concave function with $g(0) = 0$ and $g(1) = 1$; compare the equation with the formula $E[X] = \int_0^\infty P(X > x)\,dx$ (see Proposition 3). Wang [3] gives a list of distortions g that includes $g(x) = x^{1/(1+h)}$ and $g(x) = (1-e^{-hx})/(1-e^{-h})$ with $h > 0$. He then proposed in [4] another distortion $g(x) = \Phi\big(\Phi^{-1}(x) + h\big)$ with $h > 0$, now known as the *Wang transform*, where Φ is the cumulative distribution function of the standard normal distribution. See Wang [5] for a discussion of the Wang transform from an economic perspective.

This article focuses solely on the premium principle based on the Wang transform, aimed at mathematically minded people with no knowledge about the Wang transform. Section 2 gives the definition of the Wang transform; Section 3 provides a few examples; Section 4 describes some properties that can be summarized as being a coherent risk measure. Although we occasionally use results in measure theory, a reader unfamiliar with the theory should also be able to understand most of the contents without much difficulty.

2 Wang Transform

For a random variable X, we define its *cumulative distribution function* $F_X\colon \mathbb{R} \to [0,1]$ by $F_X(x) = P(X \le x)$. We sometimes conveniently extend F_X to $[-\infty,\infty] = \mathbb{R} \cup \{\pm\infty\}$ by setting $F_X(-\infty) = 0$ and $F_X(\infty) = 1$.

Let Φ denote the cumulative distribution function of the standard normal distribution. For $h \in \mathbb{R}$, we define an increasing homeomorphism $g_h\colon [0,1] \to [0,1]$ by

$$g_h(x) = \Phi\big(\Phi^{-1}(x) - h\big),$$

with the understanding that $\infty - h = \infty$ and $-\infty - h = -\infty$ (Fig. 1).

Lemma 1 *1. We have $g_{h_1+h_2} = g_{h_1} \circ g_{h_2}$ and $g_0 = \mathrm{id}_{[0,1]}$; therefore, $g_h^{-1} = g_{-h}$.*
2. We have $1 - g_h(x) = g_{-h}(1-x)$.

Proof 1. Obvious from the definition of g_h.
2. We have

$$\begin{aligned} 1 - g_h(x) &= 1 - \Phi\big(\Phi^{-1}(x) - h\big) = \Phi\big(-\Phi^{-1}(x) + h\big) = \Phi\big(\Phi^{-1}(1-x) + h\big) \\ &= g_{-h}(1-x). \end{aligned}$$

□

Definition 1 Let $h \in \mathbb{R}$ be a constant. The *Wang transform* of a random variable X is a random variable $W_{X,h}$ whose cumulative distribution function is given by

$$F_{W_{X,h}}(x) = g_h\big(F_X(x)\big).$$

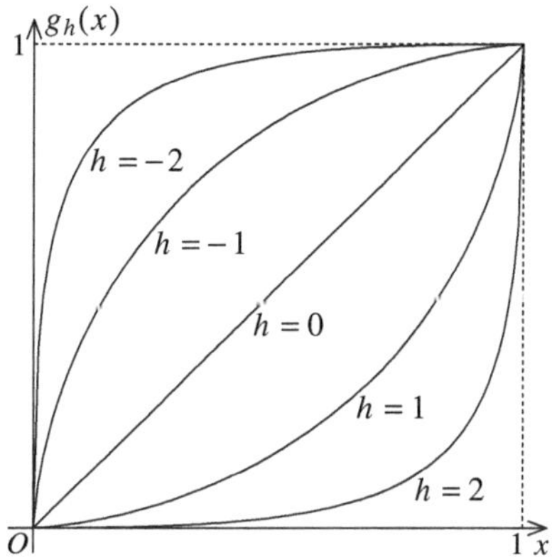

Fig. 1 Graphs of g_h for various values of h

We write $\pi(X, h) = E[W_{X,h}]$ if it exists in $[-\infty, \infty]$.

Since the definition above specifies only the distribution of $W_{X,h}$, we shall not be concerned with its correlation to other random variables.

Remark 1 We view $\pi(X, h)$ as the premium of the insurance risk modeled by the random variable X. In practice, the constant h is chosen to be non-negative because in this case we have $F_{W_{X,h}}(x) \leq F_X(x)$ and so $\pi(X, h) \geq E[X]$, which is required to ensure that the insurer does not lose money on average.

For random variables X and Y, we write $X \overset{d}{=} Y$ to mean that X and Y have the same distribution; i.e., $F_X(x) = F_Y(x)$ for all $x \in \mathbb{R}$. Definition 1 and Lemma 1 immediately tell us that we have $Y \overset{d}{=} W_{X,h}$ if and only if $X \overset{d}{=} W_{Y,-h}$.

The *support* of a random variable X, denoted by supp$X$, is the smallest closed subset C of $\mathbb{R}$ for which $P(X \in C) = 1$. Throughout what follows, equality and inequality between random variables will always mean almost sure equality and almost sure inequality, respectively.

Proposition 1 *We have* $\mathrm{supp} W_{X,h} = \mathrm{supp} X$. *In particular, if* $X \geq 0$, *then* $W_{X,h} \geq 0$ *and* $\pi(X, h) \geq 0$; *if* $a \leq X \leq b$, *then* $a \leq W_{X,h} \leq b$ *and* $a \leq \pi(X, h) \leq b$.

Proof It suffices to show that $\mathrm{supp} W_{X,h} \subset \mathrm{supp} X$, because it will imply that $\mathrm{supp} X = \mathrm{supp} W_{W_{X,h},-h} \subset \mathrm{supp} W_{X,h}$. Let $x \in \mathbb{R} \setminus \mathrm{supp} X$. Then we may choose real numbers p and q with $p < x < q$ so that $F_X(p) = F_X(q)$. This gives us $F_{W_{X,h}}(p) = F_{W_{X,h}}(q)$, which implies that $x \in \mathbb{R} \setminus \mathrm{supp} W_{X,h}$. □

Proposition 1 and the observation in Remark 1 imply that if $h \geq 0$, then we can think of $W_{X,h}$ as a random variable obtained from X by placing more emphasis on larger values in its support.

3 Examples

This section gives examples of probability distributions whose Wang transform and/or its expectation can be computed analytically.

3.1 Discrete Distributions

Example 1 Suppose that X is a constant c. Then $F_X(x) = 0$ for $x < c$ and $F_X(x) = 1$ for $x \geq c$; therefore, $F_{W_{X,h}}(x) = 0$ for $x < c$ and $F_{W_{X,h}}(x) = 1$ for $x \geq c$. It follows that $W_{X,h} = c$ and $\pi(X, h) = c$. These results also easily follow from Proposition 1.

Example 2 Suppose that X is a discrete random variable on a finite set $\{x_1, \ldots, x_n\}$, where $x_1 < \cdots < x_n$, with $P(X = x_j) = p_j$ for $j = 1, \ldots, n$. For ease of notation, set $x_0 = -\infty$, $x_{n+1} = \infty$, and $q_j = \sum_{i=1}^{j} p_i$ for $j = 0, \ldots, n$. Then for $j = 0, \ldots, n$, if $x_j \leq x < x_{j+1}$, we have $F_X(x) = q_j$ and so $F_{W_{X,h}}(x) = g_h(q_j)$. This means that $W_{X,h}$ is a discrete random variable on $\{x_1, \ldots, x_n\}$ with $P(W_{X,h} = x_j) = g_h(q_j) - g_h(q_{j-1})$ for $j = 1, \ldots, n$. We therefore have

$$\pi(X, h) = \sum_{j=1}^{n} x_j \big(g_h(q_j) - g_h(q_{j-1})\big).$$

Although very few probability distributions allow their Wang transforms to be computed analytically, Example 2 provides us with a Monte Carlo method for computing $\pi(X, h)$ numerically if sampling from the distribution of X is possible.

3.2 Normal Distribution

For $\mu \in \mathbb{R}$ and $\sigma > 0$, we write $N(\mu, \sigma^2)$ for the normal distribution with mean μ and variance σ^2.

Example 3 Suppose that X has the normal distribution $N(\mu, \sigma^2)$. Then $F_X(x) = \Phi\big((x - \mu)/\sigma\big)$, and so

$$F_{W_{X,h}}(x) = \Phi\left(\frac{x - \mu}{\sigma} - h\right) = \Phi\left(\frac{x - (\mu + h\sigma)}{\sigma}\right).$$

It follows that $W_{X,h}$ has the normal distribution $N(\mu + h\sigma, \sigma^2)$, which yields $\pi(X, h) = \mu + h\sigma = E[X] + h\sigma(X)$.

3.3 Lognormal Distribution

A random variable X has the *lognormal distribution* $LN(\mu, \sigma^2)$ if X is positive and $\log X$ has the normal distribution $N(\mu, \sigma^2)$. Since the lognormal distribution is closely connected to the normal distribution, the following proposition allows its Wang transform to be computed analytically. For a random variable X, we write $F_X(x-) = \lim_{x' \nearrow x} F_X(x') = P(X < x)$. The continuity of g_h implies that $F_{W_{X,h}}(x-) = g_h\big(F_X(x-)\big)$.

Proposition 2 *If* $\psi\colon \mathbb{R} \to \mathbb{R}$ *is increasing, then* $W_{\psi(X),h} \overset{d}{=} \psi(W_{X,h})$*; in particular, we have* $\pi\big(\psi(X), h\big) = E[\psi(W_{X,h})]$ *if they exist.*

Proof Given any $y \in \mathbb{R}$, we need to show that $F_{W_{\psi(X),h}}(y) = F_{\psi(W_{X,h})}(y)$. Set $x = \sup \psi^{-1}\big((-\infty, y]\big) \in [-\infty, \infty]$. We leave the easier cases $x = \pm\infty$ to the reader and hereafter assume that $x \in \mathbb{R}$.

The set $\psi^{-1}\big((-\infty, y]\big)$ is then either $(-\infty, x]$ or $(-\infty, x)$. In the former case, we have

$$F_{\psi(Y)}(y) = P\big(\psi(Y) \le y\big) = P(Y \le x) = F_Y(x)$$

for any random variable Y, from which it follows that

$$F_{W_{\psi(X),h}}(y) = g_h\big(F_{\psi(X)}(y)\big) = g_h\big(F_X(x)\big) = F_{W_{X,h}}(x) = F_{\psi(W_{X,h})}(y).$$

In the latter case, we have

$$F_{\psi(Y)}(y) = P\big(\psi(Y) \le y\big) = P(Y < x) = F_Y(x-)$$

for any random variable Y, from which it follows that

$$F_{W_{\psi(X),h}}(y) = g_h\big(F_{\psi(X)}(y)\big) = g_h\big(F_X(x-)\big) = F_{W_{X,h}}(x-) = F_{\psi(W_{X,h})}(y).$$

□

Corollary 1 *If a and b are constants and $a \ge 0$, then* $W_{aX+b,h} \overset{d}{=} aW_{X,h} + b$*; in particular, we have* $\pi(aX + b, h) = a\pi(X, h) + b$ *if* $\pi(X, h)$ *exists.*

Proof Apply Proposition 2 to the function $\psi(x) = ax+b$ to get $W_{aX+b,h} \overset{d}{=} aW_{X,h}+b$. It follows that

$$\pi(aX + b, h) = E[aW_{X,h} + b] = aE[W_{X,h}] + b = a\pi(X, h) + b.$$

□

If Y has the normal distribution $N(\mu, \sigma^2)$, then its moment-generating function, defined by $M_Y(t) = E[\exp(tY)]$, is given by $M_Y(t) = \exp(\mu t + \sigma^2 t^2/2)$. Therefore, if X has the lognormal distribution $LN(\mu, \sigma^2)$, then

$$E[X] = E[\exp(\log X)] = M_{\log X}(1) = \exp\left(\mu + \frac{\sigma^2}{2}\right).$$

Example 4 Suppose that X has the lognormal distribution $LN(\mu, \sigma^2)$. Then Proposition 2 shows that

$$W_{X,h} = W_{\exp(\log X),h} \stackrel{d}{=} \exp(W_{\log X,h}).$$

Since $\log X$ has the normal distribution $N(\mu, \sigma^2)$, Example 3 implies that $W_{\log X,h}$ has the normal distribution $N(\mu + h\sigma, \sigma^2)$. It follows that $W_{X,h}$ has the lognormal distribution $LN(\mu + h\sigma, \sigma^2)$ and so $\pi(X, h) = \exp(\mu + h\sigma + \sigma^2/2) = e^{h\sigma} E[X]$.

3.4 Uniform Distribution

It appears to be little known that $\pi(X, h)$ can be computed analytically when X is uniformly distributed on an interval. The computation is based on an interesting integration formula (Lemma 2) concerning the function Φ, which the author found in Owen [2]. Let $\varphi\colon \mathbb{R} \to \mathbb{R}$ denote the probability density function of the standard normal distribution:

$$\varphi(x) = \frac{1}{\sqrt{2\pi}} \exp\left(-\frac{x^2}{2}\right).$$

Lemma 2 *For $a, b \in \mathbb{R}$, we have*

$$\int_{-\infty}^{\infty} \Phi(a + bx)\varphi(x)\,\mathrm{d}x = \Phi\left(\frac{a}{\sqrt{1 + b^2}}\right).$$

Proof This proof was communicated by Tomoyuki Shirai. Let X and Y be independent random variables, both having the standard normal distribution. Then

$$\int_{-\infty}^{\infty} \Phi(a + bx)\varphi(x)\,\mathrm{d}x = \int_{-\infty}^{\infty} P(Y \le a + bx)\varphi(x)\,\mathrm{d}x = P(Y \le a + bX)$$
$$= P(Y - bX \le a) = \Phi\left(\frac{a}{\sqrt{1 + b^2}}\right),$$

where the last equality follows from the fact that $Y - bX$ has the normal distribution $N(0, 1 + b^2)$. □

Example 5 Suppose that X has a uniform distribution on the interval $[a, b]$. Let Y be a random variable having the standard normal distribution. Then $\Phi(Y)$has a uniform distribution on the unit interval [0, 1], and so $X \overset{d}{=} a + (b-a)\Phi(Y)$. Since $W_{Y,h}$ has the normal distribution $N(h, 1)$ by Example 3, we have

$$\pi\big(\Phi(Y), h\big) = E[\Phi(W_{Y,h})] = \int_{-\infty}^{\infty} \Phi(x)\varphi(x-h)\,\mathrm{d}x = \int_{-\infty}^{\infty} \Phi(x+h)\varphi(x)\,\mathrm{d}x$$
$$= \Phi\left(\frac{h}{\sqrt{2}}\right)$$

by Proposition 2 and Lemma 2. It follows from Corollary 1 that

$$\pi(X, h) = \pi\big(a + (b-a)\Phi(Y), h\big) = a + (b-a)\pi\big(\Phi(Y), h\big)$$
$$= a + (b-a)\Phi\left(\frac{h}{\sqrt{2}}\right).$$

Since $E[X] = (a+b)/2$ and $\sigma(X) = (b-a)/2\sqrt{3}$, we have

$$\pi(X, h) = E[X] + 2\sqrt{3}\left(\Phi\left(\frac{h}{\sqrt{2}}\right) - \frac{1}{2}\right)\sigma(X).$$

4 Properties

Important properties of the premium principle $X \mapsto \pi(X, h)$ can be summarized as being a *coherent risk measure*. Let $\mathscr{L}^\infty$ denote the linear space of bounded random variables.

Definition 2 A *coherent risk measure* is a map $\pi\colon \mathscr{L}^\infty \to \mathbb{R}$ with the following properties:

1. *monotonicity*: if $X \le Y$, then $\pi(X) \le \pi(Y)$;
2. *translation invariance*: if $c \in \mathbb{R}$ is a constant, then $\pi(X + c) = \pi(X) + c$;
3. *positive homogeneity*: if $c \ge 0$ is a constant, then $\pi(cX) = c\pi(X)$;
4. *subadditivity*: we have $\pi(X + Y) \le \pi(X) + \pi(Y)$.

Remark 2 Monotonicity means that if an insurance policy Y always pays more than an insurance policy X, then Y should be priced higher. Translation invariance means that the combination of an insurance policy X and a fixed amount c of payment by the insurer should be priced at the price of X plus c. Positive homogeneity means that if an insurance policy pays c times as much as an insurance policy X, then its price should be the price of X multiplied by c. Subadditivity means that it should be more reasonable to buy two insurance policies X and Y together at the price of $\pi(X + Y)$ than to buy them separately at the prices of $\pi(X)$ and $\pi(Y)$.

If X is a bounded random variable, then $W_{X,h}$ is also bounded by Proposition 1, and so $\pi(X, h)$ is defined in $\mathbb{R}$. If we take h to be non-negative, then the map $X \mapsto \pi(X, h)$ restricted to $\mathscr{L}^\infty$ turns out to be a coherent risk measure:

Theorem 1 *If $h \geq 0$, then the map $\pi(\cdot, h)\colon \mathscr{L}^\infty \to \mathbb{R}$ is a coherent risk measure.*

The rest of this section will be devoted to the proof of Theorem 1. Since translation invariance and positive homogeneity readily follow from Corollary 1, we shall verify monotonicity (Sect. 4.1) and subadditivity (Sect. 4.2).

4.1 Monotonicity

For a random variable X, we define its *complementary cumulative distribution function* $S_X\colon \mathbb{R} \to [0, 1]$ by $S_X(x) = 1 - F_X(x) = P(X > x)$. Then we have

$$S_{W_{X,h}}(x) = 1 - F_{W_{X,h}}(x) = 1 - g_h\big(F_X(x)\big) = g_{-h}\big(1 - F_X(x)\big) = g_{-h}\big(S_X(x)\big)$$

by Lemma 1.

Proposition 3 *If $X \geq 0$, then*

$$\pi(X, h) = \int_0^\infty g_{-h}\big(S_X(x)\big)\,\mathrm{d}x \in [0, \infty].$$

Proof For any non-negative random variable Y, Fubini's theorem for non-negative functions shows that

$$E[Y] = E\bigg[\int_0^Y \mathrm{d}x\bigg] = E\bigg[\int_0^\infty 1_{(0,Y)}(x)\,\mathrm{d}x\bigg] = \int_0^\infty E[1_{(0,Y)}(x)]\,\mathrm{d}x$$
$$= \int_0^\infty P(Y > x)\,\mathrm{d}x = \int_0^\infty S_Y(x)\,\mathrm{d}x.$$

Since $W_{X,h} \geq 0$ by Proposition 1, it follows that

$$\pi(X, h) = E[W_{X,h}] = \int_0^\infty S_{W_{X,h}}(x)\,\mathrm{d}x = \int_0^\infty g_{-h}\big(S_X(x)\big)\,\mathrm{d}x.$$

□

Proposition 4 (Monotonicity) *If $X, Y \in \mathscr{L}^\infty$ and $X \leq Y$, then $\pi(X, h) \leq \pi(Y, h)$.*

Proof By adding a sufficiently large constant to X and Y if necessary, we may assume that $X, Y \geq 0$. Then the claim immediately follows from Proposition 3 because g_{-h} is increasing and $S_X(x) \leq S_Y(x)$ for all $x \in \mathbb{R}$. □

4.2 Subadditivity

Lemma 3 *If $0 \leq X_1 \leq X_2 \leq \cdots \to X$, then $0 \leq \pi(X_1, h) \leq \pi(X_2, h) \leq \cdots \to \pi(X, h)$.*

Proof For each $x \in \mathbb{R}$, we have $0 \leq S_{X_1}(x) \leq S_{X_2}(x) \leq \cdots \to S_X(x)$ by the monotone convergence theorem (for measures). Therefore, the conclusion follows from Proposition 3 and the monotone convergence theorem. □

Lemma 4 *If $h \geq 0$, then the function g_h is convex.*

Proof We have

$$g_h'(x) = \frac{\varphi\big(\Phi^{-1}(x) - h\big)}{\varphi\big(\Phi^{-1}(x)\big)} = \exp\left(h\Phi^{-1}(x) - \frac{h^2}{2}\right) \geq 0,$$
$$g_h''(x) = \frac{hg_h'(x)}{\varphi\big(\Phi^{-1}(x)\big)} \geq 0,$$

and the lemma follows. □

A random variable is *simple* if it can be expressed as $\sum_{j=1}^n x_j 1_{A_j}$, where $x_1, \dots, x_n$ are constants and $A_1, \dots, A_n$ are events.

Proposition 5 (Subadditivity) *If $h \geq 0$ and $X, Y \in \mathscr{L}^\infty$, then*

$$\pi(X + Y, h) \leq \pi(X, h) + \pi(Y, h).$$

Proof By adding a sufficiently large constant to X and Y if necessary, we may assume that $X, Y \geq 0$. By Lemma 3, we may assume that X and Y are simple. We may write $X = \sum_{j=1}^n x_j 1_{A_j}$ and $Y = \sum_{j=1}^n y_j 1_{A_j}$, where $x_1, \dots, x_n, y_1, \dots, y_n$ are non-negative constants and $A_1, \dots, A_n$ are disjoint events whose union is the whole sample space. Set $p_j = P(A_j)$ for $j = 1, \dots, n$. We assume without loss of generality that our underlying probability space is the unit interval $[0, 1]$ equipped with the Lebesgue measure, for the technical reason that we will wish to divide given events into smaller pieces.

We first prove the required inequality when $p_1, \dots, p_n$ are all rational. Then by dividing the events $A_1, \dots, A_n$ further if necessary, we may assume that $p_1 = \cdots =$

$p_n = 1/n$. Let $\mathfrak{S}_n$ denote the symmetric group on the set $\{1, \ldots, n\}$. If $\sigma_0 \in \mathfrak{S}_n$ is such that $x_{\sigma_0(1)} \leq \cdots \leq x_{\sigma_0(n)}$, then Example 2 shows that

$$\pi(X, h) = \sum_{j=1}^{n} x_{\sigma_0(j)} \left(g_h\left(\frac{j}{n}\right) - g_h\left(\frac{j-1}{n}\right) \right).$$

Since g_h is convex by Lemma 4, the number $g_h(j/n) - g_h\big((j-1)/n\big)$ increases as j increases. We therefore have

$$\pi(X, h) = \max_{\sigma \in \mathfrak{S}_n} \sum_{j=1}^{n} x_{\sigma(j)} \left(g_h\left(\frac{j}{n}\right) - g_h\left(\frac{j-1}{n}\right) \right).$$

Since we similarly have

$$\pi(Y, h) = \max_{\sigma \in \mathfrak{S}_n} \sum_{j=1}^{n} y_{\sigma(j)} \left(g_h\left(\frac{j}{n}\right) - g_h\left(\frac{j-1}{n}\right) \right),$$

$$\pi(X + Y, h) = \max_{\sigma \in \mathfrak{S}_n} \sum_{j=1}^{n} (x_{\sigma(j)} + y_{\sigma(j)}) \left(g_h\left(\frac{j}{n}\right) - g_h\left(\frac{j-1}{n}\right) \right),$$

we conclude that $\pi(X + Y, h) \leq \pi(X, h) + \pi(Y, h)$.

We now turn to the general case where $p_1, \ldots, p_n$ are not necessarily rational. Let $\varepsilon > 0$ be arbitrary. Choose $M > 0$ larger than all of $x_1, \ldots, x_n, y_1, \ldots, y_n$. Since $g_{-h}\colon [0, 1] \to [0, 1]$ is uniformly continuous, we may take $\delta > 0$ so that $|g_{-h}(s) - g_{-h}(t)| < \varepsilon/M$ whenever $s, t \in [0, 1]$ and $|s - t| < \delta$. We may choose disjoint events $B_1, \ldots, B_n$ of rational measure whose union is $[0, 1]$ so that, setting $X' = \sum_{j=1}^{n} x_j 1_{B_j}$ and $Y' = \sum_{j=1}^{n} y_j 1_{B_j}$, we have $P(X = X' \text{ and } Y = Y') > 1 - \delta$. Then $|S_X(x) - S_{X'}(x)| < \delta$ for all $x \in \mathbb{R}$, and so by Proposition 3 we have

$$\begin{aligned}
|\pi(X, h) - \pi(X', h)| &= \left| \int_0^\infty g_{-h}\big(S_X(x)\big)\, \mathrm{d}x - \int_0^\infty g_{-h}\big(S_{X'}(x)\big)\, \mathrm{d}x \right| \\
&\leq \int_0^M \big| g_{-h}\big(S_X(x)\big) - g_{-h}\big(S_{X'}(x)\big) \big|\, \mathrm{d}x \\
&\leq \int_0^M \frac{\varepsilon}{M}\, \mathrm{d}x = \varepsilon.
\end{aligned}$$

Since we similarly have $|\pi(Y, h) - \pi(Y', h)| < \varepsilon$ and $|\pi(X+Y, h) - \pi(X'+Y', h)| < \varepsilon$, we obtain

$$\pi(X+Y,h) \leq \pi(X'+Y',h)+\varepsilon \leq \pi(X',h)+\pi(Y',h)+\varepsilon$$
$$\leq \pi(X,h)+\pi(Y,h)+3\varepsilon.$$

Since $\varepsilon > 0$ was arbitrary, we conclude that $\pi(X+Y,h) \leq \pi(X,h)+\pi(Y,h)$. □

5 Further Reading

For premium principles in general, see Young [7] and the references therein. For (coherent) risk measures, see Föllmer and Schied [1].

References

1. H. Föllmer, A. Schied, *Stochastic Finance*, extended edn. An Introduction in Discrete Time (Walter de Gruyter & Co., Berlin, 2011)
2. D.B. Owen, A table of normal integrals. Communications in Statistics. B. Simul. Comput. **9**(4), 389–419 (1980). doi:10.1080/03610918008812164
3. S. Wang, Premium calculation by transforming the layer premium density. ASTIN Bulletin **26**(1), 71–92 (1996). http://www.casact.org/library/astin/vol26no1/71.pdf
4. S. Wang, A class of distortion operators for pricing financial and insurance risks. J. Risk Insur. **67**(1), 15–36 (2000). doi:10.2307/253675
5. S.S. Wang, A universal framework for pricing financial and insurance risks. ASTIN Bulletin **32**(2), 213–234 (2002). http://www.casact.org/library/astin/vol32no2/213.pdf
6. S.S. Wang, V.R. Young, H.H. Panjer, Axiomatic characterization of insurance prices. Insur. Math. Econ. **21**(2), 173–183 (1997). doi:10.1016/S0167-6687(97)00031-0
7. V.R. Young, *Premium Principles*, ed. by J. Teugels, B. Sundt (eds.) Encyclopedia of Actuarial Science. (Wiley, New York, 2004), pp. 1322–1331. http://www.wiley.com/legacy/wileychi/eoas/pdfs/TAP027-.pdf

[illegible]

Further Reading

[illegible] ... see [illegible] ... see Föllmer and Schied [illegible]

[illegible]

[illegible]

Stochastic Process Models

Hiroki Masuda

Abstract A stochastic process model describes how an objective "randomly" varies over time and is typically referred to as an infinite-dimensional random variable $X = X(\omega) = \{X_t(\omega)\}_{t \in T}$ whose value is either a continuous or a càdlàg (right-continuous with left-hand limits) function of $t \in T \subset \mathbb{R}_+$. The probabilistic structure of X can be wonderfully rich, ranging from a piece-wise constant type describing a low-frequency state change to a very rapidly varying type for which we cannot define $\int f \mathrm{d}X$ pathwise as the Riemann-Stieltjes integral even for a smooth f; typical examples are a compound-Poisson process and a Wiener process, respectively. Examples of application fields include signal processing (detection, estimation, etc.), population dynamics, finance, hydrology, radiophysics, and turbulence.

Keywords Asymptotic statistics · Itô calculus · Lévy process, Stochastic differential equation · Stochastic process.

1 Lévy Process

Let $\mathscr{L}(\zeta)$ denote the distribution of a random variable ζ. We say that a real-valued stochastic process $X = (X_t)_{t \in \mathbb{R}_+}$ is a *Lévy Process* if the properties (L1) and (L2) hold true:

(L1) For any finite points $0 = t_0 < t_1 < \cdots < t_n$, the increments

$$X_{t_1} - X_{t_0},\ X_{t_2} - X_{t_1}, \ldots,\ X_{t_n} - X_{t_{n-1}}$$

are independent, and for each j,

H. Masuda (✉)
Institute of Mathematics for Industry, Kyushu University, 744, Motooka, Fukuoka, Nishiku 819-0395, Japan
e-mail: hiroki@imi.kyushu-u.ac.jp

R. Nishii et al. (eds.), *A Mathematical Approach to Research Problems of Science and Technology*, Mathematics for Industry 5,
DOI: 10.1007/978-4-431-55060-0_17,

$$\mathscr{L}(X_{t_j} - X_{t_{j-1}}) = \mathscr{L}(X_{t_j - t_{j-1}}).$$

(L2) For each $t \in \mathbb{R}_+$, $X_s \xrightarrow{p} X_t$ as $s \to t$.

Here $\xrightarrow{p}$ stands for the convergence in probability. The initial variable X_0 can be any fixed point, but here we will set $X_0 = 0$. Given any Lévy Process X, $t > 0$, and $n \in \mathbb{N}$, we can always write X_t as

$$X_t = \sum_{j=1}^{n} (X_{jt/n} - X_{(j-1)t/n}), \tag{1}$$

where, according to (L1), the summands $X_{jt/n} - X_{(j-1)t/n}$ form an array of i.i.d. random variables. Expression (1) suggests that X is a natural continuous-time analogue of the discrete-time random walk and also that $\mathscr{L}(X_t)$ is infinitely divisible for each t. In fact, there is one-to-one correspondence between Lévy Processes and infinitely divisible distributions: we can associate any Lévy Process X with an infinitely divisible distribution μ such that $\mu = \mathscr{L}(X_1)$, and conversely, for any infinitely divisible distribution μ there is a corresponding Lévy Process X such that $\mathscr{L}(X_1) = \mu$. The class of infinitely divisible distributions is reasonably large: to mention just a few, gamma, inverse-Gaussian, log-normal, normal, stable, hyperbolic, Student-t, Pareto, Meixner, logistic, negative binomial, geometric, F, Gumbel, and Weibull. Because an i.i.d. sequence can be used to construct a time series model, a Lévy Process can be a building block for constructing a much wider class of stochastic processes.

In view of the general stochastic process theory (e.g., [18], Chap. I), we may always suppose that each sample path $t \mapsto X_t(\omega)$ is càdlàg (right-continuous with left-hand limits). The convergence in probability is metrizable by several equivalent metrics, and the continuity in probability (L2) means that X_s converges to X_t for any t under one of them, say $E\{1 \wedge |X_t - X_s|\} \to 0$ as $s \to t$ for each $t \in \mathbb{R}_+$, where E denotes the expectation operator associated with the underlying probability P. More intuitively, (L2) means that a sample path of X has no prearranged jump time point; $P(\Delta X_t \neq 0) = 0$ for each $t \geq 0$, where

$$\Delta X_t := X_t - \lim_{\varepsilon \downarrow 0} X_{t-\varepsilon}$$

denotes the (directed) jump size of X at time t. For example, the process

$$X_t = \sum_{j=1}^{[t]} \xi_j$$

with i.i.d. random variables $\xi_1, \xi_2, \ldots$ is not a Lévy Process, because it has fixed points of discontinuity.

Although the definition of a Lévy process looks pretty simple, unbelievably many inherent properties follow from it. In particular, the *Lévy-Itô decomposition* of a sample path says that any Lévy Process admits the pathwise representation

$$X_t = bt + \sqrt{c}w_t + J_t, \tag{2}$$

where the ingredients are given as follows.

- $b \in \mathbb{R}$ and $c \geq 0$ are constants.
- w is a *standard Wiener process*, a Lévy Process having continuous but everywhere non-differentiable sample paths and normally distributed increments: $\mathscr{L}(w_t - w_s) = N(0, |t-s|)$ for each $s, t \in \mathbb{R}_+$.
- J is a *pure-jump Lévy Process*, a Lévy Process evolving only by jumps, which is independent of w and is characterized by the *Lévy measure* $\nu(\mathrm{d}z)$.

A Lévy measure is a σ-finite measure on the Borel space $(\mathbb{R}, \mathscr{B}(\mathbb{R}))$ such that

$$\nu(\{0\}) = 0 \quad \text{and} \quad \int_{\mathbb{R}} (1 \wedge |z|^2)\nu(\mathrm{d}z) < \infty. \tag{3}$$

For each Borel set $A \subset \mathbb{R}\backslash\{0\}$, the quantity $\nu(A)$ represents the occurrence frequency of jumps of size in A over the unit time interval $[0, 1]$. A sample path $t \mapsto J_t$ may have infinitely many "small" jumps over any nonempty interval; then we have $\nu(B) = \infty$ for any open neighborhood B of the origin. If $\int_{|z|\leq 1} |z|\nu(\mathrm{d}z) < \infty$, then J may take the form

$$J_t = \sum_{0<s\leq t} \Delta X_s \tag{4}$$

as soon as the sum is convergent a.s. (with probability 1). Although the presence of "too many" small jumps makes the sum fail to converge, we can formulate J even for such cases by means of suitable centering of small jumps. Condition (3) puts an upper limit on the occurrence frequency of small jumps; that is, we have to impose that $\int_{|z|<\varepsilon} |z|^2\nu(\mathrm{d}z) < \infty$ for any $\varepsilon > 0$.

The distribution $\mathscr{L}(X)$ can be completely characterized in terms of the non-random *generating triplet*

$$(b, c, \nu(\mathrm{d}z)).$$

In particular, the characteristic function of $\mathscr{L}(X_t)$ can be given by the so-called *Lévy-Khintchine representation*:

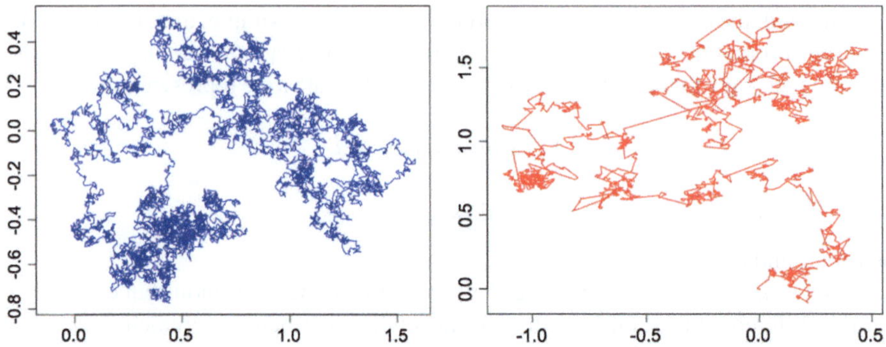

Fig. 1 Simulated sample paths of a Wiener process (*left*) and a normal inverse Gaussian Lévy Process (*right* for clarity, we connected the points by *lines*)

$$\int_{\mathbb{R}} e^{\mathrm{iux}} P^{X_t}(\mathrm{d}x)$$
$$= \exp\left[t \left\{ \mathrm{iub} - \frac{1}{2} cu^2 + \int_{\mathbb{R}} \left(e^{\mathrm{iuz}} - 1 - \mathrm{iuz}\mathbf{1}_U(z) \right) \nu(\mathrm{d}z) \right\} \right], \quad u \in \mathbb{R},$$

where $U := \{z \in \mathbb{R} : |z| \leq 1\}$ and P^{X_t} stands for the distribution of X_t. This formula can be deduced from the Lévy-Itô decomposition (2), but another proof which does not make use of (2) is possible.

There are several basic references concerning general Lévy Processes. We refer to [9, 15, 18, 30] for a detailed systematic account of Lévy Processes.

Typical sample paths of a standard Wiener process and a normal-inverse Gaussian Lévy Process in the plane are shown in Fig. 1. The latter is of the pure-jump type and belongs to the class of the generalized hyperbolic Lévy Processes that will be introduced shortly hereafter.

There exist many real phenomena for which Gaussian noise is never suitable. In stochastic process modeling, incorporating a pure-jump Lévy Process instead of only using the continuous Wiener process seems to be one of the current mainstream approaches. We may often get a significant gain in model fitting, hence in prediction too, just by replacing Gaussian noise with non-Gaussian noise. There is some empirical evidence to support this. An example is use of the generalized inverse-Gaussian and the generalized hyperbolic distributions in turbulence and econometrics.

The *Generalized Inverse-Gaussian (GIG) distribution* $\mathrm{GIG}(\lambda, \delta, \gamma)$ is the distribution on the positive half line admitting the density

$$p_{\mathrm{GIG}}(x; \lambda, \delta, \gamma) = \frac{(\gamma/\delta)^{\lambda}}{2K_{\lambda}(\gamma\delta)} x^{\lambda-1} \exp\left\{ -\frac{1}{2}\left(\frac{\delta^2}{x} + \gamma^2 x \right) \right\}, \quad x > 0. \tag{5}$$

Here $K_\lambda(x)$, $\lambda \in \mathbb{R}$, denotes the modified Bessel function of the third kind with index λ, and the region of admissible values for the parameter $(\lambda, \delta, \gamma)$ is specified as follows:

$$\begin{cases} \lambda > 0 \Rightarrow \ \delta \geq 0, \ \ \gamma > 0; \\ \lambda = 0 \Rightarrow \ \delta > 0, \ \ \gamma > 0; \\ \lambda < 0 \Rightarrow \ \delta > 0, \ \ \gamma \geq 0. \end{cases} \tag{6}$$

The GIG distribution is infinitely divisible; hence, we can choose a Lévy Process X such that $\mathscr{L}(X_1) = \mathrm{GIG}(\lambda, \delta, \gamma)$: we note that $\mathscr{L}(X_t)$ for $t \neq 1$ may or may not belong to the GIG family.

The Laplace transform of the $\mathrm{GIG}(\lambda, \delta, \gamma)$-distribution is explicitly given by

$$u \mapsto \left(\frac{\gamma^2}{\gamma^2 + 2u}\right)^{\lambda/2} \frac{K_\lambda\left(\delta\sqrt{\gamma^2 + 2u}\right)}{K_\lambda(\delta\gamma)},$$

so we can derive moments of any order in closed forms if any exist. The moments are expressed in terms of $K_{\cdot}(\cdot)$; moments of any order are finite if $\gamma > 0$, while only those of order less than $-\lambda$ exist if $\gamma = 0$. Moreover, we have $E(X^{-q}) < \infty$ for every $q > 0$ as soon as $\delta > 0$. By suitable control of the parameters, we can derive several well-known positive distributions as special cases. For example:

- the *gamma* distribution for $\delta = 0, \gamma > 0, \lambda > 0$;
- the *reciprocal gamma* distribution for $\delta > 0, \gamma = 0, \lambda < 0$;
- the *inverse-Gaussian distribution* for $\delta > 0, \gamma \geq 0, \lambda = -1/2$;
- the *delta (degenerate)* distribution as the limit for $\delta, \gamma \to \infty$ under the condition $\delta/\gamma \to c \in (0, \infty)$ (then we can prove that $p_{\mathrm{GIG}}(x; \lambda, \delta, \gamma) \to 0$ for $x \neq c$ while $p_{\mathrm{GIG}}(x; \lambda, \delta, \gamma) \to \infty$ for $x = c$).

Moreover, other distributions such as Weibull and log-normal can be derived through appropriate transformations such as power transformation.

Given constants μ, β, and random variables $\sigma \sim \mathrm{GIG}(\lambda, \delta, \gamma)$ and $\eta \sim N(0, 1)$ independent of σ, we define the normal variance-mean mixture of σ as the random variable

$$Y := \mu + \beta\sigma + \sqrt{\sigma}\eta. \tag{7}$$

Then we call $\mathscr{L}(Y)$ the *Generalized Hyperbolic (GH)* distribution, usually denoted by $GH(\lambda, \alpha, \beta, \delta, \mu)$ with the reparametrization $\alpha := \sqrt{\gamma^2 + \beta^2}$. The GH distribution is infinitely divisible: hence, we can define a Lévy Process X such that $\mathscr{L}(X_1) = GH(\lambda, \alpha, \beta, \delta, \mu)$. The admissible region of the parameters is determined according to (6): $\lambda, \mu \in \mathbb{R}$, and

$$\begin{cases} \lambda > 0 \Rightarrow \ \delta \geq 0, \ \ \alpha > |\beta|, \\ \lambda = 0 \Rightarrow \ \delta > 0, \ \ \alpha > |\beta|, \\ \lambda < 0 \Rightarrow \ \delta > 0, \ \ \alpha \geq |\beta|. \end{cases} \tag{8}$$

Note that $\mathscr{L}(X_1)$ is symmetric around μ if $\beta = 0$.

It follows from (5) and (7) that the distribution $\mathscr{L}(Y)$ admits the density

$$p_{GH}(y; \lambda, \alpha, \beta, \delta, \mu) = C(\lambda, \alpha, \beta, \delta)\{h(y; \delta, \mu)\}^{\lambda-1/2} \times K_{\lambda-\frac{1}{2}}(\alpha h(y; \delta, \mu))\, e^{\beta(y-\mu)}, \quad y \in \mathbb{R},$$

where

$$h(y; \delta, \mu) := \sqrt{\delta^2 + (y-\mu)^2},$$

$$C(\lambda, \alpha, \beta, \delta) := \frac{(\alpha^2 - \beta^2)^{\lambda/2}}{\sqrt{2\pi}\alpha^{\lambda-1/2}\delta^{\lambda} K_{\lambda}(\delta\sqrt{\alpha^2 - \beta^2})}.$$

The function p_{GH} is unimodal (and so is $p_{\mathrm{GIG}}(\cdot; \lambda, \delta, \gamma)$). The parameters λ, α, β determine the tail behavior:

$$\lim_{|y|\to\infty} \frac{p_{GH}(y; \lambda, \alpha, \beta, \delta, \mu)}{|y|^{\lambda-1}\exp\{-\alpha|y| + \beta y\}} = \frac{e^{-\beta\mu}}{\sqrt{\alpha}} C(\lambda, \alpha, \beta, \delta),$$

clarifying the semi-heavy tail property of GH distributions except for the case $\alpha = |\beta|$, which corresponds to an asymmetric variant of the Cauchy distribution. As in the GIG case, suitable control of the parameters leads to special distributions such as normal, skewed Student-t, normal-inverse Gaussian, hyperbolic, and GIG.

For more details concerning the GIG and GH distributions together with their histories, we refer to [5, 8, 12, 21] as well as the references therein.

2 Itô Calculus and Stochastic Differential Equation

2.1 Semimartingale

For later exposition, let us mention here the notion of the semimartingale. A semimartingale X is an adapted process represented as the sum of a finite-variation process A and a local martingale M, both starting from the origin; we say that a function $x : \mathbb{R}_+ \to \mathbb{R}$ is of finite variation (resp. infinite variation), if for each $t \in \mathbb{R}_+$

$$\sup_{n\in\mathbb{N}} \sum_{j=1}^{2^n} |x_{jt/2^n} - x_{(j-1)t/2^n}| < \infty \quad (\text{resp.} = \infty).$$

Furthermore, the local martingale part can be decomposed into its continuous part X^c and purely discontinuous part X^d, so that

$$X = X_0 + A + X^c + X^d.$$

In the case of Lévy Process (2), we have $X^c = \sqrt{c}w$ while the representation of X^d is really determined in conjunction with A. A typical way is to divide the jump part into small- and large-jump parts with centering for the former:

$$X_t = \left(bt + \int_0^t \int_{|z|>1} z\mu(\mathrm{d}s, \mathrm{d}z) \right) + \sqrt{c}w_t + \int_0^t \int_{|z|\le 1} z\{\mu(\mathrm{d}s, \mathrm{d}z) - \nu(\mathrm{d}z)\mathrm{d}s\}, \quad (9)$$

where μ denotes the Poisson random measure associated with the Lévy measure $\nu(\mathrm{d}z)$; for each $t > 0$ and $B \in \mathscr{B}(\mathbb{R})$, the random variable $\mu([0, t], B)$ is distributed as Poisson with intensity $t\nu(B)$, with $\nu(B)$ possibly being infinity when $0 \in B$. If J takes the form (4), then (9) can be written as

$$X_t = \left(b - \int_{|z|\le 1} z\nu(\mathrm{d}s, \mathrm{d}z) \right) t + \sqrt{c}w_t + \int_0^t \int_{\mathbb{R}} z\mu(\mathrm{d}s, \mathrm{d}z),$$

with the last term on the right-hand side corresponding to J.

2.2 Itô Integral

Consider two sequences of real-valued random variables $Y_1, Y_2, \ldots$ and $\zeta_1, \zeta_2, \ldots$ where ζ_i are independent and zero mean, and let the σ-fields

$$\mathscr{F}_n := \sigma(\zeta_i;\ i \le n) = \bigvee_{i\le n} \{\zeta_i^{-1}(B);\ B \in \mathscr{B}(\mathbb{R})\}$$

represent the information about ζ_i up to time n. Assume that each Y_i is predictable, i.e. $\mathscr{F}_{i-1}$-measurable. Then, the martingale transform $Y \cdot \zeta$ is defined by

$$(Y \cdot \zeta)_n := \sum_{i=1}^n Y_i \zeta_i. \quad (10)$$

In fact, $Y \cdot \zeta$ is an $(\mathscr{F}_i)$-martingale. The martingale transform is an easily understandable general procedure of making a martingale from predictable and independent sequences, and it corresponds to the prototype for the construction of Itô's stochastic integration, which has brought about a revolution in the theory of continuous-time martingales.

Kiyosi Itô (1915–2008) introduced the concept of stochastic integration of the form

$$\int Y \mathrm{d}X,$$

where both Y and X are stochastic processes whose sample paths are of possibly unbounded variation, and then developed the theory of a random perturbation of differential equations. This paradigm is called *Itô calculus*, a term known to everybody concerned with stochastic processes. Itô calculus is nowadays a basic analysis tool in a wide range of application fields such as system control engineering, mathematical biology, finance, and econometrics. Owing to Itô's great achievement, it has become possible to deal with, for example, the limit of the Riemann sum of the suitable predictable semimartingale Y with respect to w in a rigorous manner: there exists a unique random variable $\int_0^t Y_s \mathrm{d}w_s$ such that

$$\sum_{j=1}^{n} Y_{(j-1)t/n}(w_{jt/n} - w_{(j-1)t/n}) \xrightarrow{p} \int_0^t Y_s \mathrm{d}w_s, \tag{11}$$

with the convergence holding true uniformly over any compact time interval. The limit process $\int Y \mathrm{d}w$ is called the *Itô integral* of Y with respect to w, which defines a local martingale. We note that the notation $\int Y dw$ is just symbolic, and the random variable $\int Y dw$ cannot be defined *pathwise* (*ω-wise)* as the Riemann-Stieltjes integral. In fact, let $m(j;t,n)$ denote the midpoint between jt/n and $(j-1)t/n$, and replace the predictable weight "$Y_{(j-1)t/n}$" by "$Y_{m(j;t,n)}$". Then, the corresponding limit is still well-defined as a locally uniform limit in probability, say

$$\sum_{j=1}^{n} Y_{m(j;t,n)}(w_{jt/n} - w_{(j-1)t/n}) \xrightarrow{p} \int_0^t Y_s \circ \mathrm{d}w_s.$$

However, the limit process $\int Y \circ dw$ is essentially different from that in (11), and is called the *Stratonovich integral*. The two stochastic integrals $\int Y \mathrm{d}w$ and $\int Y \circ \mathrm{d}w$ have the explicit relation

$$\int_0^t Y_s \circ \mathrm{d}w_s = \int_0^t Y_s \mathrm{d}w_s + \frac{1}{2}\langle Y^c, w\rangle_t.$$

Here the term $\langle Y^c, w\rangle$ is the continuous part of the *quadratic variation* process $[Y, w]$, which vanishes especially if Y is of finite variation. Here, given two semimartingales X and Y, we can define the quadratic variation process $[X, Y]$ by the following limit in probability (locally uniform in time):

$$[X, Y]_t := \lim_{n\to\infty} \sum_{j=1}^{n} (X_{jt/n} - X_{(j-1)t/n})(Y_{jt/n} - Y_{(j-1)t/n}). \tag{12}$$

It admits decomposition into its continuous and discontinuous parts:

$$[X, Y]_t = \langle X^c, Y^c \rangle_t + \sum_{0<s\le t} (\Delta X_t)(\Delta Y_t).$$

The continuous part $\langle X^c, Y^c \rangle$ can be more explicit according to the specific structures of X and Y; in particular, if $Y = w$, then $\langle Y^c, w^c \rangle_t = \langle w, w \rangle_t = t$. It is the concept of quadratic variation that most clearly separates stochastic calculus from classical calculus. More than half a century has passed since Itô's invention, but the device still remains a basic requisite in continuous-time modeling.

Importantly, the Itô integral (11) can be defined with respect to much more general local martingales X instead of the Wiener process w. For example, we can construct the Itô integral $\int Y_- \mathrm{d}X$ for general local martingale X (hence automatically for general semimartingales), especially for Lévy Processes with jumps; in this case, in order to make the variable suitably defined, we have to pick the left-hand limit version $Y_{t-} := \lim_{t\uparrow s} Y_t$ for the integrand process. As a result, a broad class of stochastic processes beyond Lévy Processes can be constructed; see Sect. 2.4 below.

2.3 Itô's Formula for Semimartingale

Given a right-continuous semimartingale X and a $\mathscr{C}^2$-function f, the celebrated *Itô's formula* provides us the pretty change-of-variable formula

$$\begin{aligned} f(X_t) = f(X_0) + \int_0^t f'(X_{s-})\mathrm{d}X_s + \frac{1}{2}\int_0^t f''(X_{s-})d\langle X^c \rangle_s \\ + \sum_{0<s\le t} \left\{ f(X_s) - f(X_{s-}) - f'(X_{s-})\Delta X_s \right\}, \qquad (13) \end{aligned}$$

where we simply wrote $\langle X^c \rangle = \langle X^c, X^c \rangle$. The last term on the right-hand side of (13) is absolutely convergent locally uniformly in time, and in particular it vanishes if X has continuous samples paths. Itô's formula looks somewhat like a Taylor approximation. Indeed, we make use of the Taylor expansion in the proof of (13). Nevertheless, it is an *equality* with the third- and higher-order parts vanishing and instead having some seemingly strange second-order part. Itô's formula is an indispensable tool in studying various stochastic process models.

Here is a simple illustration. Let X be a Lévy Process and $S_0 > 0$ a constant. Then, the process $S = (S_t)$ defined by

$$S_t = S_0 \exp(X_t)$$

is called the *geometric Lévy Process*, which is one of the basic examples for describing random fluctuation of a financial price process. Applying Itô's formula, we have

$$S_t = S_0 + \int_0^t S_{s-}\mathrm{d}X_s + \frac{1}{2}\int_0^t S_{s-}\mathrm{d}\langle X^c\rangle_s + \sum_{0<s\le t} S_{s-}(e^{\Delta X_s} - 1 - \Delta X_s).$$

In particular, if X is of the form (2), we have

$$S_t = S_0 + \left(b + \frac{c}{2}\right)\int_0^t S_s\mathrm{d}s + \sqrt{c}\int_0^t S_s dw_s$$
$$+ \int_0^t S_{s-}dJ_s + \sum_{0<s\le t} S_{s-}(e^{\Delta J_s} - 1 - \Delta J_s).$$

This is an extension of the classical Black-Scholes model, where $J \equiv 0$.

2.4 Stochastic Differential Equation

The Itô calculus enables us to consider a random perturbation of the ordinary integral equation

$$x_t = x_0 + \int_0^t b(x_s)\mathrm{d}s$$

for some measurable function b. Let X be a Lévy Process X of the form (2) with $b = 0$ and $c = 1$. Given suitable measurable functions c and ζ, consider the process X of the form

$$X_t = X_0 + \int_0^t b(X_s)\mathrm{d}s + \int_0^t c(X_s)\mathrm{d}w_s + \int_0^t \zeta(X_{s-})\mathrm{d}J_s,$$

where the last two terms on the right-hand side are Itô integrals. Conventionally, this integral equation is written in the differential form:

$$\mathrm{d}X_t = b(X_t)\mathrm{d}t + c(X_t)\mathrm{d}w_t + \zeta(X_{t-})\mathrm{d}J_t, \tag{14}$$

which is called the *Stochastic Differential Equation (SDE)*. Under appropriate conditions on the coefficient functions, (14) admits a unique solution process X as a functional of (X_0, w, J); such a solution is called a strong solution. We note that solutions become non-Markovian if the coefficients of (14) depend on the past of X, e.g. $b(x.)_t = \int_{t-1}^t x_s\mathrm{d}s$. Then things become much more complicated, and we can no longer make use of Markov operator theory to study X.

We refer to [14, 29] for a comprehensive account of the theory of SDE (including non-Markovian types) as well as of stochastic integration.

2.5 Long-Term Stability

We may control various characteristics of X in (14) in terms of the ingredients b, c, ζ, and $\nu(\mathrm{d}z)$, to mention just a few:

- Heaviness of the tail of $\mathscr{L}(X_t)$;
- Auto-covariance structure (short- and long-memory properties);
- Finiteness of moments such as not only $E\{g(X_t)\}$ but also $\sup_{t\in\mathbb{R}_+} E\{g(X_t)\}$ for some unbounded measurable function g;
- Ergodicity of X.

Here, the ergodicity means that there exists a unique invariant measure $\pi(\mathrm{d}x)$ to which the distribution $\mathscr{L}(X_t)$ in total variation converges as $t \to \infty$ for every initial point:

$$\lim_{t\to\infty} \sup_{A\in\mathscr{B}(\mathbb{R})} |P(X_t \in A|X_0 = x) - \pi(A)| = 0, \quad x \in \mathbb{R}.$$

We then have the weak law of large numbers

$$\frac{1}{t}\int_0^t f(X_s)\mathrm{d}s \xrightarrow{p} \int_{\mathbb{R}} f(x)\pi(\mathrm{d}x), \quad t \to \infty, \tag{15}$$

also known as the ergodic theorem, roughly meaning that a time average converges to a space average. The convergence (15) is valid for every $g \in L^1(\pi)$.

If in particular $b(x) = -\lambda x$ with c and ζ being constant, then we can write it as

$$\mathrm{d}X_t = -\lambda X_t \mathrm{d}t + \mathrm{d}Z_t$$

for some Lévy Process Z. The solution is called the *Lévy-Ornstein-Uhlenbeck (OU) process*, admitting the closed form

$$X_t = e^{-\lambda t}X_0 + \int_0^t e^{-\lambda(t-s)}\mathrm{d}Z_s, \quad t \in \mathbb{R}_+. \tag{16}$$

This is the continuous-time counterpart to the first-order autoregressive model $\zeta_n = \rho\zeta_{n-1} + \varepsilon_n$ in the time series literature. The OU process has several inherent characteristics, which cannot be shared with a general nonlinear SDE. For example, there is a simple relation between the generating triplet of Z and the invariant distribution π, the latter being necessarily selfdecomposable; see Masuda [22] as well

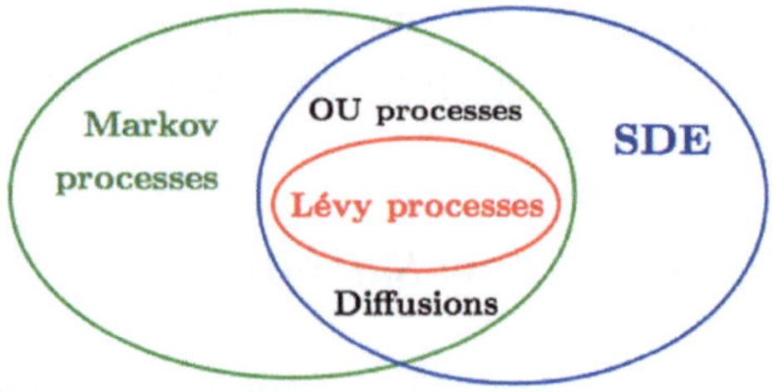

Fig. 2 Relations among Lévy Process, Markov process, and SDE models

as the references therein for details. Because of its mathematical tractability, the OU process is used in many application fields such as stochastic volatility modeling in econometrics (Barndorff-Nielsen et al. [5]) and signal estimation in diffusion leaky integrate-and-fire neuronal models (Lansky and Ditlevsen [20]).

Figure 2 shows relations among several stochastic models.

2.6 Sample Path Generation

It is important for simulation purposes to generate a sample path of $X = (X_t)_{t\in[0,1]}$ on a computer. There is a huge literature of stochastic numerics for the SDE (14). The most naive way is the Euler approximation. Suppose that we know how to (approximately) generate random numbers obeying $\mathscr{L}(J_t)$ for any $t > 0$ small enough. Then, for a sufficiently large $n \in \mathbb{N}$ we inductively generate discrete-time skeleton process $X_0, X_{1/n}, \ldots, X_{(n-1)/n}, X_1$ via the recurrence formula

$$X_{j/n} = X_{(j-1)/n} + b(X_{(j-1)/n})\frac{1}{n} + c(X_{(j-1)/n})\Delta_j^n w + \zeta(X_{(j-1)/n})\Delta_j^n J, \quad j \le n, \tag{17}$$

where and in what follows

$$\Delta_j^n \zeta := \zeta_{j/n} - \zeta_{(j-1)/n}$$

for any process ζ. Actually, there are several concrete examples of J for which an exact or approximate algorithm for generating $\mathscr{L}(J_{1/n})$-random numbers is available. As a result of (17), we get a discretized (piecewise constant) process X^n defined by

$$X_t^n := X_{[nt]/n}, \quad t \in [0, 1];$$

see Fig. 3. Under appropriate conditions, the original process X is realized as a limit of X^n under a suitable topology.

More sophisticated methods are known. For diffusion processes, the monograph [19] is a milestone. We refer to [4, 27] for numerics concerning Lévy Processes and SDE models driven by a Lévy Process.

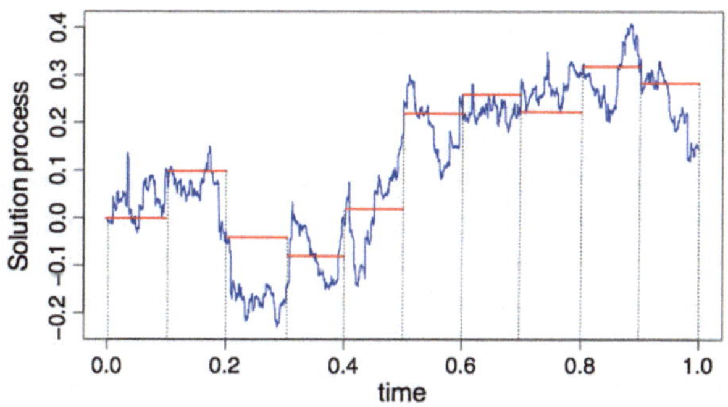

Fig. 3 Sample paths of a SDE model X (*blue*) and its discretized version X^n (red)

Finally, we note that the SDE of the form (14) may arise as a weak limit of a time series model under high-frequency sampling. For example, a diffusion approximation is valid for the famous GARCH model, which is used to model (normally daily) stock-return fluctuation; see [26]. This fact enables us to reflect good properties of the diffusion model in the limit into the original GARCH model; for example, it is easy to specify the invariant distribution for a one-dimensional diffusion process, whereas it is not easy to do so for the GARCH model.

3 Stochastic Process as a Statistical Model

Mathematical statistics is a discipline to quantify information from observed data in various fruitful ways with theoretical rationale, and then to put it to use for future decision-making. It has been penetrating deeply into a great number of research fields handling actual phenomena involving randomness, expanding its impact in an increasing number of future directions. One of the essences of mathematical statistics is distribution approximation. *Asymptotic statistics* is a collective term for analyses of the distribution behavior of statistics when the number of data increases, forming a core of mathematical statistics.

In most types of parameter estimation, we usually define an estimator $\hat{\theta}_n$ of θ, where θ stands for the parameter of interest, as the maximum point of some objective (random) function $\theta \mapsto \mathbb{M}_n(\theta)$, $\theta \in \Theta \subset \mathbb{R}^p$; for example, the penalized maximum-likelihood method, the weighted least-squares method, and the least absolute deviation method. In order to deduce the asymptotic behavior of $\hat{\theta}_n$, it is crucial to clarify that of a suitably rescaled $\mathbb{M}_n$, especially, the weak limit in the function space; see e.g. van der Varrt [33, Chap. 5] for details. When attempting to estimate a stochastic process model, we have to verify limit theorems such as the law of large numbers and the central limit theorem, building on a correct understanding of local (small-time) and global (long-term) stochastic behaviors of the underlying stochastic process. Toward that end, martingale limit theory and ergodic theory often play important roles. Concerning statistical inference for diffusion models, interested readers can consult [17, 28, 32] for an extensive review of the existing literature.

Despite its importance in application fields, a solid basis of asymptotic statistics for the SDE models with jumps such as (14) has not yet been well-established as yet in the presence of jumps. Needless to say, this is primarily because of the difficulty of handling the diversity of the driving Lévy Process. On the one hand, for a broad class of SDE with jumps the likelihood analysis has been developed when *continuous-time data* $(X_t)_{t\in[0,T]}$ is available: see Sø [31]. On the other hand, however, for the more realistic *discrete-time sampling*, many things are yet far from being well-developed; indeed, the exact maximum likelihood estimation is usually of no utility because of the lack of a closed-form transition probability, and we do not know what kind of $\mathbb{M}_n$ is universally good to use even when the class of $\mathscr{L}(J_t)$, or equivalently the class of $\nu(\mathrm{d}z)$, is limited to some extent. This area of research is now under development.

Example 1 Population Growth Dynamics

The logistic diffusion is defined by

$$\mathrm{d}X_t = r\,(1 - X_t/K)\,X_t\mathrm{d}t + X_t\sigma\mathrm{d}w_t \tag{18}$$

for some positive constants r, K, and σ. This model is a randomly perturbed version of the logistic equation (the ordinary differential equation)

$$\mathrm{d}x_t = r(1 - x_t/K)x_t\mathrm{d}t$$

by a state-dependent diffusion term "$x_t\sigma\mathrm{d}w_t$". The SDE (18) is used to describe the random dynamics of population growth, X_t denoting, say, the number of cells. A solution process is regarded as an approximate macroscopic model; it can take values in $\mathbb{R}\setminus\mathbb{N}$ too.

In this example, we can explicitly calculate the invariant distribution of X by means of an analysis of one-dimensional diffusions: if $r > \sigma^2/2$, the solution to (18) admits a chi-square invariant distribution with its degrees of freedom depending on the parameters r, K, and σ. See [3, 25] for details.

If we can estimate the parameters r, K, and σ from observed time series data, then we can also calculate the distribution of the population in which X will end up after long period. However, the random dynamical system (18) may be too simple to describe several stylized features in reality. Various extensions and generalizations of the model incorporating the features would be possible.

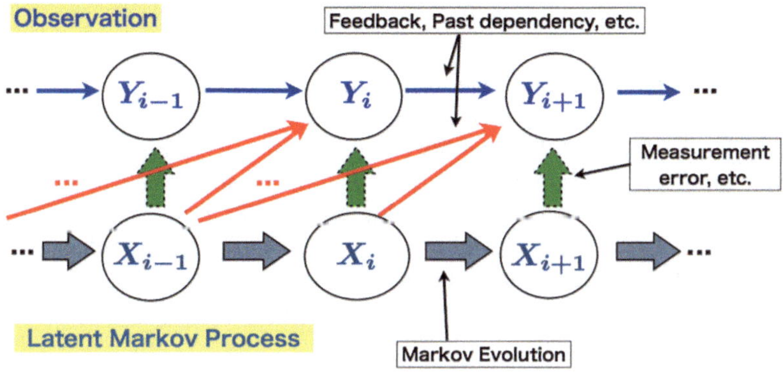

Fig. 4 Graphic illustration of HMM

Example 2 Hidden Markov Models

A *Hidden Markov model (HMM)* consists of an unobserved Markov process X and an observed process Y. The probabilistic structure is then specified in terms of:

- The transition probability of the latent Markov process;
- The conditional probability of Y given state of X.

Figure 4 shows a graphical illustration of HMM. In particular, if the model is discrete in both time and states, then the HMM $\{(X_i, Y_i); i = 1, 2, \dots\}$ can be completely characterized by the quantities

$$p^X_{i,i+1}(k, l) := P(X_{i+1} = l \mid X_i = k),$$
$$p^{Y|X}_i(y|k) := P(Y_i = y \mid X_i = k).$$

If these are parametrized by θ, a statistical problem arises: we want to estimate θ only based only on available data $(Y_i)_{i\leq n}$. If the state spaces of X and Y are not compact, then maximum-likelihood estimation is theoretically difficult and computationally heavy. See [11, 34] for a detailed account of the statistics of discrete-time HMMs.

Estimation of a continuous-time HMM $(X, Y) = \{(X_t, Y_t)\}_{t\in\mathbb{R}_+}$ is directly related to parameter estimation in a filtering problem, e.g. [1]. Instead of the maximum-likelihood method, we may resort to other practical ways to estimate a model in question. One such possibility is the method of moments discussed in [23] for a class of discretely and partially observed ergodic stochastic process models driven by a Lévy Process.

Here is an illustrative example borrowed from [23]. Assume that we observe $\{Y_{jh} : j = 0, 1 \dots .n\}$ for a given sampling step size $h > 0$, and that the latent Markov process is given by the OU process X of (16) with $\lambda > 0$, $X_0 > 0$ a.s., and $\Delta Z_t > 0$ a.s. so that we have $X_t > 0$ a.s. for each $t \in \mathbb{R}_+$. The stochastic volatility model introduced by Barndorff-Nielsen and Shephard [5, 7]) is given by

$$\mathrm{d}Y_t = (\mu + \beta X_t)\mathrm{d}t + \sqrt{X_t}\mathrm{d}w_t + \rho \mathrm{d}Z_t,$$

which describes the variation of the logarithm of a stock price; $Y = \log S$ for some stock price process S. Here the standard Wiener process w is independent of (X_0, Z). In this model, X_t describes the volatility at time t. With $\rho < 0$, this model can capture the leverage effect; that is, the volatility tends to increase after a negative shock in price. Let $y_j := Y_{jh} - Y_{(j-1)h}$ denote the log-returns. For each constants $m \in \mathbb{N}$ and $k = (k_1, \ldots, k_m) \in \mathbb{Z}_+^m$, the special nature of the OU process provides us with the closed form of the characteristic function of $\mathscr{L}(y_1, \ldots, y_m)$, so that the expressions for its kth cumulants $\kappa^{(k)} = \kappa^{(k_1,\ldots,k_m)}$ can be derived through the relation

$$\kappa^{(k)} := i^{-(k_1+\cdots+k_m)} \frac{\partial^{k_1+\cdots+k_m}}{\partial u_1^{k_1} \cdots \partial u_m^{k_m}} \log E\left\{\exp\left(i \sum_{j=1}^{m} u_j y_j\right)\right\}.$$

This enables us, under regularity conditions including the ergodicity and the stationarity of $\{(y_{j+1}, \ldots, y_{j+m})\}_{j=0}^{\infty}$, to construct an easy-to-use estimator $\hat{\theta}_n$ of the parameter $\theta_0 \in \mathbb{R}^p$ of interest, based on moment fitting such as

$$\frac{1}{n-m+1} \sum_{j=1}^{n-m+1} \prod_{l=1}^{m} y_{j+l-1}^{k_l} \approx E\left(\prod_{l=1}^{m} y_1^{k_l}\right).$$

The above type of closed-form method of moments is also applicable to the continuous-time state space model

$$\mathrm{d}Y_t = X_t \mathrm{d}t + \mathrm{d}Z_t,$$

where the OU process X describes some unseen state process, which is now not necessarily positive, and where Z represents measurement error piling up as time passes.

In either case, we can deduce the asymptotic normality of the normalized estimator: for some explicit asymptotic covariance matrix $\Sigma(\theta_0, h)$,

$$\sqrt{n}(\hat{\theta}_n - \theta_0) \xrightarrow{\mathscr{L}} N_p\left(0, \Sigma(\theta_0, h)\right).$$

To achieve this distribution approximation, the mixing property of X plays a crucial role in verification of the appropriate law of large numbers and central limit theorem, paving the way to formulation of how to construct confidence regions and model assessment devices.

Example 3 Estimation of Integrated Volatility Via Realized Multipower Variation

There is no relation between model and actual time scales. Time series data over any fixed period may be regarded as, say, a discrete-time sample from a stochastic process X over unit interval [0, 1]. According to a somewhat universal local (small-time) structure of $\mathscr{L}(X_{t+\varepsilon} - X_t)$, a high-frequency data setting allows us to estimate the integrated volatility by means of very simple statistics.

To be more specific, let us assume that a discrete-time sample $(X_{j/n})_{j=0}^{n}$ is observed from a continuous Itô semimartingale of the form

$$X_t = X_0 + \int_0^t b_s \mathrm{d}s + \int_0^t \sigma_s \mathrm{d}w_s, \tag{19}$$

where b and σ are stochastic processes satisfying mild regularity conditions. Suppose that the function σ never vanishes, and let m ($\ll n$) and $r > 0$ be given constants. The *realized Multi-Power Variation (MPV)* (with index (m, r)) is defined by the easily computable statistics

$$V_n(m, r; X) := n^{rm/2-1} \sum_{j=1}^{n-m+1} \prod_{k=1}^{m} |\Delta_{j+k-1}^n X|^r.$$

From [6], which proved the asymptotic mixed normality of the MPV for general multivariate continuous Itô semimartingales, we know that

$$V_n(m, r; X) \xrightarrow{p} \mu_r^m \int_0^1 \sigma_s^{rm} \mathrm{d}s,$$

where μ_r denotes the rth absolute moment of the standard normal distribution. Importantly, we do not need to impose any concrete structure of b and σ. Concerning (19), the *integrated volatility* over [0, 1] is the possibly random quantity

$$\int_0^1 \sigma_s^2 \mathrm{d}s.$$

To estimate this, we can make use of $V_n(m, r; X)$ for $mr = 2$, the simplest case being the realized volatility $V_n(1, 2; X)$, the sum of squared log returns:

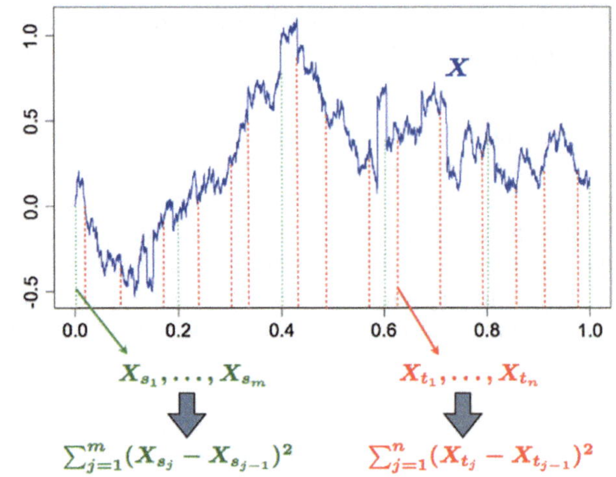

Fig. 5 Computing two time-scale realized volatilities: the accuracy of volatility estimation (20) can be improved just by finer sampling, but excessively fine sampling may lead to a heavy contamination owing to market microstructure noise

$$V_n(1, 2; X) = \sum_{j=1}^{n} (\Delta_j^n X)^2 \xrightarrow{p} \int_0^1 \sigma_s^2 \mathrm{d}s, \quad n \to \infty. \tag{20}$$

The unit-period integrated volatility appears as the key variable in the option-price formula in financial engineering. We note that the realized volatility $V_n(1, 2; X)$ is very familiar in the classical martingale theory and has been well-recognized as an estimator of the quadratic variation $[X, X]_1$ (Recall (12)): Itô calculus plays a crucial role in developing the theory of MPV.

In reality, however, using ultrahigh-frequency data, such as financial tick data (stock price data recording every change), may cause trouble in the limit theorem (20); that is, so-called market microstructure noise violates the limit. See e.g. [2, 13] for a detailed empirical analysis in this direction. In such cases, we may use the subsampling method or direct data thinning (say, using returns every 15 min). Alternatively and hopefully, we may get a more stabilized estimation and volatility prediction procedures by introducing a more sophisticated way of incorporating the effect of market microstructure noise. Moreover, when concerned with the integrated co-volatility of two return-process models, the effect of non-synchronicity of sampling times is non-negligible. Several attempts have been made so far in this direction; among others, we refer to [10] and the references therein. This area of research remains active.

In the case where X has a jump component, things become much more complicated; we refer to Jacod [16] for a detailed study of the asymptotic distribution of power variation $V_n(1, r; X)$ when X is a general Itô semimartingale with jumps (Fig. 5).

The case of $m \geq 2$, such as the bipower variation $V_n(2, 1; X)$, has the attractive feature that it can estimate the continuous part in a way somewhat (but not very much) robust to the presence of jumps. For example, suppose that X takes the form

$$X_t = X_0 + \int_0^t b_s \mathrm{d}s + \int_0^t \sigma_s \mathrm{d}w_s + J_t$$

with some pure-jump process with finite-activity jumps (only finite number of jumps can occur in each compact time interval). Then (20) becomes

$$V_n(1, 2; X) \xrightarrow{p} \int_0^1 \sigma_s^2 \mathrm{d}s + \sum_{0<s\le t} (\Delta X_s)^2, \quad n \to \infty,$$

while

$$V_n(2, 1; X) = \sum_{j=1}^{n-1} |\Delta_j^n X||\Delta_{j+1}^n X| \xrightarrow{p} \frac{2}{\pi} \int_0^1 \sigma_s^2 \mathrm{d}s, \quad n \to \infty.$$

These statistics can be used to construct a simple test procedure for the existence of jump part J. See [6, 21, 24] as well as the references therein for more information on the MPV literature.

References

1. H. Ahn, R.E. Feldman, Optimal filtering of a Gaussian signal in the presence of Lévy noise. SIAM J. Appl. Math. **60**, 359–369 (2000) (electronic)
2. Y. Aït-Sahalia, J. Yu, High frequency market microstructure noise estimates and liquidity measures. Ann. Appl. Stat. **3**, 422–457 (2009)
3. L.H.R. Alvarez, L.A. Shepp, Optimal harvesting of stochastically fluctuating populations. J. Math. Biol. **37**, 155–177 (1998)
4. S. Asmussen, J. Rosiński, Approximations of small jumps of Lévy processes with a view towards simulation. J. Appl. Probab. **38**, 482–493 (2001)
5. O.E. Barndorff-Nielsen et al., (ed.), Lévy Processes: Theory and Applications (Birkhäuser, Boston, 2001)
6. O.E. Barndorff-Nielsen, S.E. Graversen, J. Jacod, M. Podolskij, N. Shephard, *A Central Limit Theorem for Realised Power and Bipower Variations of Continuous Semimartingales. From Stochastic Calculus to Mathematical Finance* (Springer, Berlin, 2006), pp. 33–68
7. O.E. Barndorff-Nielsen, N. Shephard, Non-Gaussian Ornstein-Uhlenbeck-based models and some of their uses in financial economics. J. R. Stat. Soc. Ser. B Stat. Methodol. **63**, 167–241 (2001)
8. O.E. Barndorff-Nielsen, N. Shephard, M. Winkel, Limit theorems for multipower variation in the presence of jumps. Stoch. Process. Appl. **116**, 796–806 (2006)
9. J. Bertoin, *Lévy Processes* (Cambridge University Press, Cambridge, 1996)
10. M. Bibinger, Efficient covariance estimation for asynchronous noisy high-frequency data. Scand. J. Stat. **38**, 23–45 (2011)
11. O. Cappé, E. Moulines, T. Rydén, *Inference in Hidden Markov Models* (Springer, New York, 2005)
12. E. Eberlein, E.A. v Hammerstein, Generalized hyperbolic and inverse Gaussian distributions: limiting cases and approximation of processes, in *Seminar on Stochastic Analysis, Random Fields and Applications*, vol. IV, pp. 221–264, Progr. Probab. 58, Birkhä user, Basel (2004)

13. P.R. Hansen, A. Lunde, Realized variance and market microstructure noise. J. Bus. Econ. Stat. **24**, 127–218 (2006)
14. N. Ikeda, S. Watanabe, *Stochastic Differential Equations and Diffusion Processes*, 2nd edn. (North-Holland Publishing Co., Amsterdam; Kodansha Ltd, Tokyo, 1989)
15. K. Itô, Stochastic Processes. Lectures given at Aarhus University. Reprint of the 1969 original, ed. by with a foreword by O.E. Barndorff-Nielsen, K. Sato. (Springer-Verlag, Berlin, 2004)
16. J. Jacod, Asymptotic properties of realized power variations and related functionals of semimartingales. Stoch. Process. Appl. **118**, 517–559 (2008)
17. J. Jacod, Inference for stochastic processes. *Handbook of Financial Econometrics*, vol. 2: i, pp. 197–239, Access Online via Elsevier (2009)
18. J. Jacod, A.N. Shiryaev, *Limit Theorems for Stochastic Processes*, 2nd edn. (Springer, Berlin, 2003)
19. P.E. Kloeden, E. Platen, *Numerical Solution of Stochastic Differential Equations* (Springer, Berlin, 1992)
20. P. Lansky, S. Ditlevsen, A review of the methods for signal estimation in stochastic diffusion leaky integrate-and-fire neuronal models. Biol. Cybern. **99**, 253–262 (2008)
21. H. Masuda, Analytical properties of GIG and GH distributions (in Japanese). Proc. Inst. Statist. Math. **50**, 165–199 (2002)
22. H. Masuda, On multidimensional Ornstein-Uhlenbeck processes driven by a general Lévy process. Bernoulli **10**, 97–120 (2004)
23. H. Masuda, Classical method of moments for partially and discretely observed ergodic models. Stat. Infer. Stoch. Process. **8**, 25–50 (2005)
24. H. Masuda, Estimation of second-characteristic matrix based on realized multipower variations (in Japanese). Proc. Inst. Statist. Math. **57**, 17–38 (2009)
25. R.M. May, Stability in randomly fluctuating versus deterministic environments. Am. Nat. **107**, 621–650 (1973)
26. D.B. Nelson, ARCH models as diffusion approximations. J. Econ. **45**, 7–38 (1990)
27. E. Platen, N. Bruti-Liberati, *Numerical Solution of Stochastic Differential Equations with Jumps in Finance* (Springer, Berlin, 2010)
28. B.L.S. Prakasa Rao, *Statistical inference for diffusion type processes* (Oxford University Press, New York, 1999)
29. P.E. Protter, *Stochastic Integration and Differential Equations*, 2nd edn. Version 2.1. Corrected third printing (Springer, Berlin, 2005)
30. K. Sato, *Lévy Processes and Infinitely Divisible Distributions* (Cambridge University Press, Cambridge, 1999)
31. M. Sørensen, *Likelihood Methods for Diffusions with Jumps. Statistical Inference in Stochastic Processes*, pp. 67–105 (Dekker, New York, 1991) (Probab. Pure Appl., 6)
32. M. Uchida, Statistical inference for diffusion processes from discrete observations. Sugaku Expo. **24**, 169–181 (2011)
33. A.W. van der Vaart, *Asymptotic Statistics* (Cambridge University Press, Cambridge, 1998)
34. W. Zucchini, I.L. MacDonald, *An introduction Using R Hidden Markov Models for Time Series* (CRC Press, Boca Raton, FL, 2009)

Signal Detection and Model Selection

Yoshiyuki Ninomiya

Abstract Signal detection is a basic statistical problem in various fields including engineering, econometrics and psychometrics. It is performed by statistical testing or model selection, but we cannot apply conventional statistical theory to it. The reason is that the signal model, a statistical model for signal detection, has an irregularity, called non-identifiability. Because of this non-identifiability problem, the signal model needs to be shrunk in its geometrical representation. After drawing it, we prove there is an asymptotic property of the likelihood ratio statistics for the model, which is indicated by the geometrical representation. Then, on the basis of this asymptotic property, we introduce a criterion for model selection considering non-identifiability that is a reevaluated Akaike information criterion (AIC). We check the validity of the reevaluated AIC through simulation studies and real data analysis using a factor analysis model, which can be regarded as a kind of signal model.

Keywords Factor analysis · Information criterion · Likelihood ratio · Locally conic parameterization · Non-identifiability

1 Introduction

Consider the following experiment to detect regions of the human brain that respond to hot temperatures. Prepare lukewarm water and hot water, and have an examinee soak his or her hands in one then the other. While this is going on, capture an image of blood flow in his/her brain; the image will be darker or lighter when the blood flow is faster or slower. For example, if the image made when the examinee's hands are in

Y. Ninomiya (✉)
Institute of Mathematics for Industry, Kyushu University,
744, Motooka, Nishiku, Fukuoka 819-0395, Japan
e-mail: nino@imi.kyushu-u.ac.jp

R. Nishii et al. (eds.), *A Mathematical Approach to Research Problems of Science and Technology*, Mathematics for Industry 5,
DOI: 10.1007/978-4-431-55060-0_18,

the hot water has a region that is clearly darker than the corresponding region in the image for lukewarm water, we can say that the region responds to hot temperatures.

The problem is the criterion for determining whether the region is "clearly" darker. Can we discriminate an image that is dark by necessity from an image that is dark by chance? Various environmental and bodily factors such as biorhythms influence blood flow. That is, even if the region is not related to hotness, its blood flow when the examinee's hands are in hot water may still be faster than when his or her hands are in lukewarm water.

Such an unpredictable random variation in an image is called noise. On the other hand, the variation derived from the sensation of hotness is called a signal in the broad sense. Signal detection is the task of detecting variations in an image that are due to a signal, not noise, and its difficulty is the possibility that the variations are due to noise. For this reason, statistics is required in signal detection. The existence of signals can be determined by statistical testing, and the number of signals can be estimated by using statistical model selection. However, statistical models for signals, which hereafter are called signal models, have an irregular property. Because of it, we cannot apply conventional statistical theory to signal detection. Hereafter, we explain this irregularity especially through model selection.

2 Non-identifiability in Signal Models

We will illustrate the problem mentioned in the previous section with a simple model. Let y_t be the difference between two data at position t. In image analysis, t is usually two or three dimensional, but here, we will assume that it is one dimensional for simplicity. For this y_t, we assume the following model:

$$y_t = \alpha g_t(\beta) + \varepsilon_t, \quad \varepsilon_t \overset{\text{indep.}}{\sim} \mathrm{N}(0, \sigma_0^2), \quad t = 1, \ldots, T, \tag{1}$$

where $\alpha g_t(\beta)$ is the term for a signal with amplitude α and position β, and ε_t is the term for noise. Let us assume, for simplicity, that the variance of the noise σ_0^2 and shape of the signal $g_t(\cdot)$ are known, and so only α and β are unknown parameters ($\alpha \in \mathbb{R}, 1 \leq \beta \leq T$). From (1), the probability density function for $y = (y_1, \ldots, y_T)$ can be written as

$$f(y|\alpha, \beta) = \frac{1}{(2\pi\sigma_0^2)^{T/2}} \exp\Big[-\frac{1}{2\sigma_0^2} \sum_{t=1}^{T} \{y_t - \alpha g_t(\beta)\}^2 \Big].$$

For the time being, let us express the one-signal model as follows:

$$\mathscr{M}_1 = \{ f(y|\alpha, \beta) \mid \alpha \in \mathbb{R},\ 1 \leq \beta \leq T \}.$$

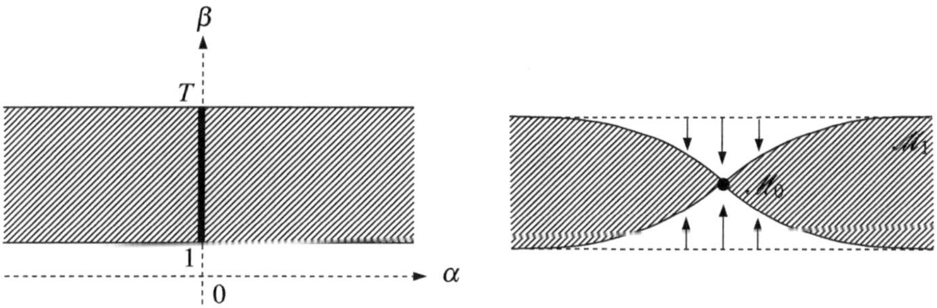

Fig. 1 *Signal model* The *left panel* shows its representation in parameter space, and the *right panel* shows its geometrical representation

The hatched area in the left panel of Fig. 1 is a representation of $\mathcal{M}_1$ in the space of parameter (α, β), but is it reasonable to regard $\mathcal{M}_1$ as this type of region? The one-signal model $\mathcal{M}_1$ becomes a no-signal one $y_t = \varepsilon_t$ independent of the value of β if $\alpha = 0$; in other words, the one-signal model becomes a no-signal model regardless of its position if its amplitude is zero. This property of $\mathcal{M}_1$ leads to a negative answer to the question. The property means that even if $\beta^\dagger \neq \beta^\ddagger$, the probability distributions $f(y|0, \beta^\dagger)$ and $f(y|0, \beta^\ddagger)$ become the same. For the time being, we will denote this no-signal model as follows:

$$\mathcal{M}_0 = \{f(y|0, \beta) \mid 1 \leq \beta \leq T\}.$$

Because we cannot discriminate among probability distributions in $\mathcal{M}_0$, $\mathcal{M}_1$ is said to be non-identifiable at $\mathcal{M}_0$. In the left panel of Fig. 1, $\mathcal{M}_0$ is drawn as a thick line, but it should be drawn as a one point because every point of the thick line corresponds to the same probability distribution. That is, $\mathcal{M}_1$ should be "shrunk" as in the right panel of Fig. 1.

Let us generalize this one-signal model. Given a set of probability distributions

$$\mathcal{M}_1 = \{f(y|\alpha, \beta, \gamma) \mid \alpha \in \mathbb{R}^p,\ \beta \in \mathbb{R}^q,\ \gamma \in \mathbb{R}^r\} \tag{2}$$

for data y, we assume that $f(y|\alpha, \beta, \gamma)$ reduces to $f(y|0, \beta, \gamma)$ if and only if $\alpha = 0$, and $f(y|0, \beta, \gamma)$ does not depend on β. Then, $\mathcal{M}_1$ is said to be non-identifiable at

$$\mathcal{M}_0 = \{f(y|0, \beta, \gamma) \mid \beta \in \mathbb{R}^q,\ \gamma \in \mathbb{R}^r\}, \tag{3}$$

and $\mathcal{M}_1$ can be drawn as in Fig. 2 (when $p = 1, q = 2$ and $r = 0$). From a viewpoint near $\mathcal{M}_0$, $\mathcal{M}_1$ can be regarded as a cone, and so Dacunha-Castelle and Gassiat [5] call $\mathcal{M}_1$ a locally conic model with a vertex $\mathcal{M}_0$. Note that $\mathcal{M}_1$ is often expressed with parameters different from (α, β, γ), and in this case, it is called a locally conic parameterization for deriving (α, β, γ).

On the other hand, let us consider the model,

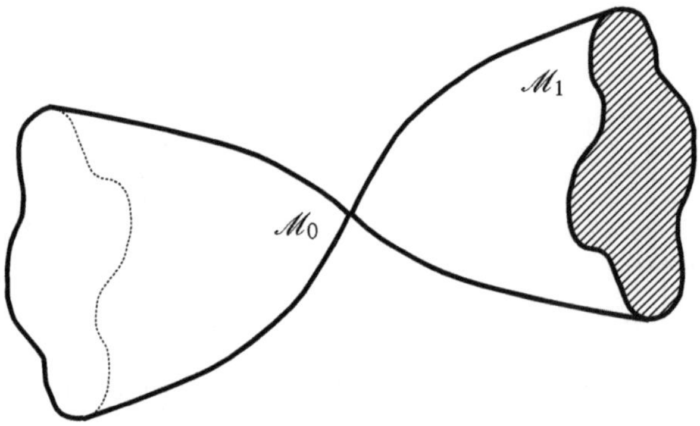

Fig. 2 Geometrical representation of locally conic model

$$y_t = \alpha_1 + \varepsilon_t, \quad \varepsilon_t \overset{\text{indep.}}{\sim} \mathrm{N}(0, \sigma_0^2), \quad t = 1, \ldots, T, \tag{4}$$

or

$$y_t = \alpha_1 + \alpha_2 t + \varepsilon_t, \quad \varepsilon_t \overset{\text{indep.}}{\sim} \mathrm{N}(0, \sigma_0^2), \quad t = 1, \ldots, T, \tag{5}$$

where $\alpha_1, \alpha_2 \in \mathbb{R}$. In this model, the probability distribution expressed by $\alpha_1 = 0$, or $\alpha_1 = \alpha_2 = 0$, does not have any other expression; that is, the model is identifiable at $\alpha_1 = 0$ or $\alpha_1 = \alpha_2 = 0$. Let us consider this model in more general form. We assume that a set of probability distributions

$$\mathcal{M}_1 = \{g(y|\alpha, \gamma) \mid \alpha \in \mathbb{R}^p, \ \gamma \in \mathbb{R}^r\} \tag{6}$$

for data y are identifiable at its subset

$$\mathcal{M}_0 = \{g(y|0, \gamma) \mid \gamma \in \mathbb{R}^r\}. \tag{7}$$

The model $\mathcal{M}_1$ with $p = 1$ and $r = 0$, which corresponds to (4), can be drawn as in the left panel of Fig. 3, and the model $\mathcal{M}_1$ with $p = 2$ and $r = 0$, which corresponds to (5), can be drawn as in the right panel of Fig. 3. These models do not need to be shrunk, and so $\mathcal{M}_1$ can be regarded as a line or plane from a viewpoint near $\mathcal{M}_0$.

3 Statistical Theory for Non-identifiable Models

First, we will describe the conventional statistical theory; then, we will show it does not to hold for non-identifiable models. Let $\hat{l}_0$ and $\hat{l}_1$ be the maximum log-likelihoods for models $\mathcal{M}_0$ and $\mathcal{M}_1$, respectively. The difference $\hat{l}_1 - \hat{l}_0$ is regarded

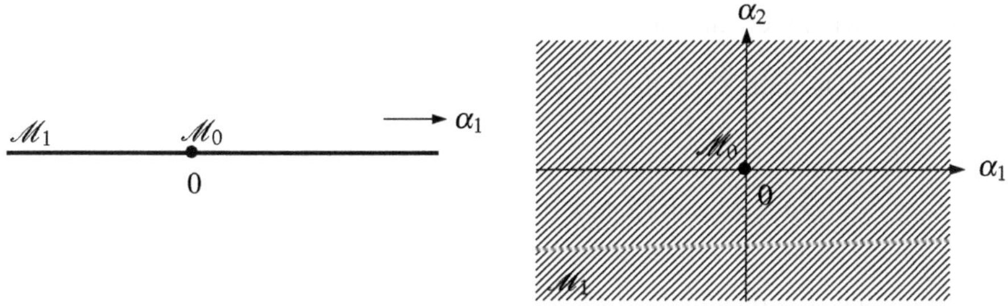

Fig. 3 Model without the non-identifiability problem. The *left panel* shows a one-dimensional model, and the *right panel* shows a two-dimensional model

as an important index for comparing $\mathcal{M}_0$ and $\mathcal{M}_1$, and twice the difference is called the likelihood ratio statistic. When they are (6) and (7), i.e., the models do not have the non-identifiability problem, the likelihood ratio statistic can be written as

$$2\hat{l}_1 - 2\hat{l}_0 = \sup_{\alpha,\gamma}\{2\log g(y|\alpha,\gamma)\} - \sup_{\gamma}\{2\log g(y|0,\gamma)\}.$$

The following is a well-known asymptotic property for this statistic.

Theorem 1 (Wilks [11]) *For models* (6) *and* (7)*, if $\mathcal{M}_1$ is identifiable at $\mathcal{M}_0$ and the true distribution exists in $\mathcal{M}_0$, the following*

$$\exists p \in \mathbb{N};\quad \exists T \sim \chi^2(p);\quad 2\hat{l}_1 - 2\hat{l}_0 \xrightarrow{\mathrm{d}} T$$

holds under certain regularity conditions.

Here, $\chi^2(p)$ denotes the chi-square distribution with p degrees of freedom, and $\xrightarrow{\mathrm{d}}$ denotes convergence in law.

The above theorem says that although there are various models without the non-identifiability problem, the asymptotic distributions of their likelihood ratio statistics are of the same type. This is attributed to the fact that it is an asymptotic property and the true distribution exists in $\mathcal{M}_0$. In fact, the asymptotic distribution is influenced by the behavior of likelihoods whose parameters are in the immediate vicinity of the true ones, and any model in the vicinity can be regarded as a hyper-plane. Note that a $\chi^2(1)$ (or $\chi^2(2)$) distribution appears when the hyper-plane is a line (or plane).

On the other hand, for models with the non-identifiability problem, $\mathcal{M}_1$ can be regarded as a cone in the vicinity of $\mathcal{M}_0$, as explained in the previous section, and so Theorem 1 does not hold. When the models are (2) and (3), the likelihood ratio statistic can be written as

$$2\hat{l}_1 - 2\hat{l}_0 = \sup_{\alpha,\beta,\gamma}\{2\log f(y|\alpha,\beta,\gamma)\} - \sup_{\beta,\gamma}\{2\log f(y|0,\beta,\gamma)\},$$

and the following theorem can be obtained in place of Theorem 1.

Theorem 2 (Dacunha-Castelle and Gassiat [5]) *For models* (2) *and* (3), *if* $\mathscr{M}_1$ *is non-identifiable at* $\mathscr{M}_0$ *and the true distribution exists in* $\mathscr{M}_0$, *the following*

$$\exists p \in \mathbb{N}; \quad \exists \{T_\beta \sim \chi^2(p)\}; \quad 2\hat{l}_1 - 2\hat{l}_0 \xrightarrow{\mathrm{d}} \sup_\beta T_\beta$$

holds under certain regularity conditions.

Note that $\mathscr{M}_1$ becomes a model without the non-identifiability problem if β is fixed, and that the likelihood ratio statistic for the case in which β is unknown is the maximum of the likelihood ratio statistic for the case in which β is fixed with respect to β, that is, $2\hat{l}_1 - 2\hat{l}_0 = \sup_\beta[\sup_{\alpha,\gamma}\{2\log f(y|\alpha,\beta,\gamma)\} - \sup_\gamma\{2\log f(y|0,\beta,\gamma)\}]$. These two facts enable us to intuitively understand the above theorem.

On the basis of Theorems 1 and 2, we can conduct statistical testing, that is, evaluate the p value for the null hypothesis where the true distribution exists in $\mathscr{M}_0$ against the alternative hypothesis where it exists in $\mathscr{M}_1 \setminus \mathscr{M}_0$. The p value is given by a tail probability of T in Theorem 1 or $\sup_\beta T_\beta$ in Theorem 2. The tail probability of $\sup_\beta T_\beta$ reduces to the exceedance probability of a Gaussian random field, and its evaluation is an important topic in probability theory because of its demands in statistics. The evaluation method requires a differential geometric tool called the tube method (Hotelling [8], Weyl [10]). The tube method has been shown to be part of the Euler characteristic method, an integral geometric approach (Adler [1]), and it continues to be a topic of interest. The topic is summarized in Adler and Taylor [2], but here we shall focus on its use in statistical model selection.

The model selection is the task of selecting an appropriate model by examining data from a set of candidates $\{\mathscr{M}^{(m)} \mid m = 0, 1, 2, \ldots\}$, and it is indispensable to the field of statistical analysis. The m-th degree polynomial regression model defined by

$$y_t = \sum_{j=0}^{m} \alpha_{j+1} t^j + \varepsilon_t, \quad \varepsilon_t \overset{\text{indep.}}{\sim} \mathrm{N}(0, \sigma_0^2), \quad t = 1, \ldots, T,$$

which is an extension of (4) and (5), is an example of $\mathscr{M}^{(m)}$, and in this case, the model selection task is to select the degree m.

One of the most frequently used model selection methods involves the Akaike information criterion (AIC; Akaike [3]). Letting $\hat{l}^{(m)}$ and $q^{(m)}$ be the maximum log-likelihood for $\mathscr{M}^{(m)}$ and the number of parameters in $\mathscr{M}^{(m)}$, AIC is defined as

$$\mathrm{AIC}^{(m)}_{\mathrm{formal}} = -2\hat{l}^{(m)} + 2q^{(m)}. \tag{8}$$

The model that yields the smallest AIC value is regarded as an optimal one. Here, we denote the AIC for $\mathscr{M}^{(m)}$ by $\mathrm{AIC}^{(m)}_{\mathrm{formal}}$ in (8) because later we reevaluate AIC for models with the non-identifiability problem.

If $\mathcal{M}^{(m)}$ does not have the non-identifiability problem and the true probability distribution exists in $\mathcal{M}^{(m)}$, it is known that $\mathrm{AIC}_{\mathrm{formal}}^{(m)}$ is an asymptotically unbiased estimator of twice the Kullback-Leibler divergence $\mathrm{KL}^{(m)}$ (minus some constant) between the best probability distribution in $\mathcal{M}^{(m)}$ and the true probability distribution. Therefore, by focusing on the comparison between just two models $\mathcal{M}^{(m)}$ and $\mathcal{M}^{(m+1)}$, we obtain the following property.

Proposition 1 *If $\mathcal{M}^{(m+1)}$ including $\mathcal{M}^{(m)}$ is identifiable at $\mathcal{M}^{(m)}$ and the true probability distribution exists in $\mathcal{M}^{(m)}$, $\mathrm{AIC}_{\mathrm{formal}}^{(m+1)} - \mathrm{AIC}_{\mathrm{formal}}^{(m)}$ is an asymptotically unbiased estimator of $2\mathrm{KL}^{(m+1)} - 2\mathrm{KL}^{(m)}$ under certain regularity conditions.*

When the true probability distribution is in or near $\mathcal{M}^{(m)}$, we can say from Proposition 1 that $\mathrm{AIC}_{\mathrm{formal}}^{(m+1)} - \mathrm{AIC}_{\mathrm{formal}}^{(m)}$ is a good estimator of $2\mathrm{KL}^{(m+1)} - 2\mathrm{KL}^{(m)}$, and so the selection based on $\mathrm{AIC}_{\mathrm{formal}}$ must be reasonable. On the other hand, when the true probability distribution is far from $\mathcal{M}^{(m)}$, $\mathrm{AIC}_{\mathrm{formal}}^{(m+1)} - \mathrm{AIC}_{\mathrm{formal}}^{(m)}$ is not a good estimator. In this case, however, $\hat{l}^{(m+1)}$, which is regarded as an index of goodness of fit for $\mathcal{M}^{(m+1)}$, becomes considerably larger than $\hat{l}^{(m)}$, which is regarded as an index of goodness of fit for $\mathcal{M}^{(m)}$, then $\mathrm{AIC}_{\mathrm{formal}}^{(m+1)} - \mathrm{AIC}_{\mathrm{formal}}^{(m)}$ will usually be less than 0. The result is that $\mathcal{M}^{(m+1)}$ is regarded as better than $\mathcal{M}^{(m)}$, and so no problem occurs. Hence, owing to the above proposition, we can say that $\mathrm{AIC}_{\mathrm{formal}}$ has good performance. On the other hand, if the proposition does not hold, we cannot assure that $\mathrm{AIC}_{\mathrm{formal}}$ will have good performance.

The difference $\mathrm{AIC}_{\mathrm{formal}}^{(m+1)} - \mathrm{AIC}_{\mathrm{formal}}^{(m)}$ can be rewritten as $-2(\hat{l}^{(m+1)} - \hat{l}^{(m)}) + 2(q^{(m+1)} - q^{(m)})$; that is, it is related to the likelihood ratio statistic. As can be imagined from this fact, Proposition 1 does not hold for models with the non-identifiability problem, and so $\mathrm{AIC}_{\mathrm{formal}}$ does not work well. For example, if $\mathcal{M}^{(m)}$ is an m-signal model defined by

$$y_t = \sum_{j=1}^{m} \alpha_j g_t(\beta_j) + \varepsilon_t, \quad \varepsilon_t \overset{\text{indep.}}{\sim} \mathrm{N}(0, \sigma_0^2), \quad t = 1, \ldots, T,$$

which is an extension of (1), $\mathcal{M}^{(m+1)}$ is non-identifiable at $\mathcal{M}^{(m)}$, and so it is not reasonable to use $\mathrm{AIC}_{\mathrm{formal}}$.

Now let us consider candidates of a locally conic model such as the signal model $\{\mathcal{M}^{(m)} \mid m = 0, 1, 2, \ldots\}$, and denote the maximum log-likelihood for $\mathcal{M}^{(m)}$ by $\hat{l}^{(m)}$. Here, we obtain

$$\exists p^{(m)} \in \mathbb{N}; \quad \exists \{T_\beta^{(m)} \sim \chi^2(p^{(m)})\}; \quad 2\hat{l}^{(m+1)} - 2\hat{l}^{(m)} \overset{\mathrm{d}}{\to} \sup_\beta T_\beta^{(m)}$$

from Theorem 2. On the basis of the above equation, we can reevaluate AIC as

$$\mathrm{AIC}_{\mathrm{proposed}}^{(m)} = -2\hat{l}^{(m)} + 2\sum_{j=1}^{m-1} \mathrm{E}\Big(\sup_\beta T_\beta^{(j)}\Big). \tag{9}$$

Accordingly, its properties can be obtained as follows.

Proposition 2 *If $\mathscr{M}^{(m+1)}$ including $\mathscr{M}^{(m)}$ is non-identifiable at $\mathscr{M}^{(m)}$ and the true probability distribution exists in $\mathscr{M}^{(m)}$, $\mathrm{AIC}^{(m+1)}_{\mathrm{proposed}} - \mathrm{AIC}^{(m)}_{\mathrm{proposed}}$ is an asymptotically unbiased estimator of $2\mathrm{KL}^{(m+1)} - 2\mathrm{KL}^{(m)}$ under certain regularity conditions.*

From this proposition, $\mathrm{AIC}_{\mathrm{proposed}}$ is expected to perform well at model selection. However, we must point out here that the expectation in (9) is for the supremum of an infinite number of chi-square variables, and so it is generally difficult to evaluate explicitly.

4 Application to Factor Analysis Model

To check the validity of $\mathrm{AIC}_{\mathrm{proposed}}$, we considered the factor analysis model, which is a basic model of psychometrics and a typical example of models with the non-identifiability problem. The factor analysis model assumes there are several latent factors behind multivariate data $\{x_i \in \mathbb{R}^p \mid 1 \le i \le n\}$. Concretely speaking, letting

$$z_i = (z_{i1}, \dots, z_{im})' \overset{\text{indep.}}{\sim} \mathrm{N}(0, \mathrm{diag}(1, \dots, 1)) \tag{10}$$

and

$$\varepsilon_i \overset{\text{indep.}}{\sim} \mathrm{N}(0, \mathrm{diag}(\psi_1, \dots, \psi_p)), \tag{11}$$

if x_i can be expressed as

$$x_i = \sum_{j=1}^{m} \lambda_j z_{ij} + \varepsilon_i = (\lambda_1, \dots, \lambda_m) z_i + \varepsilon_i, \quad i = 1, \dots, n, \tag{12}$$

this model is called the m-factor model for p-variate data. In this section, we denote it by $\mathscr{M}^{(m)}$ and consider the problem of selecting m, the number of factors. Here, for i-th sample, z_i is a vector whose components are m factors, and ε_i is noise. In addition, the unknown coefficient $\lambda_j = (\lambda_{j1}, \dots, \lambda_{jp})'$ is called factor loading vector. The m-factor model $\mathscr{M}^{(m)}$ can be rewritten as

$$x_i \overset{\text{indep.}}{\sim} \mathrm{N}\Big(0, \sum_{j=1}^{m} \lambda_j \lambda_j' + \mathrm{diag}(\psi_1, \dots, \psi_p)\Big), \qquad i = 1, \dots, n \tag{13}$$

from (10) to (12).

The $(m+1)$-factor model $\mathscr{M}^{(m+1)}$ is non-identifiable at m-factor model $\mathscr{M}^{(m)}$, and so $\mathscr{M}^{(m+1)}$ can be regarded as a cone in the vicinity of $\mathscr{M}^{(m)}$, as in Fig. 2. It also

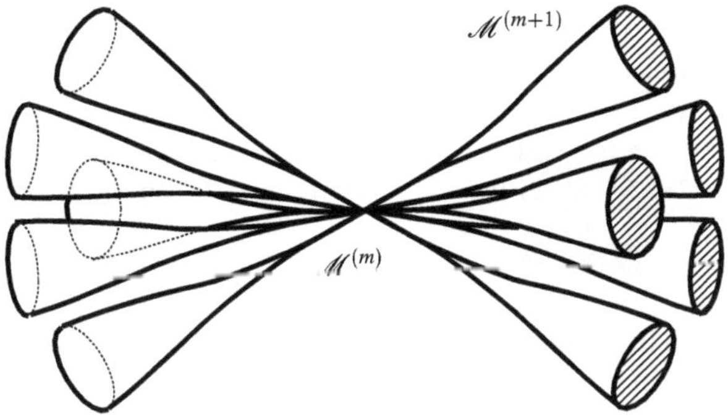

Fig. 4 Geometrical representation of factor analysis model

Table 1 Evaluation of the expectation term in $\mathrm{AIC}_{\mathrm{proposed}}$ for factor analysis model

$p-j$	4	5	6	7	8	9	10	11	12	13	14	15
$\mathrm{E}(\max_{1\le\beta\le p-j} T_\beta^{(j)})$	6.4	8.5	10.5	12.4	14.2	16.0	17.8	19.5	21.2	22.8	24.5	26.1

has the following characteristic. As in Fig. 4, in the vicinity of $\mathcal{M}^{(m)}$, $\mathcal{M}^{(m+1)}$ is divided into $(m+1)$ cones and each cone is shrunk. Using this characteristic, we obtain the following theorem.

Theorem 3 (Ninomiya et al. [9]) *Let $\mathcal{M}^{(m)}$ be the factor analysis model defined in* (13)*, and assume that the parameter space is compact and that $\lambda_j \neq 0\,(1 \le j \le m)$. Then,* (9) *reduces to*

$$\mathrm{AIC}_{\mathrm{proposed}}^{(m)} = -2\hat{l}^{(m)} + 2\sum_{j=1}^{m-1} \mathrm{E}\Big(\max_{1\le\beta\le p-j} T_\beta^{(j)}\Big), \tag{14}$$

where $T_\beta^{(j)}$ $(1 \le \beta \le p-j)$ are chi-square variables with $(p-j-1)$ degrees of freedom.

Note that it is not difficult to evaluate the expectation in (14) because it is for the maximum of a finite number of chi-square variables. The expectation can be evaluated as in Table 1, for example, and thus, model selection can be easily conducted.

In order to compare $\mathrm{AIC}_{\mathrm{proposed}}$ and $\mathrm{AIC}_{\mathrm{formal}}$, we will introduce the data analysis presented in Ninomiya et al. [9] which applies $\mathrm{AIC}_{\mathrm{proposed}}$ and $\mathrm{AIC}_{\mathrm{formal}}$ to the data in Holzinger and Swineford [7]. The data consists of 145 samples of 24 psychological variables, and it is benchmark data in psychometrics. The results are shown in Table 2. Note that $\mathrm{AIC}_{\mathrm{consistent}}$ is a criterion proposed by Bozdogan [4]. We can see that each criterion selects a different number of factors, which indicates that the differences among the criteria are not trivial. According to various analyses by psychologists,

Table 2 Application of AIC to the data in Holzinger and Swineford [7]

	1-factor	2-factor	3-factor	4-factor	5-factor	6-factor
$AIC_{proposed}$	2573.1	2433.7	2374.0	**2372.4**	2397.4	2462.1
AIC_{formal}	2589.4	2419.2	2329.7	2299.3	**2296.5**	2303.1
$AIC_{consistent}$	2780.2	2701.5	**2699.5**	2752.6	2829.4	2943.0

4-factor is regarded to be reasonable [6, p.164], and so we can verify the reasonableness of $AIC_{proposed}$ in this sense.

References

1. R.J. Adler, *The Geometry of Random Fields* (Wiley, New York, 1981)
2. R.J. Adler, J.E. Taylor, *Random Fields and Their Geometry* (Springer, New York, 2007)
3. H. Akaike, in *Information Theory and an Extension of the Maximum Likelihood Principle*, ed. by B.N. Petrov, F. Csaki. 2nd International Symposium on Information Theory (Akademiai Kiado, Budapest, 1973), pp. 716–723
4. H. Bozdogan, Model selection and Akaike's information criterion (AIC): the general theory and its analytical extensions. Psychometrika **52**, 345–370 (1987)
5. D. Dacunha-Castelle, E. Gassiat, Testing in locally conic models and application to mixture models. ESAIM Probab. Stat. **1**, 285–317 (1997)
6. H.H. Harman, *Modern Factor Analysis*, 3rd edn. (The University of Chicago Press, Chicago, 1976)
7. K.J. Holzinger, F. Swineford, A study in factor analysis: the stability of a bi-factor solution, in *Supplementary Educational Monographs*, vol. 48 (University Chicago Press, Chicago, 1939)
8. H. Hotelling, Tubes and spheres in n-space a class of statistical problems. Am. J. Math. **61**, 440–460 (1939)
9. Y. Ninomiya, H. Yanagihara, K.-H. Yuan, Selecting the number of factors in exploratory factor analysis via locally conic parameterization. ISM Research Memorandum, 1078 (2008)
10. H. Weyl, On the volume of tubes. Am. J. Math. **61**, 461–472 (1939)
11. S.S. Wilks, The large-sample distribution of the likelihood ratio for testing composite hypotheses. Ann. Math. Stat. **9**, 60–62 (1938)

Regression Analysis and Its Development

Ryuei Nishii

Abstract Regression analysis aims to predict a target variable statistically by using explanatory variables. The analysis has a long history and is utilized in various situations. We will review linear regression analysis and describe model assessment methods based on the coefficient of determination and Akaike information criterion (AIC). Furthermore, we propose a relative coefficient of determination based on AIC for general statistical modeling. Finally, we illustrate variable selection and discuss recent developments in regression analysis.

Keywords Akaike information criterion · Bayesian information criterion · Coefficient of determination · Model selection · Regression analysis

1 Introduction

Regression analysis aims to predict a target variable by using a set of explanatory variables. It is the most-frequently utilized statistical method. We will illustrate regression analysis with a simple dataset representing the postnatal development of 32 infants [1].

Figure 1(left) shows a scatter plot of the weights of 32 babies just after birth against a weight growth ratio 3 months after birth. A straight line relation can be detected between the two variables. Hence, we can fit a linear model for the training data $\{(x_i, y_i) \mid i = 1, 2, \ldots, n\}$, where x_i and y_i are the weight just after birth and weight growth ratios after 3 months of the babies $i = 1, \ldots, n(= 32)$. Our aim is to predict

R. Nishii (✉)
Institute of Mathematics for Industry, Kyushu University,
744, Motooka, Nishi-ku,
Fukuoka 819-0395, Japan
e-mail: nishii@imi.kyushu-u.ac.jp

R. Nishii et al. (eds.), *A Mathematical Approach to Research Problems of Science and Technology*, Mathematics for Industry 5,
DOI: 10.1007/978-4-431-55060-0_19,

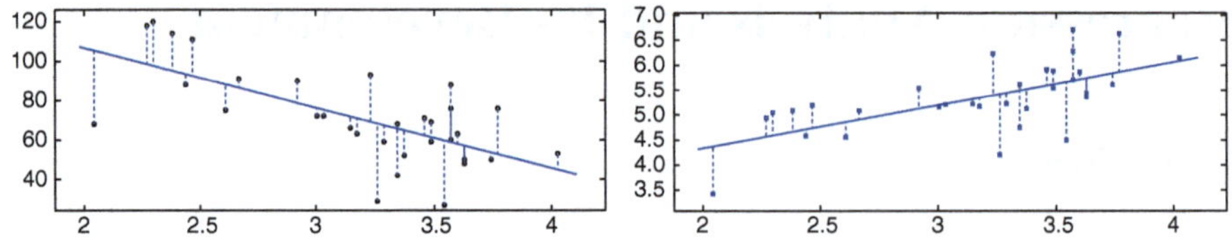

Fig. 1 Weight growth data of 32 babies. *left* weight just after birth (kg) against weight growth ratio 3 months later (%), *right* weight just after birth (kg) against weight after 3 months (kg)

the growth ratio y_i from the weight x_i at birth. Hence, these two variables are called the target and explanatory variables, respectively.

We fit the following regression model:

$$y_i = \beta_0 + \beta_1 x_i + \varepsilon_i, \quad i = 1, 2, \ldots, n \tag{1}$$

where β_0 and β_1, called **regression coefficients**, are constants, and ε_i, called **errors**, are independent random variables following a stochastic distribution with mean zero and variance σ^2. Here, β_0, β_1, and σ^2 are unknown parameters that all the data have in common.

1.1 Parameter Estimation and Significance Test

A vector of the unknown regression coefficients $\beta = (\beta_0, \beta_1)^T$ is estimated with the least squares method. Let $Q(\beta)$ be a quadratic form defined by

$$Q(\beta) \equiv \sum_{i=1}^{n} (y_i - \beta_0 - \beta_1 x_i)^2.$$

Then, β is estimated by minimizing $Q(\beta)$. It can be seen that $Q(\beta)$ is a convex function of β. Hence, the stationary point minimizes the function. The partial derivatives of the function lead to the following normal equations:

$$\frac{\partial Q(\beta)}{\partial \beta_0} = -2n(\bar{y} - \beta_0 - \beta_1 \bar{x}) = 0, \tag{2}$$

$$\frac{\partial Q(\beta)}{\partial \beta_1} = -2 \sum_{i=1}^{n} x_i (y_i - \beta_0 - \beta_1 x_i) = 0 \tag{3}$$

where $\bar{x} = \sum_{i=1}^{n} x_i / n$ and $\bar{y} = \sum_{i=1}^{n} y_i / n$. The solution of the simultaneous Eqs. (2) and (3) is given by

$$\hat{\beta} \equiv \begin{pmatrix} \hat{\beta}_0 \\ \hat{\beta}_1 \end{pmatrix} = \begin{pmatrix} \bar{y} - \hat{\beta}_1 \bar{x} \\ s_{xy}/s_{xx} \end{pmatrix} \tag{4}$$

where $s_{xy} = \sum_{i=1}^{n}(x_i - \bar{x})(y_i - \bar{y})$ and $s_{xx} = \sum_{i=1}^{n}(x_i - \bar{x})^2 > 0$. The $\hat{\beta}$ is called the **least squares estimate**.

From the assumptions placed on the error distribution, it is shown that the estimate $\hat{\beta}$ given by Eq. (4) follows a bi-variate random distribution with mean vector β and variance-covariance matrix $\sigma^2 D_1$, where

$$D_1 = \frac{1}{s_{xx}} \begin{pmatrix} s_{xx}/n + \bar{x}^2 & -\bar{x} \\ -\bar{x} & 1 \end{pmatrix}. \tag{5}$$

If the errors independently follow a normal distribution $N(0, \sigma^2)$ with mean 0 and variance σ^2, $\hat{\beta}$ follows a bi-variate normal distribution, where a probability density function of a general m-dimensional normal distribution $N_m(\mu, \Sigma)$ is given by

$$\psi(x|\mu, \Sigma) = (2\pi)^{-m/2}|\Sigma|^{-1/2} \exp\left\{-(x-\mu)^T \Sigma^{-1}(x-\mu)/2\right\}. \tag{6}$$

Let $\hat{y}_i = \hat{\beta}_0 + \hat{\beta}_1 x_i$ be the predicted values and $\hat{\varepsilon}_i = y_i - \hat{y}_i$ be the estimated errors. Accordingly, the sum of the squared errors follows a chi-squared distribution, summarized as

$$\hat{\beta} \sim N_2(\beta, \sigma^2 D_1) \text{ and } s_1 \equiv \sum_{i=1}^{n} \hat{\varepsilon}_i^2 = \sum_{i=1}^{n}(y_i - \hat{y}_i)^2 \sim \sigma^2 \chi_{n-2}^2. \tag{7}$$

Additionally, $\hat{\beta}$ and s_e are stochastically independent.

From the distributions (7), a **confidence interval and significance test** can be carried out on β by using the t-distribution. For example, the significance of β_1 is tested by using a statistic $(\hat{\beta}_1 - \beta_{10})\sqrt{s_{xx}}\big/\sqrt{s_1/(n-2)}$ following a t-distribution with the degree of freedom $n-2$ under the null hypothesis $H_0 : \beta_1 = \beta_{10}$. For residual diagnostics, the same t-distribution of $\hat{\varepsilon}_i/\sqrt{1-h_{ii}}\big/\sqrt{s_1/(n-2)}$ is used, where $h_{ii} = 1/n + (x_i - \bar{x})^2/s_{xx}$.

1.2 Model Assessment of Regression Analysis

Next, we derive a measure for model assessment. The sample correlation coefficient R between the target variable y_i and the predicted value $\hat{y}_i \equiv \hat{\beta}_0 + \hat{\beta}_1 x_i$ may be utilized for this purpose. After some calculation, one finds that

$$R^2 = 1 - \sum_{i=1}^{n}(y_i - \hat{y}_i)^2 \Big/ \sum_{i=1}^{n}(y_i - \bar{y})^2. \tag{8}$$

The squared correlation coefficient R^2 is called the coefficient of determination. It takes a value from 0 to 1, and a large R^2 implies that the linear model fits the data well.

Numerical example (Weight growth data) The estimated regression line for Fig. 1 (left) was $y = 167.8 - 30.48x$. The line segment from the observations (x_i, y_i) to the regression line gives the estimated error, and the y-value on the regression line gives the predicted value $\hat{y}_i$. In this case, the two correlation coefficients are both highly significant, and the coefficient of determination is 0.4465. When the target variable is substituted with the weight after 3 months (Fig. 1 (right)), the coefficient of determination of the second regression model is 0.4102. Hence, the first model is better than the second one.

2 Multiple Regression Analysis and Significance Test

Concerning the prediction of postnatal development, conception day is another important explanatory variable. Next, let us consider regression analysis with multiple explanatory variables.

Let y_i be a target variable of the ith sample and $x_{i1}, \ldots, x_{ip}$ be p-dimensional explanatory variables. We will assume a linear model:

$$y_i = \beta_0 + \beta_1 x_{i1} + \cdots + \beta_p x_{ip} + \varepsilon_i, \quad i = 1, \ldots, n. \tag{9}$$

The case of $p = 1$ reduces to model (1) with a single explanatory variable. We will employ vector notation for simplicity. Let y be a vector of target variables, X a design matrix of size $n \times (p+1)$ constituted by explanatory variables, β a coefficient vector, and ε an error vector defined by

$$y = \begin{pmatrix} y_1 \\ \vdots \\ y_n \end{pmatrix}, \quad X = \begin{pmatrix} 1 & x_{11} & \cdots & x_{1p} \\ \vdots & \vdots & \vdots & \vdots \\ 1 & x_{n1} & \cdots & x_{np} \end{pmatrix}, \quad \beta = \begin{pmatrix} \beta_0 \\ \vdots \\ \beta_p \end{pmatrix} \quad \text{and} \quad \varepsilon = \begin{pmatrix} \varepsilon_1 \\ \vdots \\ \varepsilon_n \end{pmatrix}. \tag{10}$$

Here, we assume that the rank of X is equal to $p + 1 (\leq n)$ and the errors ε_i independently follow $N(0, \sigma^2)$. y and X are observed, whereas β and σ^2 are unknown.

Now, the regression model (9) leads to the following joint expression,

$$y = X\beta + \varepsilon \ \text{ with } \ \varepsilon \sim N_n(0_n, \sigma^2 I). \tag{11}$$

The coefficient vector β can be estimated using the least squares method. Put the sum of squared errors as

$$Q(\beta) = \sum_{i=1}^{n} (y_i - \beta_0 - \beta_1 x_{i1} - \cdots - \beta_p x_{ip})^2 = \sum_{i=1}^{n} \varepsilon_i^2 = \varepsilon^T \varepsilon = (y - X\beta)^T (y - X\beta).$$

The derivatives of $Q(\beta)$ with respect to each coefficient of β yield the following normal equation in the general case:

$$\partial Q(\beta)/\partial \beta = -2X^T y + 2X^T X\beta = 0_{p+1}. \tag{12}$$

Equation (12) has a unique solution $\hat{\beta} = D_p X^T y$ following the normal distribution:

$$\hat{\beta} = D_p X^T y \sim N_{p+1}(\beta, \sigma^2 D_p) \tag{13}$$

$$\text{with } D_p = (X^T X)^{-1} : (p+1) \times (p+1). \tag{14}$$

The vectors of predicted target values and estimated errors are respectively given by $\hat{y} \equiv X\hat{\beta}$ and $\hat{\varepsilon} \equiv y - X\hat{\beta}$. The sum of squared errors is defined by $s_p \equiv \hat{\varepsilon}^T \hat{\varepsilon}$. The estimated error vector and the sum of squared errors independently follow a singular normal distribution and a chi-square distribution with degree of freedom $n - p - 1$:

$$\hat{\varepsilon} \sim N_n\big(0, \sigma^2 (I - H)\big) \text{ and } s_p \equiv \sum_{i=1}^{n} (y_i - \hat{y}_i)^2 \sim \sigma^2 \chi^2_{n-p-1} \tag{15}$$

where $H \equiv X D_p X^T : n \times n$. The null hypothesis $H_0 : \beta_j = \beta_{j0}$ is tested by

$$\frac{(\hat{\beta}_j - \beta_{j0})/\sqrt{d_{jj}}}{\sqrt{s_p/(n-p-1)}} \sim t_{n-p-1} \quad (j = 0, 1, \ldots, p) \tag{16}$$

where d_{jj} is the jth diagonal element of the matrix D_p defined by (14).

The tolerance of the residual $\hat{\varepsilon}_i$ is assessed by the standardized residual $\hat{\varepsilon}_i/\sqrt{1 - h_{ii}} \,/\sqrt{s_p/(n-p-1)}$, which follows a t-distribution with df $n - p - 1$, where h_{ii} is the ith diagonal element of the matrix $H = X D_p X^T$.

Regression analysis is usually carried out under the assumption that the errors are independently and identically distributed as a normal distribution. Hence, the validity of the assumption should be checked by using residual plot and outlier detection. Note also that the matrix $X^T X$ is nearly singular if there are two highly correlated explanatory variables. In such case, the estimated coefficients will be unstable. Furthermore, if $n < p$, $X^T X$ is singular. This means that variable selection is an important issue.

3 Model Assessment and Selection of Explanatory Variables

3.1 Coefficient of Determination R^2 and the Adjusted R^2

The coefficient of determination R^2 defined by (8) is used in multiple regression by substituting the predicted vector with $\hat{y} = X\hat{\beta}$. Unfortunately, R^2 is monotone increasing if additional explanatory variables are implemented in the regression model, and hence, it is of no use for variable selection. Instead, the following $\boldsymbol{R}^2$ is adjusted by the degrees of freedom of the denominator and the numerator

$$R^2 = 1 - \frac{s_p}{s_0} \text{ and } R^2_{adj} = 1 - \frac{s_p/(n-p-1)}{s_0/(n-1)} \text{ with } s_0 = \sum_{i=1}^{n}(y_i - \bar{y})^2. \quad (17)$$

Note that the adjusted R^2 may take negative values if the model is very poor.

3.2 Information Criterions AIC and BIC

Now, let us consider a general model evaluation. Consider a statistical model $f_\theta(x)$ specified by a parameter vector θ. Let $L(\theta) = \prod_{i=1}^{n} f_\theta(x_i)$ be the likelihood based on random samples $x_1, \ldots, x_n$ and $\hat{\theta}$ be the maximum likelihood estimate (MLE) which maximizes the likelihood. **Akaike Information Criterion**, **AIC,** is a measure for evaluating general statistical models; it is an unbiased estimate of the expected log likelihood of the model. Bayesian Information Criterion (BIC) has a similar form, but it is derived from a Laplace approximation of the Bayes factor [6]. AIC and BIC evaluate the losses of the model by using the following formulas:

$$\text{AIC} = -2\log L(\hat{\theta}) + 2 \times \dim\hat{\theta} \quad \Longrightarrow \quad \text{minimum}$$
$$\text{BIC} = -2\log L(\hat{\theta}) + \log n \times \dim\hat{\theta} \quad \Longrightarrow \quad \text{minimum}$$

where $L(\hat{\theta})$ denotes the maximum likelihood.

The AIC of the regression model (11) is derived as follows. The parameter vector θ and likelihood are (β, σ^2) and $L(\beta, \sigma^2) = \psi(y|X\beta, \sigma^2 I)$ from formula (6). The MLE of the parameter is given by $\hat{\beta} = D_p X^T y$ and $\hat{\sigma}^2 = s_p/n$. Thus, the maximum likelihood is expressed by $L(\hat{\beta}, \hat{\sigma}^2) = (2\pi e)^{-n/2}\hat{\sigma}^{-n}$. Therefore, the **AIC and BIC** of the regression model are given by

$$\text{AIC} = n\log(2\pi e) + n\log\hat{\sigma}^2 + 2(p+2) \quad \Longrightarrow \text{minimum} \quad (18)$$
$$\text{BIC} = n\log(2\pi e) + n\log\hat{\sigma}^2 + (p+2)\log n \Longrightarrow \text{minimum} \quad (19)$$

where $\hat{\sigma}^2 = s_p/n$. Now, $p+2$ implies the sum of the numbers of unknown parameters $p+1$ (regression coefficients) and 1 (the variance). They are typical criteria for model selection. If the complexity of the statistical model increases (equivalently, the number of parameters becomes large), then $\hat{\sigma}^2$ becomes small. AIC balances this trade-off. Actually, it aims to predict an appropriate model of future samples.

3.3 Lasso (Least Absolute Shrinkage and Selection Operator)

Finding the optimal regression model by using AIC requires us to calculate 2^p AIC values if the full model has p explanatory variables. Exhaustive searches are currently impractical for more than 20 explanatory variables, and forward selection, backward elimination and stepwise methods are used instead.

It is known that AIC has a tendency to select a large model. Therefore, BIC with model consistency is utilized when one needs to choose a relatively small model. The optimal models selected by AIC and BIC can be compared by cross-validation, where the data are divided up into training and test data, and the performance of the optimal model determined from the training data is evaluated on the test data.

LASSO [8] and adaptive lasso are variable selection and parameter estimation methods that work by minimizing the following formulas.

$$\sum_{i=1}^{n}(y_i - \beta_0 - \beta_1 x_{i1} - \cdots - \beta_p x_{ip})^2 + \lambda \sum_{j=1}^{p} |\beta_j| \quad : \text{Lasso} \tag{20}$$

$$\sum_{i=1}^{n}(y_i - \beta_0 - \beta_1 x_{i1} - \cdots - \beta_p x_{ip})^2 + \lambda \sum_{j=1}^{p} w_j |\beta_j| \quad : \text{Adaptive Lasso} \tag{21}$$

where the explanatory variables are standardized so that $\sum_{i=1}^{n} x_{ij} = 0$ and $\sum_{i=1}^{n} x_{ij}^2 = n$. Here, $\lambda > 0, w_1 > 0, \ldots, w_p > 0$ are tuning parameters. The target functions given by (20) and (21) are derived by using the least squares method with L_1 penalty against overfitting. Lasso has the ability of **variable selection**; i.e., some of the regression coefficients β_j are estimated to be exactly zero. This feature is especially useful when one choose a small model among large numbers of explanatory variables.

4 Absolute Measure of Model Assessment Based on AIC

Consider the regression model (9). Its AIC is given by (18), and the AIC of the simplest model $y_i = \beta_0 + \varepsilon_i$ is $\text{AIC}_0 = n\log(2\pi e) + n\log(s_0/n) + 4$, where s_0 is determined using (17). Now, let us define the **relative coefficients of AIC determination and of BIC determination** by

$$R_{\text{AIC}} = 1 - \exp\left(\frac{\text{AIC} - \text{AIC}_0}{n}\right) \text{ and } R_{\text{BIC}} = 1 - \exp\left(\frac{\text{BIC} - \text{BIC}_0}{n}\right). \quad (22)$$

Accordingly, the following relations hold.

$$\begin{aligned} 1 - R^2 &= \left(1 - R^2_{\text{adj}}\right) \exp\left(-\log\frac{n-1}{n-p-1}\right) \\ &= (1 - R_{\text{AIC}}) \exp\left(-\frac{2p}{n}\right) \\ &= (1 - R_{\text{BIC}}) \exp\left(-\frac{p}{n}\log n\right). \end{aligned}$$

These equalities yield the inequalities,

$$R_{\text{BIC}} < R_{\text{AIC}} < R^2_{\text{adj}} < R^2 \le 1$$

if $n > e^2 = 7.39$ and $n > p$. Therefore, R_{AIC} gives a smaller value than the adjusted R^2.

The relative coefficient R_{AIC} defined by (22) is proposed as an extension of the coefficient of determination in regression settings. Of course, AIC can be used for general model evaluations. Here, we propose the use of R_{AIC} for a general model evaluation. It gives a relative evaluation of the current model M with respect to the simplest model M_0. Maximization of R_{AIC} is equivalent to minimization of AIC, and R_{AIC} is less than 1. Furthermore, it gives an absolute scale of AIC. $R_{\text{AIC}} = 0.5$ means a 50 % improvement with respect to the simplest (baseline) model.

The choice of the simplest model is an important issue. A reasonable candidate of the simplest model would be the simplest one in the family of distributions under consideration. Note that R_{AIC} or R_{BIC} may take negative values if the current model M is poorer than the simplest model M_0.

4.1 Numerical Examples

The first example is regression analysis of "air quality data" provided by the statistical software R [13]. "Ozone" is regressed by "Solar.R", "Wind", and "Temp". Complete data of size $n = 111$ were used. Eight models were made consisting of all possible combinations of three explanatory variables, and these are denoted by M000, M100,…, and M111. Model M000 means "Ozone $= \beta_0 + \varepsilon$", M100 means "Ozone $= \beta_0 + \beta_1 \times$ Solar.R $+ \varepsilon$", and so on. Model M000 was chosen as the simplest model (the baseline model). p denotes the number of explanatory variables used in the model.

Table 1 lists the estimated parameters and four model evaluation measures for each model. R_{AIC} provides a similar value of R_{dj}, whereas R_{BIC} is smaller.

Table 1 Model assessment of "air quality" data. The target variable "Ozone" was regressed using all combinations of "Solar.R", "Wind", and "Temp". The sample size was $n = 111$

Model	p	β_0	β_1	β_2	β_3	σ	R_{BIC}	R_{AIC}	R^2_{adj}	R^2
M000	0	42.10	–	–	–	33.28	0	0	0	0
M100	1	18.60	0.1272	–	–	31.33	0.0833	0.1054	0.1133	0.1213
M010	1	99.04	–	−5.729	–	26.42	0.3481	0.3638	0.3694	0.3752
M001	1	−147.7	–	–	2.439	23.92	0.4658	0.4787	0.4833	0.4880
M110	2	77.25	0.1004	−5.402	–	24.92	0.4007	0.4293	0.4393	0.4495
M101	2	−145.7	0.0571	–	2.278	23.50	0.4670	0.4923	0.5012	0.5103
M011	2	−67.32	–	−3.295	1.828	21.73	0.5443	0.5660	0.5736	0.5814
M111	**3**	**−64.34**	**0.0598**	**−3.334**	**1.652**	**21.18**	**0.5524**	**0.5840**	**0.5948**	**0.6059**

Table 2 Model evaluation at the group stage of the 2010 FIFA World Cup. The logarithm of the Poisson mean is regressed by using all combinations of "World Ranking," "Points," and "+/− Pos". The sample size was $n = 96$

Model	p	β_0	β_1	β_2	β_3	BIC	R_{BIC}	AIC	R_{AIC}
M000	0	0.0577	–	–	–	268.94	0	266.38	0
M100	1	0.2884	−0.0101	–	–	269.02	−0.0008	263.89	0.0256
M010	**1**	**−0.7223**	–	**0.000823**	–	**267.94**	**0.0104**	**262.81**	**0.0365**
M001	1	0.0653	–	–	−0.04675	272.47	−0.0423	267.79	−0.0148
M110	2	−0.5912	−0.00162	0.000724	–	272.47	−0.0374	264.78	0.0165
M101	2	0.2888	−0.01014	–	0.00118	273.58	−0.0495	265.89	0.0051
M011	2	−0.7080	–	0.000811	−0.00931	272.48	−0.0375	264.79	0.0164
M111	3	−0.6009	−0.00138	0.000730	−0.00648	277.03	−0.0878	266.77	−0.0041

The second example is Poisson regression, one of the generalized linear models (GLM) [5]. Scores from the group stage of the 2010 FIFA World Cup [3] were analyzed using Poisson regression. Each of the 32 teams played three matches. Accordingly, we assumed that the 96 scores are independently drawn from a Poisson distribution with mean

$$\lambda_i = \exp(\beta_0 + \beta_1 x_{i1} + \beta_2 x_{i2} + \beta_3 x_{i3}) \quad (i = 1, \ldots, 96)$$

where the explanatory variables x_{i1}, x_{i2}, and x_{i3} are given by World Ranking, Points and +/− Pos (change of ranking) of the ith team. Note that the explanatory variables of each team appears three times among $i = 1, \ldots, 96$.

Table 2 lists the parameters estimated by the maximum likelihood method and model evaluation measures for the eight Poisson regression models. All criteria choose M010 as the best model, which is based only on FIFA points. Also, R_{BIC} and R_{AIC} take on negative values in many models, and the improvement of R_{AIC} is 3.65 %. This implies that the baseline model M000 is nearly optimal. Figure 2 shows the regression line estimated by the optimal model M010. The horizontal and

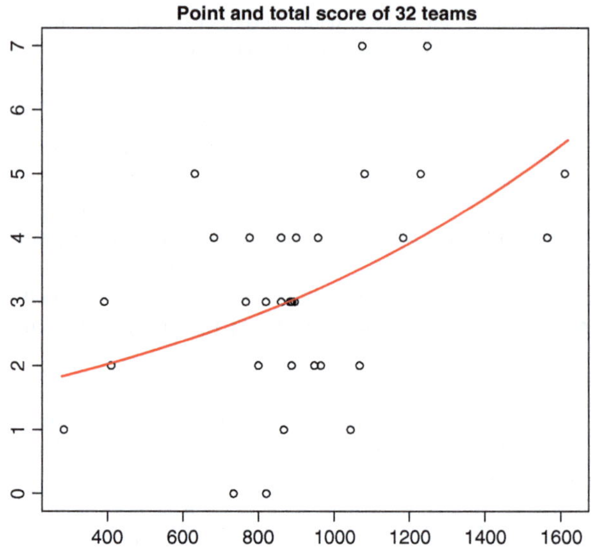

Fig. 2 FIFA points versus total scores of 32 teams at the group stage. The *curve* was estimated by Poisson regression

vertical axes correspond to FIFA points in 2010 and total scores at the group stage of 32 teams. It can be seen that teams with many points got high scores.

5 Development of Regression Analysis

The above-mentioned regression model does not always perform well in practical circumstances. The following illustrates various practical approaches for improving regression models.

1. Power transformation of the target variable and/or explanatory variables
   When the target variable is positive, the power transform is used to (a) improve the linearity of the explanatory variables and (b) make the error distribution closer to a normal distribution.

$$\phi(z;\lambda) = \begin{cases} \log z & \text{if } \lambda = 0, \\ \dfrac{z^\lambda - 1}{\lambda} & \text{if } \lambda \neq 0. \end{cases} \tag{23}$$

   The transform is also useful when the effect of the explanatory variable is not linear. Figure 3 shows that $\phi(z+1;\lambda)$ are monotone increasing functions even if λ is negative [11].

2. Expansion using basis functions

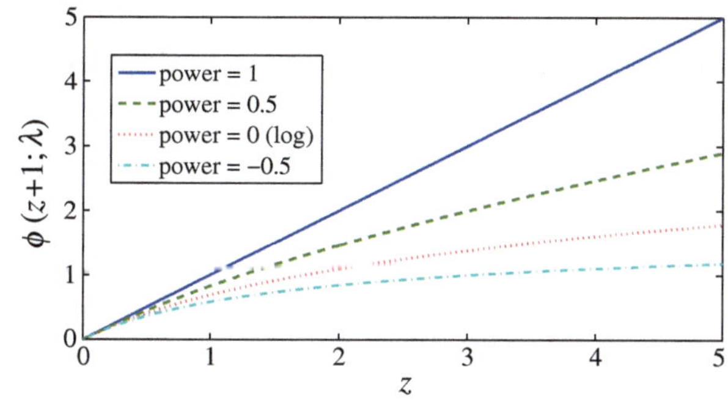

Fig. 3 Power transform of $z+1$ (λ = power)

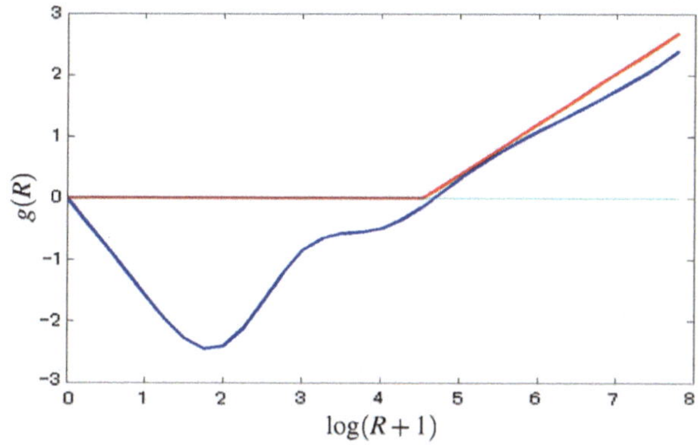

Fig. 4 Forest areal rate against relief energy R *Polygonal line* parametric model, *Curved line* natural cubic splines

The functional form of the effect of an explanatory variable on the target variable may not be a linear/monotone one. In such a case, an expansion based on basis functions is useful. The following formulas give linear models wherein the effect of x_1 is expressed in two ways. Let $\mathrm{E}(y|x)$ denote the conditional mean given explanatory variables $x = (x_1, \ldots, x_p)^T$.

$$\mathrm{E}(y \mid x) = \beta_0 + \beta_{11}x_1 + \beta_{12}x_1^2 + \cdots + \beta_{1q}x_1^q + \beta_2 x_2 + \cdots + \beta_p x_p$$
$$\mathrm{E}(y \mid x) = \beta_0 + \beta_{11}b_1(x_1) + \beta_{12}b_2(x_1) + \cdots + \beta_{1q}b_q(x_1) + \beta_2 x_2 + \cdots + \beta_p x_p.$$

The first one is a polynomial regression of x_1, and the second gives the basis function expansion using the known basis functions $b_1(\cdot), \ldots, b_q(\cdot)$ (e.g. x^2, $\log x$, $\sin x$). They are members of generalized additive models (GAM [5]).

Figure 4 shows an example of basis function expansion. The target variable: forest areal rate is expressed in terms of the relief energy [9]. Natural cubic spline functions are used as basis functions, and a non-monotone effect is detected.

3. Modeling of error variance
   The model (9) assumed a common variance σ^2. Generalized linear models (GLMs) [5] are available in the heteroscedastic case. $\log(\sigma)$ is expressed by a linear combination of explanatory variables.

$$\sigma = \exp\left(\gamma_0 + \gamma^T x\right).$$

   Of course, the mean regression and standard deviation are jointly estimated. The R package named "gamlss" [4] is useful for estimating the parameters of GLM.

4. Random effect models
   Model (9) assumed that the effect of each explanatory variable is fixed (a fixed effect model). Next, let us consider a case in which some of the effects are random. We divide the coefficient vector β^T into two subvectors (β_1^T, β_2^T): β_1 is random, and β_2 is fixed. The following is a typical random effect model.

$$\begin{aligned} y &= X_1\beta_1 + X_2\beta_2 + \varepsilon \quad \left(\beta_1 \sim N_q(H\gamma, \tau^2 D), \quad \varepsilon \sim N_n(0, \sigma^2 I)\right) \\ &\sim N_n(X_1 H\gamma + X_2\beta_2, \; \tau^2 X_1 H D H^T X_1^T + \sigma^2 I) \end{aligned}$$

   where $H : q \times r$ is known, $p \geqq q \geqq r$, and $D : q \times q$ is also a known positive-definite matrix. This regression model has a specialized covariance structure. If $\tau = 0$, β_1 is also fixed. The unknown parameters $\gamma : r \times 1$, $\tau^2 \geqq 0$, $\sigma^2 > 0$ can be estimated using the maximum likelihood principle, and AIC and BIC can be used to evaluate the model.

5. Design of experiments
   Consider a situation where explanatory variables can be arbitrarily chosen from a compact domain. Here, an optimal design by which the estimated parameters provide a good prediction on the domain is required. Suppose that the initial estimates of the parameters are derived in a preliminary experiment with a design matrix X. Then, the predictive variance at the explanatory vector x is proportional to $x^T(X^T X)^{-1}x$. The predictive variance is useful for finding additional experimental points.

6. Global models based on local linear models
   Linear models are approximations of unknown mean functions made by taking a Taylor expansion to the first degree. So, we cannot expect that such models provide a good approximation over all regions of regressors. LOLIMOT (LOcal LInear MOdel Tree) is a weighted sum of local linear models derived in subregions of regressors [10]. LOLIMOT needs to be tuned in many ways: One needs to determine the number of subregions, find a way to determine the subregions, and decide on how to tune the weights to the local models. These tunings are all mutually related.

7. Application to time series
Suppose we have a time series $\{(x_t, y_t) \mid t = 1, \ldots, T\}$ with $x_t = (x_{1t}, \ldots, x_{pt})^T$. Let us consider the following ARX model (AutoRegressive model with eXogenous variables).

$$y_t = \alpha_0 + \sum_{i=1}^{u} \alpha_i y_{t-i} + \sum_{j=1}^{p} \sum_{k=1}^{v_j} \beta_{jk} x_{j,t-k} + \varepsilon_t, \quad t = \max\{u, v_1, \ldots, v_p\} + 1, \ldots, T.$$

This model has an auto-regressive term. There are many candidate models involving different explanatory variables and different time delays $u,\ v_1, \ldots, v_p$.
The state-space models are generalizations of ARX models.

8. Weighted least squared method
In regression analysis, it is implicitly required to obtain a uniformly good prediction. However, there are situations in which a large target variable should be precisely predicted. Prediction of the tension of a winding process is a typical example [12]. In this case, the least weighted squares method can be employed even if the variance of the errors is common, and the Generalized Information Criterion GIC [6] can be used to make the model selection.

9. Model averaging method
Suppose that there are candidate models $M_1, \ldots,$ and M_m. Usually, the best model is selected according to a certain criterion, and it is used for prediction. In contrast, the model averaging method [2] uses all models and weights them according to the information criterion. For example, the model averaging method based on AIC is defined by

$$\hat{y} = \sum_{i=1}^{m} w_i \hat{y}_i \ \text{ with } \ w_i = \exp(-\text{AIC}_i/2) \Big/ \sum_{j=1}^{m} \exp(-\text{AIC}_j/2)$$

where AIC_i denotes the AIC value of model M_i, and $\hat{y}_i$ denotes a value predicted by M_i for the explanatory vector x. This method aims to derive a stable prediction.

10. Regression model of spatio-temporal data
Spatio-temporal data, e.g., the amount of precipitation at different sites and times, have spatial and time dependencies. The regression model of the spatio-temporal data is significantly improved by incorporating a spatio-temporal dependency into the error [11].

11. Zero-Inflated regression analysis
Consider the above-mentioned stochastic model for precipitation again. The model needs to specify zero probability of rainfall as well as a continuous density on a positive half plane. More precisely, let y be rainfall under weather condition x. Then, the zero-inflated model [11] is given by

$$p(y|x) = \Delta(x)\delta(y) + \{1 - \Delta(x)\} q(y|x) I(y > 0)$$

where $\delta(\cdot)$ is the Dirac delta function, $\Delta(x)$ is a probability $\Pr(y = 0|x)$, $q(y|x)$ is a conditional probability density function on the positive half plane, and $I(\cdot)$ denotes the indicator function. Here, the zero-inflated probability can be modeled by logistic regression [7] as $\Delta(x) = \{1 + \exp(\alpha_0 + \alpha^T x)\}^{-1}$.

6 Discussion

Many methods for conducting regression analyses have been proposed, and nonlinear regression analysis and multivariate regression analysis have been discussed. The latest developments and related software are available on the Web.

The word "regression" in regression analysis comes from a law discovered by Francis Galton (1822–1911), which is that the conditional expectation gets closer to the average. It is quite interesting that regression analysis continues to be one of the hot topics of statistics.

References

1. P. Armitage, G. Berry, J.N.S. Matthews, *Statistical Methods in Medical Research*, 4th edn. (Blackwell Publishing, Malden, 2002)
2. G. Claeskens, N.L. Hjort, *Model Selection and Model Averaging* (Cambridge University Press, Cambridge, 2008)
3. FIFA World Ranking, http://www.fifa.com/worldranking/rankingtable/, Score of 2010 Group stage, http://en.wikipedia.org/wiki/2010_FIFA_World_Cup/
4. Gamlss page (Generalized Additive Models for Location, Scale and Shape), http://www.gamlss.org/
5. T.J. Hastie, R.J. Tibshirani, *Generalized Additive Models* (Chapman and Hall, London, 1990)
6. S. Konishi, G. Kitagawa, *Information Criteria and Statistical Modeling* (Springer, New York, 2007)
7. T. Hastie, R. Tibshirani, J. Friedman, *The Elements of Statistical Learning*, 2nd edn. (Springer, New York, 2008)
8. Lasso page http://www-stat.stanford.edu/~tibs/lasso.html/
9. D. Miyata, R. Nishii, S. Tanaka, Nonlinear regression modeling of forest area ratios. Tokei Suri **60**(1), 109–119 (2012). (in Japanese)
10. O. Nelles, *Nonlinear System Identification* (Springer, New York, 2001)
11. R. Nishii, S. Tanaka, Modeling and inference of forest coverage ratio using zero-one inflated distributions with spatial dependence. Environ. Ecol. Stat. **20**, 315–336 (2013)
12. P. Qin, R. Nishii, Selection of ARX models estimated by the penalized weighted least squares method. Bull. Inf. Cybern. **42**, 35–43 (2010)
13. The R Project for Statistical Computing, http://www.r-project.org/

Stochastic Analytical Models in Mathematical Finance

Setsuo Taniguchi

Abstract Stochastic analysis is a key tool in the recent study of Mathematical Finance. Stochastic analytical models in Mathematical Finance are classified into two types. One is a discrete model, in which the trading time is restricted to the set of natural numbers, and moreover the underlying probability space is often a finite set. The other is a continuous model, which admits the trading time to be any non-negative real number. In a lot of continuous models, stochastic differential equations govern the time evolution of the models. A short survey on these two models will be given.

Keywords Equivalent martingale measure · Pricing formula · CRR model · Trinomial model · Stochastic integral · Black-Scholes model · Implied volatility · Stochastic volatility model · Single-factor model

1 Discrete Models

1.1 A Review on the Probability Theory on a Finite Set

Let Ω be a finite set, say $\Omega = \{\omega_1, \ldots, \omega_N\}$, and $\mathscr{F} = 2^\Omega$, the set of all subsets of Ω. In this case, a probability measure $\mathbf{P}$ on $(\Omega, \mathscr{F})$ is specified by a finite sequence $\{p_\alpha = \mathbf{P}(\{\omega_\alpha\})\}_{\alpha=1}^N$ so that

$$\mathbf{P}(A) = \sum_{\alpha:\omega_\alpha \in A} p_\alpha, \quad A \in \mathscr{F}.$$

S. Taniguchi (✉)
Faculty of Arts and Science, Kyushu University, 744 Motooka, Nishi-ku, Fukuoka 819-0395, Japan
e-mail: se2otngc@artsci.kyushu-u.ac.jp

R. Nishii et al. (eds.), *A Mathematical Approach to Research Problems of Science and Technology*, Mathematics for Industry 5,
DOI: 10.1007/978-4-431-55060-0_20,

Throughout this section, we assume that $p_\alpha > 0$ for every $\alpha = 1, \dots, N$. For a random variable $X : \Omega \to \mathbf{R}$, the expectation $\mathbf{E}[X]$ of X is given by

$$\mathbf{E}[X] = \sum_{\alpha=1}^{N} X(\omega_\alpha) p_\alpha.$$

The expectation of X on $A \in \mathscr{F}$, denoted by $\mathbf{E}[X; A]$, is defined to be the expectation of $X\mathbf{1}_A$, where $\mathbf{1}_A$ is the indicator function of A.

For a sub-σ-field $\mathscr{G} \subset \mathscr{F}$, the conditional expectation $\mathbf{E}[X|\mathscr{G}]$ of random variable X given $\mathscr{G}$ is a unique $\mathscr{G}$-measurable random variable satisfying that

$$\mathbf{E}[X; A] = \mathbf{E}[\mathbf{E}[X|\mathscr{G}]; A] \quad \text{for any } A \in \mathscr{G}.$$

With the help of the unique $\mathscr{G}_0 = \{A_1, \dots, A_n\} \subset \mathscr{G}$ satisfying that $A_i \cap A_j = \emptyset$ if $i \neq j$ and every $A \in \mathscr{G}$ is represented as a union of elements of $\mathscr{G}_0$, the conditional expectation is given by

$$\mathbf{E}[X|\mathscr{G}] = \sum_{i=1}^{n} \frac{\mathbf{E}[X; A_i]}{\mathbf{P}(A_i)} \mathbf{1}_{A_i}.$$

1.2 Fundamental Theorems of Mathematical Finance

Let $T \in \mathbf{N}$ and $\mathbf{T} = \{0, 1, \dots, T\}$, which is thought of as the set of trading times. An information structure on $(\Omega, \mathscr{F}, \mathbf{P})$ is an increasing sequence of sub-σ-fields of $\mathscr{F}$ such that $\{\emptyset, \Omega\} = \mathscr{F}_0 \subset \mathscr{F}_1 \subset \cdots \subset \mathscr{F}_T = \mathscr{F}$.

A market model is an $\mathbf{R}^{d+1}$-valued stochastic process $S = \{S_t = (S_t^0, \dots, S_t^d)\}_{t\in\mathbf{T}}$ such that every S_t^i is positive and $\mathscr{F}_t$-measurable, and $S_0^0 = 1$. S_t^0 represents the price of bond or safe security at time t, and S_t^i, $i = 1, \dots, d$, are those of stock or risky securities.

A trading strategy is a stochastic process $\theta = \{\theta_t = (\theta_t^0, \dots, \theta_t^d)\}_{t\in\mathbf{T}}$ with values in $\mathbf{R}^{d+1}$ such that $\{\theta_t\}_{t\in\mathbf{T}}$ is predictable: for each $0 < t < T$, θ_t is $\mathscr{F}_{t-1}$-measurable, and $\theta_0 = \theta_1$. The value process $\{V_t(\theta)\}_{t\in\mathbf{T}}$ of θ is given by

$$V_t(\theta) = \theta_t \cdot S_t = \sum_{i=0}^{d} \theta_t^i S_t^i \quad \text{for } t \in \mathbf{T}, \tag{1}$$

where $x \cdot y$ stands for the inner product of $x, y \in \mathbf{R}^{d+1}$. A trading strategy is self-financing if

$$V_t(\theta) - V_{t-1}(\theta) = \theta_t \cdot (S_t - S_{t-1}) \quad \text{for any } 1 \le t \le T. \tag{2}$$

An admissible trading strategy θ is a self-financing trading strategy such that $V_t(\theta) \geq 0$ for any $t \in \mathbf{T}$. An arbitrage opportunity is an admissible trading strategy θ with $V_0(\theta) = 0$ and $\mathbf{P}(V_T(\theta) > 0) > 0$. An arbitrage opportunity is a mathematical modeling of "free lunch".

A probability measure $\mathbf{Q}$ on $(\Omega, \mathscr{F})$ is called an equivalent martingale measure (EMM in short) if $\mathbf{Q}(\{\omega_\alpha\}) > 0$ for any $\alpha = 1, \ldots, N$ and every discounted stock price $\overline{S}^i = \{\overline{S}^i_t = \xi_t S^i_t\}_{t \in \mathbf{T}}$, where $\xi_t = \frac{1}{S^0_t}$, is a martingale under $\mathbf{Q}$: $\mathbf{E}_\mathbf{Q}[\overline{S}^i_t | \mathscr{F}_{t-1}] = \overline{S}^i_{t-1}$, $1 \leq t \leq T$, $\mathbf{E}_\mathbf{Q}$ being the expectation with respect to $\mathbf{Q}$.

Theorem 1 (The 1st fundamental theorem of mathematical finance) *The market model admits no arbitrage opportunity if and only if there exists an EMM.*

For the 1st fundamental theorem in the general setting, consult [5].

A contingent claim with maturity T is a non-negative random variable $F : \Omega \to \mathbf{R}$, which represents the payoff at time T. The claim F is said to be attainable if an admissible strategy θ replicates F: $V_T(\theta) = F$.

Theorem 2 (The 2nd fundamental theorem of mathematical finance) *The market model is complete, that is, every contingent claim is attainable, if and only if there exists only one EMM.*

For a contingent claim F, put

$$\pi_b(F) = \sup\{y \in \mathbf{R} \mid V_0(\theta) = -y,\ V_T(\theta) + F \geq 0 \text{ for some self-financing } \theta\}$$
$$\pi_s(F) = \inf\{z \in \mathbf{R} \mid V_0(\psi) = z,\ V_T(\psi) - F \geq 0 \text{ for some self-financing } \psi\}.$$

The first one is a price acceptable to a buyer, and the second is one to a seller.

Theorem 3 (Pricing formula) *Suppose that the market model possesses an EMM* $\mathbf{Q}$. *Then* $\pi_b(F) \leq \mathbf{E}_\mathbf{Q}[\xi_T F] \leq \pi_s(F)$. *Moreover, if F is attainable, then the identities hold:* $\pi_b(F) = \pi_s(F) = \mathbf{E}_\mathbf{Q}[\xi_T F]$.

1.3 CRR Model

Let $r > 0$, $b > a > -1$ and set $W_2 = \{1+a, 1+b\}$ and $\Omega = W_2^T$. Given $0 < p < 1$, define the probability measure $\mathbf{P}$ on $(\Omega, \mathscr{F})$ by

$$\mathbf{P}(\{\omega\}) = p^{\#\{t|\omega_t = 1+b\}}(1-p)^{\#\{t|\omega_t=1+a\}} \quad \text{for } \omega = (\omega_1, \ldots, \omega_T) \in \Omega.$$

Set $\mathscr{F}_0 = \{\emptyset, \Omega\}$ and $\mathscr{F}_t = \{A \times W_2^{T-t} | A \subset W_2^t\}$, $1 \leq t \leq T$. Let

$$S^0_t = (1+r)^t \quad \text{and} \quad S^1_t(\omega) = s_0 \prod_{s=1}^{t} \omega_s \quad \text{for } t \in \mathbf{T}, \tag{3}$$

where $s_0 > 0$ and $\prod_{s=1}^{0} \omega_s$ is understood to be 1. The market model $S = \{S_t = (S_t^0, S_t^1)\}_{t \in \mathbf{T}}$ is called the CRR (Cox-Ross-Rubinstein) model, which was introduced in 1979 [3]. In the CRR model, the stock price value S_t^1 varies according to the "coin-tossing". The CRR model satisfies that

Theorem 4 1. *There exists an EMM if and only if $a < r < b$.*
2. *Assume that $a < r < b$ and set $q = (r - a)/(b - a)$. Then the probability measure given by*

$$\mathbf{Q}(\{\omega\}) = q^{\#\{t|\omega_t=1+b\}}(1-q)^{\#\{t|\omega_t=1+a\}} \quad \text{for } \omega = (\omega_1, \dots, \omega_T) \in \Omega.$$

is the only one EMM.

In conjunction with the fundamental theorems, this theorem implies that in the CRR market model,

1. there is no arbitrage opportunity,
2. every contingent claim is attainable.

Moreover, the price $\pi(F)$ of contingent claim F is given by

$$\pi(F) = (1+r)^{-T} \sum_{\omega_1,\dots,\omega_T \in W_2} F(\omega_1, \dots, \omega_T) q^{\#\{t|\omega_t=1+b\}}(1-q)^{\#\{t|\omega_t=1+a\}}.$$

In particular, if $F(\omega)$ is of the form $f(\#\{t|\omega_t = 1 + b\})$, then

$$\pi(F) = (1+r)^{-T} \sum_{s=0}^{T} \binom{T}{s} f(s) q^s (1-q)^{T-s}.$$

A typical example of contingent claim of this kind is the call option with strike price $K > 0$: $F = (S_T - K)^+$, where $a^+ = \max\{a, 0\}$. Its price is given by:

$$\pi((S_T - K)^+) = (1+r)^{-T} \sum_{s=A}^{T} \binom{T}{s} q^s (1-q)^{T-s} \{s_0(1+b)^s(1+a)^{T-s} - K\},$$

where $A = \min\{s \in \mathbf{T} | s_0(1+b)^s(1+a)^{T-s} - K \geq 0\}$.

The put option with strike price $K > 0$ is a contingent claim with payoff $(K - S_T)^+$. Since $\mathbf{E}_{\mathbf{Q}}[\overline{S}_T^1] = s_0$, the put-call parity holds:

$$\pi((S_T - K)^+) - \pi((K - S_T)^+) = s_0 - (1+r)^{-T} K.$$

It should be noted that $(1+r)^{-T} K$ is the discounted value of the strike price at $t = 0$.

For $\omega = (\omega_1, \dots, \omega_T) \in \Omega$ and $t \leq T$, define

$$\omega^{(t)} = (\omega_1, \dots, \omega_t), \quad \omega_a^{(t)} = (\omega_1, \dots, \omega_t, 1+a), \quad \omega_b^{(t)} = (\omega_1, \dots, \omega_t, 1+b)$$

Since S_t^1 is a function of $\omega^{(t)}$ and θ_t is that of $\omega^{(t-1)}$ for each $t \in \mathbf{T}$, a self-financing trading strategy θ replicating $f(S_T^1)$, f being a non-negative function, is computed backward as follows:

$$\begin{pmatrix} \theta_T^0(\omega^{(T-1)}) \\ \theta_T^1(\omega^{(T-1)}) \end{pmatrix} = \frac{1}{b-a} \begin{pmatrix} (1+b)(1+r)^{-T} & -(1+a)(1+r)^{-T} \\ -(S_{T-1}^1(\omega^{(T-1)}))^{-1} & (S_{T-1}^1(\omega^{(T-1)}))^{-1} \end{pmatrix} \times \begin{pmatrix} f((1+u)S_{T-1}^1(\omega^{(T-1)})) \\ f((1+b)S_{T-1}^1(\omega^{(T-1)})) \end{pmatrix},$$

and, for $t < T$,

$$\begin{pmatrix} \theta_t^0(\omega^{(t-1)}) \\ \theta_t^1(\omega^{(t-1)}) \end{pmatrix} = \frac{1}{b-a} \begin{pmatrix} (1+b)(1+r)^{-t} & -(1+a)(1+r)^{-t} \\ -(S_{t-1}^1(\omega^{(t-1)}))^{-1} & (S_{t-1}^1(\omega^{(t-1)}))^{-1} \end{pmatrix} \times \begin{pmatrix} (1+r)^t\theta_{t+1}^0(\omega_a^{(t-1)}) + (1+a)S_{t-1}^1(\omega^{(t-1)})\theta_{t+1}^1(\omega_a^{(t-1)}) \\ (1+r)^t\theta_{t+1}^0(\omega_b^{(t-1)}) + (1+b)S_{t-1}^1(\omega^{(t-1)})\theta_{t+1}^1(\omega_b^{(t-1)}) \end{pmatrix}.$$

The first identity comes from replicating and the second does from being self-financing.

1.4 Trinomial Model

Let $r > 0$, $a_1 > a_2 > a_3 > -1$ and take $p_1, p_2, p_3 > 0$ with $p_1 + p_2 + p_3 = 1$. Set $W_3 = \{1+a_1, 1+a_2, 1+a_3\}$, $\Omega = W_3^T$, and $\mathscr{F}_0 = \{\emptyset, \Omega\}$, $\mathscr{F}_t = \{A \times W_3^{T-t} | A \subset W_3^t\}$, $1 \le t \le T$. Define the probability measure $\mathbf{P}$ on $(\Omega, \mathscr{F})$ by

$$\mathbf{P}(\{\omega\}) = \prod_{i=1}^{3} p_i^{\#\{t|\omega_t = 1+a_i\}} \quad \text{for } \omega = (\omega_1, \ldots, \omega_T) \in \Omega,$$

and the asset price processes $\{S_t^0\}_{t\in\mathbf{T}}$ and $\{S_t^1\}_{t\in\mathbf{T}}$ by Eq. 3. The market model $S = \{S_t = (S_t^0, S_t^1)\}_{t\in\mathbf{T}}$ is called the trinomial model.

A probability measure on $(\Omega, \mathscr{F})$ is characterized as follows:

Theorem 5 *Let $\mathscr{P}$ be the set of all probability measures on $(\Omega, \mathscr{F})$ and $\mathscr{C}$ be the set of sequences $\mathbf{f} = \{f_t^\delta | \delta \in W_3, 1 \le t \le T\}$ of functions $f_t^\delta : W_3^{t-1} \to (0, 1)$ with $\sum_{\delta\in W_3} f_t^\delta = 1$ $(1 \le t \le T)$. Then,*

1. *$Q \in \mathscr{P}$ and $\mathbf{f} \in \mathscr{C}$ are in the one-to-one correspondence through the relationship:*

$$\mathbf{Q}(\{(\omega_1, \ldots, \omega_T)\}) = \prod_{t=1}^{T} f_t^{\omega_t}(\omega_1, \ldots, \omega_{t-1}),$$

$$\mathbf{E_Q}[\mathbf{1}_{\{R_t=\delta\}}|\mathscr{F}_{t-1}] = f_t^\delta \quad \textit{for } \delta \in W_3, 1 \le t \le T, \quad \textit{where } R_t(\omega) = \omega_t.$$

2. *R_t, $1 \le t \le T$, are independent under* $\mathbf{Q}$ *if and only if every* f_t^δ, $\delta \in W_3$, $1 \le t \le T$, *is a constant function.*
3. *R_t, $1 \le t \le T$, are independent and identically distributed under* $\mathbf{Q}$ *if and only if every* f_t^δ, $\delta \in W_3$, $1 \le t \le T$, *is a constant function and* $f_t^\delta = f_s^\delta$ *for any* $\delta \in W_3$, $1 \le s, t \le T$.

The above one-to-one correspondence, and hence the derivation of f_t^δ's, is found in the following expression of $\mathbf{Q}(\{(\delta_1, \dots, \delta_T)\})$ for $\delta_1, \dots, \delta_T \in W_3$:

$$\mathbf{Q}(\{(\delta_1, \dots, \delta_T)\}) = \mathbf{E}\Big[\prod_{t=1}^{T} \mathbf{1}_{\{R_t=\delta_t\}}\Big] = \mathbf{E}\Big[\mathbf{E}[\mathbf{1}_{\{R_T=\delta_T\}}|\mathscr{F}_{T-1}]\Big(\prod_{t=1}^{T-1} \mathbf{1}_{\{R_t=\delta_t\}}\Big)\Big]$$
$$= \mathbf{E}[\mathbf{1}_{\{R_T=\delta_T\}}|\mathscr{F}_{T-1}](\delta_1, \dots, \delta_{T-1})\mathbf{E}\Big[\prod_{t=1}^{T-1} \mathbf{1}_{\{R_t=\delta_t\}}\Big].$$

Theorem 6 1. *There exists an EMM if and only if* $a_3 < r < a_1$.

2. *Suppose* $a_3 < r < a_1$. $\mathbf{Q}$ *is an EMM if and only if the corresponding* $\mathbf{f} = \{f_t^\delta | \delta \in W_3, 1 \le t \le T\}$ *satisfies that*

$$\sum_{i=1}^{3} a_i f_t^{1+a_i} = r \quad \text{for } 1 \le t \le T.$$

In particular, there exist infinitely many EMM's.

3. *If* $a_3 < r < a_1$, *then the trinomial model admits no arbitrage opportunity, but is not complete.*
4. *For* $\eta \in W_3^{T-1}$, *set* $\widehat{\eta_i} = (\eta, 1 + a_i) \in W_3^T$, $i = 1, 2, 3$. *If a contingent claim* F *is replicated by* θ, *then it holds that*

$$F(\widehat{\eta_3}) = \frac{(a_2 - a_3)F(\widehat{\eta_1}) - (a_1 - a_3)F(\widehat{\eta_2})}{a_2 - a_1} \quad \text{for } \eta \in W_3^{T-1}.$$

2 Continuous Models

2.1 Stochastic Integral and Stochastic Differential Equation

Let $T > 0$, $\mathbf{T} = [0, T]$, and $\{B_t\}_{t\in\mathbf{T}}$ be a one-dimensional Brownian motion on the probability space $(\Omega, \mathscr{F}, \mathbf{P})$: (i) $B_0(\omega) = 0$ and $\mathbf{T} \ni t \mapsto B_t(\omega) \in \mathbf{R}$ is continuous for every $\omega \in \Omega$, and (ii) for any $0 = t_0 < t_1 < \dots < t_n \le T$, $B_{t_{i+1}} - B_{t_i}$, $i = 0, \dots, n-1$, are independent and each $B_{t_{i+1}} - B_{t_i}$ obeys the normal distribution with mean 0 and variance $t_{i+1} - t_i$. Let $\mathscr{F}_t$ be the smallest σ-field containing all null sets and $\{B_s \le a\}$, $s \le t$, $a \in \mathbf{R}$, and assume that $\mathscr{F} = \mathscr{F}_T$.

A stochastic process $\{\theta_t\}_{t\in\mathbf{T}}$ is said to be progressively measurable if for each t, the mapping $[0,t]\times\Omega\ni(s,\omega)\mapsto\theta_s(\omega)$ is $\mathscr{B}([0,t])\times\mathscr{F}_t$-measurable, where $\mathscr{B}([0,t])$ is the Borel σ-field of $[0,t]$ and $\mathscr{B}([0,t])\times\mathscr{F}_t$ the product σ-field of $\mathscr{B}([0,t])$ and $\mathscr{F}_t$.

Let $\mathscr{L}^2_{\text{loc}}$ be the totality of progressively measurable $\{\theta_t\}_{t\in\mathbf{T}}$ with $\mathbf{P}(\int_0^T\theta_s^2ds<\infty)=1$. Denote by $\mathscr{L}_0$ the set of all $\{\theta_t\}_{t\in\mathbf{T}}\in\mathscr{L}^2_{\text{loc}}$ such that for some $0=t_0<t_1<\ldots,<t_n=T, \theta_t=\theta_{t_i}$ if $t\in[t_i,t_{i+1}), i=0,\ldots,n-1$, and $\sup_{t\in\mathbf{T},\omega\in\Omega}|\theta_t(\omega)|<\infty$. The stochastic integral $\int_0^t\theta_s dB_s$ of $\{\theta_t\}_{t\in\mathbf{T}}\in\mathscr{L}^2_{\text{loc}}$ is defined as follows: (1) if $\{\theta_t\}_{t\in\mathbf{T}}\in\mathscr{L}_0$, then

$$\int_0^t\theta_s\mathrm{d}B_s=\sum_{i=0}^{n-1}\theta_{t_i}\{B_{t\wedge t_{i+1}}-B_{t\wedge t_i}\},\quad t\in\mathbf{T},$$

(2) for general $\{\theta_t\}_{t\in\mathbf{T}}\in\mathscr{L}^2_{\text{loc}}$, taking a sequence $\{\{\theta_t^{(n)}\}_{t\in\mathbf{T}}\}_{n=1}^\infty\subset\mathscr{L}_0$ such that $\int_0^T|\theta_s^{(n)}-\theta_s|^2ds\to0$ in probability ($n\to\infty$), that is, $\mathbf{E}[\{\int_0^T|\theta_s^{(n)}-\theta_s|^2ds\}\wedge1]\to0$, define $\{\int_0^t\theta_s dB_s\}_{t\in\mathbf{T}}$ to be the stochastic process such that

$$\sup_{t\in\mathbf{T}}\left|\int_0^t\theta_s^{(n)}\mathrm{d}B_s-\int_0^t\theta_s\mathrm{d}B_s\right|\to0\ \text{ in probability.}$$

A stochastic process $\{\xi_t\}_{t\in\mathbf{T}}$ represented as

$$\xi_t=\xi_0+\int_0^t\theta_s\mathrm{d}B_s+\int_0^t\beta_s\mathrm{d}s,\quad t\in\mathbf{T}$$

for some $\{\theta_t\}_{t\in\mathbf{T}}$, $\{\beta_t\}_{t\in\mathbf{T}}$ is often called an Itô process, and symbolically denoted by

$$\mathrm{d}\xi_t=\theta_t\mathrm{d}B_t+\beta_t\mathrm{d}t.$$

For continuous $\sigma, b:\mathbf{R}\to\mathbf{R}$, a solution to the stochastic differential equation (SDE in short)

$$\mathrm{d}X_t=\sigma(X_t)\mathrm{d}B_t+b(X_t)\mathrm{d}t \tag{4}$$

with initial condition $X_0=x$ is an $\{X_t\}_{t\in\mathbf{T}}\in\mathscr{L}^2_{\text{loc}}$ such that

$$X_t=x+\int_0^t\sigma(X_s)\mathrm{d}B_s+\int_0^t b(X_s)\mathrm{d}s,\quad t\in\mathbf{T}.$$

It is known that if σ, b are both globally Lipschitz continuous, then the SDE Eq. 4 possesses a unique solution. The SDE describes the time evolution of random events, and hence is thought of as a probabilistic counterpart to the Newton equation.

For details of stochastic integrals and SDE's, see [11, 13, 16]. The above definition of stochastic integrals is different from the original one due to K. Itô [12], and is found in [13].

2.2 Black-Scholes Model

2.2.1 Market Model

The Black-Scholes market model (BM model in short) consists of two assets, one is safe and the other is risky, as the CRR model.

Let $r, \mu \geq 0, \sigma > 0$. The price processes $\{S_t^0\}_{t\in\mathbf{T}}$ and $\{S_t^1\}_{t\in\mathbf{T}}$ of the safe and risky assets, respectively, are given by

$$S_t^0 = e^{rt}, \quad S_t^1 = s_0 \exp\left(\left\{\mu - \frac{\sigma^2}{2}\right\} t + \sigma B_t\right), \quad t \in \mathbf{T},$$

where $s_0 > 0$. They obey the SDE

$$\mathrm{d}S_t^0 = rS_t^0 \mathrm{d}t, \quad \mathrm{d}S_t^1 = \mu S_t^1 \mathrm{d}t + \sigma S_t^1 \mathrm{d}B_t.$$

Hence r is the force of return of the safe asset, μ is a true value of the force of return of the risky asset, which is perturbed randomly by the Brownian motion. The constant σ is called a volatility and used to evaluate the level of risk.

The idea of using the Brownian motion as the price process goes back to 1900 [1], and the price process of the above form does to 1960s [2, 14, 17, 18].

2.2.2 Trading Strategy

A trading strategy $\theta = \{\theta_t\}_{t\in\mathbf{T}}$ is a progressively measurable $\mathbf{R}^2$-valued process and its value process $\{V_t(\theta)\}_{t\in\mathbf{T}}$ is defined by Eq. 1:

$$V_t(\theta) = \theta_t^0 S_t^0 + \theta_t^1 S_t^1, \quad t \in \mathbf{T},$$

where we represented θ_t as (θ_t^0, θ_t^1). A trading strategy $\theta = \{\theta_t\}_{t\in\mathbf{T}}$ is said to be self-financing if its components $\{\theta_t^i\}_{t\in\mathbf{T}}$ are in $\mathscr{L}^2_{\mathrm{loc}}$ and it holds that

$$\mathrm{d}V_t(\theta) = \theta_t^0 \mathrm{d}S_t^0 + \theta_t^1 \mathrm{d}S_t^1 = \{r\theta_t^0 S_t^0 + \mu\theta_t^1 S_t^1\}\mathrm{d}t + \sigma\theta_t^1 S_t^1 \mathrm{d}B_t.$$

The totality of all self-financing strategy is denoted by $\mathscr{S}_{\rm sf}$. As is easily guessed, the formula is obtained via a "limiting procedure" from Eq. 2: take $\{0, \frac{1}{n}, \frac{2}{n}, \dots, \frac{nT}{n} = T\}$ instead of $\{0, 1, \dots, T\}$, and then let $n \to \infty$. It should be also mentioned that

Proposition 1 *Given $\{\theta_t^1\}_{t\in\mathbf{T}} \in \mathscr{L}^2_{\rm loc}$ and $a \in \mathbf{R}$, there exists $\{\theta_t^0\}_{t\in\mathbf{T}} \in \mathscr{L}^2_{\rm loc}$ such that $\{(\theta_t^0, \theta_t^1)\}_{t\in\mathbf{T}} \in \mathscr{S}_{\rm sf}$ and $V_0(\theta) = a$.*

A self-financing trading strategy $\{\theta_t\}_{t\in\mathbf{T}}$ is an arbitrage opportunity if $V_0(\theta) = 0$, $V_T(\theta) \geq 0$, and $\mathbf{P}(V_T(\theta) > 0) > 0$. Let $\mathscr{S}_{\rm arb}$ be the set of all arbitrage opportunity.

Example 1 Suppose that $r = \mu = 0$ and $a > 0$. Under this assumption, $\mathrm{d}S_t^0 = 0$ and $\mathrm{d}S_t^1 = \sigma S_t^1 \mathrm{d}B_t$. Set $Y_t = \int_0^t \frac{1}{\sqrt{T-s}} \mathrm{d}B_s$, $t < T$, and $\tau_a = \inf\{t | Y_t = a\}$. Then the time change argument yields that $\mathbf{P}(\tau_a < T) = 1$.

Let $\theta_t^1 = \frac{1}{\sigma S_t^1 \sqrt{T-t}} \mathbf{1}_{[0,\tau_a)}(t)$, $t \in \mathbf{T}$. On account of Proposition 1, take $\{\theta_t^0\}_{t\in\mathbf{T}} \in \mathscr{L}^2_{\rm loc}$ so that $\theta = \{(\theta_t^0, \theta_t^1)\}_{t\in\mathbf{T}} \in \mathscr{S}_{\rm sf}$ and $V_0(\theta) = 0$. Since $dS_t^0 = 0$, it then holds that

$$V_t(\theta) = \int_0^t \theta_s^1 \mathrm{d}S_s^1 = Y_{t\wedge\tau_a}.$$

In particular, $V_T(\theta) = Y_{\tau_a} = a > 0$. Thus $\theta \in \mathscr{S}_{\rm arb}$ and hence $\mathscr{S}_{\rm arb} \neq \emptyset$.

2.2.3 EMM

A probability measure $\mathbf{Q}$ on $(\Omega, \mathscr{F})$ is said to be an equivalent martingale measure (EMM in short) if $\{\overline{S}_t^1 = \xi_t S_t^1\}_{t\in\mathbf{T}}$, where $\xi_t = \frac{1}{S_t^0}$, is a martingale under $\mathbf{Q}$. In the BS model, one can directly construct an EMM as follows:

Theorem 7 *Let $\alpha = \frac{r-\mu}{\sigma}$ and define the probability measure $\mathbf{Q}$ on $(\Omega, \mathscr{F})$ by*

$$\mathbf{Q}(A) = \mathbf{E}[e^{\alpha B_T - \frac{\alpha^2 T}{2}}; A], \quad A \in \mathscr{F}.$$

Then,

1. *$\{\overline{B}_t = B_t - \alpha t\}_{t\in\mathbf{T}}$ is a Brownian motion under $\mathbf{Q}$,*
2. *$\mathbf{Q}$ is an EMM,*
3. *$\{\overline{V}_t(\theta) = \xi_t V_t(\theta)\}_{t\in\mathbf{T}}$ is a local martingale under $\mathbf{Q}$ for any $\theta \in \mathscr{S}_{\rm sf}$.*

As in the discrete case, the existence of EMM implies that the BS model possesses no arbitrage opportunity in some restricted classes of self-financing trading strategies (cf. Example 1). To state this, we introduce two classes of admissible strategies:

$$\mathscr{S}_{\text{adm},2} = \left\{ \{\theta_t\}_{t\in\mathbf{T}} \in \mathscr{S}_{\text{sf}} \middle| \mathbf{E}_{\mathbf{Q}} \left[\int_0^T (\theta_t^1 S_t^1)^2 dt \right] < \infty \right\},$$

$$\mathscr{S}_{\text{adm},0} = \left\{ \{\theta_t\}_{t\in\mathbf{T}} \in \mathscr{S}_{\text{sf}} \middle| \inf_{t\le T, \omega\in\Omega} V_t(\theta)(\omega) > -\infty \right\}.$$

The BS model possesses no arbitrage opportunity in the sense that

$$\mathscr{S}_{\text{arb}} \cap \mathscr{S}_{\text{adm},2} = \emptyset, \quad \mathscr{S}_{\text{arb}} \cap \mathscr{S}_{\text{adm},0} = \emptyset.$$

See [6, 16].

2.2.4 Pricing Formula

A contingent claim is an $\mathscr{F}$-measurable function, which is bounded from below. $\mathscr{C}_0$ stands for the totality of all contingent claims, and $\mathscr{C}_2$ does the set of all $F \in \mathscr{C}_0$ with $\mathbf{E}_{\mathbf{Q}}[F^2] < \infty$. For $\alpha = 0$ or 2, $F \in \mathscr{C}_\alpha$ is said to be attainable if there is a $\theta \in \mathscr{S}_{\text{adm},\alpha}$ replicating F: $V_T(\theta) = F$. If F is in $\mathscr{C}_2$ or $\mathscr{C}_0 \cap L^1(\mathbf{Q})$, then it is replicated by θ with $V_0(\theta) = \mathbf{E}_{\mathbf{Q}}[\xi_T F] = e^{-rT}\mathbf{E}_{\mathbf{Q}}[F]$.

For $\alpha = 0$ or 2, two prices of contingent claim F by buyers and sellers are defined by

$$\pi_b^\alpha(F) = \sup\{y \in \mathbf{R} \mid V_0(\theta) = -y,\ V_T(\theta) + F \ge 0 \text{ for some } \theta \in \mathscr{S}_{\text{adm},\alpha}\}$$
$$\pi_s^\alpha(F) = \inf\{z \in \mathbf{R} \mid V_0(\psi) = z,\ V_T(\psi) - F \ge 0 \text{ for some } \psi \in \mathscr{S}_{\text{adm},\alpha}\}.$$

Theorem 8 1. *For $F \in \mathscr{C}_0$, $\pi_b^0(F) \le \pi_s^0(F)$. If, in addition, $F \in L^1(\mathbf{Q})$, then*

$$\pi_b^0(F) = \pi_s^0(F) = e^{-rT}\mathbf{E}_{\mathbf{Q}}[F].$$

2. *If $F \in \mathscr{C}_2$, then*

$$\pi_b^2(F) = \pi_s^2(F) = e^{-rT}\mathbf{E}_{\mathbf{Q}}[F].$$

In conjunction with the representation $S_T = s_0 e^{rT}\exp(\sigma\overline{B}_T - \frac{\sigma^2}{2}T)$, this yields that the price $\pi(f(S_T))$ of the contingent claim $f(S_T)$ is given by

$$\pi(f(S_T)) = e^{-rT} \int_{\mathbf{R}} f\left(s_0 e^{(r-\frac{\sigma^2}{2})T} e^x\right) \frac{1}{\sqrt{2\pi\sigma^2 T}} e^{-\frac{x^2}{2\sigma^2 T}} \mathrm{d}x. \quad (5)$$

2.2.5 Implied Volatility

The payoff C of the European call option with strike price $K > 0$ is $(S_T - K)^+$. By Eq. 5, it holds that

$$\pi(C) = s_0\Phi(d_+) - Ke^{-rT}\Phi(d_-),$$

where

$$\Phi(x) = \int_{-\infty}^{x} \frac{1}{\sqrt{2\pi}} e^{-\frac{y^2}{2}} dy, \quad d_\pm = \frac{1}{\sigma\sqrt{T}}\left(\log\left(\frac{s}{K}\right) + \left(r \pm \frac{\sigma^2}{2}\right)T\right).$$

Denote $\pi(C;\sigma)$ instead of $\pi(C)$ in order to emphasize the dependence on σ. Then $\frac{\partial}{\partial\sigma}\pi(C;\sigma) > 0$, that is, $\pi(C;\sigma)$ is a strictly increasing function of σ. Thus, for $\gamma \geq 0$, one finds the unique σ_γ so that $\pi(C;\sigma_\gamma) = \gamma$. This σ_γ is called an implied volatility.

The payoff P of the European call option with strike price $K > 0$ is $(K - S_T)^+$. By Eq. 5, it holds that

$$\pi(P) = Ke^{-rT}\Phi(-d_-) - s_0\Phi(-d_+).$$

Since $\Phi(-x) = 1 - \Phi(x)$, the put-call parity holds:

$$\pi(C) - \pi(P) = s_0 - Ke^{-rT}.$$

Denoting $\pi(P;\sigma)$ instead of $\pi(P)$ to emphasize the dependence on σ, it is also seen that $\pi(P;\sigma)$ is a strictly increasing function of σ. Thus, for $\gamma \geq 0$, via the option pricing formula for the put option, one finds the implied volatility σ_γ so that $\pi(P;\sigma_\gamma) = \gamma$.

Observe that the pricing formula Eq. 5 involves only σ, but not μ. Thus if one knows the volatility σ, the pricing formula works. Hence from the practical point of view, it is necessary to determine the σ through the market data, and for this purpose the implied volatility is used: once the market price of the call (or put) option is observed and the implied volatility is computed, all the above pricing formulas start working.

2.3 *Stochastic Volatility Model*

In the BS market model, the volatility is a fixed constant. Pragmatically speaking, it seems more natural to assume that the volatility varies randomly. The models of this kind are called stochastic volatility models and are determined by the SDE of the form

$$\begin{cases} dS_t = \mu(S_t, t)dt + \sigma_t S_t dB_t, \\ d\sigma_t = a(\sigma_t, t)dt + b(\sigma_t, t)dW_t, \end{cases}$$

where $\{B_t\}_{t\in\mathbf{T}}$, $\{W_t\}_{t\in\mathbf{T}}$ are both one-dimensional Brownian motions, which may not be independent.

There are several famous stochastic volatility models. Wiggins [21] considered the model

$$\begin{cases} dS_t = \mu S_t dt + \sigma_t dB_t, \\ d\sigma_t = a(\sigma_t)dt + \theta\sigma_t dW_t, \end{cases}$$

and Heston [8] did, putting $\nu_t = \sigma_t^2$,

$$\begin{cases} dS_t = \mu S_t dt + \sqrt{\nu_t} S_t dB_t, \\ d\nu_t = \kappa(\nu - \nu_t)dt + \eta\sqrt{\nu_t} dW_t, \end{cases}$$

where μ, κ, ν, η are all constants. In both models, S_t possesses concrete expressions:

$$S_t = s_0 e^{\mu t} \int_0^t e^{-\mu s} \sigma_s dB_s \qquad \text{(the Wiggins model)},$$

$$S_t = s_0 \exp\left(\int_0^t \sqrt{\nu_s} dB_s + \left\{ \mu - \frac{1}{2} \int_0^t \nu_s ds \right\} \right) \qquad \text{(the Heston model)},$$

while neither σ_t nor ν_t is represented explicitly.

2.4 Short-Term Rate Models: Single-Factor Models

There is a lot of derivatives whose prices vary in response to interest rates, like bond options, swaps, swaptions, and so on. In this subsection, an application of SDE's to the study of interest rates is shown.

Let $T^* > 0$ be a horizon date for all market activities, and $B(t, T)$ be the price at time t of a zero coupon bond, whose holder is repaid one unit cash at maturity time $T \le T^*$. The yield of the zero coupon bond is given by $Y(t, T) = -\frac{1}{T-t} \log B(t, T)$, and the term structure of interest rates (or the yield curve) is the function $T \mapsto Y(t, T)$. The instantaneous forward rate $f(t, T)$ and the instantaneous short-term rate r_t, $0 \le t \le T \le T^*$, are defined by

$$f(t, T) = -\frac{\partial \log B(t, T)}{\partial T}, \quad r_t = f(t, t),$$

respectively.

While r_t has less information than $f(t, T)$ and hence $B(t, T)$, there are several models where $\{B(t, T)\}_{t \leq T}$ can be recovered by $\{r_t\}_{t \leq T^*}$: namely, suppose that there exist a progressively measurable process $r = \{r_t\}_{t \leq T^*}$ on a filtered probability space $(\Omega, \mathscr{F}, \mathbf{P}, \{\mathscr{F}_t\}_{t \leq T^*})$ and a probability measure $\mathbf{P}^*$ such that (i) $\mathbf{P}^*$ is equivalent to $\mathbf{P}$, and (ii) the discounted price process $\{\xi_t B(t, T)\}_{t \leq T}$, where $\xi_t = \exp(-\int_0^t r_s ds)$, is a martingale under $\mathbf{P}^*$. Then the family $\{B(t, T)\}_{t \leq T \leq T^*}$ is called an arbitrage-free family of bond prices relative to $r = \{r_t\}_{t \leq T^*}$, and it holds that

$$B(t, T) = \mathbf{E}_{\mathbf{P}^*}\left[e^{-\int_t^T r_s \mathrm{d}s} \middle| \mathscr{F}_t\right], \quad t \leq T.$$

From this it follows that $r_t = -\frac{\partial}{\partial T}\Big|_{T=t} \log B(t, T)$, i.e., r_t is the instantaneous short-term rate.

In the remainder, several arbitrage-free families of bond prices, whose short-term rate processes $\{r_t\}_{t \leq T^*}$ obey SDE's governed by a one-dimensional Brownian motion $\{B_t\}_{t \leq T^*}$, will be given. Since the Brownian motion is one-dimensional, the models are called single-factor models.

- **Merton model**: [14] $\{r_t\}_{t \leq T^*}$ obeys the SDE

$$\mathrm{d}r_t = a\mathrm{d}t + \sigma \mathrm{d}B_t,$$

where $a, \sigma > 0$ are constants. In this case,

$$r_t = r_0 + at + \sigma B_t.$$

The unwelcome event $\{r_t < 0\}$ occurs with positive probability.

- **Vasicek model**: [20] $\{r_t\}_{t \leq T^*}$ obeys the SDE

$$\mathrm{d}r_t = (a - br_t)\mathrm{d}t + \sigma \mathrm{d}B_t,$$

where $a, b, \sigma > 0$ are constants. In this case,

$$r_t = r_0 e^{-bt} + \frac{a}{b}\left(1 - e^{-bt}\right) + \sigma \int_0^t e^{-b(t-u)} \mathrm{d}B_u.$$

The event $\{r_t < 0\}$ also occurs with positive probability.

- **Cox-Ingersoll-Ross (CIR) model**: [4] $\{r_t\}_{t \leq T^*}$ obeys the SDE

$$\mathrm{d}r_t = (a - br_t)\mathrm{d}t + \sigma \sqrt{r_t}\, \mathrm{d}B_t,$$

where $a, b, \sigma > 0$ are constants. Using the Bessel processes, one can pull out several properties of $\{r_t\}_{t \in \mathbf{T}}$. In particular, one can show the positivity of r_t. It should be noted that the SDE for the CIR model is exactly the same as the one determining the stochastic volatility in the Heston model.

- **Hull-White model**: [10] $\{r_t\}_{t\le T^*}$ obeys the SDE

$$\mathrm{d}r_t = (a(t) - b(t)r_t)\mathrm{d}t + \sigma(t)\mathrm{d}B_t,$$

where a, b, σ are functions from $[0, \infty)$ to $(0, \infty)$. The solution is given by

$$r_t = e^{-\ell(t)}\left(r_0 + \int_0^t e^{\ell(u)}a(u)\mathrm{d}u + \int_0^t e^{\ell(u)}\sigma(u)\mathrm{d}B_u\right), \quad \ell(t) = \int_0^t b(u)\mathrm{d}u.$$

Again the probability $\mathbf{P}(r_t < 0)$ may be positive.

Bibliographical comment

Those who are interested in the details of the subjects, see for example the books by Duffie [6], Elliott-Kopp [7], Musiela-Rutkowski [15], Øksendal [16], and Shreve [19]. For mathematical financial overview, see the book by Hull [9].

References

1. L. Bachelier, Théorie de la spéculation. Ann. Sci. Ecole Norm. Super. **17**, 21–86 (1900)
2. F. Black, M. Scholes, The pricing of options and corporate liabilities. J. Polit. Econ. **81**, 637–654 (1973)
3. J.C. Cox, S.A. Ross, M. Rubinstein, Option pricing: a simplified approach. J. Finan. Econ. **7**, 229–263 (1979)
4. J.C. Cox, J.E. Ingersoll, S.A. Ross, A theory of the term structure of interest rates. Econometrica **53**, 385–407 (1985)
5. F. Delbean, W. Schachermayer, *The Mathematics of Arbitrage* (Springer, New York, 2006)
6. D. Duffie, *Dynamic Asset Pricing Theory*, 2nd edn. (Princeton University Press, Princeton, 1996)
7. E. Elliott, P.E. Kopp, *Mathematics of Financial Markets* (Springer, New York, 1999)
8. S.L. Heston, A closed-form solution for options with stochastic volatility with applications to bond and currency options. Rev. Financ. Stud. **6**, 327–343 (1993)
9. J. Hull, *Introduction to Futures and Option Markets*, 3rd edn. (Prentice Hall, Upper Saddle River, 1998)
10. J. Hull, A. White, Pricing interest-rate derivative securities. Rev. Financ. Stud. **3**, 573–592 (1990)
11. N. Ikeda, S. Watanabe, *Stochastic Differential Equations and Diffusion Processes*, 2nd edn. (North Holland, Amsterdam, 1989)
12. K. Itô, Differential equations determining Markov processes. Zenkoku-Shijo-Danwakai **244**, 1352–1400 (1942). (in Japanese)
13. H.P. McKean, *Stochastic Integrals*, (Academic Press, New York, 1969)
14. R. Merton, Theory of rational option pricing. Bell J. Econ. Manage. Sci. **4**, 141–183 (1973)
15. M. Musiela, M. Rutkowski, *Martingale Methods in Financial Modeling*, 2nd edn. (Springer, New York, 2005)
16. B. Øksendal, *Stochastic Differential Equations. An Introduction with Applications*, 6th edn. (Springer, New York, 2003)

17. P. Samuelson, Proof that properly anticipated prices fluctuate randomly. Ind. Manage. Rev. **6**, 41–49 (1965)
18. P. Samuelson, Rational theory of warrant pricing, Ind. Manage. Rev. **6**, 13–39 (1965)
19. S. Shreve, *Stochastic Calculus for Finance I: The Binomial Asset Pricing Model, II: Continuous-Time Models* (Springer, New York, 2004)
20. O. Vasicek, An equilibrium characterisation of the term structure. J. Financ. Econ. **5**, 177–188 (1977)
21. J.B. Wiggins, Option values under stochastic volatility; theory and empirical estimates. J. Financ. Econ. **19**, 351–372 (1987)

An Introduction to the Minimum Description Length Principle

Jun'ichi Takeuchi

Abstract We give a brief introduction to the minimum description length (MDL) principle. The MDL principle is a mathematical formulation of Occam's razor. It says '*simple explanations of a given phenomenon are to be preferred over complex ones.*' This is recognized as one of basic stances of scientists, and plays an important role in statistics and machine learning. In particular, Rissanen proposed MDL criterion for statistical model selection based on information theory in 1978. After that, much literature has been published and the notion of MDL principle was founded in the 1990s. In this article, we review some important results on the MDL principle.

Keywords Bayes mixture · Laplace estimator · MDL · Model selection · Minimax regret · Universal code

1 Introduction

The minimum description length (MDL) principle [10, 11, 18] is one of key concepts in information science. It is a mathematical formulation of Occam's razor from William of Ockham (see [16], e.g.,) in the fourteenth century, '*simple explanations of a given phenomenon are to be preferred over complex ones* [10].' This idea is applicable in particular to machine learning and statistical estimation, where the simpler hypothesis is recognized as better one among competing ones. In particular, Rissanen formulated it based on information theory [18] under an influence of Akaike's Information Criterion (AIC) [1]. He founded a quantitative formalization of Occam's razor using the codelength in universal data compression and named it the MDL principle.

J. Takeuchi (✉)
Kyushu University, 744, Motooka, Nishiku, Fukuoka 819-0395, Japan
e-mail: tak@inf.kyushu-u.ac.jp

R. Nishii et al. (eds.), *A Mathematical Approach to Research Problems of Science and Technology*, Mathematics for Industry 5,
DOI: 10.1007/978-4-431-55060-0_21,

The MDL principle says that we should pursuit the shortest codelength, to obtain a good performance in various problems in information science including machine learning. The most basic recognition here is that there is essentially one to one correspondence between a code for data compression and a probability distribution of the data. Shannon found that if a data string $x^n \in \mathscr{X}^n$ is distributed according to a probability function $p(x^n)$, that is, the information source is p, the decodable code which minimizes the expected codelength has the code length $L(x^n) = -\log_2 p(x^n)$ bit, where we neglect values less than 1 bit. Based on this recognition, we can think of a probability distribution as a code, hence we often refer to a probability function $p(x)$ as code.

In particular for statistical model selection, the MDL principle takes a form "use the model by which the observed data is encoded into the shortest code." We note that "shortest code" is defined in terms of *regret* in universal data compression [20]. Here, the regret of a universal code q designed for the considered model for a data string $x^n = x_1 x_2 \ldots x_n$ is the difference between the codelength by q and the codelength by the hindsight best code for x^n in the considered model. Further, the MDL principle says that we should design the code q so that q minimizes the worst case regret for various data strings. For an observed data, the codelength by such a code is called *stochastic complexity* (SC) [19] of the data x^n, which is the most important notion in the MDL principle.

The remainder of this article is organized as follows. In Sects. 2 and 3, we review basic concepts of data compression and related issues in statistics. In Sect. 4, we introduce the two part code MDL, which is the first version of the MDL proposed by Rissanen for model selection. Then, we introduce the refined MDL, which is based on the notion of *stochastic complexity* in Sect. 5. In Sect. 6, we argue Bayesian approach to achieve the stochastic complexity with some application to statistical inference.

2 Data Compression

In this section, we review the basic notion of data compression and universal coding. See [9].

Let $\mathscr{X}$ be a finite set, which is called alphabet. Let x^n denote a string $x_1 x_2 \ldots x_n \in \mathscr{X}^n$ $(n = 1, 2, \ldots)$, and $X^n = X_1 X_2 \ldots X_n$ be a random variable on $\mathscr{X}^n$. In general, we let x_t denote a realized value of X_t. We also refer to $\mathscr{X}^n$ as an enlarged alphabet.

Let p be a probability function on $\mathscr{X}^n$, that is $\forall x^n \in \mathscr{X}^n,\ p(x^n) \geq 0$ and

$$\sum_{x^n \in \mathscr{X}^n} p\left(x^n\right) = 1.$$

In this article, we usually assume that p defines an i.i.d. stochastic process, that is $p(x^n) = \prod_{t=1}^n p(x_t)$ for $n = 1, 2, \ldots$. Let $H(X) = H(X|p)$ denote the entropy of $X \sim p$:

$$H(X|p) = -E_p \log p(X) = -\sum_x p(x) \log p(x),$$

where E_p denotes the expectation provided X is drawn from p and log is the natural logarithm (the unit of entropy is 'nat'). A function $\phi : \mathcal{X}^n \to \{0, 1\}^*$ is referred to as a code for the alphabet $\mathcal{X}^n$, where A^* denotes the set of all strings of finite length over an alphabet A. Here, each $\phi(x^n)$ is referred to as code word. We refer to a function $L : \mathcal{X}^n \to [0, \infty)$ as codelength function. Let $|\phi(x^n)|$ denote the length of the string $\phi(x^n)$ then the function which maps x^n to $|\phi(x^n)|$ is a codelength function. We refer to it as the codelength function of the code ϕ.

Assume that ϕ is a *prefix code*, which is a code satisfying the prefix condition: for each x^n, $\phi(x^n)$ is not a prefix of any other code words. Note that 'NULL', 'a', and 'ab' are prefixes of the string 'ab' for example. Then, ϕ has a property of 'instantly decodable'. See [9] for detail.

Note that the codelength function $L(x^n) = |\phi(x^n)|$ of any prefix code satisfies the inequality

$$\sum_{x^n \in \mathcal{X}^n} 2^{-L(x^n)} \leq 1,$$

which is known as Kraft's inequality. Conversely for any codelength function L satisfying Kraft's inequality, construction of prefix code ϕ with $|\phi(x^n)| = \lceil L(x^n) \rceil$ is possible, where $\lceil x \rceil$ is the minimum integer not less than x.

From now on, we change the unit of codelength from 'bit' to 'nat', i.e., let $L(x^n) = |\phi(x^n)|_e = |\phi(x^n)| / \log_2 e$. Then Kraft's inequality becomes

$$\sum_{x^n \in \mathcal{X}^n} e^{-L(x^n)} \leq 1.$$

The following Proposition is known as the source coding theorem.

Proposition 1 *Assume that X^n is drawn from p. For any codelength function satisfying Kraft's inequality,*

$$\frac{1}{n} E_p L\left(X^n\right) \geq H(X|p).$$

holds, where equality holds, if and only if $L(x^n) = -\log p(x^n)$.

The proof is possible via several ways. For example, the inequity $Ef(X) \geq f(EX)$ for strictly convex function $f(x)$ suffices, where equality holds if and only if X is a constant. This is known as Jensen's inequality. See [9].

Define q's redundancy with respect to p as

$$R_n(p, q) = E_p \left(\log \frac{1}{q(X^n)} - \log \frac{1}{p(X^n)} \right).$$

Then, we have $R_n(p, q) \geq R_n(p, p) = 0$. Note that $R_1(p, q)$ is referred to as Kullback-Leibler divergence (KL-divergence for short), which we denote by

$$D(p|q) = E_p \log \frac{p(X)}{q(X)}.$$

The source coding theorem provides a relation between a prefix code and a sub-probability function. Here, a function $f(x) \geq 0$ with $\sum_x f(x) \leq 1$ is referred to as sub-probability function. Given a prefix code ϕ, define

$$q\left(x^n\right) = e^{-|\phi(x^n)|_e}.$$

Then,

$$\sum_{x^n} q\left(x^n\right) \leq 1$$

holds. Here, q is referred to as a sub-probability function corresponding to the prefix code ϕ. The source coding theorem means that p is optimal in terms of expected codelength, among all sub-probability functions when $X^n \sim p$. Based on this recognition, we regard a probability function as a source code. This is our fundamental recognition that we understand probabilistic notions in terms of data compression.

The above discussion concerns data compression in the situation that the probability distribution of the data is known. Such situation is somewhat restricted, and in fact we can design source codes provided X^n's distribution is unknown but belongs to a known set $\mathcal{M}$. Strictly speaking, a code q which satisfies

$$\forall p \in \mathcal{M}, \quad \lim_{n\to\infty} \frac{1}{n} E_p\left(-\log q\left(X^n\right)\right) = H(X|p)$$

is referred to as a universal code with respect to $\mathcal{M}$. When $\mathcal{M}$ corresponds to the set of all stationary ergodic processes for finite alphabet $\mathcal{X}$, there exists a universal code [9]. For example, well-known applications such as zip, compress, etc., are implementations of universal coding algorithms.

We can extend the notions introduced in this section to the case in which the set $\mathcal{X}$ is not discrete. Typically, we assume $\mathcal{X} = \Re^d$, for which we introduce a reference measure $\nu(\mathrm{d}x)$. Then we assume $p(x)$ is a probability density function with respect to $\nu(\mathrm{d}x)$ and the sum in the above is replaced by integral with respect to $\nu(\mathrm{d}x)$. In particular, Kraft's inequality is

$$\int e^{-L(x^n)} \nu\left(\mathrm{d}x^n\right) \leq 1.$$

Note that we need quantization of $\mathcal{X}$ to have the corresponding code to p.

3 Statistical Preliminaries

Let $\mathcal{M}$ be a parametric model of probability densities of $x \in \mathcal{X}$ with respect to the reference measure ν:

$$\mathcal{M} = \{p(\cdot|\theta) : \theta \in \Theta\},$$

where $\Theta \subset \Re^k$ is a k-dimensional parameter space. Let $\hat{\theta} = \hat{\theta}(x^n)$ denote the maximum likelihood estimate (MLE) of θ given x^n, that is, $p(x^n|\hat{\theta}) = \max_\theta p(x^n|\theta)$.

We introduce the empirical Fisher information $\hat{J}(\theta, x^n)$ given x^n and the Fisher information $J(\theta)$:

$$\hat{J}_{ij}(\theta) = \hat{J}_{ij}\left(\theta, x^n\right) = \frac{-1}{n} \frac{\partial^2 \log p\left(x^n|\theta\right)}{\partial\theta^i \partial\theta^j},$$
$$J_{ij}(\theta) = E_\theta \hat{J}_{ij}\left(\theta, X^n\right).$$

The exponential family is defined as follows [2, 3, 7].

Definition 1 (*Exponential Family*) Given a Borel measurable function $T : \mathcal{X} \to \Re^k$, define

$$\Theta = \left\{\theta : \theta \in \Re^k, \int_{\mathcal{X}} \exp\left(\theta \cdot T(x)\right) \nu(\mathrm{d}x) < \infty\right\},$$

where $\theta \cdot T(x)$ denotes the inner product of θ and $T(x)$. Define a function ψ and a probability density p on $\mathcal{X}$ with respect to ν by $\psi(\theta) = \log \int \exp(\theta \cdot T(x))\nu(\mathrm{d}x)$ and $p(x|\theta) = \exp(\theta \cdot T(x) - \psi(\theta))$. We refer to the set $\mathcal{M} = \{p(\cdot|\theta)|\theta \in \Theta\}$ as an exponential family of densities.

Note that exponential families include many common statistical models such as Gaussian, Poisson, Bernoulli models, and so on.

For the exponential family above, we refer to θ as the canonical parameter (or θ-coordinates). We define the expectation parameter (or η-coordinates) as $\eta_i = E_\theta(T_i)$.

Note that $\partial\psi(\theta)/\partial\theta_i = E_\theta(T_i(X)) = \eta_i$ and $\partial^2\psi(\theta)/\partial\theta_j\partial\theta_i = E_\theta((T_i(X) - \eta_i)(T_j(X) - \eta_j))$ hold on Θ°. Here, this $E_\theta((T_i(X) - \eta_i)(T_j(X) - \eta_j))$ is an entry of the Fisher information matrix with respect to θ.

Given a data string x^n, we have $p(x^n|\theta) = \prod_{\tau=1}^n p(x_\tau|\theta) = \prod_{\tau=1}^n \exp(\theta \cdot T(x_\tau) - \psi(\theta)) = \exp(n(\theta \cdot \bar{T} - \psi(\theta)))$, where $\bar{T} = \bar{T}(x^n) = (1/n)\sum_{\tau=1}^n T(x_\tau)$. It is easy to see $\hat{\eta}(x^n) = \bar{T}(x^n)$ holds for $x^n \in \{x^n : \bar{T}(x^n) \in \mathcal{H}^\circ\}$.

Also, we have

$$\hat{J}_{ij}\left(\theta, x^n\right) = \frac{-1}{n} \frac{\partial^2 \log p\left(x^n|\theta\right)}{\partial\theta_i \partial\theta_j} = J_{ij}(\theta). \tag{1}$$

This is a remarkable property of the canonical parameter of exponential families.

Further note that the following holds. When we employ other parametrization than the canonical parameter θ by $\theta = \phi(u)$ and denote $\bar{p}(x|u) = p(x|\phi(u))$, where u is k-dimensional and ϕ is one to one. Then, (1) does not generally holds, but we still have

$$\hat{J}_u \left(x^n, \hat{u}\right) = J_u \left(\hat{u}\right), \tag{2}$$

where $\hat{J}_u$ and J_u denote the empirical Fisher information and the Fisher information of u, respectively.

4 Two Part Code MDL

The MDL was first proposed as an information criterion for statistical model selection [18], which is given as

$$L\left(x^n|\gamma\right) = -\log p(x^n|\hat{\theta}_\gamma, \gamma) + \frac{k_\gamma}{2}\log n. \tag{3}$$

Here $p(\cdot|\theta_\gamma, \gamma)$ is an element of a parametric model $\mathcal{M}_\gamma$ specified by an index γ, k_γ is θ_γ's dimension, and $\hat{\theta}_\gamma$ is the MLE of θ_γ given x^n. This is derived as the codelength of a universal code designed for $\mathcal{M}_\gamma$. The first term of (3) is referred to as data description length and the second is referred to as model description length (or parameter description length). In this section, we discuss this derivation and properties of inference using the MDL criterion.

4.1 MDL Criterion

Define

$$\mathcal{M} = \bigcup_{\gamma\in\Gamma} \mathcal{M}_\gamma$$
$$\mathcal{M}_\gamma = \left\{p\left(\cdot|\theta_\gamma, \gamma\right) : \theta_\gamma \in \Theta_\gamma\right\},$$

where Γ is a countable set, Θ_γ is a subset of $\Re^{k_\gamma}$, and k_γ is a natural number depending on γ.

We can design a universal code which corresponds to the MDL criterion (3), i.e., there exists a universal code for $\mathcal{M}_\gamma$ of which codelength is given by (3). The parameter description length here is codelength for quantized value of θ_γ, since almost every real number cannot be encoded. Here, precision of quantization is important. The more precise quantization becomes, the longer the parameter description length becomes. As for the data description length, the more precise quantization becomes, the shorter it becomes, since the maximum likelihood estimate minimizes the data description length and in general the MLE cannot be realized by the quantized values of parameters. Hence, we have a trade-off between data description length and parameter description length depending on precision of quantization.

Let us find (sub) optimal precision for data strings of size n in an informal discussion, assuming θ is one-dimensional for simplicity. In the discussion below, we omit the subscript γ. We also assume that $\Theta = [0, 1]$. Then to encode θ, quantize Θ into

$$\bar{\Theta}_n = \bar{\Theta}_n(\delta) = \left\{\delta, 2\delta, \ldots, \lceil\delta^{-1}\rceil\delta\right\}$$

and let $L(\bar{\theta}|\bar{\Theta}_n) = -\log\delta + \log 2$ be the codelength for $\bar{\theta} \in \bar{\Theta}_n$. Then, we want to optimize δ so as to minimize

$$L\left(x^n|\bar{\Theta}_n\right) = \min_{\bar{\theta}\in\bar{\Theta}_n}\left(-\log p\left(x^n|\bar{\theta}\right) + L\left(\bar{\theta}|\bar{\Theta}_n\right)\right),$$

depending on n. Here, the corresponding universal code is constructed as follows. Given x^n, first find $\ddot{\theta} = \ddot{\theta}(x^n)$ which minimizes the sum $-\log p(x^n|\bar{\theta}) + L(\bar{\theta}|\bar{\Theta}_n)$, then encode $\ddot{\theta}$ with the code $q(\bar{\theta}|\bar{\Theta}_n) = \exp(-L(\bar{\theta}|\bar{\Theta}_n))$, and then encode x^n with the code $p(x^n|\ddot{\theta})$. This code can be decodable by the reverse process and $L(x^n|\bar{\Theta}_n)$ satisfies Kraft's inequality. It can be directly confirmed as

$$\sum_{x^n} e^{-L(x^n|\bar{\Theta}_n)} = \sum_{x^n} p\left(x^n|\ddot{\theta}\right) e^{-L(\ddot{\theta}|\bar{\Theta}_n)} \leq \sum_{\bar{\theta}\in\bar{\Theta}_n}\sum_{x^n} p\left(x^n|\bar{\theta}\right) e^{-L(\bar{\theta}|\bar{\Theta}_n)} \tag{4}$$

$$= \sum_{\bar{\theta}\in\bar{\Theta}_n} e^{-L(\bar{\theta}|\bar{\Theta}_n)} \leq 1. \tag{5}$$

Out task is to optimize typical values of $L(x^n|\bar{\Theta}_n)$. Let $q(x^n|\bar{\Theta})$ denote the sub-probability function corresponding to $L(x^n|\bar{\Theta}_n)$, i.e., $q(x^n|\bar{\Theta}) = e^{-L(x^n|\bar{\Theta}_n)}$. If we knew the value $\hat{\theta}(x^n)$ prior to the compression process, we did not need to encode θ. Then, the optimal codelength was $-\log p(x^n|\hat{\theta}(x^n))$. We think it as a reference for universal coding, that is, we are to minimize *regret* of the code $q(\cdot|\bar{\Theta}_n)$, which is defined as

$$r\left(x^n, q\left(\cdot|\bar{\Theta}_n\right)\right) = -\log p(x^n|\ddot{\theta}) + L(\ddot{\theta}|\bar{\Theta}_n) - (-\log p(x^n|\hat{\theta}(x^n))).$$

This regret is a performance measure for universal code and sometimes called point-wise redundancy. Note that regret is defined for individual strings, while redundancy is defined in terms of expectation. Performance measures like regret are often used in machine learning community.

Since the regret depends on the data x^n, we minimize its worst case value. Assuming $|\ddot{\theta} - \hat{\theta}|$ is small, by Taylor expansion, we have

$$r\left(x^n, q\left(\cdot|\bar{\Theta}_n\right)\right) \approx -\frac{1}{2}\left.\frac{\partial^2 \log p\left(x^n|\theta\right)}{\partial\theta^2}\right|_{\theta=\hat{\theta}} (\ddot{\theta}-\hat{\theta})^2 + L\left(\ddot{\theta}|\bar{\Theta}_n\right).$$

Recalling $-\partial^2 \log p\left(x^n|\theta\right)/\partial\theta^2 = n\hat{J}(\theta)$, we have

$$r\left(x^n, q\left(\cdot|\bar{\Theta}_n\right)\right) \approx \frac{n\hat{J}(\hat{\theta})(\ddot{\theta}-\hat{\theta})^2}{2} + L\left(\ddot{\theta}|\bar{\Theta}_n\right). \tag{6}$$

As for the first term of (6), we replace $(\ddot{\theta}-\hat{\theta})^2$ by its worst value $\delta^2/4$, that is,

$$r\left(x^n, q\left(\cdot|\bar{\Theta}_n\right)\right) \approx \frac{n\hat{J}(\hat{\theta})\delta^2}{8} - \log\delta + \log 2.$$

The right-hand side is minimized when $\delta = 2(n\hat{J}(\hat{\theta}))^{-1/2}$. Ignoring dependency of $\hat{J}(\hat{\theta})$ on x^n, we have $L(\ddot{\theta}|\bar{\Theta}_n) = (1/2)\log n + C$, where C is a certain constant. When θ is k-dimensional, we have $L(\bar{\theta}) = (k/2)\log n + C'$. This yields the MDL criterion (3).

Now we redefine $L(x^n|\gamma)$ as

$$L\left(x^n|\gamma\right) = -\log p\left(x^n|\ddot{\theta}_\gamma\right) + \frac{k_\gamma}{2}\log n + C_\gamma, \tag{7}$$

where we introduce the constant C_γ so that $L(x^n|\gamma)$ satisfies Kraft's inequality. Then the information criterion (3) is slightly modified to (7). Let $\hat{\gamma} = \hat{\gamma}(x^n)$ denote the minimizer for $L(x^n|\gamma)$. Then, we should note that it is impossible to decode x^n from the codeword with the codelength $L(x^n|\hat{\gamma}(x^n))$, since we do not identify $\hat{\gamma}$ in advance. Hence, we need a code to encode γ prior to encoding θ_γ. Let $L(\gamma)$ be a codelength function for a code for $\gamma \in \Gamma$. Then, we employ the codelength below for model selection.

$$L\left(x^n, \gamma\right) = -\log p\left(x^n|\ddot{\theta}_\gamma, \gamma\right) + \frac{k_\gamma}{2}\log n + C_\gamma + L(\gamma). \tag{8}$$

Note that the term $L(\gamma)$ is significant in the proof for the consistency of the model selection by MDL, in particular when Γ is not a finite set. When we employ (8) for our information criterion for model selection, $\hat{\gamma}$ is redefined as the minimizer for $L(x^n, \gamma)$ as well. Let $L(x^n) = L(x^n, \hat{\gamma}(x^n))$, then $L(x^n)$ satisfies Kraft's inequality and corresponds to a universal code for $\mathcal{M}$.

4.2 Convergence of MDL Estimator

Using the two part code MDL (8), we can design an estimator which maps x^n to an element of $\mathcal{M}$. We refer to it as the MDL estimator based on the two part code. Here, we discuss its convergence property.

First we introduce new notations for two part code MDL and define a variant of two part code MDL. Let $\ddot{\mathcal{M}}_n$ denote a quantized set of $\mathcal{M}$ and encode $p \in \ddot{\mathcal{M}}_n$ with $\dot{L}_n(p)$ nats. Assume that $\dot{L}_n(p)$ satisfies Kraft's inequality. Then we have two part code MDL

$$L_{2\text{-}p}^{(n)}(x^n) = \min_{p \in \ddot{\mathcal{M}}_n} \left(-\log p\left(x^n\right) + \dot{L}_n(p)\right).$$

By a technical reason, define β two part code MDL assuming $\beta \geq 1$ [4]:

$$L_{\beta 2\text{-}p}^{(n)}(x^n) = \min_{p \in \ddot{\mathcal{M}}_n} \left(-\log p\left(x^n\right) + \beta \dot{L}_n(p)\right) \tag{9}$$

$$= -\log \ddot{p}\left(x^n\right) + \beta \dot{L}_n\left(\ddot{p}\right), \tag{10}$$

where we let $\ddot{p}$ denote the value of p to achieve $L_{\beta 2\text{-}p}^{(n)}(x^n)$.

Let $p_{\beta 2\text{-}p}(x^n) = \exp(-L_{\beta 2\text{-}p}^{(n)}(x^n))$, then $p_{\beta 2\text{-}p}$ is a sub-probability density.

Now, we give a theorem to guarantee the convergence of the MDL estimator. To state it, we use a notion of Rényi divergence [17]. The following is the Rényi divergence of order λ (> 0) from p to q.

$$\bar{d}_\lambda(p|q) = -\frac{1}{1-\lambda} \log \int p^\lambda(x) q^{1-\lambda}(x) \nu(\mathrm{d}x),$$

with $\bar{d}_1(p|q) = \lim_{\lambda \to 1-0} \bar{d}_\lambda(p|q)$, which equals the KL-divergence. Also note that $\bar{d}_\lambda(p|q)$ is increasing in $\lambda \in (0, 1)$. For the proof, see [11] for example.

Noting $-\log t \geq 1 - t$, we have

$$\bar{d}_\lambda(p|q) \geq \frac{1}{1-\lambda}\left(1 - \int p^\lambda(x) q^{1-\lambda}(x) \nu(\mathrm{d}x)\right) \tag{11}$$

$$= \frac{1}{1-\lambda} \int \left(1 - \left(\frac{q(x)}{p(x)}\right)^{1-\lambda}\right) p(x) \nu(\mathrm{d}x). \tag{12}$$

The last expression times λ^{-1} equals the α-divergence [2, 3]:

$$D^{(\alpha)}(p|q) = \frac{4}{1-\alpha^2} \int \left(1 - \left(\frac{q(x)}{p(x)}\right)^{(1+\alpha)/2}\right) p(x) \nu(\mathrm{d}x)$$

with $\lambda = (1-\alpha)/2$. Note that $\lim_{\alpha \to -1} D^{(\alpha)}(p|q) = D(p|q)$. When $\alpha = 0$, the α-divergence equals the squared Hellinger distance:

$$d_H^2(p, q) = 4 \int \left(1 - \left(\frac{q(x)}{p(x)}\right)^{1/2}\right) p(x) \nu(\mathrm{d}x) = 2 \int \left(\sqrt{p(x)} - \sqrt{q(x)}\right)^2 \nu(\mathrm{d}x) \tag{13}$$

which is a popular distance measure in statistics and information theory.

The first form of Theorem 1 below was given by Barron and Cover [4] in 1991. Theorem 1 is a variant of Theorem 4 of [4] and noted in [11] as Theorem 15.3. The sophisticated proof stated below is quoted from [11].

Theorem 1 *The following holds for any* $\lambda \in (0, 1-\beta^{-1})$ *with* $\beta > 1$.

$$E_{X^n \sim p^*} \bar{d}_\lambda \left(p^* | \ddot{p}\right) \le \frac{1}{n} R_n \left(p^*, p_{\beta 2\text{-}p}\right).$$

Proof The proof for $\lambda = 1 - \beta^{-1}$ is sufficient, since $\bar{d}_\lambda(p^*|\ddot{p})$ is increasing in $\lambda \in (0, 1-\beta^{-1}]$. Define

$$A_\lambda(p_0|p_1) = \int p_1^\lambda(x) p_0^{1-\lambda}(x) \nu(\mathrm{d}x).$$

Then

$$\begin{aligned}
\bar{d}_\lambda \left(p^* | \ddot{p}\right) &= -\frac{1}{1-\lambda} \log A_\lambda \left(p^* | \ddot{p}\right) \\
&= \frac{1}{n} \log \frac{\ddot{p}\left(x^n\right) e^{-\beta L(\ddot{p})}}{p_{\beta 2\text{-}p}\left(x^n\right)} + \frac{\beta}{n} \log \frac{1}{A_\lambda^n \left(p^* | \ddot{p}\right)} \\
&= \frac{1}{n} \log \frac{p^*\left(x^n\right)}{p_{\beta 2\text{-}p}\left(x^n\right)} + \frac{\beta}{n} \log \frac{\left(\frac{\ddot{p}(x^n)}{p^*(x^n)}\right)^{1/\beta} e^{-L(\ddot{p})}}{A_\lambda^n \left(p^* | \ddot{p}\right)} \\
&= \frac{1}{n} \log \frac{p^*\left(x^n\right)}{p_{\beta 2\text{-}p}\left(x^n\right)} + \frac{\beta}{n} \log \frac{\left(\frac{\ddot{p}(x^n)}{p^*(x^n)}\right)^{1-\lambda} e^{-L(\ddot{p})}}{A_\lambda^n \left(p^* | \ddot{p}\right)} \\
&\le \frac{1}{n} \log \frac{p^*\left(x^n\right)}{p_{\beta 2\text{-}p}\left(x^n\right)} + \frac{\beta}{n} \log \sum_{q \in \ddot{\mathcal{M}}_n} \frac{\left(\frac{q(x^n)}{p^*(x^n)}\right)^{1-\lambda} e^{-L(q)}}{A_\lambda^n (p^* | q)}.
\end{aligned}$$

Taking expectation with respect to the data string, we have

$$E_{X^n \sim p^*} \bar{d}_\lambda \left(p^* | \ddot{p}\right) \le \frac{1}{n} R_n(p^*, p_{\beta 2\text{-}p}) + \frac{\beta}{n} E_{X^n \sim p^*} \log \sum_{q \in \ddot{\mathcal{M}}_n} \frac{\left(\frac{q(x^n)}{p^*(x^n)}\right)^{1-\lambda} e^{-L(q)}}{A_\lambda^n \left(p^* | q\right)}.$$

By Jensen's inequality, the second-term's expectation in the right-hand side is not greater than

$$\log E_{p^*} \sum_{q \in \ddot{\mathcal{M}}_n} \frac{\left(\frac{q(x^n)}{p^*(x^n)}\right)^{1-\lambda} e^{-L(q)}}{A_\lambda^n(p^*|q)} = \log \frac{\int \sum_q (q(x^n))^{1-\lambda} (p^*(x^n))^\lambda \nu(\mathrm{d}x^n) e^{-L(q)}}{A_\lambda^n(p^*|q)} \tag{14}$$

$$= \log \sum_q e^{-L(q)} \leq 0. \tag{15}$$

This completes the proof.

Assume that $p^* \in \mathcal{M}$, then this theorem says that the better the code $p_{\beta 2\text{-}p}(x^n)$ is as a universal code for $\mathcal{M}$, the tighter upper bound on $\ddot{p}$'s convergence to p^* is obtained. We would like to let $\beta = 1$, since it corresponds to the genuine two part code MDL $L_{2\text{-}p}^{(n)}(x^n)$. However, $\beta = 1$ means that we must have $\lambda = 0$, for which the proof of the theorem does not hold. The corresponding convergence result to $\beta = 1$ has been an open problem for over 20 years.

5 Refined MDL

Here, we refine the codelength derived in the previous section. There, we minimized the maximum regret $\max_{x^n} r(x^n, q(\cdot|\bar{\Theta}_n))$ by choosing the code for the parameter θ. It means that the objective universal code was restricted to two part codes. In this section, we remove this restriction and obtain the shorter codelength.

5.1 Stochastic Complexity and Parametric Complexity

First we introduce the notion of minimax regret with respect to a target class $\mathcal{M} = \{p(\cdot|\theta) : \theta \in \Theta\}$ [20]. For a subset $K \subset \Theta$, let $\mathcal{M}_K = \{p(\cdot|\theta) : \theta \in K\}$ and $\mathcal{X}^n(K) = \{x^n : \hat{\theta}(x^n) \in K\}$. Here, we assume K satisfies $\bar{K} = \bar{K^\circ}$. (For A, $\bar{A}$ is A's closure and A° is A's interior.)

We consider a problem to find the following quantity:

$$\bar{r}_n(\mathcal{M}_K) = \inf_q \sup_{x^n \in \mathcal{X}^n(K)} r(x^n, q) = \inf_q \sup_{x^n \in \mathcal{X}^n(K)} \left(\log \frac{1}{q(x^n)} - \log \frac{1}{p(x^n|\hat{\theta}(x^n))} \right),$$

where infimum is taken for all probability densities over $\mathcal{X}^n$.

We refer to the value $\bar{r}_n(\mathcal{X}^n(K))$ as the minimax regret for the target class $\mathcal{M}_K$ and the minimizer as the minimax code for $\mathcal{M}_K$.

The minimax regret is achieved by the normalized maximum likelihood [20] defined as

$$\hat{m}_n(x^n) = \frac{p(x^n|\hat{\theta}(x^n))}{\int_{\mathcal{X}^n(K)} p(x^n|\hat{\theta}(x^n))\nu(\mathrm{d}x^n)}.$$

Fig. 1 Computation of parametric complexity of Bernoulli model [14]

Input: n, d
1. $G(1,n) := 1$
2. $G(2,n) := \sum_{k=0}^{n} C(n,k)(k/n)^k(1-k/n)^{n-k}$
3. for $i = 1$ to $d-2$
   $G(i+2,n) := G(i+1,n) + (n/i)G(i,n)$

This can be confirmed by noting

$$r\left(x^n, q\right) = \log \frac{p(x^n|\hat{\theta}(x^n))}{q(x^n)} = \log \frac{\hat{m}_n\left(x^n\right)}{q(x^n)} + \log \int_{\mathscr{X}^n(K)} p(x^n|\hat{\theta}(x^n))\nu(\mathrm{d}x^n).$$

We have

$$\max_{x^n \in \mathscr{X}^n(K)} \frac{\hat{m}_n\left(x^n\right)}{q(x^n)} \geq 1,$$

where equality holds when $q = \hat{m}_n$. It implies

$$\max_{x^n \in \mathscr{X}^n(K)} r\left(x^n, q\right) \geq \max_{x^n \in \mathscr{X}^n(K)} r\left(x^n, \hat{m}_n\right) = \log \int_{\mathscr{X}^n(K)} p(x^n|\hat{\theta}(x^n))\nu(\mathrm{d}x^n).$$

Note that the codelength $-\log \hat{m}_n(x^n)$ is referred to as *stochastic complexity* (SC) of x^n with respect to $\mathscr{M}_K$ [11, 19], and $\log \int p(x^n|\hat{\theta}(x^n))\nu\,(\mathrm{d}x^n)$ is referred to as *parametric complexity* of $\mathscr{M}_K$ [11].

It is an important issue to determine the SC of various statistical models, since the SC is what we should use for statistical inference including model selection in the view point of the MDL principle. In general, computation of normalized maximum likelihood is intractable, since the expression of parametric complexity contains exponentially many number of sum (integral) in n. However, for the multinomial Bernoulli case, an $O(n+d)$ time algorithm is known [14].

Let $p(x|\theta) = \theta_x$ for $x \in \mathscr{X} = \{1, 2, \ldots, d\}$. Then, $\sum_{x^n} p(x^n|\hat{\theta}(x^n))$ can be obtained as $G(d,n)$, which is computed by code in Fig. 1. Here $C(n,k)$ is the binomial coefficient.

This algorithm is very important, but is a special one for the multinomial Bernoulli model. Efforts to extend it to other classes have been done, but it is not successful to date, except for a few classes like the Bayesian networks, which have the essentially same structure as the Bernoulli model [21].

Note that the regret is the sum of the incremental regrets of prediction:

$$r\left(x^n, q\right) = \sum_{i=1}^{n-1} \left(\log \frac{1}{q(x_{i+1}|x^i)} - \log \frac{1}{p(x_{i+1}|x^i, \hat{\theta}(x^n))}\right).$$

Hence, the minimax code is also the minimax prediction strategy.

As mentioned above, strict evaluation of stochastic complexity is difficult in general. Instead, asymptotic evaluation for various cases is known. When $\mathscr{M}$ is the multinomial Bernoulli model, Xie and Barron [28] gave an asymptotic evaluation

$$\log \int_{\mathscr{X}^n} p(x^n|\hat{\theta}(x^n))\nu\left(\mathrm{d}x^n\right) = \frac{k}{2}\log\frac{n}{2\pi} + \log\int |J(\theta)|^{1/2}\,\mathrm{d}\theta + o(1), \tag{16}$$

where k equals the size of alphabet minus 1, $|J(\theta)|$ denotes the determinant of the Fisher information matrix $J(\theta)$, and $o(1)$ is a quantity which converges to 0 as n goes to infinity. For the stationary Markov model with finite alphabet, the analogous result is known [12, 27]. As for these cases, the stochastic complexity is evaluated for $K = \Theta$, but it is difficult in general, since the parametric complexity is infinite for many natural models. For a compact $K \subset \Theta^{\circ}$, the parametric complexity for various target models is evaluated as

$$\log \int_{\mathscr{X}^n(K)} p(x^n|\hat{\theta}(x^n))\nu(\mathrm{d}x^n) = \frac{k}{2}\log\frac{n}{2\pi} + \log\int_K |J(\theta)|^{1/2}\,\mathrm{d}\theta + o(1), \tag{17}$$

using universal codes based on Bayes mixtures. In the next section, we review some results about it.

6 Bayesian Approach

For most cases, the asymptotic expression (17) has been shown by Bayesian methods. This section provides a brief review about it.

6.1 Bayes Codes

The universal code based on the probability density function of mixture

$$m_w\left(x^n\right) = \int_K p\left(x^n|\theta\right) w(\theta)\,\mathrm{d}\theta$$

is referred to as Bayes code. Here $w(\theta) \geq 0$ is a prior density over $K \subset \Theta$: $\int_K w(\theta)\,\mathrm{d}\theta = 1$.

We are interested in the regret of Bayes codes. In this context, the Jeffreys prior [8, 13] is important, which is the prior density proportional to $|J(\theta)|^{1/2}$. The value $C_J(K) = \int_K |J(\theta)|^{1/2}\,\mathrm{d}\theta$ is the normalization constant for the Jeffreys prior over K. We refer to the mixture with Jeffreys prior as the Jeffreys mixture.

For the exponential family including the stationary Markov model with finite alphabet, it is known that a sequence of Jeffreys mixtures achieves the minimax regret asymptotically [6, 22, 23, 27, 28]. For the multinomial exponential family case except for multinomial Bernoulli and Markov models, these facts are proven under the condition that K is a compact subset included in the interior of Θ.

We briefly review outline of the proof of these results. Let $\{K_n\}$ be a sequence of subsets of Θ such that $K_n^\circ \supset K$. Suppose that K_n reduces to K as $n \to \infty$. Let $m_{J,n}$ denote the Jeffreys mixture over K_n. If the rate of that reduction is sufficiently slow, then we can prove

$$\log \frac{p\left(x^n|\hat{u}\right)}{m_{J,n}\left(x^n\right)} = \frac{k}{2}\log\frac{n}{2\pi} + \log C_J(K) + o(1), \tag{18}$$

where the remainder $o(1)$ tends to zero uniformly over all sequences with MLE in K. This implies that the sequence $\{m_{J,n}\}$ is asymptotically minimax.

The asymptotic (18) can be shown as follows. Let m_w denote the Bayes mixture with a prior $w(\theta)$. First by Taylor expansion of $\log p\,(x^n|\theta)$ around $\hat{\theta}$, we have

$$\log p\left(x^n|\theta\right) = \log p\left(x^n|\hat{\theta}\right) - \frac{1}{2}\left(\theta-\hat{\theta}\right)^T n\hat{J}(\theta')\left(\theta-\hat{\theta}\right).$$

where θ' is a certain point between θ and $\hat{\theta}$. We have used the fact that $\partial \log p(x^n|\theta)/\partial\theta_i = 0$ at $\theta = \hat{\theta}$ and that $\partial^2 \log p(x^n|\theta)/\partial\theta_i\partial\theta_j = -n\hat{J}(\theta)$. Hence we have

$$\frac{p(x^n|\theta)}{p(x^n|\hat{\theta})} = \exp\!\left(-\frac{1}{2}(\theta-\hat{\theta})^T n\hat{J}(\theta')(\theta-\hat{\theta})\right).$$

Let B_n denote a sphere with radius $n^{-1/2}\log n$ centered at $\hat{\theta}$, then we have

$$\frac{m_w\left(x^n\right)}{p(x^n|\hat{\theta})} \sim \int_{B_n} \exp\!\left(\frac{-n\theta^T\hat{J}(\hat{\theta})\theta}{2}\right)w(\theta)\,d\theta \sim \frac{(2\pi)^{k/2}w(\hat{\theta})}{n^{k/2}|\hat{J}(\hat{\theta})|^{1/2}}.$$

Hence, we have the following asymptotic for the regret of $m_{J,n}$:

$$r\left(x^n, m_{J,n}\right) = \frac{k}{2}\log\frac{n}{2\pi} + \log\frac{|\,\hat{J}(\hat{\theta},x^n)\,|^{1/2}\,C_J(K_n)}{|J(\theta)|^{1/2}} + o(1). \tag{19}$$

When $\mathscr{M}$ is an exponential family, $\hat{J}(\hat{\theta},x^n) = J(\hat{\theta})$ holds. Hence, the above expression asymptotically equals the minimax regret.

If K is the entire space for the statistical model, we cannot define the superset of K and need a different technique, which was established for the cases of multinomial Bernoulli model, Markov model, and a certain type of one-dimensional exponential families. See [23, 25, 27, 28].

6.2 Beyond Exponential Families

When the target class $\mathcal{M}$ is not an exponential family, we have a difficulty that the empirical Fisher information $\hat{J}(\hat{\theta}, x^n)$ differs from the Fisher information $J(\hat{\theta})$ in general. It means that we do not have the cancelation in the second term of the regret (19). Conversely, if $|\hat{J}(\theta) - J(\theta)|$ is small, then the regret approximately equals minimax level. Our difficulty is in the case that $|\hat{J}(\theta) - J(\theta)|$ is large. Hence, we form an enlargement of the target class $\mathcal{M}$ by an exponential tilting using linear combinations of the entries of the differences $V(x^n|\theta) = \hat{J}(\theta) - J(\theta)$, which has a point with larger likelihood than $p(x^n|\hat{\theta})$, when $|V(x^n|\theta)|$ is large. Let $\mathcal{B} = (-b/2, b/2)^{k\times k}$ for some $b > 0$. The enlargement is formed as

$$\bar{p}\left(x^n|u\right) = p\left(x^n|\theta\right) e^{n(\beta\cdot V(x^n|\theta) - \psi_n(u))}, \tag{20}$$

where u denotes the pair (θ, β), β is a matrix in $\mathcal{B}$, $V(x^n|\theta)\cdot\beta$ denotes $\mathrm{Tr}(V(x^n|\theta)\beta^t) = \sum_{ij} V_{ij}(x^n|\theta)\beta_{ij}$, and $\Psi_n(u) = \log \int p(x^n|\theta)e^{n\beta\cdot V(x^n|\theta)}\nu(dx^n)$. Note that this enlarged family is an example of local exponential family bundle in Amari's information geometry [3]. Then the mixture defined below asymptotically achieves the minimax regret under certain regularity conditions, where r is a certain small positive constant.

$$\bar{m}_n\left(x^n\right) = \left(1 - n^{-r}\right) m_{J,n}\left(x^n\right) + n^{-r}\int \bar{p}\left(x^n|u\right) w(u)\, du. \tag{21}$$

In particular for the general mixture family [2, 3] defined below, it is shown that the regularity conditions are satisfied for $\mathcal{M}_K$ with a compact $K \subset \Theta^\circ$ [24].

Definition 2 (*Mixture Family*) For $i = 0, 1, \ldots, k$, let $p_i(x)$ be a probability density function over $\mathcal{X}$. Define

$$p\left(x|\theta\right) = \sum_{i=0}^{k} \theta_i p_i(x),$$

where $\theta \in \Theta = \{\theta \in \Re^k : 0 \le \sum_{i=1}^{k} \theta_i \le 1 \text{ and } \forall i \ge 1,\ \theta_i \ge 0\}$ and $\theta_0 = 1 - \sum_{i=1}^{k} \theta_i$. Then, the set $\{p(\cdot|\theta) : \theta \in \Theta\}$ is referred to as a mixture family of densities.

Further, some refinement of the above strategy was recently proposed [25], which achieves the parametric complexity for $K = \Theta$.

Note that the idea for this enlargement in addressing minimax regret originates in preliminarily form in [6, 23] as informally discussed in [5]. The literature [26] gives discussion in the context of information geometry [3].

6.3 Prediction by Bayes Codes

An advantage of the Bayesian approach is in the following form of conditional probabilities for prediction.

$$m_w\left(x_{n+1}|x^n\right) = \frac{m_w\left(x^{n+1}\right)}{m_w\left(x^n\right)} = \int p\left(x_{n+1}|\theta\right) w\left(\theta|x^n\right)\, \mathrm{d}\theta,$$

where $w(\theta|x^n)$ is the posterior density defined as

$$w\left(\theta|x^n\right) = \frac{p\left(x^n|\theta\right) w(\theta)}{\int p\left(x^n|\theta\right) w(\theta)\, \mathrm{d}\theta}.$$

For example, consider the case of Bernoulli model $p(x|\theta) = \theta^x(1-\theta)^{1-x}$ ($x = 0, 1, \theta \in \Theta = [0, 1]$). Assume that we have a data string x^n with $\sum_{t=1}^n x_i = n_1$. If we employ the uniform prior $w(\theta) = 1$, we have

$$m_w\left(1|x^n\right) = \frac{n_1+1}{n+2}.$$

This is known as the *rule of succession* by Laplace from the nineteenth century. If we employ the Jeffreys prior $w(\theta) \propto \theta^{-1/2}(1-\theta)^{-1/2}$, then

$$m_w\left(1|x^n\right) = \frac{n_1+0.5}{n+1}$$

holds. This is a special case of the Kritchevsky-Trofimov (KT) estimator for the multinomial Bernoulli model [15]. In the view point of MDL principle, the KT estimator is better than the Laplace estimator, since the former is approximately equal to the minimax estimator.

We can generalize the KT estimator to the case of the stationary Markov models with finite alphabet based on the result from [27]. Let us see an example for the first order with binary alphabet case. Let $p(x^n|x_0, \eta)$ be the density of first-order stationary Markov model on the alphabet $\mathscr{X} = \{0, 1\}$, where x_0 is the initial symbol and $\eta = (\eta_{0|1}, \eta_{1|0})$. The parameter $\eta_{i|j}$ denotes the conditional probability that $x_{t+1} = i$ is observed after $x_t = j$. Suppose that $x_n = 0$, then the following holds:

$$m_J\left(1|x_0^n\right) \approx \frac{n_{01}+0.5}{n_0+1} + \frac{(n_{00}/n_0)(\hat{\mu}_0-0.5)}{n_0+1} \approx \frac{n_{01}+\hat{\mu}_0}{n_0+0.5+\hat{\mu}_0}, \tag{22}$$

where n_{ij} is the number of occurrences of pattern ij in the string $x_0^n = x_0x_1\ldots x_n$, $n_j = n_{0j}+n_{1j}$, $\hat{\eta}_{i|j} = n_{ji}/n_i$, $\hat{\mu}_i = n_i/n$, and the residual term is of order $o(1/n_0)$. From this, difference between the minimax predictor and the simple KT estimator $(n_{00}+0.5)/(n_0+1)$ is of order $\Omega(1/n_0)$, which is statistically significant. Also, it is reported in [27] that this estimator performs better in terms of regret than the simple KT estimator in numerical simulation.

7 Conclusions

We reviewed important notions of the MDL principle and some recent results. We did not mention applications at all. The readers can find many examples in [10, 11] and reach more literature on recent applications via the web site "http://www.mdl-research.org/".

References

1. H. Akaike, A new look at the statistical model identification. IEEE Trans. Autom. Control **19**(6), 716–723 (1974)
2. S. Amari, *Differential-Geometrical Methods in Statistics*, 2nd edn. (Springer, Berlin Heidelberg, 1990)
3. S. Amari, H. Nagaoka, *Methods of Information Geometry* (AMS & Oxford University Press, Oxford 2000)
4. A.R. Barron, T.M. Cover, Minimum complexity density estimation. IEEE Trans. Inf. Theory **37**(4), 1034–1054 (1991)
5. A.R. Barron, J. Rissanen, B. Yu, The minimum description length principle in coding and modeling. IEEE Trans. Inf. Theory **44**(6), 2743–2760 (1998)
6. A.R. Barron, J. Takeuchi, in *Proceedings of 1998 Information Theory Workshop*. Mixture models achieving optimal coding regret (1998), p. 16
7. L. Brown, *Fundamentals of Statistical Exponential families* (Institute of Mathematical Statistics, Hayward, 1986)
8. B. Clarke, A.R. Barron, Jeffreys prior is asymptotically least favorable under entropy risk. JSPI **41**, 37–60 (1994)
9. T.M. Cover, J.A. Thomas, in *Elements of Information Theory*, 2nd edn. Wiley Series in Telecommunications and Signal Processing (Wiley-Interscience, New York, 2006)
10. P. Grünwald, I.J. Myung, M. Pitt, *Advances in Minimum Description Length: Theory and Applications* (MIT Press, Cambridge, 2005)
11. P. Grünwald, *The Minimum Description Length Principle* (MIT Press, Cambridge, 2007)
12. P. Jacquet, W. Szpankowski, Markov types and minimax redundancy for Markov sources. IEEE Trans. Inf. Theory **50**(7), 1393–1402 (2004)
13. H. Jeffreys, *Theory of Probability*, 3rd edn. (University of California Press, Berkeley, 1961)
14. P. Kontkanen, P. Myllymäki, A linear-time algorithm for computing the multinomial stochastic complexity. Inf. Process. Lett. **103**, 227–233 (2007)
15. R.E. Krichevsky, V.K. Trofimov, The performance of universal encoding. IEEE Trans. Inf. Theory **27**(2), 199–207 (1981)
16. A.A. Maurer, in *Medieval Philosophy*. Etienne Gilson Series (Pontifical Instutite of Medieval Studies, Toronto, 1982)
17. A. Rényi, On measures of entropy and information, in *Proceedings of the Fourth Berkeley Symposium on Mathematical Statistics and Probability*, vol. 1 (1961), pp. 547–561
18. J. Rissanen, Modeling by shortest data description. Automatica **14**, 465–471 (1978)
19. J. Rissanen, Fisher information and stochastic complexity. IEEE Trans. Inf. Theory **40**(1), 40–47 (1996)

20. Y.M. Shtar'kov, Universal sequential coding of single messages. Probl. Inf. Transm. **23**, 3–17 (1988)
21. T. Silander, T. Roos, P. Myllymäki, Learning locally minimax optimal Bayesian networks. Int. J. Approx. Reason. **51**(5), 544–557 (2010)
22. J. Takeuchi, A.R. Barron, Asymptotically minimax regret for exponential families, in *Proceedings of the 20th Symposium on Information Theory and Its Applications (SITA'97)*, (1997), pp. 665–668
23. J. Takeuchi, A.R. Barron, Asymptotically minimax regret by Bayes mixtures, in *Proceedings of 1998 IEEE International Symposium on Information Theory*, (1998), p. 318
24. J. Takeuchi, A.R. Barron, Asymptotically minimax regret by Bayes mixtures for non-exponential families, in *Proceedings of 2013 IEEE Information Theory Workshop*, (2013a), pp. 204–208
25. J. Takeuchi, A.R. Barron, Asymptotically minimax prediction for mixture families, in *Proceedings of the 36th Symposium on Information Theory and Its Applications (SITA'13)*, (2013b), pp. 653–657
26. J. Takeuchi, A.R. Barron, T. Kawabata, Statistical curvature and stochastic complexity, in *Proceedings of the 2nd Symposium on Information Geometry and Its Applications* (2006), pp. 29–36
27. J. Takeuchi, T. Kawabata, A.R. Barron, Properties of Jeffreys mixture for Markov sources. IEEE Trans. Inf. Theory **59**(1), 438–457 (2013)
28. Q. Xie, A.R. Barron, Asymptotic minimax regret for data compression, gambling and prediction. IEEE Trans. Inf. Theory **46**(2), 431–445 (2000)

An Introduction to Ergodic Theory

Khanh Duy Trinh

Abstract Ergodic theory concerns with the study of the long-time behavior of a dynamical system. An interesting result known as Birkhoff's ergodic theorem states that under certain conditions, the time average exists and is equal to the space average. The applications of ergodic theory are the main concern of this note. We will introduce fundamental concepts in ergodic theory, Birkhoff's ergodic theorem and its consequences.

Keywords Benford's law · Ergodic theorem · Markov chain · Measure-preserving transformation · Poincaré's recurrence theorem · Strong-mixing · Weak-mixing

1 Ergodic Transformations and Examples

1.1 Measure-Preserving Transformations and Poincaré's Recurrence Theorem

Let $(\Omega, \mathscr{F}, \mathbf{P})$ be a probability space. A transformation $T\colon \Omega \to \Omega$ is called measurable if $T^{-1}(\mathscr{F}) \subset \mathscr{F}$.

Definition 1 A measurable transformation $T\colon \Omega \to \Omega$ is said to be measure preserving if $\mathbf{P}(T^{-1}A) = \mathbf{P}(A)$ for all $A \in \mathscr{F}$.

Let T be a measure-preserving transformation on $(\Omega, \mathscr{F}, \mathbf{P})$. Then $(\Omega, \mathscr{F}, \mathbf{P}, T)$ is called a dynamical system. The orbit or trajectory of ω under T is the set $\{T^n\omega\}_{n\ge 0}$, where T^n denotes the nth iterate of T.

K. D. Trinh(✉)
Institute of Mathematics for Industry, Kyushu University, 744, Motooka, Nishiku, Fukuoka 819-0395, Japan
e-mail: trinh@imi.kyushu-u.ac.jp

R. Nishii et al. (eds.), *A Mathematical Approach to Research Problems of Science and Technology*, Mathematics for Industry 5,
DOI: 10.1007/978-4-431-55060-0_22, © Springer Japan 2014

Consider a physical phenomena whose state changes over time. Ω denotes all possible states which are distributed according to a probability measure $\mathbf{P}$. The evolution over time of states is described by a transformation T from Ω to itself. A function $f : \Omega \to \mathbb{R}$ is regarded as value of the physical quantity, which can be observed in consecutive periods. In fact, we can only measure the value of a physical quantity over a single orbit $\{T^n\omega\}_{n\geq 0}$. From the measured data $\{f(T^n\omega)\}_{n\geq 0}$, we wish to derive properties of the physical quantity itself. Note that when T is a measure-preserving transformation, the sequence $f_n = f(T^n)$ is a stationary sequence; i.e., a sequence whose distribution is the same as that of the shifted sequence $\{f_{k+n}\}_{n\geq 0}$.

Here are some examples of measure-preserving transformations.

Example 1 (Circle rotation) Let $\Omega = [0,1)$, $\mathscr{F}$ = Borel subsets, $\mathbf{P}$ = Lebesgue measure. For fixed $\theta \in (0, 1)$, define

$$R_\theta : \omega \mapsto \omega + \theta \bmod 1.$$

If we identify $[0, 1)$ with the unit circle $\mathbb{T} := \{z \in \mathbb{C} : |z| = 1\}$ on the complex plane by mapping $[0, 1) \ni x \mapsto \exp(2\pi i x)$, then R_θ acts as a rotation of angle $2\pi\theta$. This identification makes it clear that R_θ is measure preserving.

Example 2 (Bernoulli shift) The probability space is the same as in the previous example. Let

$$T : [0, 1) \to [0, 1), \quad \omega \mapsto 2\omega \bmod 1 = \begin{cases} 2\omega, & \text{if } \ 0 \leq \omega < 1/2, \\ 2\omega - 1, & \text{if } \ 1/2 \leq \omega < 1. \end{cases}$$

It is clear that

$$T^{-1}(a, b) = \left(\frac{a}{2}, \frac{b}{2}\right) \cup \left(\frac{1+a}{2}, \frac{1+b}{2}\right),$$

which is a disjoint union. Thus, T is measure preserving.

Let

$$d_1 = d_1(\omega) = \begin{cases} 0, & \text{if } \ 0 \leq \omega < 1/2, \\ 1, & \text{if } \ 1/2 \leq x < 1, \end{cases}$$

and let $d_n := d_1(T^n)$. Then

$$\omega = \sum_{n=1}^{\infty} \frac{d_n}{2^n}$$

gives the binary expansion for a number $\omega \in [0, 1)$. In fact, the sequence $\{d_n\}_{n\geq 1}$ can be shown to be an independent identically distributed sequence with the common distribution $\mathbf{P}(d_n = 0) = \mathbf{P}(d_n = 1) = 1/2$, which is a Bernoulli process. The transformation T shifts $\{d_n\}_{n\geq 1}$ one step to the left, thus we may call it a Bernoulli shift.

The following property of a measure-preserving transformation, called recurrence, may be considered the starting point of ergodic theory.

Theorem 1 (Poincaré's recurrence theorem) *Let T be a measure-preserving transformation on a probability space $(\Omega, \mathscr{F}, \mathbf{P})$, and let $A \in \mathscr{F}$ be measurable with $\mathbf{P}(A) > 0$. Then for almost all $\omega \in A$, $\{T^n(\omega)\}_{n \geq 0}$ returns infinitely often to A (i.e., there exists $B \subset A$ with $\mathbf{P}(B) = \mathbf{P}(A)$ such that for each $\omega \in B$, there is a sequence $n_1 < n_2 < \cdots$ of natural numbers with $T^{n_i}(\omega) \in B$ for each i).*

1.2 Ergodicity

Let T be a measure-preserving transformation on a probability space $(\Omega, \mathscr{F}, \mathbf{P})$. A measurable set $A \in \mathscr{F}$ is called invariant (more precisely, T-invariant) if $T^{-1}(A) = A$.

Definition 2 A measure-preserving transformation T is called ergodic if every invariant set A has probability 0 or 1.

Here are several equivalent ways to state ergodicity.

Theorem 2 *Let T be a measure-preserving transformation on $(\Omega, \mathscr{F}, \mathbf{P})$. Then the following statements are equivalent:*

(i) *T is ergodic;*
(ii) *for $B \in \mathscr{F}$ with $\mathbf{P}(T^{-1}B \Delta B) = 0$, we have $\mathbf{P}(B) = 0$, or $\mathbf{P}(B) = 1$, where $A \Delta B := (A \setminus B) \cup (B \setminus A)$ denotes the symmetric difference of A and B;*
(iii) *for $A \in \mathscr{F}$ with $\mathbf{P}(A) > 0$, we have $\mathbf{P}(\bigcup_{n=1}^{\infty} T^{-n}A) = 1$;*
(iv) *for $A, B \in \mathscr{F}$ with $\mathbf{P}(A) > 0, \mathbf{P}(B) > 0$, there exists $n > 0$ such that $\mathbf{P}(T^{-n}A \cap B) > 0$.*

Theorem 3 *Let T be a measure-preserving transformation on $(\Omega, \mathscr{F}, \mathbf{P})$. Then the following statements are equivalent:*

(i) *T is ergodic;*
(ii) *if f is measurable and $f(T\omega) = f(\omega)$ for all (or almost surely) $\omega \in \Omega$, then f is constant almost surely (a.s. for short);*
(iii) *if $f \in L^2(\Omega, \mathscr{F}, \mathbf{P})$ and $f(T\omega) = f(\omega)$ for all (or a.s.) $\omega \in \Omega$, then f is constant almost surely.*

A measurable function $f\colon (\Omega, \mathscr{F}, \mathbf{P}) \to \mathbb{C}$ with $f(T\omega) = f(\omega)$ (for all or almost surely $\omega \in \Omega$) is called an invariant function. The ergodicity is equivalent to the statement that any invariant function is constant. In condition (iii), we can replace $L^2(\Omega, \mathscr{F}, \mathbf{P})$ by $L^p(\Omega, \mathscr{F}, \mathbf{P})$ for any $p \geq 1$. Here

$$L^p(\Omega, \mathscr{F}, \mathbf{P}) := \{f\colon \Omega \to \mathbb{C} | f \text{ is measurable and } \int_\Omega |f(\omega)|^p d\mathbf{P}(\omega) < \infty\}.$$

Example 3 (Circle rotation—revisited)

(i) If $\theta = \frac{p}{q}$ is rational, then $f(\omega) = \exp(2\pi i q\omega)$ satisfies

$$f(R_{p/q}\omega) = \exp(2\pi i q(\omega + \frac{p}{q})) = \exp(2\pi i q\omega)\exp(2\pi i p) = f(\omega).$$

However, $f(\omega)$ is not constant. Therefore, by the previous criterion (Theorem 3(ii)), the circle rotation R_θ is not ergodic if θ is rational.

(ii) We claim that if θ is irrational, then R_θ is ergodic. Indeed, let $f \in L^2([0, 1))$ satisfying $f(R_\theta\omega) = f(\omega)$ almost surely. We will show that f is constant almost surely; hence, Theorem 3(iii) implies that R_θ is ergodic. Let

$$f(\omega) = \sum_{n=-\infty}^{\infty} c_n \exp(2\pi i n\omega)$$

be the Fourier series of f. Then

$$f(R_\theta\omega) = \sum_{n=-\infty}^{\infty} c_n \exp(2\pi i n\theta)\exp(2\pi i n\omega)$$

is the Fourier series of $f(R_\theta)$. Now $f = f(R_\theta)$ implies that their Fourier coefficients are the same; i.e., $c_n = c_n \exp(2\pi i n\theta)$ for all n. Therefore $c_n = 0$, if $n \neq 0$. This means that f is constant almost surely.

Example 4 (Bernoulli shift—revisited) The Bernoulli shift is ergodic. Indeed, let $f \in L^2([0, 1))$ with Fourier series

$$f(\omega) = \sum_{n=-\infty}^{\infty} c_n \exp(2\pi i n\omega)$$

and $f(T\omega) = f(\omega)$. Then

$$f(T\omega) = f(2\omega) = \sum_{n=-\infty}^{\infty} c_n \exp(2\pi i(2n)\omega)$$

is the Fourier series of $f(T\omega)$. The invariance of f implies that $c_n = c_{2n}$ for all n. It follows that

$$c_n = c_{2n} = c_{2^2 n} = \cdots \to 0,$$

if $n \neq 0$, because the Fourier coefficients c_n tend to zero as $|n| \to \infty$. Therefore, f is constant almost surely, hence the Bernoulli shift T is ergodic.

2 Birkhoff's Ergodic Theorem

Theorem 4 (Birkhoff's ergodic theorem) *Let T be a measure-preserving transformation on $(\Omega, \mathscr{F}, \mathbf{P})$. Then for any $X \in L^1(\Omega, \mathscr{F}, \mathbf{P})$,*

$$\lim_{n\to\infty} \frac{1}{n} \sum_{m=0}^{n-1} X(T^m\omega) =: X^* \text{ a.s. and in } L^1.$$

Moreover, X^* is invariant, $X^*(T) = X^*$, and $\mathbf{E}[X^*] = \mathbf{E}[X]$. In addition, if the transformation T is ergodic, then $X^* = \mathbf{E}[X]$, and the above limit can be rewritten as

$$\lim_{n\to\infty} \frac{1}{n} \sum_{m=0}^{n-1} X(T^m\omega) = \mathbf{E}[X] \text{ a.s.}$$

Here, $\mathbf{E}[X] = \int_\Omega X(\omega) d\mathbf{P}(\omega)$ denotes the expectation of X.

In his original paper, Birkhoff considered only the case of indicator functions. Then Khintchine extended Birkhoff's result to arbitrary integrable functions on a finite measure space. For this reason, this result is also called the Birkhoff-Khintchine theorem. We also call it the pointwise ergodic theorem because of the type of convergence. In the ergodic case, its meaning is that the time average exists (almost surely) and is equal to the space average.

Let $\mathscr{I}$ be the sub-σ-algebra of $\mathscr{F}$ consisting of all T-invariant subsets $A \in \mathscr{F}$. We claim that the limit $X^* = \mathbf{E}[X|\mathscr{I}]$, where $\mathbf{E}[X|\mathscr{I}]$ denotes the conditional expectation of X given $\mathscr{I}$; i.e., the unique $\mathscr{I}$-measurable function Y with the property that

$$\mathbf{E}[Y\mathbf{1}_A] = \mathbf{E}[X\mathbf{1}_A] \text{ for all } A \in \mathscr{I}.$$

Here, $\mathbf{1}_A$ denotes the indicator function of the set A, and $\mathbf{1}_A(\omega) = 1$, if $\omega \in A$; $\mathbf{1}_A(\omega) = 0$, otherwise. Indeed, for any $A \in \mathscr{I}$, by the ergodic theorem and the T-invariance of $\mathbf{1}_A$,

$$\lim_{n\to\infty} \frac{1}{n} \sum_{m=0}^{n-1} (X\mathbf{1}_A)(T^m\omega) = \mathbf{1}_A(\omega) \lim_{n\to\infty} \frac{1}{n} \sum_{m=0}^{n-1} X(T^m\omega) = \mathbf{1}_A(\omega)X^*(\omega) \text{ a.s.}$$

and

$$\mathbf{E}[X^*\mathbf{1}_A] = \mathbf{E}[X\mathbf{1}_A].$$

In addition, the limit X^* is invariant, thus X^* is $\mathscr{I}$-measurable. These imply that $X^* = \mathbf{E}[X|\mathscr{I}]$.

As a consequence of the ergodic theorem, we give another criterion for the ergodicity.

Theorem 5 *Let T be a measure-preserving transformation on $(\Omega, \mathscr{F}, \mathbf{P})$. T is ergodic if and only if for $A, B \in \mathscr{F}$,*

$$\lim_{n\to\infty} \frac{1}{n} \sum_{m=0}^{n-1} \mathbf{P}(T^{-m}A \cap B) = \mathbf{P}(A)\mathbf{P}(B).$$

Proof Assume that T is ergodic. Then, by applying the ergodic theorem to the indicator function $\mathbf{1}_A$, we have

$$\lim_{n\to\infty} \frac{1}{n} \sum_{m=0}^{n-1} \mathbf{1}_A(T^m\omega) = \mathbf{P}(A) \text{ a.s.}$$

Multiply both sides with $\mathbf{1}_B$, and we have

$$\lim_{n\to\infty} \frac{1}{n} \sum_{m=0}^{n-1} \mathbf{1}_A(T^m\omega)\mathbf{1}_B(\omega) = \mathbf{P}(A)\mathbf{1}_B(\omega) \text{ a.s.}$$

The function on the left-hand side is bounded by 1, so by the bounded convergence theorem, we obtain

$$\lim_{n\to\infty} \frac{1}{n} \sum_{m=0}^{n-1} \mathbf{P}(T^{-m}A \cap B) = \mathbf{P}(A)\mathbf{P}(B).$$

Conversely, assume that

$$\lim_{n\to\infty} \frac{1}{n} \sum_{m=0}^{n-1} \mathbf{P}(T^{-m}A \cap B) = \mathbf{P}(A)\mathbf{P}(B)$$

for any $A, B \in \mathscr{F}$. Then, in particular, for an invariant set A, $T^{-1}A = A$, take $B = A$, we have

$$\lim_{n\to\infty} \frac{1}{n} \sum_{m=0}^{n-1} \mathbf{P}(A) = \mathbf{P}(A)^2.$$

It follows that $\mathbf{P}(A) = \mathbf{P}(A)^2$; hence, $\mathbf{P}(A) = 0$ or $\mathbf{P}(A) = 1$. Therefore T is ergodic.

The convergence in the previous theorem can be changed to give the following.

Definition 3 Let T be a measure-preserving transformation on $(\Omega, \mathscr{F}, \mathbf{P})$.

(i) T is weak-mixing if for $A, B \in \mathscr{F}$,

$$\lim_{n\to\infty}\frac{1}{n}\sum_{m=0}^{n-1}|\mathbf{P}(T^{-m}A\cap B)-\mathbf{P}(A)\mathbf{P}(B)|=0.$$

(ii) T is strong-mixing if for $A, B \in \mathscr{F}$,

$$\lim_{n\to\infty}\mathbf{P}(T^{-n}A\cap B)=\mathbf{P}(A)\mathbf{P}(B).$$

We provide some remarks:

(i) In practice, to verify ergodic, weak-mixing or strong-mixing properties, we only need to check corresponding conditions for A, B belonging to a generating semi-algebra of $\mathscr{F}$.

(ii) A strong-mixing transformation is weak-mixing and a weak-mixing transformation is ergodic. Indeed, for a real sequence $\{a_n\}$, $\lim_{n\to\infty} a_n = 0$ implies

$$\lim_{n\to\infty}\frac{1}{n}\sum_{m=0}^{n-1}|a_m|=0,$$

and this condition itself implies

$$\lim_{n\to\infty}\frac{1}{n}\sum_{m=0}^{n-1}a_m=0.$$

(iii) If θ is irrational, then the circle rotation R_θ is ergodic but is not weak-mixing. Indeed, we can see this roughly as follows. If A and B are small intervals, then $T^{-m}A$ will be disjoint from B for at least half of the value of m so that $(1/n)\sum_{m=0}^{n-1}|\mathbf{P}(T^{-m}A\cap B)-\mathbf{P}(A)\mathbf{P}(B)|\geq(1/2)\mathbf{P}(A)\mathbf{P}(B)$ for large n.

(iv) There are examples of weak-mixing which are not strong-mixing.

(v) Intuitive descriptions of ergodicity and strong-mixing can be given as follows. T is strong-mixing if the sequence of sets $T^{-n}A$ becomes, asymptotically, independent of any other set B. Ergodicity means that $T^{-n}A$ becomes independent of B on average.

(vi) For a real sequence $\{a_n\}$, the condition

$$\lim_{n\to\infty}\frac{1}{n}\sum_{m=0}^{n-1}|a_m|=0$$

is equivalent to a condition that there is a subset J of zero density; i.e., $\frac{1}{n}\#(J\cap\{0,1,\ldots,n-1\})\to 0$, such that

$$\lim_{J\not\ni n\to\infty}a_n=0.$$

Thus, T is weak-mixing if for $A, B \in \mathscr{F}$, the sequence $T^{-n}A$ becomes independent of B provided that we ignore a few instants of time.

3 Some Applications of Ergodic Theorems

3.1 Uniform Distribution Modulo One

Definition 4 A real sequence $(x_n)_{n\geq 1}$ is said to be uniformly distributed modulo 1 if for all a, b with $0 \leq a < b \leq 1$, we have

$$\lim_{n\to\infty} \frac{1}{n}\#\{1 \leq m \leq n : \{x_m\} \in [a, b)\} = b - a,$$

where $\{x_n\}$ denotes the fractional part of x_n.

It follows that if a sequence $(x_n)_{n\geq 1}$ is uniformly distributed modulo 1, then for any Riemann integrable function $f : [0, 1) \to \mathbb{C}$,

$$\lim_{n\to\infty} \frac{1}{n} \sum_{m=1}^{n} f(\{x_n\}) = \int_0^1 f(x)\,\mathrm{d}x,$$

because a Riemann integrable function can be approximated by linear combinations of $\{\mathbf{1}_{[a,b)} : 0 \leq a < b \leq 1\}$, where $\mathbf{1}_{[a,b)}$ is the indicator function of the set $[a, b)$. Note that this property is not true for Lebesgue integrable functions because if f is a Lebesgue integrable function, then changing the values of f on a countable set $(\{x_n\})_{n\geq 1}$ does not change the value of the integral.

A useful criterion for testing the uniform distribution modulo one is

Theorem 6 (Weyl's criterion) *A real sequence $(x_n)_{n\geq 1}$ is uniformly distributed modulo 1 if, and only if, for any integer $k \neq 0$,*

$$\lim_{n\to\infty} \frac{1}{n} \sum_{m=1}^{n} \exp(2\pi i k x_m) = 0.$$

We now show that the sequence $(R_\theta^n \omega)_n = (\omega + n\theta \bmod 1)_n$ is uniformly distributed for all $\omega \in [0, 1)$ provided that θ is irrational. For irrational θ, recall that R_θ is ergodic. Let f be a Lebesgue integrable function on $[0, 1)$. Then, by the ergodic theorem,

$$\lim_{n\to\infty} \frac{1}{n} \sum_{m=1}^{n} f(\omega + m\theta \bmod 1) = \int_{[0,1)} f(\omega)\,\mathbf{dP}(\omega), \text{ a.s.}$$

In particular, take $f = \exp(2\pi i x)$, and we obtain

$$\lim_{n\to\infty} \frac{1}{n} \sum_{m=1}^{n} \exp(2\pi i k(\omega + n\theta)) = \int_0^1 \exp(2\pi i k x)\,\mathrm{d}x = 0 \text{ a.s.}$$

It is clear that the left-hand side does not depend much on ω. Consequently, the convergence holds for all $\omega \in [0, 1)$. Thus $(\omega + n\theta \bmod 1)_n$ is uniformly distributed by Weyl's criterion.

3.2 Benford's Law

Benford's law refers to statistical data whose leading digits $k \in \{1, \ldots, 9\}$ occur with probabilities $\log_{10}(1 + \frac{1}{k})$. The numerical values of the probabilities are

1	2	3	4	5	6	7	8	9
0.3010	0.1761	0.1249	0.0969	0.0792	0.0669	0.0580	0.0512	0.0458

The law was discovered by Newcomb (1881) and named after Benford (1938), who tested it on data from 20 different domains. His data set included the surface areas of 335 rivers, the sizes of 3259 US populations, 104 physical constants, 1,800 molecular weights, and so on. (Source: Wikipedia.)

Let us show below that the frequencies of the leading digits of the sequence $(2^n)_{n\geq 0}$ follow Benford's law. The leading digits of 2^n are k, $k \in \{1, 2, \ldots, 9\}$ if there is an $m \in \{0, 1, \ldots\}$ such that

$$10^m k \leq 2^n < 10^m (k + 1),$$

which is equivalent to

$$\{n \log_{10} 2\} \in [\log_{10} k, \log_{10}(k + 1)).$$

Since $\log_{10} 2$ is irrational, the sequence $(n \log_{10} 2)_n$ is uniformly distributed modulo 1. Therefore,

$$\lim_{n\to\infty} \frac{1}{n} \#\{1 \leq m \leq n : \{m \log_{10} 2\} \in [\log_{10} k, \log_{10}(k + 1))\}$$
$$= \log_{10}(k + 1) - \log_{10} k = \log_{10}(1 + \frac{1}{k}).$$

In this proof, we only need the fact that $\log_{10} 2$ is irrational. Thus, the result is true for any sequence $(\alpha^n)_n$ provided that $\log_{10} \alpha$ is positive irrational. Note that it is the average over time, while statistical data means the average over space. If an ergodic dynamical system can be constructed to model some phenomenon, then the time average and the space average can be related by the ergodic theorem. Thus, it is plausible that statistical data of exponentially growing quantities will agree with Benford's law.

3.3 Markov Chains

In this section, we consider a Markov chain on finite state space. For simplicity, the state space is taken as $S = \{1, 2, \dots, k\}$, $(k \geq 2)$. A vector $\lambda = (\lambda_i)_{i \in S}$ is called a distribution (on S) if $\lambda_i \geq 0$, $(i \in S)$, and $\sum_{i \in S} \lambda_i = 1$. A matrix $P = (p_{ij})$ is said to be stochastic if every row of P is a distribution; i.e., $p_{ij} \geq 0$, $(i, j \in S)$, and $\sum_{j \in S} p_{ij} = 1$, for all i.

Definition 5 A sequence of S-valued random variables $\{X_n\}_{n \geq 0}$ (defined on some probability space $(\Omega', \mathscr{F}', \mathbf{Pr})$) is said to be a Markov chain with initial distribution λ and transition matrix P if

(i) X_0 has distribution λ; i.e., $\mathbf{Pr}(X_0 = i) = \lambda_i$, $(i \in S)$;
(ii) for $n \geq 0$, conditional on $X_n = i$, X_{n+1} has distribution $(p_{ij} : j \in S)$, and is independent of $X_0, \dots, X_{n-1}$; i.e.,

$$\mathbf{Pr}(X_{n+1} = j | X_n = i, X_{n-1} = i_{n-1}, \dots, X_0 = i_0) = p_{ij}.$$

We say that $\{X_n\}_{n \geq 0}$ is Markov(λ, P) for short. Here, λ is a distribution and P is a stochastic matrix.

We list here some properties of Markov chains. Let $\{X_n\}_{n \geq 0}$ be Markov(λ, P). We then have the following:

- Joint probability distribution

$$\mathbf{Pr}(X_0 = i_0, X_1 = i_1, \dots, X_n = i_n) = \lambda_{i_0} p_{i_0 i_1} \cdots p_{i_{n-1} i_n}.$$

- The distribution of X_n is λP^n.
- n-step probability transition

$$\mathbf{Pr}(X_n = j | X_0 = i) = p_{ij}^{(n)},$$

  where $p_{ij}^{(n)}$ denotes the (i, j) element of matrix P^n.
- A distribution π is called invariant if $\pi P = \pi$. Let $\{X_n\}_{n \geq 0}$ be Markov(π, P) with invariant distribution π. Then the distribution of X_n is $\pi P^n = \pi$. Moreover, $\{X_n\}_{n \geq 0}$ becomes a stationary sequence.
- *Ergodic theorem for Markov chain.* P is called irreducible if for any $i, j \in S$, there exists an $n > 0$ such that $p_{ij}^{(n)} > 0$. Let P be an irreducible stochastic matrix.

Then there is a unique invariant distribution π with $\pi_i > 0$, $(i \in S)$. Let $\{X_n\}_{n\geq 0}$ be Markov(λ, P) with λ being any initial distribution. Then, for any function $f : S \to \mathbb{R}$, we have

$$\lim_{n\to\infty} \frac{1}{n} \sum_{m=0}^{n-1} f(X_m) = \bar{f} = (\sum_{i\subset S} f_i \pi_i) \text{ almost surely.}$$

- *Convergence to equilibrium.* A state i is called aperiodic if the greatest common divisor of $\{m \geq 0 : p_{ii}^{(m)} > 0\}$ equals one. The Markov chain (or the matrix P) is called aperiodic if every state is aperiodic. Assume that P is irreducible and aperiodic, and π is the invariant distribution. Let $\{X_n\}_{n\geq 0}$ be Markov(λ, P). Then we have
$$\lim_{n\to\infty} \mathbf{Pr}(X_n = j) = \pi_j, \text{ for all } j.$$
In particular,
$$\lim_{n\to\infty} p_{ij}^{(n)} = \pi_j, \text{ for all } i, j.$$

Let us see how the last two results are related to ergodic theory. Actually, we will deal with a one-sided Markov shift. Let P be a stochastic matrix and π be an invariant distribution with $\pi_i > 0$, $(i \in S)$.

Consider the measurable space $(S, 2^S)$, where 2^S denotes the collection of all subsets of S. Let $(\Omega, \mathscr{F})$ be the direct product space:

$$(\Omega, \mathscr{F}) = \prod_0^\infty (S, 2^S).$$

Let T denote the shift transformation defined by

$$T((\omega_0, \omega_1, \ldots)) = (\omega_1, \omega_2, \ldots).$$

The probability measure $\mathbf{P}$ is defined on the semi-algebra of measurable elementary rectangles by

$$\mathbf{P}(\{(\omega_i) \in \Omega | \omega_0 = i_0, \omega_1 = i_1, \ldots, \omega_n = i_n\}) := \pi_{i_0} p_{i_0 i_1} \cdots p_{i_{n-1} i_n},$$

then $\mathbf{P}$ can be extended to a probability measure on $(\Omega, \mathscr{F})$ and T preserves the probability measure $\mathbf{P}$. The transformation T is called a one-sided (π, P)-Markov shift.

Let ξ_n denote the projection from Ω onto S, which maps $\omega = (\omega_i)$ to $\omega_n \in S$. Then regard $\{\xi_n\}_{n\geq 0}$ as a sequence of random variables defined on $(\Omega, \mathscr{F}, \mathbf{P})$, the sequence $\{\xi_n\}_{n\geq 0}$ is Markov(π, P). Note that $\xi_n = \xi_0(T^n)$.

The Markov shift T has the following properties.

(i) If P is irreducible, then T is ergodic.
(ii) If P is irreducible and aperiodic, then T is strong-mixing.

These results are related to the above results of Markov chain on *ergodic theorem for Markov chain* and *convergence to equilibrium*. Indeed, recall that $\{\xi_n\}_{n\ge 0}$ is Markov(π, P). Let $f\colon S \to \mathbb{R}$ be any function. Then $f \circ \xi_0$ is a random variable on $(\Omega, \mathscr{F}, \mathbf{P})$ and

$$\mathbf{E}[f \circ \xi_0] = \sum_{i\in S} f_i \mathbf{P}(\xi_0 = i) = \sum_{i \in S} f_i \pi_i.$$

If P is irreducible, then the Markov shift T is ergodic. Applying Birkhoff's ergodic theorem to $f \circ \xi_0$, we then obtain

$$\lim_{n\to\infty} \frac{1}{n} \sum_{m=0}^{n-1} f \circ \xi_0(T^m) = \lim_{n\to\infty} \frac{1}{n} \sum_{m=0}^{n-1} f(\xi_m) = \mathbf{E}[f \circ \xi_0] = \sum_{i\in S} f_i \pi_i.$$

Now if P is irreducible and aperiodic, then T is strong-mixing. Let $A = \{\omega = (\omega_i) : \omega_0 = j\}$, $B = \{\omega = (\omega_i) : \omega_0 = i\}$. Then

$$T^{-n} A \cap B = \{\xi_n = j, \xi_0 = i\}.$$

Consequently, by the strong-mixing property, we have

$$\lim_{n\to\infty} \mathbf{P}(\xi_n = j, \xi_0 = i) = \mathbf{P}(\xi_0 = j)\mathbf{P}(\xi_0 = i) = \pi_j \pi_i.$$

Since $\mathbf{P}(\xi_n = j, \xi_0 = i) = \pi_i p_{ij}^{(n)}$, it follows that

$$\lim_{n\to\infty} p_{ij}^{(n)} = \pi_j.$$

Note that for results above regarding a Markov chain, the Markov chain can start at any initial distribution. However, when we consider the Markov shift, we assume that the Markov chain starts at the invariant distribution.

4 Conclusion

This note is mainly taken from [1] (Chap. 7) and [4] (Chap. 1). Refer to [2, 3] for further details on the topics of the distribution modulo one of sequences and Markov chains, respectively.

References

1. R. Durrett, *Probability: Theory and Examples*, 4th edn. (Cambridge Series in Statistical and Probabilistic Mathematics. Cambridge University Press, Cambridge, 2010)
2. L. Kuipers, H. Niederreiter, *Uniform Distribution of Sequences*. (Dover Publishing, New York, 2006)
3. J.R. Norris, *Markov Chains*. (Cambridge University Press, Cambridge, 1997)
4. P. Walters, *An Introduction to Ergodic Theory*. *Graduate Texts in Mathematics*, vol. 79. (Springer, New York, 1982)

Part V
Applied Mathematics

Discrete Optimization: Network Flows and Matchings

Naoyuki Kamiyama

Abstract In this paper, we give a brief introduction to network flow problems and matching problems that are representative problems in discrete optimization. Network flow problems are used for modeling, e.g., car traffic and evacuation. Matching problems are used when we allocate jobs to workers and assign students to laboratories, and so on. Especially, we focus on mathematical models that are used in these problems.

Keywords Discrete optimization · Matching · Network flow

1 Introduction

Optimization is a branch of mathematics that studies problems of finding the optimal object from a set of objects. Especially, discrete optimization is the study of optimization problems with certain discrete structures. In this paper, we give a brief introduction to network flow problems and matching problems that are representative problems in discrete optimization. Network flow problems are used for modeling, e.g., car traffic and evacuation. Matching problems are used when we allocate jobs to workers and assign students to laboratories, and so on. Especially, we focus on mathematical models that are used in these problems. See references given in each section for theory and algorithms.

In the rest of this paper is organized as follows. In Sect. 2, we explain graphs that play an important role in discrete optimization. In Sect. 3, we consider network flow problems. In Sect. 4, we consider matching problems.

N. Kamiyama (✉)
Institute of Mathematics for Industry, Kyushu University, 744 Motooka, Nishi-ku, Fukuoka 819-0395, Japan
e-mail: kamiyama@imi.kyushu-u.ac.jp

R. Nishii et al. (eds.), *A Mathematical Approach to Research Problems of Science and Technology*, Mathematics for Industry 5,
DOI: 10.1007/978-4-431-55060-0_23, © Springer Japan 2014

Throughout this paper, we denote by $\mathbb{R}$, $\mathbb{R}_+$ and $\mathbb{Z}_+$ the sets of real numbers, nonnegative real numbers and nonnegative integers, respectively.

2 Graphs

In this section, we explain basic concepts related to graphs. Intuitively speaking, graphs model "links" connecting "objects" (e.g., people, countries, papers and words). In graph theory, "vertices" and "arcs" (or "edges") correspond to objects and links, respectively. See, e.g., [2, 17] for coverage of concepts related to graph theory. Here we define two kinds of graphs. The first one is a directed graph and the second one is an undirected graph.

A *directed graph* $D = (V, A)$ is a pair of a vertex set V and an arc set A, where every arc in A is an ordered pair of vertices in V. See Fig. 1a for an example of a directed graph. Notice that in a directed graph, we take the direction of each arc into consideration. For each arc $a = (v, w)$ in A, we call v and w the *tail* and *head* of a, respectively. For each vertex v in V, we denote by $\Delta^+(v)$ and $\Delta^-(v)$ the sets of arcs of A whose tails and heads are v, respectively. That is, $\Delta^+(v)$ represents the set of arcs "leaving" v, and $\Delta^-(v)$ represents the set of arcs "entering" v. For example, in Fig. 1a, $\Delta^+(v) = \{a_3\}$ and $\Delta^-(v) = \{a_1, a_2\}$.

An *undirected graph* $G = (V, E)$ is a pair of a vertex set V and an edge set E, where every edge e in E is a subset of V with $|e| = 2$. See Fig. 1b for an example of an undirected graph. Notice that in an undirected graph, we do not take the direction of each edge into consideration. For each subset F of E and each vertex v in V, we denote by $F(v)$ the set of edges e in F with $v \in e$. For example, in Fig. 1b, $E(v) = \{e_1, e_2, e_3\}$. An undirected graph $G = (V, E)$ is called a *bipartite graph*, if V is partitioned into two subsets P and Q, and every edge in E connects a vertex in P and a vertex in Q. See Fig. 6 for an example of a bipartite graph.

3 Network Flows

In this section, we consider network flow problems that are used for modeling, e.g., car traffic and evacuation. See [1] for applications of network flow problems. In the first half of this section, we consider an ordinary network flow model, called a static network flow. In the second half, we consider a dynamic network flow in which we take an important factor "time" into consideration. See, e.g., [1, 5, 17, 18] for coverage of concepts related to networks flows.

3.1 Static Network Flows

In this section, we explain an ordinary network flow model, called a static network flow model. Intuitively speaking, we do not take into account "time," i.e., objects "ceaselessly" flow.

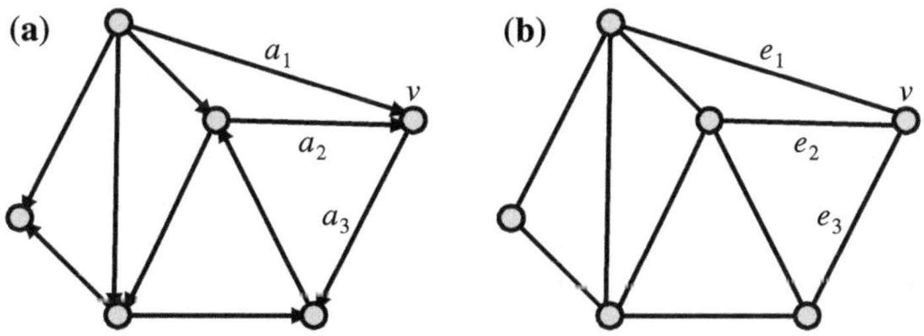

Fig. 1 **a** A directed graph. **b** An undirected graph

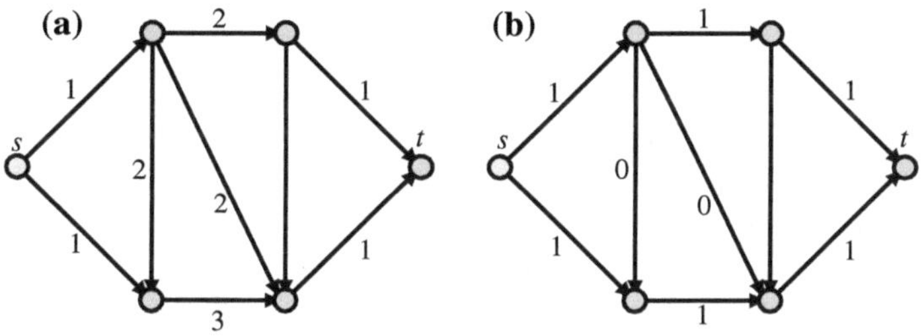

Fig. 2 **a** A static network. **b** A static flow

More formally, in a static network flow model, we are given a directed graph $D = (V, A)$ with specified vertices $s, t \in V$ and a capacity function $c\colon A \to \mathbb{R}_+$. We call the pair of D and c a *static network*. A function $\xi : A \to \mathbb{R}_+$ is called a *static flow*, if it satisfies the following two condition.

Capacity constraint. For every arc a in A,

$$\xi(a) \leq c(a).$$

Flow conservation. For every vertex v in V with $v \neq s, t$,

$$\sum_{a \in \Delta^+(v)} \xi(a) = \sum_{a \in \Delta^-(v)} \xi(a).$$

The *value* of a static flow ξ is defined as

$$\sum_{a \in \Delta^-(t)} \xi(a) - \sum_{a \in \Delta^+(t)} \xi(a).$$

See Fig. 2 for an example of a static flow model and a static flow. In Fig. 2a, the numbers attached to arcs represent their capacities. In Fig. 2a, the numbers attached to arcs represent a static flow.

Here we explain two representative problems in a static flow model. The first one is the *maximum-flow problem*. The goal of this problem is to find a static flow with maximum value. That is, this problem models the situation in which we want to send objects as much as much possible from s to t. For example, in the static network illustrated in Fig. 2a, the static flow in Fig. 2b is a solution of the maximum-flow problem. It is known that this problem can be efficiently solved. See, e.g., [9] for an efficient algorithm for the maximum-flow problem.

The second problem is the *minimum-cost flow problem*. In this problem, we are given a demand $d \in \mathbb{R}_+$ and a cost function $k\colon A \to \mathbb{R}_+$. The *cost* of a static flow $\xi\colon A \to \mathbb{R}_+$ is defined as

$$\sum_{a \in A} k(a) \cdot \xi(a).$$

The goal of the minimum-cost flow problem is to find a static flow whose cost is minimum among all static flows whose value is equal to d. That is, this problem models the situation in which a penalty incurs when we send objects on arcs. This problem can be efficiently solved. See, e.g., [15] for an efficient algorithm for the minimum-cost flow problem.

3.2 Dynamic Network Flows

In this section, we consider a dynamic network flow in which we take an important factor "time" into consideration. That is, in this model, the time required to transit an arc plays an important role.

More formally, in a dynamic network flow model, we are given a directed graph $D = (V, A)$ with a terminal subsets S of V partitioned into S^+ and S^-, a capacity function $c\colon A \to \mathbb{R}_+$, a transit time function $\tau\colon A \to \mathbb{Z}_+$ and a time horizon $T \in \mathbb{Z}_+$. The value $\tau(a)$ represent the time required to transit from the tail of a to the head of a. We call the triple D, c and τ a *dynamic network*. A function $f\colon A \times \mathbb{Z}_+ \to \mathbb{R}_+$ is called a *dynamic flow*, if it satisfies the following two conditions.

Capacity constraint. For each arc a in A and each nonnegative integer θ,

$$f(a, \theta) \le c(a).$$

Flow conservation. For each vertex v in $V \setminus S$ and each nonnegative integer θ,

$$\mathsf{ex}_f(v, \theta) \begin{cases} \ge 0 & \text{if } \theta = 0, 1, \ldots, T-1 \\ = 0 & \text{if } \theta \ge T, \end{cases}$$

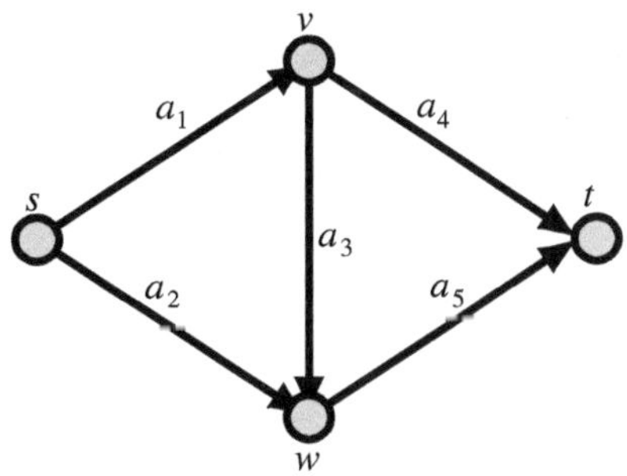

Fig. 3 A dynamic network. In this figure, $S^+ := \{s\}$ and $S^- := \{t\}$. Furthermore, we set $c(a) := 1$ for every arc a in A, and $\tau(a_1) = \tau(a_5) = 0$, $\tau(a_3) = 1$, $\tau(a_2) = \tau(a_4) = 3$ and $T = 5$

where define

$$\mathsf{ex}_f(v,\theta) := \sum_{a\in\Delta^-(v)} \sum_{t=0}^{\theta-\tau(a)} f(a,t) - \sum_{a\in\Delta^+(v)} \sum_{t=0}^{\theta} f(a,t).$$

Intuitively speaking, $f(a,\theta)$ represents the value of flow entering the tail of a at the time θ. The value $\mathsf{ex}_f(v,\theta)$ represents the excess of supplies on the vertex v until the time θ. Notice that in the flow conservation constraint, we allow supplies to stay at vertices. See Fig. 3 for an example of a dynamic network.

Here we explain two representative problems in a dynamic flow model. The first one is the *maximum dynamic flow problem*. Intuitively speaking, this problem is a dynamic version of the maximum-flow problem in a static flow model. In this problem, S^+ and S^- consists of single vertices s^+ and s^-, respectively. The goal of the maximum dynamic flow problem is to find a dynamic flow maximizing $\mathsf{ex}_f(s^-,T)$, i.e., we want to send objects from s^+ to s^- as much as possible within the time limit T. This problem can be efficiently solved. See, e.g., [6] for an efficient algorithm for the maximum dynamic flow problem.

The second problem is the *dynamic transshipment problem*. There exists no corresponding problem in a static flow model. In this problem, we are given a demand function $d\colon S \to \mathbb{R}$ such that

$$d(s)\begin{cases} \le 0 & s \in S^+ \\ \ge 0 & s \in S^-. \end{cases}$$

The dynamic transshipment problem asks for discerning where there exists a dynamic flow f such that

$$\forall s \in S\colon\ \mathsf{ex}_f(s,T) = d(s),$$

and find it, if one exists. That is, we want to send objects from S^+ to S^- so that all supplies and demands are satisfied. This problem can be efficiently solved. See, e.g., [11] for an efficient algorithm for the dynamic transshipment problem.

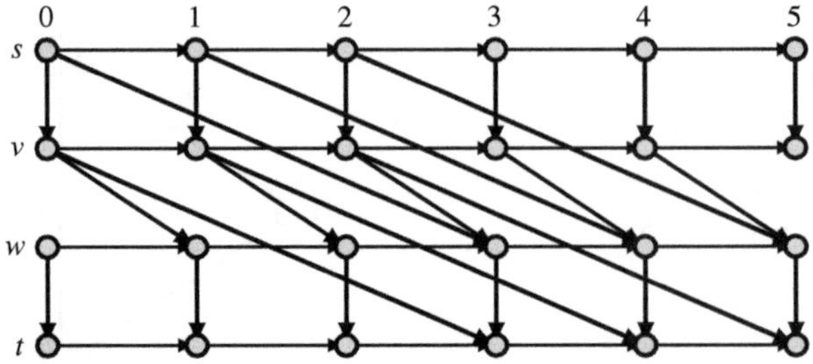

Fig. 4 The time-expanded network of the dynamic network in Fig. 3

From now on, we show that problems in a dynamic flow model can be reduced to those in a static flow model. For this, we define the *time-expanded graph* $\mathscr{T}$ of a dynamic network D with c and τ, which is a static network. See Fig. 4 for an example of a time-expanded network. The vertex set of $\mathscr{T}$ consists of a new vertex v_θ for each vertex v in V and each $\theta = 0, 1, 2, \ldots, T$. The arc set of $\mathscr{T}$ consists of the following two parts. The first part consists of an arc $a_\theta = (v_\theta, w_{\theta+\tau(a)})$ for each arc $a = (v, w)$ in A and each $\theta = 0, 1, \ldots, T - \tau(a)$. Furthermore, the capacity of a_θ is equal to $c(a)$. The second part consists of an arc $(v(\theta), v(\theta + 1))$ for each vertex v in V and each $\theta = 0, 1, \ldots, T - 1$. Furthermore, the capacity of $(v(\theta), v(\theta + 1))$ is infinite.

Let ξ be a static flow in the time-expanded network $\mathscr{T}$. By defining $f(a, \theta) := \xi(a_\theta)$ for each arc a in A and each $\theta = 0, 1, 2 \ldots, T - \tau(a)$, we can construct a dynamic flow f in D with c and τ. Conversely, we can construct a static flow from a dynamic flow in the similar way. This observation implies that by defining $s = s_0$ and $t = t_T$, we can reduce the maximum dynamic flow problem to the maximum-flow problem. It should be noted that the size of the time-expanded network is exponentially larger than that of the input dynamic network.

3.3 Other Problems

In this section, we give other problems in a dynamic flow model.

Similarly to a static flow model, it is natural to consider the problem of finding a dynamic flow with minimum cost. More precisely, we are given a cost function $k: A \to \mathbb{R}_+$ and a demand $d \in \mathbb{R}_+$. Furthermore, we assume that S^+ and S^- consists of single vertices s^+ and s^-, respectively. For each dynamic flow f, we define its cost as

$$\sum_{a \in A} \sum_{\theta=0}^{T} k(a) \cdot f(a, \theta).$$

The goal of this problem is to find a dynamic flow f whose cost is minimum among all dynamic flows f' such that $\mathsf{ex}_{f'}(s^-, T) = d$. Unlike the minimum-cost flow problem in a static flow model, this problem is very hard. See [12] for details.

Furthermore, it is practically important to consider the case where there exist many kinds of objects. For example, there exist several kinds of people in evacuation situations. We can model this problem by using "multi-commodity" flow. There are many papers considering multicommodity flow problems in a static flow model. Thus, it is natural to consider a dynamic version of a multicommodity flow problem in a dynamic flow model. See [10] for details.

In the above problems, we implicitly assume that we can control movement of objects. However, if objects are people, then it is natural to consider that objects selfishly move. That is, it is natural to consider problems from the game theoretical viewpoint. There are many papers considering network flow problems in a static flow model from the game theoretical viewpoint. In a dynamic flow model, e,g, the paper [13] considers a dynamic flow problem from the game theoretical viewpoint.

4 Matchings

In this section, we consider matching problems that are used when we allocate jobs to workers and assign students to laboratories, and so on. See [16] for applications of matching problems. In the first half of this section, we consider an ordinary matching problem, called the maximum-size matching problem. In the second half of this section, we consider the stable matching problem in which each agent has a preference list edges, i.e., a matching problem in a strategic situation. See, e.g., [14, 16, 17] for coverage of concepts related to matching problems.

4.1 Maximum-Size Matchings

In this section, we consider the *maximum-size matching problem*. Intuitively speaking, in this problem, we try to find "pairs" as many as possible.

More formally, in the maximum-size matching problem, we are give an undirected graph $G = (V, E)$. A subset M of E is called a *matching*, if

$$\forall e, f \in M \text{ s.t. } e \neq f\colon\ e \cap f = \emptyset.$$

The maximum-size matching problem asks for finding a matching with maximum cardinality. See Fig. 5 for an example of a maximum-size matching. The matching in Fig. 5b is a maximum-size matching in the undirected graph illustrated in Fig. 5a. This problem can be efficiently solved. See, e.g., [3] for a efficient algorithm for the maximum-size matching problem.

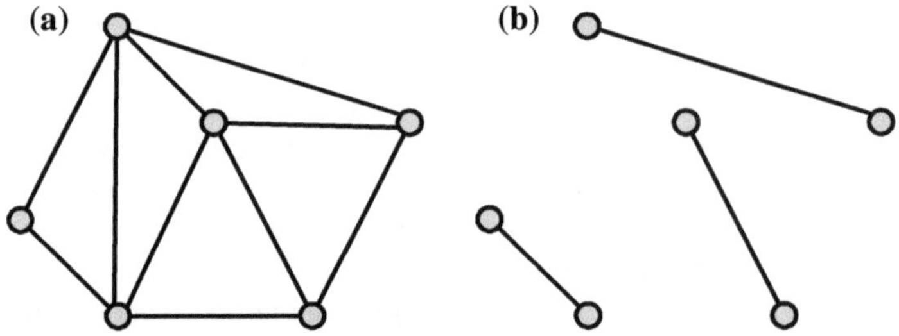

Fig. 5 **a** An undirected graph. **b** A maximum-size matching

In the maximum-size matching, the goal is to find a matching with maximum cardinality. If a "profit" is given for each edge, then it is natural to maximize the profit of a matching. This problem is called the *maximum-weight matching problem*. More formally, in the maximum-weight matching problem, we are given an undirected graph $G = (V, E)$ and a weight function $\sigma : E \to \mathbb{R}_+$. The *weight* of a matching M is defined as

$$\sum_{e \in M} \sigma(e).$$

The goal of the maximum-weight matching problem is to find a matching with maximum weight. This problem can be efficiently solved. See, e.g., [17] for details.

4.2 Stable Matchings

In this section, we explain the *stable matching problem*. Intuitively speaking, in this model, there exist two groups of agents and each agent has a preference ranking over members of the other group. The goal is to find a matching between these two groups with some specified properties.

More formally, the stable matching problem is defined as follows. We are given a bipartite graph $G = (V, E)$. We assume that V is partitioned into P and Q. For each vertex v in V, we are given a strict linear order $>_v$ that represents the preference of v. If $e >_v f$ for some edges e, f in $E(v)$, then v prefers e to f. See Fig. 6a for an example of the stable matching problem. In this example, we assume that

$$\{u, y\} >_u \{u, x\}$$
$$\{v, x\} >_v \{v, y\} >_v \{v, z\}$$
$$\{v, x\} >_x \{u, x\}$$
$$\{v, y\} >_y \{w, y\} >_y \{u, y\}.$$

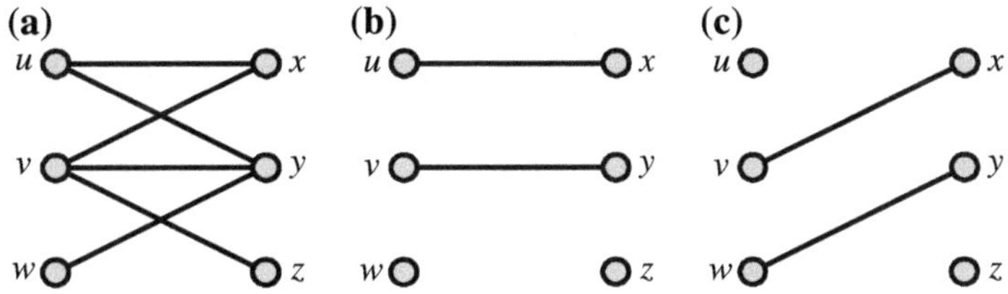

Fig. 6 **a** A bipartite graph. **b** An unstable matching. **c** A stable matching

Let M be a matching (see Sect. 4.1 for the definition of a matching). An edge e in $E \backslash M$ is said to be *free* on its end-vertex $v \in e$, if

- $M(v)$ is empty, or
- $e >_v f$, where $M(v) = \{f\}$.

We say that an edge e in $E \backslash M$ *blocks* M, if e is free on both vertices in e. A matching M is said to be *stable*, if there exists no edge in $E \backslash M$ blocking M. That is, if a given matching is not stable, then there is incentive for some pair to break a current matching.

It is not clear that there exists a stable matching in every instance of the stable matching problem. Gale and Shapley [7] proved that there always exists a stable matching and we can efficiently find it. For example, the matching in Fig. 6b is not stable since $\{v, x\}$ is a blocking pair. On the other hand, the matching in Fig. 6c is stable.

4.3 Other Problems

Here we explain other problems related to the stable matching problem.

In the stable matching problem, we try to find one-to-one matching. However, when we consider the problem of allocating jobs to workers or assigning residents to hospitals, it is natural to consider that one agent can be matched to more than one partners. That is, we consider a many-to-one or many-to-many matching. This problem is called that *hospital/residents problem*. More formally, in this problem, we are given the same input as the stable matching problem. Furthermore, we are given a capacity function $c\colon Q \to \mathbb{Z}_+$. A subset F of E is called an *assignment*, if $|F(p)| \leq 1$ for every vertex p in P and $|F(q)| \leq c(q)$ for every vertex q in Q. Let F be an assignment. An edge $e = \{p, q\}$ in $E \backslash F$ is said to be *free* on $q \in Q$, if

- $|M(q)| < c(q)$, or
- there exists an edge f in $M(q)$ with $e >_q f$.

We say that an edge e in $E \backslash F$ *blocks* F, if e is free on both vertices in e. An assignment F is said to be *stable*, if there exists no edge in $E \backslash F$ blocking F. It is known that there always exists a stable matching and we can efficiently find it. See [7] for details.

Next we consider a problem related to preference lists. In the stable matching problem, we are given a "strict" linear order as a preference list of each agent. That is, some agent v strictly prefer some edge in $E(v)$ to other edge. However, in many practical situations, it is natural to think that agents has preference lists with "ties." That is, in this case, it is possible that some agent v is "indifferent" between some edge in $E(v)$ and other edge. It is known that in this case, there always exists a stable matching, but a new issue arises. This is "Pareto efficiency" of a matching. This concept means that there exists no other matching improving some agent without hurting everyone else. See [4] for this topic.

Finally, we consider the popular matching problem introduced by Gärdenfors [8]. Recall that the concept of stability of a matching is "locally" defined. Thus, it is natural to consider a "global" fairness. The concept of popular matching is one of such concepts of global fairness. In the popular matching problem, we decide the order over matchings by "voting." More precisely, when we are given two matchings, we conclude that a matching for which much people vote is preferable. It is not clear that there always exists a popular matching, but Gärdenfors [8] proved that there always exists a popular matching. In fact, a stable matching is also a popular matching. Thus, the existence of a popular matching follows from that of a stable matching.

References

1. R.K. Ahuja, T.L. Magnanti, J.B. Orlin, *Network Flows: Theory, Algorithms, and Applications* (Prentice Hall, Englewood Cliffs, 1993)
2. J. Bang-Jensen, G.Z. Gutin, *Digraphs: Theory, Algorithms and Applications* (Springer, New York, 2009)
3. J. Edmonds, Paths, trees, and flowers. Can. J. Math. **17**, 449–467 (1965)
4. A. Erdil, H. Ergin, What's the matter with tie-breaking? Improving efficiency in school choice. Am. Econ. Rev. **98**(3), 669–689 (2008)
5. L.R. Ford, D.R. Fulkerson, *Flows in Networks* (Princeton University Press, New Jersey, 1962)
6. L.R. Ford, D.R. Fulkerson, Constructing maximal dynamic flows from static flows. Oper. Res. **6**(3), 419–433 (1958)
7. D. Gale, L.S. Shapley, College admissions and the stability of marriage. Am. Math. Monthly **69**(1), 9–15 (1962)
8. P. Gärdenfors, Match making: assignments based on bilateral preferences. Behav. Sci. **20**(3), 166–173 (1975)
9. A.V. Goldberg, R.E. Tarjan, A new approach to the maximum-flow problem. J. ACM **35**(4), 921–940 (1988)
10. A. Hall, S. Hippler, M. Skutella, Multicommodity flows over time: efficient algorithms and complexity. Theoret. Comput. Sci. **379**(3), 387–404 (2007)
11. B. Hoppe, É Tardos, The quickest transshipment problem. Math. Oper. Res. **25**(1), 36–62 (2000)
12. B. Klinz, G.J. Woeginger, Minimum-cost dynamic flows: the series-parallel case. Networks **43**(3), 153–162 (2004)
13. R. Koch, M. Skutella, Nash equilibria and the price of anarchy for flows over time, in *Proceedings of tje 2nd International Symposium on Algorithmic Game Theory*, vol. 5814, Lecture Notes in Computer Science, pp. 323–334 (2009)
14. D. Manlove, *Algorithmics of matching under preferences* (World Scientific Publishing, Singapore, 2013)

15. J.B. Orlin, A faster strongly polynomial minimum cost flow algorithm. Oper. Res. **41**(2), 338–350 (1993)
16. A.E. Roth, A.O. Sotomayor, *Two-Sided Matching: A Study in Game-Theoretic Modeling and Analysis* (Cambridge University Press, Cambridge, 1992)
17. A. Schrijver, *Combinatorial Optimization: Polyhedra and Efficiency* (Springer, Berlin, 2003)
18. M. Skutella, An introduction to network flows over time, in *Research Trends in Combinatorial Optimization* (Springer, Berlin, 2009), pp. 451–482

Strict Feasibility of Conic Optimization Problems

Hayato Waki

Abstract A conic optimization problem (COP) is the problem of minimizing a given linear objective function over the intersection of an affine space and a closed convex cone. Conic optimization problem is often used for solving nonconvex optimization problems. The strict feasibility of COP is important from the viewpoint of computation. The lack of the strict feasibility may cause the instability of computation. This article provides a brief introduction of COP and a characterization of the strict feasibility of COP. We also explain a facial reduction algorithm (FRA), which is based on the characterization. This algorithm can generate a strictly feasible COP which is equivalent to the original COP, or detect the infeasibility of COP.

Keywords Conic optimization problem · Strong duality · Strict feasibility · Facial reduction

1 Introduction

A conic optimization problem (COP) is the problem of minimizing a given linear objective function over the intersection of an affine space and a closed convex cone, and is one of the convex optimization problems. For instance, linear program (LP), second-order cone program (SOCP), and semidefinite program (SDP) are convex optimization problems and can be representable as COP. It is known that these problems are solved by primal-dual interior-point methods (PDIPMs) in practice. In contrast, nonconvex optimization problem is NP-hard, and thus, it is difficult to find a global optimal solution for such a problem in general.

H. Waki (✉)
Institute of Mathematics for Industry, Kyushu University, 744, Motooka, Nishi-ku, Fukuoka 819-0395, Japan
e-mail: waki@imi.kyushu-u.ac.jp

R. Nishii et al. (eds.), *A Mathematical Approach to Research Problems of Science and Technology*, Mathematics for Industry 5,
DOI: 10.1007/978-4-431-55060-0_24, © Springer Japan 2014

Conic optimization problem is used in convex relaxations, which are effective approaches to solve nonconvex optimization problems. For instance, one can often obtain an optimal solution by combining a convex relaxation with a branch and bound algorithm. Convex relaxations generate COP from a given nonconvex optimization problem. Then, the resulting COP has the property that the optimal value is always equal to or smaller than the optimal value of the original. In particular, LP, SOCP, and SDP are used well in convex relaxations since one can compute the optimal values by using software in which PDIPMs are implemented.

The purpose of this article is to provide a brief introduction on COP and *the strict feasibility* of COP. The strict feasibility is required to solve LP, SOCP, and SDP efficiently by PDIPMs. In fact, under the assumption that a given problem is strictly feasible, the convergence of PDIPMs is proved theoretically. If the assumption fails, PDIPMs may not converge to an optimal solution. Furthermore, in that case, it may have no optimal solutions, i.e., the optimal value is finite, but it may not be attainable. Consequently, it is hopeless to find a good approximation of an optimal solution numerically if it is not strictly feasible.

We describe a characterization of the strict feasibility of COP in this article. One can detect whether a given COP is strictly feasible or not by using the characterization. Furthermore, one can reduce the size of COP by using the certificate of nonstrict feasibility and generate a strictly feasible COP which is equivalent to the original. The characterization plays an essential role in a facial reduction algorithm (FRA) proposed by Borwein and Wolkowicz [2]. We also provide a short survey of FRA. See [11] for more details.

The organization of this article is as follows: the formulation and examples of COP are provided in Sect. 2. Here, we describe the strong duality for COP. This is closely related to the strict feasibility for COP. In Sect. 3, we describe the characterization of the strict feasibility and FRA proposed by Borwein and Wolkowicz. This algorithm is simplified by Pataki. Our explanation of FRA is based on Pataki's simplification.

1.1 Notation and Symbols

The set of real numbers is denoted $\mathbb{R}$. $\langle\cdot,\cdot\rangle_1$ and $\langle\cdot,\cdot\rangle_2$ denote inner products of $\mathbb{R}^m$ and $\mathbb{R}^n$, respectively. For a set $X \subset \mathbb{R}^n$, $\mathrm{int}(X)$ and $\mathrm{relint}(X)$ denote the interior and the relative interior of X, respectively. For a convex set X, if the affine hull of X is $\mathbb{R}^n$, then $\mathrm{relint}(X) = \mathrm{int}(X)$. For a given function $f : \mathbb{R}^n \to \mathbb{R}$ and a set $X \in\subset \mathbb{R}^n$, we denote the minimization of f over X by

$$\inf_x \{f(x) : x \in X\}.$$

x is variable in the minimization. For the maximization, we use $\sup_x$. We remark that this problem may not have any optimal solutions even if the optimal value is finite.

2 Conic Optimization Problem

2.1 Formulation and the Duality

Let $K \subseteq \mathbb{R}^n$ and $A : \mathbb{R}^n \to \mathbb{R}^m$ be a closed convex cone and a surjective linear transformation, respectively. Let us choose $b \in \mathbb{R}^m$ and $c \in \mathbb{R}^n$. We consider the following optimization problem:

$$\sup_y \{\langle b, y\rangle_1 : c - A^* y \in K, y \in \mathbb{R}^m\}, \tag{1}$$

where $A^* : \mathbb{R}^m \to \mathbb{R}^n$ denotes the adjoint of A, which satisfies $\langle Ax, y\rangle_1 = \langle x, A^* y\rangle_2$ for all $x \in \mathbb{R}^n$ and $y \in \mathbb{R}^m$. Let p^* be the optimal value of (1). If the feasible region is empty, then p^* is set to be $-\infty$.

We introduce the dual of (1). To this end, we define the function $L(y, x) : \mathbb{R}^m \times \mathbb{R}^n \to \mathbb{R}$ by $L(y, x) = \langle b, y\rangle_1 + \langle x, c - A^* y\rangle_2$. Let $K^* \subseteq \mathbb{R}^n$ be the dual cone of K, i.e., $K^* = \{s \in \mathbb{R}^n : \langle x, s\rangle_2 \geq 0 \ (\forall x \in K)\}$. Since $\langle b, y\rangle_1 \leq L(y, x)$ for all $x \in K^*$ and $y \in \mathbb{R}^m$ satisfying $c - A^* y \in K$, we have

$$\begin{aligned} p^* &\leq \sup_y \{L(y, x) : c - A^* y \in K, y \in \mathbb{R}^m\} \leq \sup_y \{L(y, x) : y \in \mathbb{R}^m\} \\ &= \langle x, c\rangle_2 + \sup_y \{\langle b - Ax, y\rangle_1 : y \in \mathbb{R}^m\} \end{aligned}$$

for $x \in K^*$. Here we use $A^{**} = A$. For the above inequality, we consider the case where $b - Ax = 0$. Otherwise, the last value is $+\infty$. Consequently, for $x \in \mathbb{R}^n$ such that $x \in K^*$ and $Ax = b$, we have $p^* \leq \langle x, c\rangle_2$, and thus the dual problem can be formulated as follows:

$$\inf_x \{\langle c, x\rangle_2 : Ax = b, x \in K^*\}. \tag{2}$$

Let d^* be the optimal value of (2). If the feasible region is empty, then d^* is set to be $+\infty$.

It follows from the introduction of the dual of (1) that $p^* \leq d^*$ holds. This inequality is known as *the weak duality* for (1) and (2). The equality $p^* = d^*$ is called *the strong duality* or *a zero duality gap* for (1) and (2). This equality does not hold in general. In fact, some instances where the gap is positive are known. For instance, see [18]. We need to impose an assumption for the strong duality. The assumption is called the *constraint qualification* (CQ). Various CQs are proposed for the strong duality for more general convex programming problems. For the strong duality for (1) and its dual (2), *Slater's CQ* is used well. Slater's CQ for (1) is that there exists a feasible solution y of (1) such that $c - A^* y \in \text{int}(K)$. Analogously, Slater's CQ for (2) is that there exists a feasible solution x of (2) such that $x \in \text{int}(K^*)$.

Theorem 1 (Strong duality; see [1]) *If Slater's CQ for* (1) *holds and* (2) *is feasible, then* $p^* = d^*$ *and* (2) *has an optimal solution. Analogously, if Slater's CQ for* (2) *holds and* (1) *is feasible, then* $p^* = d^*$ *and* (1) *has an optimal solution.*

It is known that the strong duality holds under a weaker CQ than Slater's CQ. *The generalized Slater's CQ* for (1) is that there exists a feasible solution y of (1) such that $c - A^*y \in \text{relint}(K)$. (1) is said to be *strictly feasible* if the generalized Slater's CQ holds for (1). Analogously, Slater's CQ for (2) is that there exists a feasible solution x of (2) such that $x \in \text{relint}(K^*)$. The strict feasibility for (2) is also defined in a similar manner to (1).

Theorem 2 ([21, Corollary 4.8]) *If* (1) *satisfies the generalized Slater's CQ and* (2) *is feasible, then* $p^* = d^*$ *and* (2) *has an optimal solution. Analogously, if* (2) *satisfies the generalized Slater's CQ and* (1) *is feasible, then* $p^* = d^*$ *and* (1) *has an optimal solution.*

2.2 Examples of Conic Optimization Problems

We give some typical examples of COP. The first example is LP, which is obtained by choosing the standard inner products of Euclidean spaces and K to be the nonnegative orthant, i.e., $K = \mathbb{R}^n_+$. The strong duality for LP holds under a weaker assumption than the strict feasibility.

Theorem 3 (Strong duality for LP; see [15]) *Suppose that both an LP problem and its dual are feasible. Then, the optimal values are the same and both problems have optimal solutions.*

It is well-known that LP can be solved in polynomial time by the ellipsoid method. However, the ellipsoid method is not so fast in practice. Instead of it, the simplex method and interior-point methods are used well practically. See [1, 15] and references therein.

The second one is SOCP, which is obtained by setting

$$K = \left\{ x = (x_1, x_2, x_3, \ldots, x_n) \in \mathbb{R}^n : x_1 \geq \sqrt{x_2^2 + x_3^2 + \cdots x_n^2} \right\},$$

and the standard inner products of Euclidean spaces. Unlike LP, the Slater's CQ is required for the strong duality for SOCP. See [8] for the applications.

The third example is SDP. Unlike LP and SOCP, SDP deals with symmetric matrix variables. $\mathbb{S}^n$ and $\mathbb{S}^n_+$ denote the sets of $n \times n$ real symmetric and positive semidefinite matrices, respectively. In general, the set $\mathbb{S}^n$ can be identified with $\mathbb{R}^{n(n+1)/2}$. Let K be the set of $n \times n$ positive semidefinite matrices. We fix $\langle X, S \rangle_2 = \sum_{i,j=1}^n X_{ij} S_{ij}$ for $X, S \in \mathbb{S}^n$ and $\langle y, z \rangle_1 = \sum_{i=1}^m y_i z_i$ for $y, z \in \mathbb{R}^m$. Furthermore, for $L_1, \ldots, L_m \in \mathbb{S}^n$, we define a linear map A as follows:

$$AX = \begin{pmatrix} \langle L_1, X\rangle_2 \\ \vdots \\ \langle L_m, X\rangle_2 \end{pmatrix} \quad (\forall X \in \mathbb{S}^n).$$

Then it is easy to prove $A^*y = \sum_{i=1}^m L_i y_i$ for all $y \in \mathbb{R}^m$. For $L_0 \in \mathbb{S}^n$, the SDP problem is the following problem[1]:

$$\sup_y \left\{ \langle b, y\rangle_1 : L_0 - \sum_{i=1}^m y_i L_i \in \mathbb{S}^n_+ \right\}. \tag{3}$$

For (3), the dual is

$$\inf\{\langle L_0, X\rangle_2 : \langle L_i, X\rangle_2 = b_i \ (i = 1, \ldots, m), X \in \mathbb{S}^n_+, y \in \mathbb{R}^m\}. \tag{4}$$

It should be noted that for these three cases, we have $\mathrm{int}(K) = \mathrm{relint}(K)$, and thus the generalized Slater's CQ is equivalent to the Slater's CQ.

The ellipsoid method can solve LP, SOCP, and SDP with rational coefficients in polynomial time to any fixed given precision. However, it is not effective in practice. Practically, PDIPMs are effective for solving SOCP and SDP problems. In fact, under the assumption that both SDP problem (3) and its dual (4) are strictly feasible, both problems have optimal solutions and PDIPMs converge to a pair of primal and dual optimal solutions in polynomially many iterations. From the viewpoint of computation for SDP problems, the assumption is necessary because PDIPMs often become numerically unstable if the assumption fails. See [1, 18–20] for more details on the strong duality and PDIPM, and [24, 25] for instability of computation.

A common feature among LP, SOCP, and SDP is self-dual, i.e., $K^* = K$. We provide examples of non-self-dual case. $X \in \mathbb{R}^{n\times n}$ is a *nonnegative matrix* if all elements of X are nonnegative. We define the following sets:

$$\begin{aligned}
\mathscr{C} &= \{X \in \mathbb{S}^n : x^T X x \ge 0 \ (x \in \mathbb{R}^n_+)\}, \\
\mathscr{C}^* &= \{X \in \mathbb{S}^n : X = YY^T \text{ for some nonnegative matrix } Y\}, \\
\mathscr{N} &= \{X \in \mathbb{S}^n : X \text{ is nonnegative matrix}\}, \\
\mathscr{D} &= \mathscr{N} \cap \mathbb{S}^n_+.
\end{aligned}$$

We remark that $\mathscr{C}^* \subseteq \mathscr{D} \subseteq \mathbb{S}^n_+ \subseteq \mathscr{C}$ and $\mathscr{C}^*$ is the dual of $\mathscr{C}$ with the standard inner product in $\mathbb{S}^n$.

---

[1] We remark that the formulation (3) of SDP in this article is different from one in the article of Dr. Fujisawa. However, one can obtain the form in the article of Dr. Fujisawa by applying the following replacement:

$$b \to -c, \ L_i \to -F_i.$$

Then, (3) can be reformulated as a minimization problem on y.

Conic optimization problems (1) with $K = \mathscr{C}, \mathscr{C}^*and\mathscr{D}$ are called *copositive optimization problem, completely positive optimization problem* and *doubly nonnegative optimization problem*, respectively. Since we have $\mathscr{C} \neq \mathscr{C}^*$ for $n > 1$ and $\mathscr{D}^* = \mathscr{N} + \mathbb{S}^n_+$, all problem are not self-dual.

It is known that all problems are convex programming, but copositive and completely positive optimization problems are NP-hard in general. In contrast, doubly nonnegative optimization problem can be reformulated as SDP and is solvable by PDIPMs. See [3, 4] and references therein for the applications.

3 Strict Feasibility

3.1 Characterization of Strict Feasibility

We give a characterization of the strict feasibility of COP. When the strict feasibility for (1) fails, the intersection of the relative interior of the closed convex cone K and the affine space $\mathscr{L} := \{c - A^*y : y \in \mathbb{R}^m\}$ is empty. It is natural to use the separation theorem. To give the characterization, we use the following separation theorem since the set $\mathscr{L}$ is polyhedral.

Theorem 4 (Separation theorem in [14, Theorem 20.2]) *Assume that C and D are nonempty convex sets and C is polyhedra. Then, the following statements are equivalent:*

1. *There exists a hyperplane $H \not\supseteq D$ which separates C and D.*
2. $C \cap relint(D) = \emptyset$.

In Proposition 1, we give the characterization of (1) where the strict feasibility fails by using the separation theorem 4. To this end, we explain how we apply Theorem 4 to our case for the clarity of this article. See [24] for the details of the proof. Pataki gives another shorter proof in [11].

Assume that COP (1) is not strictly feasible. Then we have

$$\mathscr{L} \cap \text{relint}(K) = \emptyset.$$

It follows from Theorem 4 that there exist $w \in \mathbb{R}^n \setminus \{0\}$ and $\delta \in \mathbb{R}$ such that

1. $\langle w, s\rangle_2 \leq \delta \leq \langle w, f\rangle_2$ $(\forall s \in \mathscr{L}, \forall f \in K)$, and
2. exists $\bar{f} \in K$ satisfying $\langle w, \bar{f}\rangle_2 > \delta$.

We remark that the first is obtained from the fact that a hyperplane H separates C and D in Theorem 4. The second one is from the fact that the hyperplane $H \not\supseteq D$ in Theorem 4. These imply that $\langle w, c\rangle_2 \leq 0$, $A^*w = 0$ and we can choose $\delta = 0$. Since $\langle w, f\rangle_2 \geq 0$ for all $f \in K$, we have $w \in K^*$.

If $\langle w, c\rangle_2 < 0$, then COP (1) is infeasible. In fact, let $\bar{y}$ be a feasible solution of COP (1). Then, we have

$$0 \leq \langle w, c - A^*\bar{y}\rangle_2 = \langle w, c\rangle_2 < 0,$$

which implies that COP (1) is infeasible.

Assume that $\langle w, c\rangle_2 = 0$. Here $\{w\}^\perp$ denotes the set $\{s \in \mathbb{R}^n : \langle s, w\rangle_2 = 0\}$. Then it is easy to see that $\mathscr{L} \subseteq \{w\}^\perp$, and thus

$$\mathscr{L} \cap K = \mathscr{L} \cap \{w\}^\perp \cap K.$$

Since there exists $\bar{f} \in K$ satisfying $\langle w, \bar{f}\rangle_2 > \delta$, we obtain $K \subsetneq K \cap \{w\}^\perp$.

We make a summary of the these result in Proposition 1.

Proposition 1 *Suppose $\mathscr{L} \cap relint(K)$ is empty. Then, there exists a nonzero $w \in K^*$ such that $Aw = 0$ and $\langle c, w\rangle_2 \leq 0$. Furthermore, if $\langle c, w\rangle_2 < 0$, then (1) is infeasible. Otherwise, $F := K \cap \{w\}^\perp \subsetneq K$ and $\mathscr{L} \cap F = \mathscr{L} \cap K$.*

It should be noted that if COP (1) is feasible, but not strictly feasible, then it is equivalent to the following optimization problem (5):

$$\sup_y\{\langle b, y\rangle_1 : c - A^*y \in F, y \in \mathbb{R}^m\}. \tag{5}$$

One can reformulate (2) as the form of (1). See [24] for the reformulation. Consequently, Proposition 1 for the dual (2) holds and is provided as follows:

Corollary 1 *Let $\mathscr{M} = \{x \in \mathbb{R}^n : Ax = b\}$. Suppose $\mathscr{M} \cap relint(K^*)$ is empty. Then, there exists a nonzero $v \in \mathbb{R}^m$ such that $-A^*v \in K$ and $\langle b, v\rangle_1 \geq 0$. Furthermore, if $\langle b, v\rangle_1 > 0$, then (2) is infeasible. Otherwise, $G := K^* \cap \{-A^*v\}^\perp \subsetneq K^*$ and $\mathscr{M} \cap G = \mathscr{M} \cap K^*$. Therefore, if (2) is feasible, then (2) is equivalent to the following problem:*

$$\inf_x\{\langle c, x\rangle_2 : Ax = b, x \in G\}. \tag{6}$$

3.2 Facial Reduction Algorithm for Conic Optimization Problems

Facial reduction algorithm (FRA) is proposed by Borwein and Wolkowicz, which works on COP. FRA for (1) is based on Proposition 1 and generates a finite sequence of COPs which are equivalent (1). If the original problem is feasible, the final problem is strictly feasible. Otherwise, FRA detects the infeasibility of the problem. As we have mentioned, the strong duality for (1) and (2) requires the Slater's CQ. On the other hand, if one applies FRA to both (1) and (2), the strong duality holds without assuming any CQs.

We describe FRA for (1) in Algorithm 1.

Algorithm 1: Facial Reduction Algorithm for COP (1)

Input: COP (1)
Output: Return a strictly feasible COP which is equivalent to (1) or detect the infeasibility

```
begin
    F ⟵ K;
    while ∃ nonzero w ∈ F* such that Aw = 0 and ⟨c, w⟩_2 ≤ 0 do
        if ⟨c, w⟩_2 < 0 then
            COP (1) is infeasible and stop;
        else
            F ⟵ F ∩ {w}^⊥;
        end
    end
    Return sup_y{⟨b, y⟩_2 : c − A*y ∈ F, y ∈ ℝ^m}
end
```

We remark that FRA for (1) terminates at finitely many iterations. See [11, 24] for the details.

All resulting subsets F in Algorithm 1 are faces[2] of K and FRA can reduce the cone K in the original COP (1). This is why this algorithm is called FRA.

One can also apply FRA to (2). The FRA is based on Corollary 1 and described in Algorithm 2. Here we use $K^{**} = K$ since K is a closed convex cone.

Algorithm 2: Facial Reduction Algorithm for COP (2)

Input: COP (2)
Output: Return a strictly feasible COP which is equivalent to (2) or detect the infeasibility

```
begin
    G ⟵ K*;
    while ∃ nonzero v ∈ ℝ^m such that −A*v ∈ G* and ⟨b, v⟩_1 ≥ 0 do
        if ⟨b, v⟩_1 > 0 then
            COP (2) is infeasible and stop;
        else
            G ⟵ G ∩ {−A*v}^⊥;
        end
    end
    Return inf_x{⟨c, x⟩_2 : Ax = b, x ∈ G}
end
```

We here give some literature on FRA. More detailed one is described in Pataki [11]. After FRA is proposed by Borwein and Wolkowicz, Ramana [12] proposes

[2] For a given convex set $C \subset \mathbb{R}^n$, a face of C is a convex subset D of C such that every $x, y \in C$, $x + y \in D$ implies that $x, y \in D$. If C is polyhedral, then the definition is simpler. In fact, a face of polyhedral set C is the intersection with a hyperplane and C. Such a face is called *exposed face*. See [10, 14] for more details.

the extended Lagrange–Slater dual to obtain a strictly feasible SDP problem for a given SDP problem. The resulting problem is a strictly feasible SDP problem with polynomially many variables and constraints. Ramana et al. [13] show the relationship between FRA and extended Lagrange–Slater dual. Pataki extends FRA and Ramana's approach by simplifying FRA in [11].

The other approach is proposed by Luo et al. [9] and Sturm [16, 17]. Their approach also achieves a strictly feasible COP for a given COP by expanding cones K and its dual K^*. The relationship between FRA by Borwein and Wolkowicz and it is revealed in [24].

3.3 Discussion About Facial Reduction Algorithm

The difficulty in FRA in Algorithm 1 is to find w such that $w \in K^* \setminus \{0\}$, $Aw = 0$ and $\langle c, w\rangle_2 \leq 0$. For instance, one may be able to find such a vector w by solving following problem:

$$\inf_{w}\{0 : w \in K^*, Aw = 0, \langle c, w\rangle_2 \leq 0, \langle \bar{f}, w\rangle_2 = 1\},$$

where $\bar{f} \in \mathrm{relint}(K)$ is added to obtain nonzero vector w. We remark that if this problem is infeasible, then FRA terminates since we have no nonzero w. Moreover, COP is infeasible if w satisfies $\langle c, w\rangle_2 < 0$.

However, the computational time for solving this problem will be almost the same as one for (1) because the scale is the almost same as (1). In this sense, applying FRA directly to (1) may be not effective in practice. Furthermore, one cannot obtain the exact solution w by numerical computation since it usually contains round-off errors.

Instead of the full application of FRA, some partial ones are proposed in [5–7, 23, 26, 27]. They exploit some structure in problem which they deal with. For instance, approaches proposed in [7, 23] reduce to the size of polynomial optimization problems by using a property of polynomials. Waki and Muramatsu [22] prove that they are partial applications of FRA. Although their approaches have no guarantees for the strict feasibility of the resulting COPs, they are simpler and more robust from viewpoint of computation. In fact, the computational results which have been already reported in these papers outperform computational one in solving original COPs. The use of partial applications of FRA will increase in the future.

Acknowledgments The author was supported by a Grant-in-Aid for JSPS Fellow 20003236 and a Grant-in-Aid for Young Scientists (B) 22740056.

References

1. A. Ben-Tal, A. Nemirovski, *Lectures on Modern Convex Optimization* (Society for Industrial and Applied Mathematics, Philadelphia, 2001)
2. M.J. Borwein, H. Wolkowicz, Facial reduction for a cone-convex programming problem. J. Aust. Math. Soc. **30**, 369–380 (1981)
3. S. Burer, On the copositive representation of binary and continuous non convex quadratic programs. Math. Programm. **120**, 479–495 (2009)
4. S. Burer, in *Copositive Programming*, ed. by M.F. Anjos, J.B. Lasserre. Handbook on Semidefinite, Conic and Polynomial Optimization (Springer, New York, 2012), pp 201–218
5. F. Burkowski, Y. Cheung, H. Wolkowicz, Efficient use of semidefinite programming for selection of rotamers in protein conformations, preprint (2013)
6. N. Krislock, H. Wolkowicz, Explicit sensor network localization using semidefinite representations and facial reductions. SIAM J. Optim. **20**, 2679–2708 (2010)
7. M. Kojima, S. Kim, H. Waki, Sparsity in sums of squares of polynomials. Math. Program. **103**, 45–62 (2005)
8. M. Lobo, L. Vandenberghe, S. Boyd, H. Lebret, Applications of second-order cone programming. Linear Algebra Appl. **284**, 193–228 (1998)
9. Z.-Q. Luo, F.J. Sturm, S. Zhang, Duality results for conic convex programming, Econometric institute report no. 9719/A. Econometric Institute, Erasmus University Rotterdam (1997)
10. G. Pataki, in *The Geometry of Cone-LP's*, ed. by H. Wolkowicz, R. Saigal, L. Vandenberghe. The Handbook of Semidefinite Programming (Springer, Berlin, 2000), pp. 29–65
11. G. Pataki, Strong duality in conic linear programming: facial reduction and extended dual, arXiv:1301.7717 (2013)
12. V.M. Ramana, An exact duality theory for semidefinite programming and its complexity implications. Math. Program. **77**, 129–162 (1997)
13. V.M. Ramana, L. Tunçel, H. Wolkowicz, Strong duality for semidefinite programming. SIAM J. Optim. **7**, 641–662 (1997)
14. R.T. Rockafellar, in *Convex Analysis*. Princeton Landmarks in Mathematics and Physics (Princeton University Press, Princeton, 1970)
15. A. Schrijver, *Theory of Linear and Integer Programming* (Wiley, New York, 1979)
16. F.J. Sturm, Primal-Dual Interior Point Approach to Semidefinite Programming, Ph.D. Thesis, Erasmus University Rotterdam (1997)
17. F.J. Sturm, in *Theory and Algorithms of Semidefinite Programming*, ed. by H. Frenk, K. Roos, T. Terlaky. High performance optimization, pp. 1–194 (Kluwer Academic Publishers, Dordrecht, 2000)
18. M.J. Todd, Semidefinite optimization. Acta Numerica **10**, 515–560 (2001)
19. L. Tunçel, On the Slater condition for the SDP relaxations of nonconvex sets. Oper. Res. Lett. **29**, 181–186 (2001)
20. L. Tunçel, in *Polyhedral and Semidefinite Programming Methods in Combinatorial Optimization*. Fields Institute Monographs (2010)
21. L. Tunçel, H. Wolkowicz, Strong duality and minimal representations for cone optimization. Comput. Optim. Appl. **53**, 619–648 (2013)
22. H. Waki, M. Muramatsu, Facial reduction algorithms for finding sparse SOS representations. Oper. Res. Lett. **38**, 361–365 (2010)
23. H. Waki, M. Muramatsu, An extension of the elimination method for a sparse SOS polynomial. J. Oper. Res. Soc. Jpn **54**, 161–190 (2011)
24. H. Waki, M. Muramatsu, Facial reduction algorithms for conic optimization problems. J. Optim. Theor. Appl. **158**, 188–215 (2013)

25. H. Waki, M. Nakata, M. Muramatsu, Strange behaviors of interior-point methods for solving semidefinite programming problems in polynomial optimization. Comput. Optim. Appl. **53**, 823–844 (2012)
26. H. Wolkowicz, Q. Zhao, Semidefinite programming relaxations for the graph partitioning problem. Discrete Appl. Math. **96–97**, 461–479 (1999)
27. Q. Zhao, S.E. Karisch, F. Rendl, H. Wolkowicz, Semidefinite programming relaxations for the quadratic assignment problem. J. Comb. Optim. **2**, 71–109 (1998)

Theory of Automata, Abstraction and Applications

Yoshihiro Mizoguchi

Abstract We introduce computational models, such as sequential machines and automata, using the category theory. In particular, we introduce a generalized theorem which states the existence of the most efficient finite state automaton, called the minimal realization. First, we introduce set theoretical elementary models using sets and functions. We then consider a category of sequential machines which is an abstract model of finite automata. In the category theory, we consider several properties of compositions of morphisms. When we look at the category of sets and functions, we describe properties using equations of compositions of functions. Since the theory of category is a general theory, we can have many concrete properties from a general theorem by assigning it to specific categories such as sets and functions, linear space and linear transformations, etc.

Keywords Sequential machine · Automaton · Category theory

1 Introduction

There is a close relationship between the theory of computing and computers, but the theory of computing predates the appearance of an electric computer. It is said that mathematics is a study of "number," "figure" and "motion." The theory of computing is a study of "motion." Similarly, calculus is a typical mathematical subject studying "motion." When we study calculus, we consider the motions of numbers which represent physical phenomena. In the theory of computing, we focus our attention towards discrete objects such as strings or formulas, rather than towards numbers. An object constructed by strings has a structure of a (formal) language, and it can

Y. Mizoguchi (✉)
Institute of Mathematics for Industry, Kyushu University,
744, Motooka, Nishiku, Fukuoka 819-0395, Japan
e-mail: ym@imi.kyushu-u.ac.jp

R. Nishii et al. (eds.), *A Mathematical Approach to Research Problems of Science and Technology*, Mathematics for Industry 5,
DOI: 10.1007/978-4-431-55060-0_25,

represent a computation process which may treat itself. An object in the theory of computing is not only a varying motion object, but also it may refer to itself to move. This leads to one of the difficulties of the theory of computing.

In this paper, we introduce computational models, such as sequential machines and automata. Of particular note, we introduce a theorem which states the existence of the most efficient finite state automaton, which is called the minimal realization. We also show the construction of the minimal realization. We omit general introductions of automata theory, which readers can obtain from several excellent textbooks. We focus on introducing an abstract theory of automata, which were introduced by Arbib and Manes [1, 2]. We pay closer attention to the theory of automata rather than to abstract theories of categories. The theorem of the minimal realization of an automaton is generalized as an abstract theorem of the decomposition of morphisms. Once we have an abstract theorem, we can interpret it with several different categories. When we use a category of discrete systems, we have the theorem of the minimal realization of an automaton. When we interpret it with a category of nondiscrete linear spaces, we can consider a minimal realization of a linear system represented by differential equations. This means that we can prove different kinds of theorems using the same abstract proof through the theory of categories. We sometimes consider optimalities where we are not aware of the notion of their formalizations. The abstract formalization may guide us in formalizing properties such as vague optimalities. When we can formulate a property which we want to minimize, it could be easy to find a way to minimize it. An abstraction using the theory of categories is the first step to solving such problems.

In the 1930s, Alan Turing introduced a formal computational model which is now called a "Turing Machine." He studied computabilities and the notion of the universality of computations. A string written in a tape of a computational model represents a number. Computations are manipulated by represented strings. The study of a finite automaton, which is a kind of computational model using strings, started in the 1950s. The first publication related to this was a collected paper published in 1956 by McCarthy and Shannon, who is now famous as a founder of information theory [3]. At that time, they investigated it as an abstract model of sequential circuits and considered several properties between input strings and output strings [4]. After the notion of an accepting using a subset of a state set was introduced, the theory of automata was established with the formal language theory. The first paper on a finite automaton was written by Rabin and Scott in 1959 [5, 6], who received the ACM Turing award in 1976 for their contributions. Nowadays, formal language theory has expanded the areas covered by their work to include "machine translation," "database theory," and "artificial intelligence."

The annual Language and Automata Theory and Applications (LATA) conference now provides a popular forum for discussion of these concepts [7]. Studies of cellular automata as a model of parallel computations are attracting much interest and theoretical studies of cellular automata are published at the international workshop on Cellular Automata and Discrete Complex Systems supported by the TC-1 working group 1.5 of the International Federation for Information Processing [8]. Further application areas of cellular automata include modeling biological, physical, or chemical systems, image processing, pattern recognition, parallel computing, hardware circuits and architectures, and traffic control. Such applications were recently discussed at the annual Cellular Automata for Research and Industry (ACRI) conference [9].

A theory of automata as a formalism of computation is used to verify programs using inference rules of symbolic logic. Extensions of automata models and logical systems are investigated. When we consider a system such as engineering system, social system, economic system or environmental system, we formalize preconditions of an object, expressions of behaviors, and states of an object. We describe the relations of those formalized objects as a function or a relation between structured sets. Such theoretical study of systems using formalism is called the general systems theory. We prove the correctness of a system by inducing formally described system behaviors through formally described preconditions. Usually, we use a specific formal system for corresponding problems. However, it is also important to arrange results and properties induced by an abstract system using the category theory. This article is an introduction to describing a system using terms of the category theory. In particular, we introduce the theory of automata using it.

2 Sequential Machines

Figure 1 shows an example of a primitive element of sequential circuits, an RS-flipflop ("reset-set" type flipflop). Outputs change sequentially according to inputs. The set of outputs is $Y = \{0, 1\}$, where 0 and 1 represent "On" and "Off," respectively. There are two input lines "S(et)" and "R(eset)." We denote inputs as one string "SR" and consider it as a binary number. That is, an input is an element of $X = \{0 = (00), 1 = (01), 2 = (10)\}$. The set of states is $Q = \{a, b\}$ and the circuit system is defined by a state transition function $\delta : Q \times X \to Q$ and an output function $\beta : Q \to Y$ in the table shown in Fig. 2. The RS-flipflop is modeled as a sequential machine defined by X, Q, Y, δ, and β. When an input is x, a state q changes to the state $\delta(q, x)$. The above Fig. 1 shows inputs "02021021" and outputs "0110010" from the first state a. The state transition diagram of the sequential machine is shown on the right side of Fig. 2. A vertex with an output represents a state and an edge with an input represents a transition. For an input string, we follow edges with input symbols and get transitions of states and output symbols.

A sequential circuit is considered as a function from an input string to an output string; i.e., a function $f : X^* \to Y^*$ from X^* to Y^*, where X^* is a set of strings over X including an empty string ε. A sequential circuit is formalized as a sequential machine defined by a state set, a transition function, and an output function. We consider problems such as 'Which kind of string functions are represented by a sequential machine ?' and 'How many states do we need to represent a string function?'.

Proposition 1 (surjection-injection factorization) *Let $F : X \to Y$ be a function from a set X to a set Y. There exists a surjection $e : X \to Z$ and an injection*

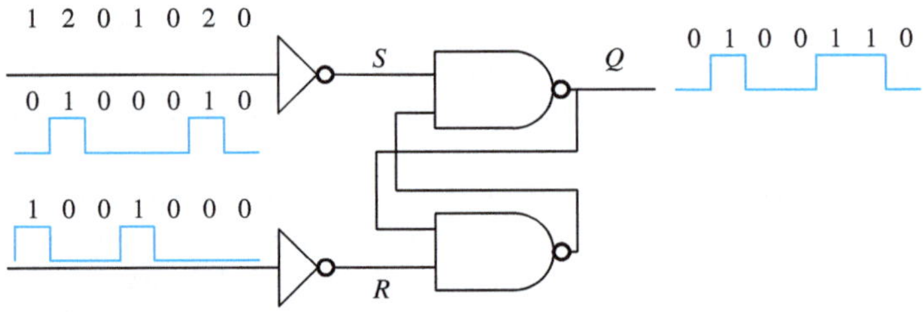

Fig. 1 Example: RS-flipflop circuit

q	x	$\delta(q,x)$
a	0	a
a	1	a
a	2	b
b	0	b
b	1	a
b	2	b

q	$\beta(q)$
a	0
b	1

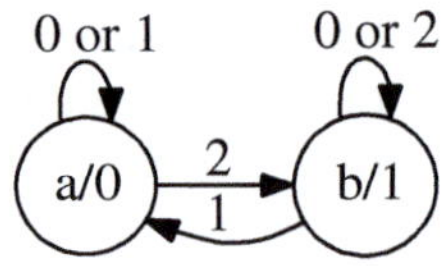

Fig. 2 A state transition diagram of a sequential machine with a state transition function δ and an output function β

$m : Z \to Y$ such that $F(x) = m(e(x))$ $(x \in X)$.[1] *The set Z is uniquely defined without isomorphic set and $Z \cong F(X) = \{F(x) \mid x \in X\}$. In addition, $Z \cong X/\sim$ $= \{[x] \mid x \in X\}$ where the equivalent relation $\sim$ over X is defined by $[x \sim x'$ iff $F(x) = F(x')]$ and $[x]$ is an equivalent class $\{x' \in X \mid x \sim x'\}$* (Fig. 3).

Definition 1 Let Q be a state set, Y a finite set of input symbols, $\delta : Q \times X \to Q$ a state transition function, $\beta : Q \to Y$ an output function and $q_0 \in Q$ an initial state. The sextuple $M = (Q, X, Y, \delta, \beta, q_0)$ is a finite state sequential machine.[2]

A state transition function $\delta : Q \times X \to Q$ can be extended to a function $\delta^* : Q \times X^* \to Q$ by defining $\delta^*(q, \varepsilon) = q$ and $\delta^*(q, xw) = \delta^*(\delta(q, x), w)$ $(q \in Q, x \in X, w \in X^*)$. A function $f : X^* \to Y$ can be generalized to a function $f_* : X^* \to Y^*$ by defining $f_*(\varepsilon) = f(\varepsilon)$ and $f_*(wx) = f_*(w)f(wx)$ $(x \in X, w \in X^*)$. Let $M = (Q, X, Y, \delta, \beta, q_0)$ be a sequential machine. A generalized function $f_{M*} : X^* \to Y^*$ of a function $f_M : X^* \to Y$ defined by $f_M(w) = \beta(\delta^*(q_0, w))$ $(w \in X^*)$ is representing a relation between the input string and the output string.

---

[1] $e : X \to Z$ is a surjection if for any element $z \in Z$, there exists an element $x \in X$ such that $e(x) = z$. $m : Z \to Y$ is an injection if $m(z_1) \neq m(z_2)$ for any elements $z_1, z_2 \in Z$ and $z_1 \neq z_2$.

[2] We follow the definition of the *Moore type* sequential machine. The *Mealy type* sequential machine uses an output function $\lambda : Q \times X \to Y$ instead of β. These two models are equivalent. If we omit the output for an initial state, they are mutually transformable. We note that there is no output of a Mealy-type sequential machine for an initial state. We can define a sequential machine as a pentad without an initial state.

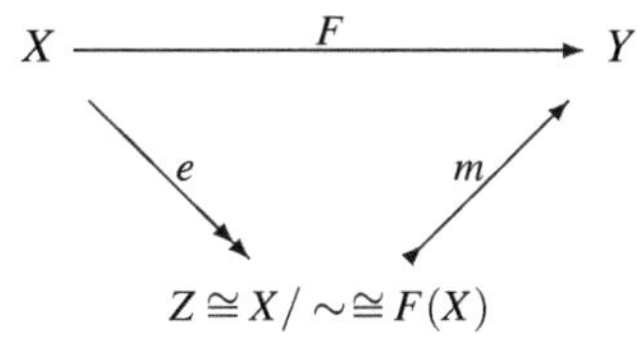

Fig. 3 Surjection–injection factorization

A function $t : X^* \to Y^*$ is *realizable* if there exists a sequential machine M such that $t = f_{M*}$. A necessary condition for a realizable function $t : X^* \to Y^*$ is that there exists a function $f : X^* \to Y$ such that $t = f_*$. The condition is equivalent to the condition such that for any $w \in X^*$, $x \in X$ there exists a $y \in Y$ such that $t(wx) = t(w)y$. That is, the first part of $t(wx)$ is dependent on just w and independent of x. A function $t : X^* \to Y^*$ satisfying the condition is called a *sequential function*.

Next we construct two sequential machines M satisfying $f = f_M$ for a given function $f : X^* \to Y$. The first machine is the most symbolical sequential machine. The state set of the machine is the set of all input strings. That is $M_I = (X^*, X, Y, \delta_I, f, \varepsilon)$ where $\delta_I(w, x) = wx$ $(w \in X^*, x \in X)$. The second machine is the most abstract sequential machine. The state set of the machine is the set of all functions between input strings to output symbols. That is $M_T = (Y^{X^*}, X, Y, \delta_T, \beta_T, f)$ where Y^{X^*} is a set $\{f \mid f : X^* \to Y\}$ of all functions from X^* to Y. The function $\delta_T(f, x) : X^* \to Y$ is defined by $\delta_T(f, x)(w) = f(xw)$ $(x \in X, w \in X^*)$ and $\beta_T(f) = f(\varepsilon)$ $(f \in Y^{X^*})$. For those two sequential machines, we have $f = f_{M_I} = f_{M_T}$. Note, however, that neither M_I nor M_T is a *finite* sequential machine.

What kind of function $f : X^* \to Y$ is realized by a *finite* sequential machine?

A solution is inside the sequential machine M_T. We do not use all states in M_T. We use a state $\delta_T(f, w)$ for some string w, where f is an initial state of M_T. The finiteness of an above state set is important. If a set $Z = \{\delta_T^*(f, w) \in Y^{X^*} \mid w \in X^*\}$ is finite, then M is realized by a finite sequential machine. Let $F : X^* \to Y^{X^*}$ be a function defined by $F(w) = \delta_T^*(f, w)$ $(w \in X^*)$. We then have $Z = F(X^*)$ and $Z = X^*/\sim$ by Proposition 1. The equivalent relation($\sim$) is $[w \sim w'$ iff $\delta_T^*(f, w) = \delta_T^*(f, w')]$. For any $z \in X^*$, $\delta_T^*(f, w)(z) = \delta_T^*(f, w')(z)$; i.e., $f(wz) = f(w'z)$. The number of equivalent classes is finite, so f is realized by a finite sequential machine. Shrinking the state set to Z, we have a finite state sequential machine. Furthermore, we can show that the number of states is minimal.

We assume a function $f : X^* \to Y$ is realized by a sequential machine $M = (Q, X, Y, \delta, \beta, q_0)$. That is, $f = f_M$. The generalized function $F : X^* \to Y^{X^*}$ of $F(w) = \delta_T^*(f_M, w)$ can be decomposed into a composition of $f_e : X^* \to Q$ and $f_m : Q \to Y^{X^*}$ such that $F(w) = f_m(f_e(w))$, where $f_e(w) = \delta^*(q_0, w)$ and

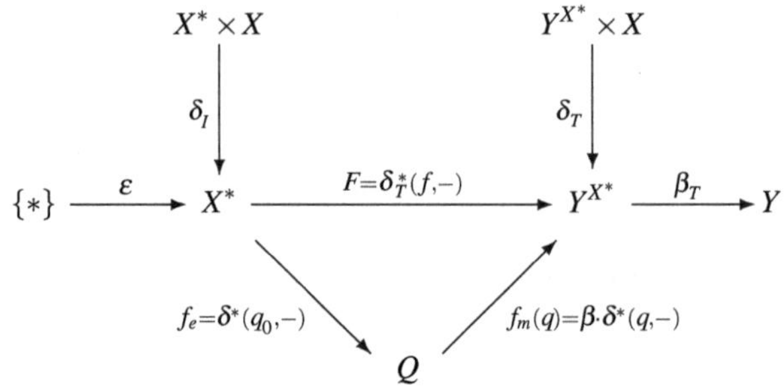

Fig. 4 $M_I = (X^*, X, Y, \delta_I, f, \varepsilon)$ and $M_T = (Y^{X^*}, X, Y, \delta_T, \beta_T, f)$

$f_m(q)(w) = \beta(\delta^*(q, w))$ $(w \in X^*, q \in Q)$.[3] If f_e is a surjection, then we say M is *reachable*. If f_m is an injection, then we say M is *observable* or *reduced*. We can show that a reachable and observable sequential machine M is a minimal state realization of $f = f_M$.

If $f_m : Q \to Y^{X^*}$ is not an injection, we can construct a minimal state sequential machine by decomposing f_m into a composition of a surjection and an injection. The equivalent relation $\sim$ on Q is $[q \sim q'$ iff $f_m(q) = f_m(q')]$. That is, $f_m(q)(w) = f_m(q')(w)$ for any $w \in X^*$. This means $\beta(\delta^*(q, w)) = \beta(\delta^*(q', w))$ for any $w \in X^*$. When the size n of the state set Q is finite, it is sufficient to check $\beta(\delta^*(q, w)) = \beta(\delta^*(q', w))$ only for $w \in X^*$ and $|w| \leq n$. Since we can check $q \sim q'$ in finite steps, we have an algorithm to construct a minimal state sequential machine from a given finite state sequential machine (Fig. 4).

3 Minimal Realization of a Sequential Machine

In this section, we consider a category of sequential machine which is an abstract model of finite automata. In the category theory, we consider several properties of compositions of morphisms. When we see the category of sets and functions, we describe properties using equations of the compositions of functions. Since the theory of category is a general theory, we can have many concrete properties from a general theorem assigning it to specific categories such as sets and functions, linear space and linear transformations, etc. A list for further reading on the category theory is given in another textbook [2, 10].

[3] Note that $f_m(q) : X^* \to Y$ $(q \in Q)$.

3.1 Category and Image Factorization System

Definition 2 A category C is a pair (Obj(C), Mor(C)) of a class of objects Obj(C) and a class of morphisms Mor(C). A set of morphisms Mor(C)(A, B) is defined by two objects A and B. We also denote Mor(C)(A, B) as $C(A, B)$. A morphism $f \in C(A, B)$ is denoted by $f : A \to B$ and we refer to A as the domain of f and B as the codomain of f. We assume the following properties for morphisms.

Property 1 *A function $C(A, B) \times C(B, C) \to C(A, C)$ is given for any objects A, B, and C. We denote the function value for a pair (f, g) of $f : A \to B$ and $g : B \to C$ as $g \cdot f$ and we call it the composition of f and g. We assume the associative axiom $h \cdot (g \cdot f) = (h \cdot g) \cdot f$ for any objects A, B, C and D, and any morphisms $f : A \to B$, $g : B \to C$ and $h : C \to D$.*

Property 2 *There exists a special morphism* id_A *in $C(A, A)$ for any object A. The function* id_A *is called as the identity of A. We assume that* $\mathrm{id}_A \cdot g = g$, $f \cdot \mathrm{id}_A = f$ *for any object B and morphisms $f : A \to B$ and $g : B \to A$.*

Example 1 The category **Set** of sets and functions, the category **Vect** of linear spaces and linear transformations and the category **Poset** of partial ordered sets and order preserving functions are examples of categories.

A morphism $f : A \to B$ is an *epimorphism* if ?$g_1 \cdot f = g_2 \cdot f \Leftrightarrow g_1 = g_2$? for any object $C \in$ Obj and morphisms $g_1 : B \to C$ and $g_2 : B \to C$. An epimorphism is denoted as $f : A \twoheadrightarrow B$.

A morphism $f : B \to C$ is an *monomorphism* if?$f \cdot g_1 = f \cdot g_2 \Leftrightarrow g_1 = g_2$? for any object $A \in$ Obj and morphisms $g_1 : A \to B$ and $g_2 : A \to B$. A monomorphism is denoted as $f : B \rightarrowtail C$.

Example 2 In the category of **Set**, an injection is equivalent to a monomorphism, and a surjection is equivalent to an epimorphism.

Definition 3 A morphism $f : A \to B$ is an *isomorphism* if there exists a morphism $g : B \to A$ such that $g \cdot f = \mathrm{id}_A$ and $f \cdot g = \mathrm{id}_B$. We call such a g an *inverse* of f and A and B are *isomorphic* and denote $A \cong B$.

Definition 4 A pair of classes of morphisms (**E**, **M**) is an *image factorization system* for a category C if the following properties hold.

Property 1 *For any $e : A \to B$ and $e' : B \to C$ in* **E**, *$e' \cdot e : A \to C$ is in* **E**. *For any $m : A \to B$ and $m' : B \to C$ in* **M**, *$m' \cdot m : A \to C$ is in* **M**.

Property 2 *Any element in* **E** *is an epimorphism. Any element in* **M** *is a monomorphism.*

Property 3 *An isomorphism is in both* **E** *and* **M**.

Property 4 *Any morphism $f : A \to B$ is decomposed into a composition of an element $e : A \to C$ in* **E** *and an element $m : C \to B$ in* **M** *such that $f = m \cdot e$. The decomposition is unique up to isomorphism. That is, if there exists another decomposition $f = m' \cdot e'$ of $e' : A \to C'$ in* **E** *and $m : C' \to B$ in* **M** *then $\psi \cdot e = e'$ and $m' \cdot \psi = m$ for some isomorphism $\psi : C \to C'$.*

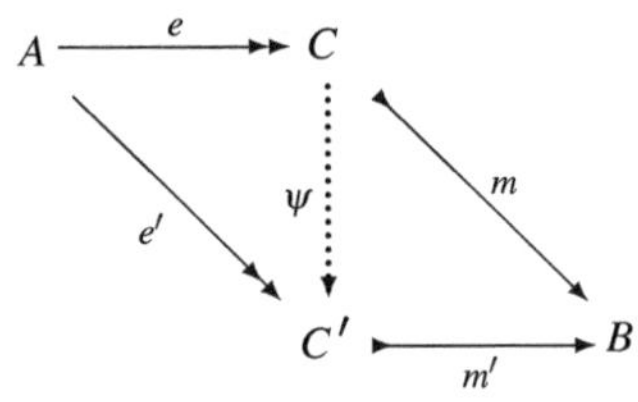

Example 3 Let $(\mathbf{E}, \mathbf{M})$ be a pair of morphisms where $\mathbf{E} = \{f \mid f \text{ is a surjection}\}$ and $\mathbf{M} = \{f \mid f \text{ is an injection}\}$. $(\mathbf{E}, \mathbf{M})$ is then an image factorization system for the category **Set**.

Theorem 1 *Let* $(\mathbf{E}, \mathbf{M})$ *be an image factorization system. Let* $e \in \mathbf{E}$ *and* $m \in \mathbf{M}$*. Consider the following commutative diagram. If* $g \cdot e = m \cdot f$ *then there exists a unique morphism* $h : B \to C$ *such that* $h \cdot e = f$ *and* $m \cdot h = g$.

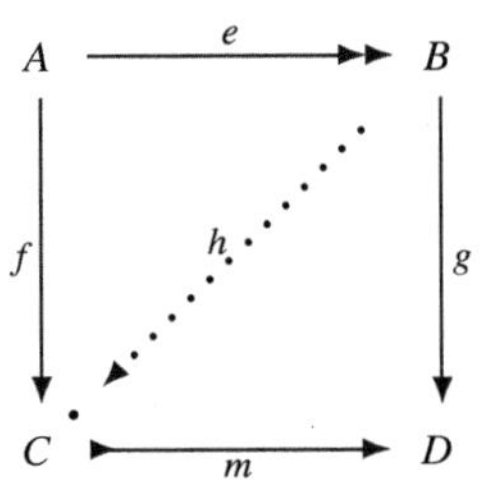

3.2 Category of Sequential Machines

Let Σ be a set of inputs, Q a set of states, $\delta : Q \times \Sigma \to Q$ a state transition function, $q_0 : 1 \to Q$ a function for an initial state, Y a set of outputs and $\beta : Q \to Y$ an output function where 1 is the terminal object in the category **Set**; i.e, a one point set $1 = \{*\}$. We call a sextuple $M = (\Sigma, Q, \delta, q_0, Y, \beta)$ as a *sequential machine*.

A function $q_0 : 1 \to Q$ can be identified as an element $q_0(*)$ in Q, so we call q_0 as an initial state. If $Y = \{0, 1\}$, then $\beta : Q \to Y$ can be considered a subset $F_\beta = \{q \in Q | \beta(q) = 1\}$ of Q. We call the subset F_β as the set of final states. Thus, a sequential machine is a generalization of a finite state automaton. We denote a state at time 0 as $q(0) = q_0$. Let an input symbol and a state at time t be $x(t)$ and $q(t)$, respectively. Then the output at time t is settled $y(t) = \beta(q(t))$ and the state at time $t + 1$ is settled $q(t + 1) = \delta(q(t), x(t))$.

Definition 5 Let Σ be a set of inputs. An object of the category $\mathbf{Dyn}(\Sigma)$ is a pair (Q, δ) of a set Q and a function $\delta : Q \times \Sigma \to Q$. We refer to the objects of the category $\mathbf{Dyn}(\Sigma)$ as Σ*-dynamics*. Let (Q', δ') be another Σ-dynamics. A morphism

$h : (Q, \delta) \rightarrow (Q', \delta')$ of the category $\mathbf{Dyn}(\Sigma)$ is a function $h : Q \rightarrow Q'$ satisfy the following commutative diagram.

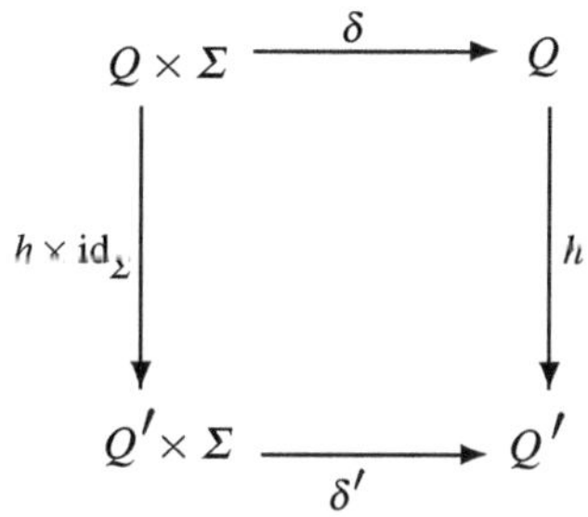

We call a morphism of $\mathbf{Dyn}(\Sigma)$ a *dynamorphism*. The composition of dynamorphisms is defined by the composition in the category **Set**.

It is easy to see that $\mathbf{Dyn}(\Sigma)$ is a category.

Definition 6 Let Σ be a set of inputs and Y a set of outputs. An object of the category $\mathbf{Mach}(\Sigma, Y)$ is a sequential machine $M = (\Sigma, Q, \delta, q_0, Y, \beta)$. Let $M' = (\Sigma, Q', \delta', q_0', Y, \beta')$ be another object. A morphism $h : M \rightarrow M'$ of the category $\mathbf{Mach}(\Sigma, Y)$ is a dynamorphism $h : (Q, \delta) \rightarrow (Q', \delta')$ satisfy the following commutative diagram.

We call a morphism of $\mathbf{Mach}(\Sigma, Y)$ a *simulation*. We say for such M and M' that M simulates M'.

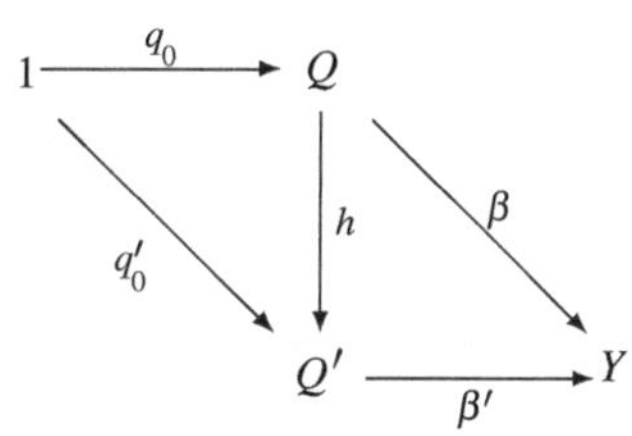

Definition 7 Let Q be a set, $\mu_0 Q : (Q \times \Sigma^*) \times \Sigma \rightarrow Q \times \Sigma^*$ a function defined by $\mu_0 Q((q, w), x) = (q, wx)$. Then, $(Q \times \Sigma^*, \mu_0 Q)$ is a Σ-dynamics. We call $(Q \times \Sigma^*, \mu_0 Q)$ the *free dynamics*.

Theorem 2 (Left Adjoint) *Let $\eta Q_0 : Q_0 \rightarrow Q_0 \times \Sigma^*$ be a function defined by $\eta Q_0(q) = (q, \varepsilon)$. For any Σ-dynamics (Q, δ) and a function $f : Q_0 \rightarrow Q$, there exists a unique dynamorphism $r_f : (Q \times \Sigma^*, \mu_0 Q) \rightarrow (Q, \delta)$ that satisfies the following commutative diagrams.*

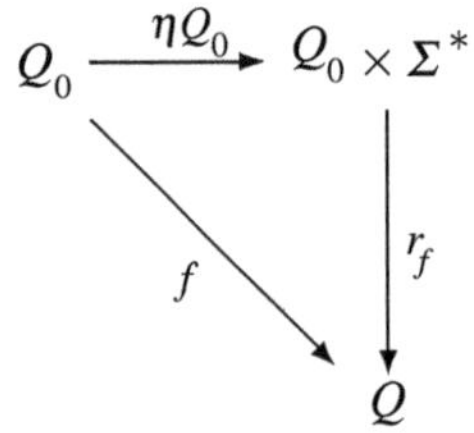

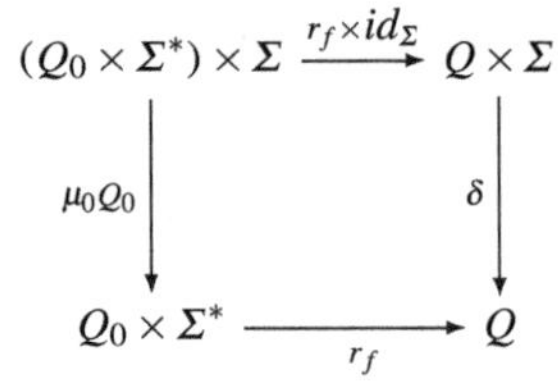

Definition 8 The function r_{id_Q} defined in Theorem 2 is denoted by $\delta^* : Q \times \Sigma^* \to Q$ and is called the *run map*. The function r_{q_0} is called the *reachability map* for an initial state $q_0 : 1 \to Q$. If the reachability map is an epimorphism, then we say Σ-dynamics (Q, δ) is *reachable*.

Definition 9 Let Y be an output set, $LY : Y^{\Sigma^*} \times \Sigma \to Y^{\Sigma^*}$ a function defined by $LY(f, x)(w) = f(xw)$. Then, (Y^{Σ^*}, LY) is a Σ-dynamics. We call (Y^{Σ^*}, LY) the *cofree dynamics* on Y.

Theorem 3 (Right Adjoint) *Let $\Lambda Y : Y^{\Sigma^*} \to Y$ be a function defined by $\Lambda Y(f) = f(\varepsilon)$. For any Σ-dynamics (Q, δ) and a function $\beta : Q \to Y$, there exists a unique dynamorphism $\sigma_\beta : (Q, \delta) \to (Y^{\Sigma^*}, LY)$ that satisfies the following commutative diagrams.*

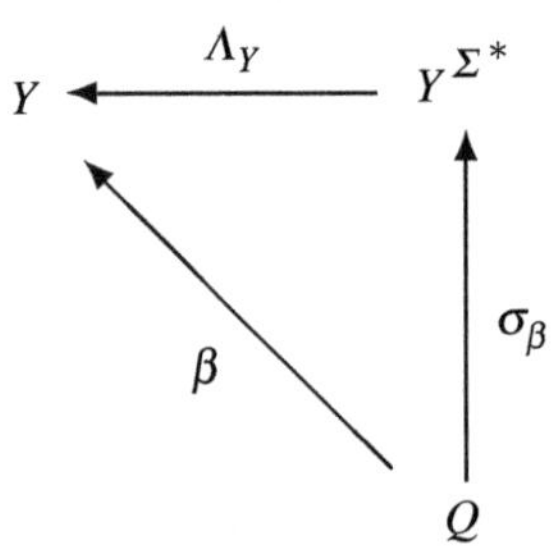

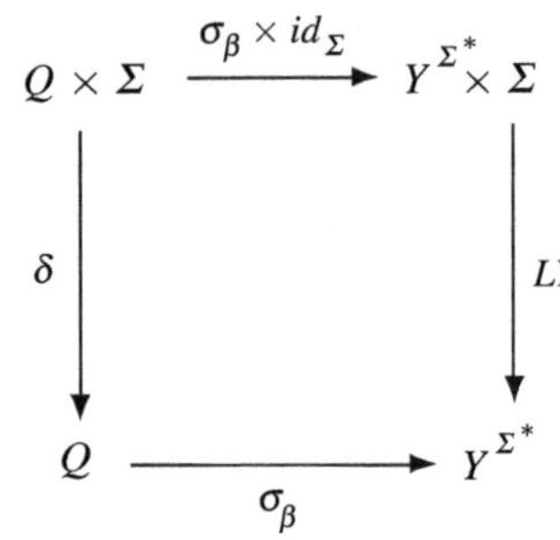

The function $\sigma_\beta : Q \to Y^{\Sigma^*}$ defined in Theorem3 for an output function $\beta : Q \to Y$ is called the *observable map*. If an observable map is an monomorphism, then we say Σ-dynamics (Q, δ) is *observable*. Let $M = (\Sigma, Q, \delta, q_0, Y, \beta)$ be a sequential machine, $r_{q_0} : \Sigma^* \to Q$ a reachable map and $\sigma_\beta : Q \to Y^{\Sigma^*}$ an observable map. We call the composition $\tau_M = \sigma_\beta^* \cdot r_{q_0} : \Sigma^* \to Y^{\Sigma^*}$ the *total response map*.[4] If a dynamorphism $\tau : (\Sigma^*, \mu_0 1) \to (Y^{\Sigma^*}, LY)$ is equal to a total response map τ_M of a sequential machine M, we say M is a *realization* of τ. If a realization M_0 of a dynamorphism $\tau : (\Sigma^*, \mu_0 1) \to (Y^{\Sigma}, LY)$ is reachable and observable, we say it is a *minimal realization*.

Example 4 Let $\tau : (\Sigma^*, \mu_0 1) \to (Y^{\Sigma^*}, LY)$ be a dynamorphism. $M_I = (\Sigma, \Sigma^*, \mu_0 1, \eta 1, Y, \Lambda Y \cdot \tau)$ and $M_T = (\Sigma, Y^{\Sigma^*}, LY, \tau \cdot \eta 1, Y, \Lambda Y)$ are realizations. M_I is reachable and M_T is observable.

[4] Note that the total response map is a dynamorphism. $\tau_M : (\Sigma^*, \mu_0 1) \to (Y^{\Sigma^*}, LY)$.

Proposition 2 *If there exists a simulation $h : M \to M'$ in the category* $\mathbf{Mach}(\Sigma, Y)$, *then the total response maps of M and M' are equal.*

Theorem 4 *Let $h : (Q, \delta) \to (Q', \delta')$ be a morphism in the category* $\mathbf{Dyn}(\Sigma)$ *and $h = m \cdot e$ is an image factorization of a function $h : Q \to Q'$ in the category* **Set**. *Then there exists a unique Σ-dynamics $(h(Q), \delta'')$ that satisfies the following commutative diagram. That is $e : (Q, \delta) \to (h(Q), \delta'')$ and $m : (h(Q), \delta'') \to (Q', \delta')$ are dynamorphisms.*

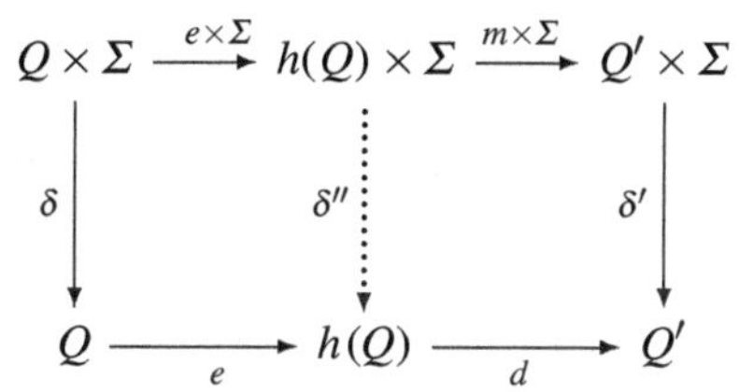

Corollary 1 *Let D be a class of dynamorphisms,* $\mathbf{E}_D = \{f \in D \mid f \text{ is a surjection}\}$ *and* $\mathbf{M}_D = \{f \in D \mid f \text{ is a injection}\}$. *Then* $(\mathbf{E}_D, \mathbf{M}_D)$ *is an image factorization system for the category* $\mathbf{Dyn}(\Sigma)$.

Theorem 5 (Simulation) *Let $\tau : (\Sigma^*, \mu_0 1) \to (Y^{\Sigma^*}, LY)$ be a dynamorphism. There exists a unique simulation $h : M \to M'$ from a reachable realization M of τ to an observable realization M' of τ.*

Proof Let $\tau_M = \sigma \cdot r$ and $\tau_{M'} = \sigma' \cdot r'$. There exists a unique morphism $h : Q \to Q'$ that satisfies the following diagram by Theorem 1.

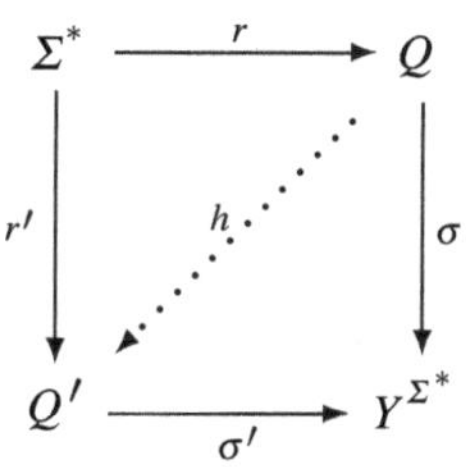

Since r is an epimorphism, h is a dynamorphism and $h : M \to M'$ is a simulation. □

Theorem 6 (Minimal Realization) *There exists a minimal realization M' for any dynamorphism $\tau : (\Sigma^*, \mu_0 1) \to (Y^{\Sigma^*}, LY)$. For any reachable realization M there exists a unique simulation $h : M \to M'$. For any observable realization M'' there exists a unique simulation $h' : M' \to M''$.*

4 Conclusion

We can consider Theorem 6 as a general theorem for a category with an image factorization system. The minimal realization theorem of finite automata is considered as the theorem for the category of sets and functions. The minimal realization of linear systems described by differential equations is considered in the category of linear spaces and linear transformations.

When we find properties of new vague objects in a specific problem, it may be ambiguous to describe the property itself. If you turn your eyes to use generalizations and you describe problems using abstract notions of the category theory, you may see some way to describe properties and objects in a specific problem. It is important to investigate an image factorization system of dynamic morphisms, adjoint functors, free dynamic systems, and cofree dynamic systems when considering extensions of automata. When we can formulate a property which we want to solve, it can become easy to find a way to solve it.

References

1. M.A. Arbib, E.G. Manes, Machines in a category, an expository introduction. SIAM Rev. **16**, 163–192 (1974)
2. M.A. Arbib, E.G. Manes, *Arrows, Structures, and Functors: The Categorical Imperative* (Academic Press, New York, 1975)
3. C.E. Shannon, J. McCarthy (eds.), *Automata Studies* (Princeton University Press, Princeton, 1956)
4. T.L. Booth, *Sequential Machines and Automata Theory* (John Wiley & Sons, New York, 1967)
5. E.F. Moore (ed.), Sequential Machines: Selected Papers (Addison-Wesley, Reading, 1964).
6. M.O. Rabin, D. Scott, Finite automata and their decision problems. IBM J. **3**, 114–125 (1959)
7. A-H. Dediu, C. Martin-Vide (ed.), in Language and Automata Theory and Applications, 6th International Conference, LATA2012, Lecture Notes in Computer Science, vol. 7183 (2012).
8. E. Formenti (ed.), in Proceedings of 18th International Workshop on Cellular Automata and Discrete Complex Systems (Automata2012) and 3rd International Symposium Journées Automates Cellulaires (JAC2012), Electronic Proceedings in Theoretical Computer Science, vol. 90 (2012).
9. G.C. Sirakoulis, S. Bandini (ed.), in Cellular Automata, 10th International Conference on Cellular Automata for Research and Industry, ACRI2012, Lecture Notes in Computer Science, vol. 7495 (2012).
10. S. Mac Lane, Categories for the Working Mathematicians. (Springer, New York, 1972).

Markov Chain Monte Carlo Algorithms

Osamu Maruyama

Abstract Markov chain Monte Carlo (MCMC) methods are a general framework of algorithms for generating samples from a specified probability distribution. They are useful when direct sampling from the distribution is unknown. This article describes theory of MCMC, presents two typical MCMC algorithms (Metropolis-Hastings and Gibbs sampling) and three tempering methods (simulated tempering, parallel tempering, and simulated annealing), and discusses the application of MCMC methods to a prediction problem in systems biology.

Keywords Markov chain Monte Carlo · MCMC · Metropolis-Hastings · Gibbs sampling · Simulated tempering · Parallel tempering · Simulated annealing

1 Introduction

Markov chain Monte Carlo (MCMC) methods are a family of algorithms for sampling from a probability distribution, using the property that a Markov chain converges to a unique stationary distribution under some condition (see, for example, [12, 20]). In physics, statistics, computer science, and other research fields, sampling from particular probability distributions is often required. However, direct sampling algorithms are unknown for many of these distributions. MCMC methods can generally be used for generating the sample. In this article, we focus on MCMC algorithms and describe two representative ones and their extensions.

MCMC sampling algorithms can also be used to optimize a scoring function $f(\mathbf{x})$ for which the optimal value is the minimum of f. For a probability distribution formulated as

O. Maruyama (✉)
Institute of Mathematics for Industry, Kyushu University, 744, Motooka, Nishiku, Fukuoka 819-0395, Japan
e-mail: om@imi.kyushu-u.ac.jp

R. Nishii et al. (eds.), *A Mathematical Approach to Research Problems of Science and Technology*, Mathematics for Industry 5,
DOI: 10.1007/978-4-431-55060-0_26, © Springer Japan 2014

$$\pi(\mathbf{x}) \propto \exp\left(-\frac{f(\mathbf{x})}{T}\right)$$

from $f(\mathbf{x})$ and temperature parameter $T > 0$, the higher the $\pi(\mathbf{x})$ the more optimal the $f(\mathbf{x})$. This means that the optimization problem can be solved by finding the state $\mathbf{x}^*$ with the highest probability,

The rest of the article is organized as follows. Section 2 describes theory of MCMC. The Metropolis-Hastings algorithm and the Gibbs sampling algorithm are presented in Sect. 3. Section 4 presents three tempering techniques that are useful for multimodal probability distributions. Lastly, an application of MCMC is discussed in Sect. 5.

2 Theory of Markov Chain Monte Carlo

2.1 Markov Chain

Let $X^{(n)}$ be random variables for $n = 0, 1, \ldots$, and let $P(X^{(n)})$ represent the probability distribution of $X^{(n)}$. A sequence of random variables, $(X^{(0)}, X^{(1)}, \ldots)$, is called a *Markov chain* if

$$P(X^{(n+1)} = x|X^{(0)} = x^{(0)}, X^{(1)} = x^{(1)}, \ldots, X^{(n)} = x^{(n)}) = P(X^{(n+1)} = x|X^{(n)} = x^{(n)})$$

for all n and $x, x^{(0)}, x^{(1)}, \ldots, x^{(n)}$. This property is called the *Markov property*.

A possible value taken by the random variables is called a *state*. We assume here that the set of all states is finite and that the states are indexed from one to $|S|$, respectively, for the sake of simplicity. This set is often called a *state space*, and here we denote it by S.

A Markov chain is said to be time-homogeneous if

$$P(X^{(n+1)} = j|X^{(n)} = i) = P(X^{(n)} = j|X^{(n-1)} = i)$$

for all n. The Markov chain discussed here is a time-homogeneous one.

Since the state space is finite, the transition probability distribution can be represented by an $|S|$-dimensional matrix in which the (i, j)th element, which is equal to

$$W(i, j) = P\left(X^{(n+1)} = j|X^{(n)} = i\right),$$

is called the *transition probability* from state i to j. The corresponding $|S|$-dimensional square matrix, $W = [W(i, j)]$, is called a transition matrix. Note that

$$\sum_j W(i, j) = 1.$$

2.2 Convergence to Stationary Distribution

A probability distribution, π, over S is called a *stationary distribution* if

$$\pi(j) = \sum_{i=1}^{|S|} \pi(i) W(i, j)$$

for any state $j \in S$. A Markov chain with transition matrix W, is said to be *ergodic* if there exists a positive integer M such that

$$W^M(i, j) > 0$$

for an arbitrary pair of states, $i, j \in S$, where W^M is the Mth matrix power of W. For an ergodic Markov chain of W, the stationary distribution of W is known to be unique.

We now consider the infinite sequence of probability distributions, $\pi^{(n)}$ with $\pi^{(n)}(j) = P(X^n = j)$, for $n = 0, 1, \ldots$, that are generated by repeatedly applying W to $\pi^{(n-1)}$ in the following way:

$$\pi^{(n)}(j) = \sum_{i=1}^{|S|} \pi^{(n-1)}(i) W(i, j), \tag{1}$$

where $\pi^{(0)}$ is an initial probability distribution. This can be also expressed as

$$\pi^{(n)}(j) = \sum_{i=1}^{|S|} \pi^{(0)}(i) W^n(i, j).$$

Theorem 1 *Let W be a transition matrix over a state space S and for which the Markov chain is ergodic, and let π be the stationary distribution of W. The infinite sequence of probability distributions, $\pi^{(0)}, \pi^{(1)}, \ldots$, converges to π:*

$$\lim_{n \to \infty} \pi^{(n)}(i) = \pi(i).$$

We now prove this theorem in a simple way [9]. The ergodicity of W implies that there exists a positive integer M such that any two states are reachable to each other with exactly M transitions by W. Thus, W is replaced with the transition matrix of the Mth matrix power of W. The resulting matrix satisfies $W(i, j) > 0$ for every pair of (i, j). This property gives the next lemma.

Lemma 1 *Suppose that π is a probability distribution over S and that W is a transition matrix on S and its Markov chain is ergodic. For π and W, there exists a* *real number, c, in* (0, 1) *such that, for an arbitrary probability distribution, U, we have*

$$\sum_i U(i)W(i,j) = c\pi(j) + (1-c)R(j),$$

where R is a probability distribution.

Proof For the sake of simplicity, the resulting probability distribution on the left side of the above equation is denoted by $V(j)$, i.e.,

$$V(j) = \sum_i U(i)W(i,j).$$

For each state j, a lower bound on $V(j)$ can be obtained as follows:

$$V(j) \geq \min_k W(k,j) \sum_i U(i) = \min_k W(k,j)$$

Recall that

$$\min_k W(k,j) > 0.$$

Thus, there exists a constant $c_j > 0$ such that

$$\min_k W(k,j) > c_j \pi(j).$$

Then the next is obtained:

$$\min_k W(k,j) > c\pi(j)$$

for any state j, where

$$c = \min_j c_j.$$

Note that $c > 0$. If $c \geq 1$, c is replaced with $c < 1$. Even then, the above equation still holds. Thus, we have

$$V(j) \geq c\pi(j). \tag{2}$$

We then define

$$R(j) = \frac{1}{1-c}\left(V(j) - c\pi(j)\right).$$

From Eq. (2) and $1 - c > 0$, we have $R(j) \geq 0$. From the above equation, we have

$$V(j) = c\pi(j) + (1-c)R(j).$$

After the summations of both sides, R satisfies

$$\sum_j R(j) = 1.$$

In addition to the above equation, we have already shown that $R(j) \geq 0$. Both mean that $R(j) \leq 1$. Thus, R is a probability distribution over S. □

We are now ready to prove Theorem 1 directly. Recall that $\pi^{(0)}$ is the initial probability distribution. Applying Lemma 1 to $\pi^{(0)}$ gives

$$\pi^{(1)}(j) = \sum_i \pi^{(0)}(i)W(i, j) = c\pi(j) + (1-c)R^{(1)}(j)$$

with some probability distribution $R^{(1)}$. Recall that c depends only on π and W. By using the above equation and the fact that π is the stationary distribution of W, the probability distribution of $\pi^{(2)}$, which is Eq. (1) with $n = 2$, can be calculated:

$$\begin{aligned}\pi^{(2)}(j) &= \sum_i \pi^{(1)}(i)W(i, j)\\ &= c\sum_i \pi(i)W(i, j) + (1-c)\sum_i R^{(1)}(i)W(i, j)\\ &= c\pi(j) + (1-c)\sum_i R^{(1)}(i)W(i, j).\end{aligned}$$

Note that in the last equation, the assumption that π is the stationary distribution of W is applied. Furthermore, the last term on the rightmost side of the equation can be transformed by using Lemma 1 as follows:

$$\sum_i R^{(1)}(i)W(i, j) = c\pi(j) + (1-c)R^{(2)}(j)$$

with some probability distribution, $R^{(2)}$. As a result, the two above equations give

$$\pi^{(2)}(j) = (1-(1-c)^2)\pi(j) + (1-c)^2 R^{(2)}(j).$$

In the same way, we can show that, for $n = 3, 4, \ldots,$

$$\pi^{(n)}(j) = (1-(1-c)^t)\pi(j) + (1-c)^t R^{(n)}(j)$$

with some probability distribution $R^{(n)}$. As a result, it can be shown that as $0 < c < 1$ and $R^{(n)}(j) \leq 1$,

$$\lim_{n\to\infty} \pi^{(n)}(j) = \pi(j).$$

The proof of Theorem 1 is thereby completed.

2.3 Detailed Balance

Given Theorem 1, we can see that, given a transition matrix W for which the Markov chain is ergodic, the states resulting from repeatedly taking the transitions from an arbitrary initial state can be considered to be samples from the stationary distribution of W. However, making the W for a particular probability distribution π is still not trivial.

Definition 1 Let W be a transition matrix over a finite state space S and π be a probability distribution over S. A Markov chain with W satisfies *the detailed balance condition* w.r.t. π if

$$\pi(i)W(i, j) = \pi(j)W(j, i) \tag{3}$$

for every (i, j).

The detailed balance condition is sufficient for π to be the stationary distribution with W because, by taking the summations of both sides of Eq. 3 over S, respectively, we have

$$\sum_i \pi(i)W(i, j) = \sum_i \pi(j)W(j, i) = \pi(j)$$

for every $j \in S$. Note that, in the case where $W(i, j) = 0$, if $W(j, i) = 0$, the detailed balance condition at (i, j) trivially holds for any π.

The discussion so far is valid on a finite state space. It is also basically valid on continuous state spaces (see, for example, [18]).

The Metropolis-Hastings algorithm, which is presented in the next section, satisfies the detailed balance condition.

3 MCMC Sampling Algorithms

Two typical MCMC methods are the Metropolis-Hastings algorithm and the Gibbs sampling algorithm.

3.1 Metropolis-Hastings Algorithm

As shown in Algorithm 1, the Metropolis-Hastings algorithm [7] repeatedly moves the current state to another state probabilistically. Distribution $Q(\mathbf{x}, \mathbf{x}')$ is formulated over states $\mathbf{x}'$, conditioned on current state $\mathbf{x}$. It thus holds that $\sum_{\mathbf{x}'} Q(\mathbf{x}, \mathbf{x}') = 1$ for every $\mathbf{x}$. The distribution is called a proposal distribution. In each iteration, a candidate for the next state, $\mathbf{x}'$, is randomly proposed accordance with $Q(\mathbf{x}, \mathbf{x}')$. Then, with a probability of

$$r = \min\left\{1, \frac{\pi\left(\mathbf{x}'\right) Q\left(\mathbf{x}', \mathbf{x}\right)}{\pi\left(\mathbf{x}\right) Q\left(\mathbf{x}, \mathbf{x}'\right)}\right\},$$

Algorithm 1 Metropolis-Hastings algorithm. The target probability distribution from which samples are generated is π. N is the number of iterations.

Let $\mathbf{x}^{(0)}$ be the initial state
$\mathbf{x} = \mathbf{x}^{(0)}$
for $n = 0$ to N **do**
  $\mathbf{x}' \sim Q(\mathbf{x}, \mathbf{x}')$
  $r = \min\left\{1, \dfrac{\pi(\mathbf{x}')\,Q(\mathbf{x}',\mathbf{x})}{\pi(\mathbf{x})\,Q(\mathbf{x},\mathbf{x}')}\right\}$
  $R \sim \mathrm{Uniform}(0, 1)$
  if $r > R$ **then**
    $\mathbf{x} = \mathbf{x}'$
  end if
end for

$\mathbf{x}$ is replaced with the proposed state, $\mathbf{x}'$. With the remaining probability, nothing is done ($\mathbf{x}$ is used again as the current state in the next iteration). The algorithm repeatedly performs this iteration. A sequence of states generated by the Metropolis-Hastings algorithm satisfies the Markov property, because the next state is probabilistically determined from only the previous one.

One of the advantages of the Metropolis-Hastings algorithm is that, even if the partition function of a target probability distribution, π, is unknown, the algorithm works because π is used only in the form $\pi(\mathbf{x}')/\pi(\mathbf{x})$. This feature is quite helpful because it is often difficult to calculate the partition function from a probability density function without the partition function.

It can be easily shown that every iteration of the Metropolis-Hastings algorithm satisfies the detailed balance condition. Suppose that $\mathbf{x}$ is the current state and that candidate state $\mathbf{x}'$ is proposed for the next iteration. We denote by $W(\mathbf{x}, \mathbf{x}')$ the resulting transition probability of moving from $\mathbf{x}$ to $\mathbf{x}'$, which can be described as follows. It is enough to consider the case in which $\pi(\mathbf{x}) > 0$, $\pi(\mathbf{x}') > 0$, and $\mathbf{x} \neq \mathbf{x}'$.

If $r \geq 1$, we have

$$W(\mathbf{x}, \mathbf{x}') = Q(\mathbf{x}, \mathbf{x}'), \tag{4}$$

$$W(\mathbf{x}', \mathbf{x}) = Q(\mathbf{x}', \mathbf{x}) \times \frac{\pi(\mathbf{x})Q(\mathbf{x}, \mathbf{x}')}{\pi(\mathbf{x}')Q(\mathbf{x}', \mathbf{x})}, \tag{5}$$

and otherwise

$$W(\mathbf{x}, \mathbf{x}') = Q(\mathbf{x}, \mathbf{x}') \times \frac{\pi(\mathbf{x}')Q(\mathbf{x}', \mathbf{x})}{\pi(\mathbf{x})Q(\mathbf{x}, \mathbf{x}')},$$

$$W(\mathbf{x}', \mathbf{x}) = Q(\mathbf{x}', \mathbf{x}).$$

In the first case, the next equation can be obtained by respectively dividing both sides of Eq. (4) by both sides of Eq. (5):

$$\frac{W(\mathbf{x},\mathbf{x}')}{W(\mathbf{x}',\mathbf{x})} = \frac{\pi(\mathbf{x}')}{\pi(\mathbf{x})}.$$

This means that π and W satisfy the detailed balance condition. The same result can be shown for the second case. Consequently, π is guaranteed to be the stationary distribution of W.

It is almost trivial that there is no need that a single transition of the Metropolis-Hastings should be ergodic. It is sufficient that a finite sequence of transitions is ergodic for the corresponding Markov chain to converge to the stationary distribution. It would be easy to formulate such transitions in most cases.

Furthermore, it can be seen that there is no need to repeatedly use the same proposal distribution, $Q(\mathbf{x},\mathbf{x}')$, for every transition. Let $Q_1, Q_2, \ldots, Q_K$ be K different "base" proposal distributions, and let $W_{Q_1}, W_{Q_2}, \ldots, W_{Q_K}$ be the resulting transitions. Suppose, at every iteration, the Metropolis-Hastings algorithm is allowed to choose Q_k with probability α_k. The resulting transition probability distribution is expressed as

$$W(\mathbf{x},\mathbf{x}') = \sum_{k=1}^{K} \alpha_k W_{Q_k}(\mathbf{x},\mathbf{x}').$$

Alternatively, the algorithm can use the base proposal distributions successively. The transition probability distribution with this scheme is

$$W(\mathbf{x},\mathbf{x}') = \sum_{\mathbf{x}_1} \cdots \sum_{\mathbf{x}_{K-1}} W_{Q_1}(\mathbf{x},\mathbf{x}_1) W_{Q_2}(\mathbf{x}_1,\mathbf{x}_2) \ldots W_{Q_K}(\mathbf{x}_{K-1},\mathbf{x}'),$$

which is equivalent to

$$W = W_{Q_1} W_{Q_2} \ldots W_{Q_K}.$$

Typical base proposal distributions for a D-dimensional space propose a new value for only one coordinate. For example, for state, $\mathbf{x} = (x_1, x_2, \ldots, x_D)^T$, the proposal distribution, $Q_d(x_d, x'_d)$, proposes a new value, x'_d, for the dth coordinate ($d = 1, 2, \ldots, D$). If the D-dimensional space is continuous, a typical proposal distribution is $Q_d(x_d, x'_d) = N(x'_d | x_d, \sigma^2)$ where $N(x'_d | x_d, \sigma^2)$ is the Gaussian distribution with mean x_d and variance σ^2. The resulting Markov chain is trivially ergodic.

The resulting probability distributions simulated by transitions converge to π. Therefore, the states chosen at each transition can be considered samples from π. By choosing states with a sufficiently long interval, we can use the resulting states as samples from P.

If a proposal distribution is symmetric, i.e., $Q(\mathbf{x},\mathbf{x}') = Q(\mathbf{x}',\mathbf{x})$, the Metropolis-Hastings algorithm is equivalent to the Metropolis algorithm [17].

3.2 Gibbs Sampling

Algorithm 2 Gibbs sampling algorithm. $\mathbf{x}_{-i}$ is the set of all elements except x_i.

Let $\mathbf{x} = (x_1, x_2, \ldots, x_D)^T$ be a variable representing a state in a D-dimensional space
Let $\mathbf{x}^{(0)}$ be the initial state
$\mathbf{x} = \mathbf{x}^{(0)}$
for $n = 1$ to N **do**
  for $i \in \{1, 2, \ldots, D\}$ **do**
    {choose i randomly or in sequential order}
    $x_i^{(n)} \sim \pi\left(x_i^{(n)} \middle| \mathbf{x}_{-i}\right)$
    $x_i = x_i^{(n)}$
  end for{New sample $\mathbf{x}^{(n)} = \left(x_1^{(n)}, x_2^{(n)}, \ldots, x_D^{(n)}\right)^T$ has been created}
end for

The Gibbs sampling algorithm [4], shown in Algorithm 2, is also an MCMC algorithm for sampling from a given *multivariate* probability distribution. The state space considered here is D-dimensional. Let π be the probability distribution from which samples are generated. The algorithm repeats the following procedure for current state $\mathbf{x} = (x_1, x_2, \ldots, x_D)^T$. For each $i \in \{1, 2, \ldots, D\}$, x_i is replaced with a sample $x_i^{(n)} \sim \pi\left(x_i^{(n)} \middle| \mathbf{x}_{-i}\right)$, where $\mathbf{x}_{-i}$ is the set of all elements except x_i. That is, $x_i^{(n)}$ is sampled from the distribution of the ith element, x_i, conditioned on all other variables. The value of x_i is then instantly updated with the latest sample, $x_i^{(n)}$.

The Gibbs sampling algorithm is a special case of the Metropolis-Hastings algorithm for the following reason. A proposal distribution can be formulated for the Gibbs sampling algorithm although it does not explicitly appear in the algorithm. We denote it by $Q(\mathbf{x}, \mathbf{x}')$. Because the algorithm chooses $i \in \{1, 2, \ldots, D\}$ and proposes a candidate value for the ith element, Q can be represented as

$$Q(\mathbf{x}, \mathbf{x}') = Q_{index}(i)\pi\left(x_i' | \mathbf{x}_{-i}\right),$$

where $Q_{index}(i)$ is the probability of choosing the ith element. Note that for every $j (\neq i)$, $x_j' = x_j$, where $\mathbf{x} = (x_1, x_2, \ldots, x_D)^T$ and $\mathbf{x}' = (x_1', x_2', \ldots, x_D')^T$. Using this proposal distribution, we can show that the term of r in the Metropolis-Hastings algorithm is always equal to one:

$$r = \frac{\pi(\mathbf{x}')Q(\mathbf{x}', \mathbf{x})}{\pi(\mathbf{x})Q(\mathbf{x}, \mathbf{x}')} = \frac{\pi(\mathbf{x}')Q_{index}(i)\pi\left(x_i | \mathbf{x}'_{-i}\right)}{\pi(\mathbf{x})Q_{index}(i)\pi\left(x_i' | \mathbf{x}_{-i}\right)} = \frac{\pi(\mathbf{x}')\pi\left(x_i | \mathbf{x}_{-i}\right)}{\pi(\mathbf{x})\pi\left(x_i' | \mathbf{x}_{-i}\right)} = \frac{\pi(\mathbf{x}')\pi\left(\mathbf{x}\right)}{\pi(\mathbf{x})\pi\left(\mathbf{x}'\right)} = 1.$$

This is why the Gibbs sampling algorithm always accepts a proposed value.

It is appropriate to use Gibbs sampling if sampling from the conditional distribution, $\pi(x_i | \mathbf{x}_{-i})$, is easy. Otherwise, an alternative method like the Metropolis-Hastings algorithm should be used.

4 Tempering

A probability distribution from which samples are generated is often designed in the form

$$\pi(\mathbf{x}) \propto \exp\left(-\frac{f(\mathbf{x})}{T}\right),$$

where $f(\mathbf{x})$ is a scoring function for which the optimal value is the minimum, and $T > 0$ is a temperature parameter. This formulation is called a *Boltzmann* or *Gibbs* distribution. In physics, $f(\mathbf{x})$ corresponds to the energy of state $\mathbf{x}$. The higher the T, the more uniform the resulting probability distribution, and the lower the T, the spikier the distribution.

The target probability distribution, π, is often *multimodal*. If it is, an MCMC sampler is likely to be trapped in local modes if the temperature is low. Conversely, an MCMC sampler increases the chance of moving to the main body of π if the temperature is high. This observation led to various techniques of multi-level sampling.

4.1 Simulated Tempering

Algorithm 3 Simulated tempering algorithm.

```
Let (x^(0), i^(0)) be the initial state
(x, i) = (x^(0), i^(0))
for n = 1 to N do
  u ~ Uniform(0, 1)
  if u ≤ α_0 then
    i^(n) = i
    x^(n) is updated via the MCMC transition for π_i from x
  else
    x^(n) = x
    j ~ Q_st(i, j)
    r_st = min{1, c_j π_j(x) Q_st(j, i) / (c_i π_i(x) Q_st(i, j))}
    R ~ Uniform(0, 1)
    if r_st > R then
      i^(n) = j
    else
      i^(n) = i
    end if
  end if
  (x, i) = (x^(n), i^(n))
end for
```

In the simulated tempering algorithm [6, 15], shown in Algorithm 3, the value of the temperature parameter in a Boltzmann distribution is also sampled from among I predetermined temperatures, $T_1 < T_2 < \cdots < T_I$. The state space of the algorithm is thereby augmented to $S \times \{1, 2, \ldots, I\}$, where S is the original state space. For a temperature T_i, the corresponding probability distribution is given as

$$\pi_i(\mathbf{x}) = \frac{1}{Z_i} \exp\left(-\frac{f(\mathbf{x})}{T_i}\right),$$

where Z_i is the partition function. The target probability distribution,

$$\pi(\mathbf{x}, i) \propto c_i \pi_i(\mathbf{x}),$$

is then defined on the augmented state space, where c_i is the weight of the ith distribution, π_i.

In each iteration of the algorithm, with some probability α_0, the algorithm simulates an MCMC transition for π_i while the current temperature, T_i, remains unchanged. With the remaining probability, $j (\in I) \sim Q_{\mathrm{st}}(i, j)$ is proposed, and j is accepted with probability

$$\min\left\{1, \frac{c_j \pi_j(\mathbf{x}) Q_{\mathrm{st}}(j, i)}{c_i \pi_i(\mathbf{x}) Q_{\mathrm{st}}(i, j)}\right\}$$

and is rejected otherwise. Usually, $Q_{\mathrm{st}}(i, j)$ proposes a neighboring value, $i \pm 1$. Trivially, this transition also satisfies the detailed balance condition. Thus, as a whole, this algorithm also satisfies the condition.

4.2 Parallel Tempering

The parallel tempering algorithm [5] shown in Algorithm 4 is also known as the exchange Monte Carlo algorithm [8]. Instead of the original state being augmented to $S \times \{1, 2, \ldots, I\}$ as in simulated tempering, it is augmented to product space S^I, and the I Markov chains on the space in parallel. In addition, instead of a temperature transition as in simulated tempering, the original states in two neighboring Markov chains are swapped.

Let $(\mathbf{x}_1, \mathbf{x}_2, \ldots, \mathbf{x}_I) \in S^I$ be a state in parallel tempering. The target probability distribution is given in the form

$$\pi(\mathbf{x}_1, \mathbf{x}_2, \ldots, \mathbf{x}_I) = \prod_{i=1}^{I} \pi_i(\mathbf{x}_i),$$

where $\pi_i(\mathbf{x}_i)$ is a probability distribution over S. Looking at Algorithm 4, we can easily see that the detailed balance condition is satisfied.

This algorithm is called "parallel tempering" because the ith probability distribution over $\mathbf{x}_i$ is typically tempered with a temperature, T_i,

Algorithm 4 Parallel tempering algorithm.

Let $\left(\mathbf{x}_1^{(0)}, \mathbf{x}_2^{(0)}, \ldots, \mathbf{x}_I^{(0)}\right)$ be the initial state in S^I
$(\mathbf{x}_1, \mathbf{x}_2, \ldots, \mathbf{x}_I) = \left(\mathbf{x}_1^{(0)}, \mathbf{x}_2^{(0)}, \ldots, \mathbf{x}_I^{(0)}\right)$
for $n = 1$ to N **do**
  $u \sim \text{Uniform}(0, 1)$
  if $u \leq \alpha_0$ **then**
    every individual state $\mathbf{x}_i$ is updated via the ith MCMC transition for π_i
  else
    $i \sim \text{Uniform}[1, 2, \ldots, \text{I} - 1]$
    $r_{\text{pt}} = \min\left\{1, \dfrac{\pi_i(\mathbf{x}_{i+1})\pi_{i+1}(\mathbf{x}_i)}{\pi_i(\mathbf{x}_i)\pi_{i+1}(\mathbf{x}_{i+1})}\right\}$
    $R \sim \text{Uniform}(0, 1)$
    if $r_{\text{pt}} > R$ **then**
      the values of $\mathbf{x}_i$ and $\mathbf{x}_{i+1}$ are exchanged
    end if
  end if
end for

$$\pi_i(\mathbf{x}) \propto \exp\left(-\frac{f(\mathbf{x})}{T_i}\right),$$

where $T_1 < T_2 < \ldots < T_I$. Probability r_{pt} can thus be expressed as

$$r_{\text{pt}} = \min\left\{1, \exp\left(\left(\frac{1}{T_i} - \frac{1}{T_{i+1}}\right)(\text{f}(\mathbf{x_i}) - \text{f}(\mathbf{x_{i+1}}))\right)\right\}.$$

It is interesting to compare the performance of these two tempering methods. For example, Zhang and Ma [25] reported that simulated tempering is superior to parallel tempering under certain condition.

4.3 Simulated Annealing

Simulated annealing, proposed by Kirkpatrick et al. [11], is a sampling method specialized for solving optimization problems. Let $T_1 > T_2 > \cdots > T_I$ be monotonically decreasing temperatures. For each $i = 1, 2, \ldots, I$, an MCMC simulation is conducted on $\pi_i \propto \exp\left(-\frac{f(\mathbf{x})}{T_i}\right)$ where the initial state is the last state sampled in the previous simulation. With a high temperature, the algorithm moves globally. With a low temperature, the algorithm moves locally. The whole simulation is not a valid MCMC simulation; it is simply a sequence of MCMC simulations with different temperatures.

5 Application of MCMC

As seen in the discussion of simulated annealing, MCMC methods can be applied to optimization problems. Successful results for the problem of predicting protein complexes from protein-protein interactions (PPIs) have been obtained by designing search algorithms based on the Metropolis-Hastings algorithm [22, 23]. Here we discuss, as an example, PPSampler2 [23], a protein complex prediction tool based on the Metropolis-Hastings algorithm.

The formulations of the state space, the scoring function, and the proposal distribution can be described as follows. Let V be the set of proteins under consideration. The input is a collection of PPIs that can be considered a subset of protein pairs from V. Each pair has a weight representing the reliability of the interaction. Let C be a partition of V:

$$C = \left\{ c_1, \ldots, c_n \subseteq V \;\middle|\; \begin{array}{l} \forall i, c_i \neq \emptyset, \\ \cup_{1 \leq i \leq n} c_i = V, \\ \forall i, j(\neq i), c_i \cap c_j = \emptyset \end{array} \right\}.$$

We call an element of C a cluster of proteins. Every partition of V is used as a state in PPSampler2.

The complete scoring function

$$f(C) = -(g_1(C) + g_2(C) + g_3(C)),$$

gives the target probability distribution,

$$P(C) \propto \exp\left(-\frac{f(C)}{T}\right).$$

The first function, $g_1(C)$, is an optimization term, and the other two are regularization terms (i.e., they control the relative frequency of the sizes of the predicted clusters and the number of proteins in the predicted clusters, respectively). The $g_2(C)$ function is designed on the basis of the observation that the frequency of sizes of known protein complexes obeys a power-law. It can be observed in CYC2008 (a comprehensive catalog of manually curated 408 heteromeric protein complexes in *S. cerevisiae*) [19] for yeast and in CORUM (a comprehensive database of mammalian protein complexes) [21] for humans.

The $g_1(C)$ function is defined as $g_1(C) = \sum_{c \in C} g_1(c)$ where

$$g_1(c) = \begin{cases} 0 & \text{if} |c| = 1, \\ -\infty & \text{else if} |c| > N \text{or} \\ & \exists u \in c, \forall v(\neq u) \in c, w(\{u, v\}) = 0, \\ \displaystyle\sum_{u,v(\neq u) \in c} \frac{w(u, v)}{\sqrt{|c|}} & \text{otherwise}, \end{cases}$$

and N is the upper bound on the size of a cluster in C.

The $g_2(C)$ function is formulated as

$$g_2(C) = -\sum_{i=2}^{N} \frac{(\psi_C(i) - \psi(i))^2}{2\sigma_{2,i}^2},$$

where $\psi(i)$ is a power-law function,

$$\frac{1}{\sum_{j=2}^{N} j^{-\gamma}} \cdot i^{-\gamma},$$

of power-law parameter γ, and $\psi_C(i)$ is the relative frequency of clusters of size i in C for size $i = (2, 3, \ldots, N)$.

The $g_3(C)$ function is formulated as

$$g_3(C) = -\frac{(s(C) - \lambda)^2}{2\sigma_3^2},$$

where $s(C)$ is the number of proteins in clusters of size two or more in C, i.e.,

$$s(C) = \sum_{c \in C \text{s.t.} |c| \geq 2} |c|,$$

and λ is the target value of $s(C)$.

The proposal distribution used is a rather naive one. It first randomly chooses a protein, p. It then moves p to another cluster with the probability proportional to the sum of the weights between p and the proteins in the cluster.

PPSampler2 returns the partition with the minimum score among the partitions sampled. All clusters of size two or more in the found partition are used as predicted ones.

The performance of PPSampler2 was compared with that of seven common prediction tools (MCL [3], MCODE [2], DPClus [1], CMC [13], COACH [24], RRW [14], NWE [16]) as well as with that of PPSampler [22], the previous version of PPSampler2 [23]. The input PPIs were all PPIs in WI-PHI (a weighted yeast interactome enriched for direct physical interactions) [10]. PPSampler2 is reported to outperform the other tools in terms of the F-measure [23] due to its more sophisticated scoring function. While the solution obtained by PPSampler2 might not be optimal, it is superior to the those of the other tools [23].

References

1. M. Altaf-Ul-Amin, Y. Shinbo, K. Mihara, K. Kurokawa, S. Kanaya, Development and implementation of an algorithm for detection of protein complexes in large interaction networks. BMC Bioinform. **7**, 207 (2006)
2. G.D. Bader, C.W.V. Hogue, An automated method for finding molecular complexes in large protein interaction networks. BMC Bioinform. **4**, 2 (2003)
3. A.J. Enright, S. Van Dongen, C.A. Ouzounis, An efficient algorithm for large-scale detection of protein families. Nucleic Acids Res. **30**, 1575–1584 (2002)
4. S. Geman, D. Geman, Stochastic relaxation, Gibbs distributions, and the Bayesian restoration of images. IEEE Trans. Pattern Anal. Mach. Intell. **6**, 721–741 (1984)
5. C.J. Geyer, Markov chain Monte Carlo maximum likelihood, in *Computing Science and Statistics: Proceedings of the 23rd Symposium on the Interface*, ed. by E.M. Keramides (Fairfax Station, Interface Foundation, 1991), pp. 156–163
6. C.J. Geyer, E.A. Thompson, Annealing Markov chain Monte Carlo with applications to ancestral inference. J. Am. Stat. Ass. **90**, 909–920 (1995)
7. W.K. Hastings, Monte Carlo sampling methods using Markov chains and their applications. Biometrika **57**, 97–109 (1970)
8. K. Hukushima, K. Nemoto, Exchange Monte Carlo method and application to spin glass simulations. J. Phys. Soc. Jpn. **65**, 1604–1608 (1996)
9. Yukito Iba, *Computational Statistics II (in Japanese), Chapter Introduction to Markov Chain Monte Carlo* (Iwanami Shoten, Tokyo, 2005)
10. L. Kiemer, S. Costa, M. Ueffing, G. Cesareni, WI-PHI: a weighted yeast interactome enriched for direct physical interactions. Proteomics **7**, 932–943 (2007)
11. S. Kirkpatrick, C.D. Gelatt, M.P. Vecchi, Optimization by simulated annealing. Science **220**, 671–680 (1983)
12. J.S. Liu, *Monte Carlo Strategies in Scientific Computing* (Springer, New York, 2008)
13. G. Liu, L. Wong, H.N. Chua, Complex discovery from weighted PPI networks. Bioinformatics **25**, 1891–1897 (2009)
14. K. Macropol, T. Can, A.K. Singh, RRW: Repeated random walks on genome-scale protein networks for local cluster discovery. BMC Bioinform. **10**, 283 (2009)
15. E. Marinari, G. Parisi, Simulated tempering: a new monte carlo scheme. Europhys. Lett. **19**, 451–458 (1992)
16. O. Maruyama, A. Chihara, NWE: Node-weighted expansion for protein complex prediction using random walk distances. Proteome Sci. **9**(Suppl 1), S14 (2011)
17. N. Metropolis, A.W. Rosenbluth, M.N. Rosenbluth, A.H. Teller, E. Teller, Equations of state calculations by fast computing machines. J. Chem. Phys. **21**, 1087–1092 (1953)
18. S.P. Meyn, R.L. Tweedie, *Markov Chains and Stochastic Stability* (Springer, Berlin, 1993)
19. S. Pu, J. Wong, B. Turner, E. Cho, S.J. Wodak, Up-to-date catalogues of yeast protein complexes. Nucleic Acids Res. **37**, 825–831 (2009)
20. C. Robert, G. Casella, *Monte Carlo Statistical Methods* (Springer, New York, 2004)
21. A. Ruepp, B. Waegele, M. Lechner, B. Brauner, I. Dunger-Kaltenbach, G. Fobo, G. Frishman, C. Montrone, H.W. Mewes, CORUM: the comprehensive resource of mammalian protein complexes-2009. Nucleic Acids Res. **38**, D497–D501 (2010)
22. D. Tatsuke, O. Maruyama, Sampling strategy for protein complex prediction using cluster size frequency. Gene **518**, 152–158 (2013)
23. C.K. Widita, O. Maruyama. Ppsampler2: Predicting protein complexes more accurately and efficiently by sampling. BMC Syst. Biol. (2013) (To appear)
24. M. Wu, X. Li, C.K. Kwoh, S.K. Ng, A core-attachment based method to detect protein complexes in PPI networks. BMC Bioinform. **10**, 169 (2009)
25. C. Zhang, J. Ma, Comparison of sampling efficiency between simulated tempering and replica exchange. J Chem. Phys. **129**, 134112 (2008)

References

[illegible]

21. A. Ruepp, B. Waegele, M. Lechner, B. Brauner, I. [illegible], G. [illegible], man, C. Montrone, H.W. Mewes, CORUM: the [illegible] complexes—2009. Nucleic Acids Res. 38, D497–[illegible] (2010)
22. [illegible] Sampling strategy for [illegible] Neurosci[illegible] 578, 132–[illegible] (2014)
23. [illegible] Mitsuyama, [illegible] BMC Syst. Biol. (2013) [illegible]
24. M. Wu, X. Li, C.K. Kwoh, S.K. Ng, A core-attachment based method to detect protein complexes in PPI networks. BMC Bioinformatics 10, 169 (2009)
25. C. Zhang, J. Ma, Comparison of sampling efficiency between simulated tempering and replica exchange. J. Chem. Phys. 129, 134112 (2008)

Modeling of Fluid Flows by Nonlinear Schrödinger Equation

Yasuhide Fukumoto

Abstract Fluid flows exhibit diverse ways of their behavior, from ordered to chaotic and turbulent motion. The Navier-Stokes or Euler equations governing such motion are formidable as they are, and even the highest performance computers have difficulty in producing accurate and therefore useful solutions. Effort has constantly been made for mathematically modeling flow phenomena by simplified equations, deriving them from the Navier-Stokes equations and solving them. In this note, we illustrate how we model the nonlinear modulation of a traveling wave, as observed in water waves, by the nonlinear Schrödinger equation. The wave modulation is captured as the instability and bifurcation of a plane-wave solution. Behind this lies the Hamiltonian structure of the Euler equations, and Krein's theory of the Hamiltonian spectra is applicable to it. We build on it a striking aspect of dissipation and diffusion that drives instability for an otherwise stable solution.

Keywords Nonlinear Schrödinger equation · Ginzburg-Landau equation · Stokes wave · Benjamin-Feir instability · Dissipation induced instability

1 Introduction

Fluid mechanics is a subject for understanding flows of liquids and gases and the force exerted on bodies moving in them, and then for utilizing and further controlling the flows. Given the boundary shape, the flow of a fluid through it is ordered when its velocity is slow but it becomes disordered when the velocity is increased. There are cases where disturbances are not amplified when their amplitude is small but are amplified when their amplitude is large. The field of hydrodynamic stability ranges

Y. Fukumoto (✉)
Institute of Mathematics for Industry, Kyushu University, 744 Motooka, Nishi-ku, Fukuoka 819-0395, Japan
e-mail: yasuhide@imi.kyushu-u.ac.jp

R. Nishii et al. (eds.), *A Mathematical Approach to Research Problems of Science and Technology*, Mathematics for Industry 5,
DOI: 10.1007/978-4-431-55060-0_27,

from engineering, planetary-geophysics, physics to mathematics. As its theory is developed, the frontier of hydrodynamic stability overlaps with dynamical system, pattern formation and singularity theory etc.

In this note, we give an exposition of how to make a mathematical modeling of phenomena governed by the hydrodynamic equations, using the nonlinear Schrödinger and the Ginzburg-Landau equations, with illustration of the concrete procedure for stability of water waves. Using the nonlinear Schrödinger equation, we discuss stability and bifurcation of a solution. Behind this lies the Hamiltonian structure of the Euler equations, and Krein's theory of the Hamiltonian spectra is applicable to it [1, 9]. We build on it a striking aspect of the effect of dissipation and diffusion causing instability [7, 8].

2 Sketch of Derivation of the Navier-Stokes Equations

Motion of a simple fluid, like the water and the air, is governed by the Navier-Stokes equations. Their derivation is a long process [2]. In this section, we give only a perfume of it.

Suppose that the fluid is incompressible with its density ρ being uniform. We think of the fluid as a collection of infinitely many fluid particles. A fluid particle is not a molecule constituting the fluid, but means an infinitesimal material parcel which includes a number of molecules. This parcel is regarded as a point in the macroscopic description of the fluid. We introduce the Cartesian coordinates (x, y, z) and write the position of a fluid particle at the time t as $\boldsymbol{x}(t) = (x(t), y(t), z(t)) = (x_1(t), x_2(t), x_3(t))$. The velocity of the fluid at the time t and the position $\boldsymbol{x}(t)$ is defined by

$$\frac{d\boldsymbol{x}}{dt} = \boldsymbol{u}(\boldsymbol{x}, t) = (u_1(\boldsymbol{x}, t), u_2(\boldsymbol{x}, t), u_3(\boldsymbol{x}, t)) . \tag{1}$$

The kth component of acceleration is

$$\frac{d^2 x_k}{dt^2} = \frac{\partial u_k}{\partial t} + \sum_{j=1}^{3} \frac{\partial u_k}{\partial x_j}\frac{dx_j}{dt} = \frac{\partial u_k}{\partial t} + \sum_{j=1}^{3} u_j \frac{\partial u_k}{\partial x_j} = \frac{\partial u_k}{\partial t} + \sum_{j=1}^{3} (\boldsymbol{u} \cdot \nabla) u_k . \tag{2}$$

Inside the fluid, the material is in a state of tension; fluid parcels push and/or pull each other. The pressure and the viscous stress are the force, per unit area, acting on a lump of the fluid through its boundary by the surrounding fluids. The pressure is the normal force, per unit area, acting trough the boundary of adjacent fluid parcels pushing each other. The summation of the x-component of the pressure force experienced by a short cylindrical region, of length δx, with the top and the bottom faces of area S oriented normal to the x-axis is

$$p(x, y, z)S - p(x + \delta x, y, z)S = -\left\{\frac{\partial p}{\partial x}(x, y, z)\delta x + O\big((\delta x)^2\big)\right\} S. \quad (3)$$

If we ignore the forces other than the pressure, Newton's second law for the short cylindrical region under question reads, in the limit of infinitesimal δx,

$$\rho\delta V\frac{d^2x}{dt^2} = -\frac{\partial p}{\partial x}\delta V \quad (\delta V = S\delta x) \iff \rho\frac{d^2x}{dt^2} = -\frac{\partial p}{\partial x}. \quad (4)$$

The viscosity is a manifestation of resistance of a fluid parcel against distortion by adjacent fluid parcels. This may be regarded as an internal friction. The viscous stress acts to diffuse the momentum from a high velocity region to the neighboring smaller velocity region, with the diffusion coefficient measured by the kinematic viscosity ν. Other than these area forces acted on the boundary by adjacent fluid parcels, there are body forces, like the gravity force, directly acting on each infinitesimal material parcel, and we denote the body force per unit mass by $\boldsymbol{b}$. With these forces taken into account, we are led from Newton's law to the following equations governing the velocity field.

$$\rho\left[\frac{\partial \boldsymbol{u}}{\partial t} + (\boldsymbol{u}\cdot\nabla)\boldsymbol{u}\right] = -\nabla p + \rho\nu\nabla^2\boldsymbol{u} + \rho\boldsymbol{b}. \quad (5)$$

These are called the *Navier-Stokes equations*. The Navier-Stokes equations with the viscous term being removed are called the *Euler equations*. In the case of the incompressible fluid, these are supplemented by the condition of constancy in time of the volume of an arbitrary material region V_t

$$\frac{d}{dt}\int_{V_t} dV = \int_{V_t} \nabla\cdot\boldsymbol{u}dV = 0 \iff \nabla\cdot\boldsymbol{u} = 0. \quad (6)$$

3 Hydrodynamic Stability and Ginzburg-Landau Equation

The body force as exemplified by the gravity force is in most cases expressed by use of a potential function $\Phi(\boldsymbol{x})$ as $\boldsymbol{b} = -\nabla\Phi$. For a fluid of constant density ρ, the body force $\rho\boldsymbol{b}$ is absorbed into the pressure term $-\nabla p$ in the Navier-Stokes equations (5). We denote the velocity and pressure fields of the base flow to be $\boldsymbol{U}_0(\boldsymbol{x}, t)$ and $P_0(\boldsymbol{x}, t)$, and explore its stability. These fulfill the incompressibility condition $\nabla\cdot\boldsymbol{U}_0(\boldsymbol{x}, t) = 0$ and the Navier-Stokes equations (5).

$$\rho\left[\frac{\partial \boldsymbol{U}_0}{\partial t} + (\boldsymbol{U}_0\cdot\nabla)\boldsymbol{U}_0\right] = -\nabla P_0 + \rho\nu\nabla^2\boldsymbol{U}_0. \quad (7)$$

As a typical setting, we take a steady base flow: $\partial\boldsymbol{U}_0/\partial t = 0$.

We superpose a disturbance on it and denote the disturbance velocity and pressure fields by $\tilde{\boldsymbol{u}}(\boldsymbol{x},t)$ and $\tilde{p}(\boldsymbol{x},t)$. The total velocity field $\boldsymbol{u}(\boldsymbol{x},t)$ and the pressure field $p(\boldsymbol{x},t)$ are $\boldsymbol{u} = \boldsymbol{U}_0 + \tilde{\boldsymbol{u}}$ and $p = P_0 + \tilde{p}$, and they satisfy he Navier-Stokes equations (5). By subtracting Eq. (7) for the base flow from the equations for the total field, we are left with evolution equations governing the disturbance field as

$$\frac{\partial \tilde{\boldsymbol{u}}}{\partial t} + (\boldsymbol{U}_0 \cdot \nabla)\tilde{\boldsymbol{u}} + (\tilde{\boldsymbol{u}} \cdot \nabla)\boldsymbol{U}_0 + (\tilde{\boldsymbol{u}} \cdot \nabla)\tilde{\boldsymbol{u}} = -\frac{1}{\rho}\nabla \tilde{p} + \nu \nabla^2 \tilde{\boldsymbol{u}}. \tag{8}$$

Crudely viewing, Eq. (8) look alike the time-dependent Ginzburg-Landau equation governing a complex valued function $\psi(x,t)$ ($\in \mathbb{C}$) of the one-dimensional space variable x and the time t [6],

$$\frac{\partial \psi}{\partial t} + U\frac{\partial \psi}{\partial x} = \mu\psi + \alpha\frac{\partial^2 \psi}{\partial x^2} - \beta|\psi|^2\psi, \tag{9}$$

where U, μ, α and β are constants. In case α and β are complex numbers, (9) is called the *complex Ginzburg-Landau equation*. With a simpler form, (9) serves as a one-dimensional toy model for the disturbance Eq. (8). Since the middle of 1960s, an effort has been made for deriving (9) from (8) in a systematic manner, and thereby the hydrodynamic stability theory has been substantially deepened. In the context of fluid mechanics, (9) is called the *Landau-Stuart* or *Stuart-Stewartson equation*.

When $\mu = 0$ and α and β are pure imaginary, (9) is reduced to the nonlinear Schrödinger equation. Relevant to the motion of an inviscid fluid is this case. Subsequently, we shall illustrate that the stability of a water wave is described by the nonlinear Schrödinger equation.

4 Water Wave

Let us consider waves excited on the surface $z = 0$ of a liquid filling the lower half space $z < 0$ [11]. We take the motionless sate $\boldsymbol{U}_0 = \boldsymbol{0}$ as the base flow in the absence of waves and restrict the wave amplitude to be infinitely small. Under this restriction, the nonlinear term is ruled out from (5), and the viscous term may be ignored except for small-scale phenomena. Taking account of gravity $-g\boldsymbol{e}_z$ as an external force per unit mass, the Euler equations (5) are approximated by

$$\frac{\partial \boldsymbol{u}}{\partial t} = -\frac{1}{\rho}\nabla p - g\boldsymbol{e}_z, \tag{10}$$

where g is the gravity acceleration, and $\boldsymbol{e}_z$ is the unit vector along the z-axis, being directed vertically upward. We recall the assumptions that the density ρ is constant and that the solenoidal condition (6) is imposed on the velocity. We readily see that, when $\rho =$ const., an irrotational flow, being characterized by $\nabla \times \boldsymbol{u} = \boldsymbol{0}$, satisfies the curl of (5). Subsequently, we assume that the flow is irrotational. Under this assumption, there a potential function ϕ for the velocity to be expressed by $\boldsymbol{u} = \nabla\phi$. In view of the condition (6), ϕ is ruled by the Laplace equation

$$\nabla \cdot \nabla\phi = \nabla^2\phi = 0. \tag{11}$$

Upon substitution from the relation $\boldsymbol{u} - \nabla\phi$, (10) becomes

$$\nabla\left(\frac{\partial\phi}{\partial t} + \frac{p}{\rho} + gz\right) = 0, \tag{12}$$

which is integrated to yield

$$\rho\frac{\partial\phi}{\partial t} + p + \rho gz = f(t), \tag{13}$$

where $f(t)$ is an arbitrary function of t. Equation (13) is no other than the Bernoulli theorem for a potential flow. The requirement that the pressure is equal to the atmospheric pressure p_{a} at the liquid surface ($z = 0$) selects $f(t) = p_{\mathrm{a}}$.

Supposing that the displacement of the liquid surface does not depends on y, we write the infinitesimal displacement of the liquid surface as $\zeta(x, t)$. Equating the pressure of the liquid to the atmospheric pressure p_{a} at the displaced surface $z = \zeta(x, t)$ yields one of the boundary conditions

$$\frac{\partial\phi}{\partial t} + g\zeta = 0 \quad \text{at } z = \zeta(x, t) \approx 0. \tag{14}$$

Within the linear approximation, we may take the free surface to be $z = 0$. As the other boundary condition, we equate, at the surface ($z = \zeta$), the normal component of the velocity of the liquid surface $z = \zeta(x, t)$ to the normal component of the liquid velocity. From the representation $F(x, z, t) = z - \zeta(x, t) = 0$ of the liquid surface, the normal vector at the surface is found to be $\boldsymbol{n} = \nabla F = (-\partial\zeta/\partial x, 0, 1)$. The velocity of the surface is $u_{\mathrm{s}} = (0, 0, \partial\zeta/\partial t)$ and the velocity of the liquid is $\nabla\phi$. With this form, the second boundary condition is expressed as $(\boldsymbol{u}_{\mathrm{s}} - \nabla\phi) \cdot \boldsymbol{n} = 0$, at $z = \zeta$, which is reduced in the linear approximation to

$$\frac{\partial\zeta}{\partial t} = \frac{\partial\phi}{\partial z} \quad \text{at } z = \zeta(x, t) \approx 0. \tag{15}$$

Elimination the variable ζ, the two boundary conditions (14) and (15) collapse to a single boundary condition on ϕ

$$\frac{\partial^2\phi}{\partial t^2} = -g\frac{\partial\phi}{\partial z} \quad \text{at } z = 0. \tag{16}$$

The motion of a liquid filling the region $z < \zeta(x,t)$ below the surface is found by solving (11). We pose a monochromatic traveling wave of the form $\zeta(x,t) = \text{Re}[A\exp\{i(kx - \omega t)\}]$ for the shape of the liquid surface. The amplitude A is a complex number, and Re[·] designates the symbol taking the real part. The real constants k and ω are referred to as the wavenumber and the angular frequency. They have a link with the wavelength $\lambda = 2\pi/k$ and the frequency $f = \omega/(2\pi)$. This wave travels, without change of form, in the positive x-direction with speed $c_p = \omega/k$. This is called the phase velocity. In keeping with the deformation of the surface, the disturbance of the potential for the liquid velocity takes the form $\phi = \Phi(z)\exp\{i(kx - \omega t)\}$. For the case of infinite depth, the condition of ϕ being bounded at the bottom ($z = -\infty$) gives $\Phi \propto \exp(kz)$, whence the general solution of (11) is obtained, using an arbitrary constant C, as $\phi = Ce^{kz}e^{i(kx-\omega t)}$. Upon substitution from (4), the boundary condition (16) produces, under the condition $C \neq 0$ a relation between ω and k as

$$\omega^2 = gk \quad i.e.\ \omega = \sqrt{gk}. \tag{17}$$

This is referred to as the *dispersion relation.*

It follows from (17) that the phase velocity is $c_p = \sqrt{g/k}$. On the other hand, the derivative $c_g = d\omega/dk = \sqrt{g/k}/2$ of the dispersion relation $\omega = \omega(k)$ with respect to k is referred to as the group velocity. This velocity signifies the propagating velocity of the wave energy. In the next section which exposes modulation of a wave, it turns out that c_g signifies the propagation velocity of a bundle of waves.

5 Wave Modulation and Nonlinear Schrödinger Equation

By an intuitive argument, the modulation of the amplitude and the phase of a wave is shown to be described by the nonlinear Schrödinger equation. We denote the complex function representing a wave by $\psi = \psi(x,t)$. It would be reasonable to pose the following nonlinear dispersion relation modified by the wave amplitude

$$\omega - \omega_0(k) + \gamma|\psi|^2 = 0, \tag{18}$$

where γ is a constant [5]. Regarding the linear part, we can take $\omega_0(k) = \sqrt{gk}$ for the case of infinite depth.

In the preceding section, we took the wave amplitude A = const. Here instead, we suppose that the amplitude A varies (= modulates) slowly in space and in time around $k = k_0$ and $\omega = \omega_0(k_0)$, and set

$$\psi = A(x,t)e^{i(k_0x - \omega_0 t)}; \quad \omega_0 := \omega_0(k_0), \tag{19}$$

where the amplitude $A(x,t)$ is a complex-valued function slowly varying in x and t. The nonlinear dispersion relation (18) becomes, when expanded in k about $k = k_0$,

$$\omega-\omega_0(k_0)-\omega_0'(k_0)(k-k_0)-\frac{1}{2}\omega_0''(k_0)(k-k_0)^2+\gamma|\psi|^2+O\left((k-k_0)^3\right)=0. \quad (20)$$

Here the superscript $(\,)'$ stands for the differentiation with respect to the argument, and the remaining terms of the Taylor expansion are ignored. As a rule, the frequency ω and the wavenumber k are replaced by derivative operators $\omega = i\partial/\partial t, \quad k = -i\partial/\partial x$. Rewriting (20) with this displacement and substituting (19) into the resulting equation, we arrive at the evolution equation for the amplitude A

$$i\frac{\partial A}{\partial t} + i\omega_0'\frac{\partial A}{\partial x} + \frac{1}{2}\omega_0''\frac{\partial^2 A}{\partial x^2} + \gamma|A|^2 A = 0. \quad (21)$$

where we have employed short-hand notation $\omega_0' = \omega_0'(k_0)$ and $\omega_0'' = \omega_0''(k_0)$. This is the *nonlinear Schrödinger equation*, and is now know to describe various phenomena of nonlinear waves. The foregoing description implies that the wave modulation is appropriately described by the nonlinear dispersion relation (18).

Compare (21) with the Ginzburg-Landau equation (9). They take the same form except for the imaginary unit i. The second term of the nonlinear Schrödinger equation indicates that a part of amplitude variation propagates with the group velocity $\omega_0'(k_0)$. In other words, the group velocity is literally the propagation velocity of a group of waves. The second term is eliminated in the coordinate frame moving with the group velocity $\omega_0'(k_0)$ in the x-direction.

The nonlinear Schrödinger equation (21) has a distinguishing feature of being a completely integrable evolution equation. It has soliton solutions. In the case of $\gamma < 0$, the localized solution of permanent form is called the dark soliton. It is only recent that the dark soliton of a water wave is observed in a laboratory experiment [4].

6 Benjamin-Feir Instability

We consider the stability of a traveling wave governed by the nonlinear Schrödinger equation of a general form

$$iA_t + \alpha A_{xx} + \gamma|A|^2 A = 0, \quad (22)$$

where $A = A(x,t)$ is a complex valued function of x and t, and the subscript stands for the derivative with respect to the indicated variable. We take α and γ as real constants. Returning to the definition (19), A is the amplitude function. Accordingly, we inquire into the amplitude modulation [3] (See also [5]).

6.1 Stokes Traveling Wave

At the outset, we seek a traveling-wave solution of (22). Inserting the solution form $A = A_0 \exp\{i(\kappa x - \varpi t + \theta_0)\}$, with $A_0(\neq 0)$ and θ_0 being constants, into (22), we have

$$\varpi = \alpha\kappa^2 - \gamma|A_0|^2. \tag{23}$$

The wave (19) with the frequency provided by (23) is called the *Stokes traveling wave* or simply the *Stokes wave*.

When rewritten for the original variable ψ, (19) becomes

$$\psi = A_0 e^{i[(k_0+\kappa)x-(\omega_0+\varpi)t+\theta_0]}, \tag{24}$$

and the dispersion relation for the Stokes wave is therefore

$$\omega_0 + \varpi = \omega_0(k_0) + \alpha\kappa^2 - \gamma|A_0|^2. \tag{25}$$

In case the wavenumber increment $\kappa = 0$, (25) is reduced the nonlinear dispersion relation (18) as expected. Hereinafter, we generalize to the cases $\kappa \neq 0$.

Though the calculation procedure may be elegant if we treat the complex function A as it is, the treatment of A as a pair of real functions has an advantage of exploiting the Hamiltonian structure [3]. We express the amplitude function, in terms of two real-valued functions $u_1(x,t)$ and $u_2(x,t)$ of the independent variables x and t, as $A = u_1 + iu_2$. With this change of variables, the nonlinear Schrödinger equation (22) is converted into a vector equation for the vector-valued function $\boldsymbol{u} = {}^t(u_1, u_2)$

$$\mathsf{J}\boldsymbol{u}_t + \alpha\boldsymbol{u}_{xx} + \gamma||\boldsymbol{u}||^2\boldsymbol{u} = \boldsymbol{0}; \quad \mathsf{J} := \begin{pmatrix} 0 & -1 \\ 1 & 0 \end{pmatrix}. \tag{26}$$

Here superscript t designates the transpose of the following vector or matrix, therefore $\boldsymbol{u}$ is a column vector, and $||\boldsymbol{u}|| = (u_1^2 + u_2^2)^{1/2}$ is an ordinary norm of the vector $\boldsymbol{u}$. The Stokes traveling wave (24) is written, in terms of the amplitude $A_0 = u_{01} + iu_{02}$ and the phase $\theta = (k_0 + \kappa)x - (\omega_0 + \varpi)t + \theta_0$, as

$$\begin{pmatrix} u_1 \\ u_2 \end{pmatrix} = \mathsf{R}_\theta \begin{pmatrix} u_{01} \\ u_{02} \end{pmatrix}; \quad \mathsf{R}_\theta := \begin{pmatrix} \cos\theta & -\sin\theta \\ \sin\theta & \cos\theta \end{pmatrix}. \tag{27}$$

6.2 Linear Stability of Stokes Wave

We look into time evolution of a disturbance $\mathsf{R}_\theta\boldsymbol{v}(x,t)$; $\boldsymbol{v} = {}^t(v_1, v_2)$, of infinitesimal amplitude, superimposed on the Stokes wave. The amplitude vector augmented with a disturbance

$$u(x,t) = \mathsf{R}_\theta\,[u_0 + v(x,t)] \tag{28}$$

is substituted into (26). Ignoring the quadratic term in disturbance amplitude and simplifying, with the aid of $\mathsf{J}\mathsf{R}_{\theta t} + \alpha \mathsf{R}_{\theta xx} + \gamma ||u_0|| \mathsf{R}_\theta = 0$, and $\mathsf{R}_{\theta x} = \kappa \mathsf{J}\mathsf{R}_\theta$, which are derived from the dispersion relation (23), we deduce, after multiplication from the left by ${}^t\mathsf{R}_\theta$, the linearized equation governing the disturbance as

$$\mathsf{J}v_t + 2\alpha\kappa\mathsf{J}v_x + \alpha v_{xx} + 2\gamma(u_0 \cdot v)u_0 = \mathbf{0}. \tag{29}$$

Here we have used the relations ${}^t\mathsf{R}_\theta \mathsf{R}_\theta = \mathsf{I}$ (I: 2×2 unit matrix) and ${}^t\mathsf{R}_\theta \mathsf{J} \mathsf{R}_\theta = \mathsf{J}$.

We substitute

$$v(x,t) = v(t)\cos\sigma x + w(t)\sin\sigma x \tag{30}$$

into (29). Abuse of the same notation v in (30) is to be kept in view. The spatial dependence lies only in the trigonometric functions $\cos\sigma x$ and $\sin\sigma x$. Collecting terms with $\cos\sigma x$, we obtain

$$\mathsf{J}\dot{v} + 2\alpha\kappa\sigma\mathsf{J}w - \alpha\sigma^2 v + 2\gamma(u_0 \cdot v)u_0 = \mathbf{0}, \tag{31}$$

where a dots stands for the derivative in the time t. Likewise, collecting the terms with $\sin\sigma x$, we obtain

$$\mathsf{J}\dot{w} - 2\alpha\kappa\sigma\mathsf{J}v - \alpha\sigma^2 w + 2\gamma(u_0 \cdot w)u_0 = \mathbf{0}. \tag{32}$$

The solution takes the form $v = {}^t(q_1, q_2)e^{\lambda t}$, $w = {}^t(p_1, p_2)e^{\lambda t}$, and, with this form, (31) and (32) are reduced to a matrix equation

$$\mathsf{A}\begin{pmatrix} q_1 \\ q_2 \\ p_1 \\ p_2 \end{pmatrix} = \mathbf{0};$$

$$\mathsf{A} = \begin{pmatrix} -\alpha\sigma^2 + 2\gamma u_{01}^2 & -\lambda + 2\gamma u_{01}u_{02} & 0 & -2\alpha\kappa\sigma \\ \lambda + 2\gamma u_{01}u_{02} & -\alpha\sigma^2 + 2\gamma u_{02}^2 & 2\alpha\kappa\sigma & 0 \\ 0 & 2\alpha\kappa\sigma & -\alpha\sigma^2 + 2\gamma u_{01}^2 & -\lambda + 2\gamma u_{01}u_{02} \\ -2\alpha\kappa\sigma & 0 & \lambda + 2\gamma u_{01}u_{02} & -\alpha\sigma^2 + 2\gamma u_{02}^2 \end{pmatrix}. \tag{33}$$

The necessary and sufficient condition for (33) to have a nontrivial solution ${}^t(q_1, q_2, p_1, p_2) \neq \mathbf{0}$ furnishes an algebraic equation for the spectral parameter λ as

$$\det A = \lambda^4 + 2(p^2 + 4\kappa^2\alpha^2\sigma^2)\lambda^2 + (p^2 - 4\kappa^2\alpha^2\sigma^2)^2 = 0; \tag{34}$$
$$p^2 := \alpha^2\sigma^4 - 2\gamma\alpha||u_0||\sigma^2.$$

The solution is readily found to be

$$\lambda^2 = -(p \pm 2\kappa\alpha\sigma)^2, \quad \text{or} \quad \lambda = \pm i(p \pm 2\kappa\alpha\sigma). \tag{35}$$

It is a characteristic feature of a Hamiltonian system that the spectra come as a quartet [1, 7, 9]. If λ is a spectrum, not only its complex conjugate $\bar{\lambda}$ but also $-\lambda$ belong to the spectra. The latter spectra reflects the time-reversal symmetry of the Hamiltonian system.

If at least one of the roots of (34) has positive real part, a disturbance growing in t can be built, that is, the Stokes wave is spectrally unstable. When the parameter p is real, all the four roots of (35) are pure imaginary, indicating that the Stokes wave is spectrally stable. In contrast, when p becomes a purely imaginary number, two among (35) have positive real part, with which the disturbance is amplified exponentially in time. As far as the amplitude of the Stokes wave is small enough ($||\boldsymbol{u}_0|| < \sqrt{\alpha\sigma^2/2\gamma}$), p is a real number, but as the amplitude is increased and exceeds the critical value $||\boldsymbol{u}_0|| = \sqrt{\alpha\sigma^2/2\gamma}$, the spectral parameter acquires positive real part, indicating the exponential growth of the disturbance. This instability is called the *Benjamin-Feir instability*. Until the middle of the twentieth century, there were trials to embody the Stokes wave in a water tank, but the excited traveling waves were necessarily disrupted. This failure could be attributable to the Benjamin-Feir instability. This pertains to the *side-band instability*. For the Stokes wave (24) with wavenumber $k = k_0 + \kappa$, the disturbances whose wavenumber is located in a narrow band, around k, with width 2σ ($\sigma^2 < 2\gamma||\boldsymbol{u}_0||^2/\alpha$) is amplified. Recalling the linear stability analysis in this subsection, a disturbance is given to the amplitude A_0 of the Stokes wave (24) with wavenumber $k = k_0 + \kappa$ as

$$\psi = [A_0 + B(x,t)]\, e^{i[(k_0+\kappa)x-(\omega_0+\varpi)t+\theta_0]}; \quad B = B_0 e^{i\sigma x+\lambda t}. \tag{36}$$

The wavenumber $k = k_0 + \kappa$ of the Stokes wave is altered, by experiencing the disturbance, to $k_\pm = k_0 + \kappa \pm \sigma$. The linearized term originating from the nonlinear term $\gamma|A|^2A$ in the nonlinear Schrödinger equation (22) couples a wave with the fundamental wave number k and that with k_+ to create a wave with $2k - k+ = k_-$ and couples also the waves with k and k_- to create a wave with $2k - k_- = k_+$. As a consequence, the sideband waves with k_+ and k_- go through resonant amplification via the base field (24).

It is illuminating to view this instability from the standpoint of the Hamiltonian spectra. When the amplitude $||\boldsymbol{u}_0||$ of the Stokes wave is small, the four spectra all confined on the pure imaginary axis in the complex λ plane. As the amplitude is raised, the two spectra on the positive imaginary axis approach each other and the same is true for the two spectra, being the mirror images, on the negative imaginary axis. At the critical amplitude $||\boldsymbol{u}_0|| = \sqrt{\alpha\sigma^2/2\gamma}$, the spectrum parameters execute pairwise collisions at $\lambda = 2i\kappa\sigma$ and $\lambda = -2i\kappa\sigma$, and a further increase of amplitude gives rise to real parts in the spectra. This is the *Hamiltonian-Hopf bifurcation*. Krein's theory dictates that a necessary condition for the spectra to escape from the imaginary axis is that the signs of the energy of the corresponding eigenmodes are opposite from each other [1, 9]. It is shown that, before the collision of spectra ($p > 0$), the eigenmode associated with $\lambda = \pm i(2\kappa\alpha\sigma + p)$, being located on the far side from $\lambda = 0$, has positive energy, but that with $\lambda = \pm i(2\kappa\alpha\sigma - p)$, being located closer to $\lambda = 0$, has negative energy, being consistent with the scenario of Krein's theory [3].

7 Dissipation Induced Instability

Dissipation is an agent to convert the kinetic energy into the thermal energy (= heat), with a tendency of subsiding the moving state to a steady state. Seemingly contrary to an intuition, there are cases in which the dissipation turns an ordered motion to unstable one [7, 8]. As argued at the end of the last section, instability of a Hamiltonian system is typically driven by a collision of two spectra with positive and negative energy. A positive-energy mode subsides down when its energy is lost by the dissipation, while on the contrary a negative-energy mode is amplified by losing its energy. Instability stemming from dissipative forces is not uncommon. In nature, there are an abundance of negative-energy modes. This section gives an account for the instability of weak diffusion origin that the Stokes wave undergoes, which occurs in the parameter region of absence of the Benjamin-Feir instability [3].

7.1 Nonlinear Schrodinger Equation with Diffusion Effect

We augment the nonlinear Schrödinger equation (22) with weak dissipation and diffusion as

$$iA_t + (\alpha - ia)A_{xx} + ibA + (\gamma + ic)|A|^2 A = 0, \tag{37}$$

where a, b and c are positive constants. If we retain only these new terms in (37) by taking the coefficient in the original form as $\alpha = \gamma = 0$, we are left with

$$A_t = aA_{xx} - (b + c|A|^2)A. \tag{38}$$

The first term on the right-hand side signifies the diffusion term with the diffusion coefficient $a (> 0)$. The rests are linear and nonlinear dissipation terms.

A comprehensive analysis is referred to [3]. Here, for the sake of simplicity, we focus only on the diffusion effect by putting $b = c = 0$. The dispersion relation (37) of the Stokes wave $A = A_0 \exp\{i(\kappa x - \varpi t + \theta_0)\}$ is modified to

$$\varpi = \alpha\kappa^2 - \gamma|A_0|^2 - ia\kappa^2. \tag{39}$$

The modification of (23) comes as the presence of the last term with the effect of damping the Stokes wave.

7.2 Diffusion Effect on Linear Stability of Stokes Wave

With an introduction of a pair of real functions for the complex amplitude function $A = u_1 + iu_2$, the generalized nonlinear Schrödinger equation (37), with restriction $b = c = 0$, admits the following real representation for the vector (u_1, u_2)

$$\mathsf{J}\boldsymbol{u}_t + \alpha \boldsymbol{u}_{xx} - a\mathsf{J}\boldsymbol{u}_{xx} + \gamma ||\boldsymbol{u}||\boldsymbol{u} = \boldsymbol{0}. \tag{40}$$

This is substituted from superposition of the Stokes wave and a disturbance $\boldsymbol{v}$, like (28), resulting in, after linearization in disturbance amplitude $||\boldsymbol{v}||$,

$$\mathsf{J}\boldsymbol{v}_t + 2\left(\alpha\kappa\mathsf{J} + a\kappa\right)\boldsymbol{v}_x + (\alpha - a\mathsf{J})\,\boldsymbol{v}_{xx} + 2\gamma(\boldsymbol{u}_0 \cdot \boldsymbol{v})\boldsymbol{u}_0 = \boldsymbol{0}. \tag{41}$$

Substitution from the disturbance (30) with wavenumber σ into (41) yields for the $\cos\sigma x$ and the $\sin\sigma x$ terms, respectively

$$\mathsf{J}\dot{\boldsymbol{v}} + 2\kappa\sigma\,(\alpha\mathsf{J} + a)\,\boldsymbol{w} - \sigma^2\,(\alpha - a\mathsf{J})\,\boldsymbol{v} + 2\gamma(\boldsymbol{u}_0 \cdot \boldsymbol{v})\boldsymbol{u}_0 = \boldsymbol{0}, \tag{42}$$
$$\mathsf{J}\dot{\boldsymbol{w}} - 2\kappa\sigma\,(\alpha\mathsf{J} + a)\,\boldsymbol{v} - \sigma^2\,(\alpha - a\mathsf{J})\,\boldsymbol{w} + 2\gamma(\boldsymbol{u}_0 \cdot \boldsymbol{w})\boldsymbol{u}_0 = \boldsymbol{0}. \tag{43}$$

The solution takes the form $\boldsymbol{v} = {}^t(q_1, q_2)e^{\lambda t}$, $\boldsymbol{w} = {}^t(p_1, p_2)e^{\lambda t}$, and (42) and (43) gives the algebraic equation (33) with the matrix A provided by

$$\mathsf{A} = \begin{pmatrix} -\alpha\sigma^2 + 2\gamma u_{01}^2 & -\lambda - a\sigma^2 + 2\gamma u_{01}u_{02} & 2a\kappa\sigma & -2\alpha\kappa\sigma \\ \lambda + a\sigma^2 + 2\gamma u_{01}u_{02} & -\alpha\sigma^2 + 2\gamma u_{02}^2 & 2\alpha\kappa\sigma & 2a\kappa\sigma \\ -2a\kappa\sigma & 2\alpha\kappa\sigma & -\alpha\sigma^2 + 2\gamma u_{01}^2 & -\lambda - a\sigma^2 + 2\gamma u_{01}u_{02} \\ -2\alpha\kappa\sigma & -2a\kappa\sigma & \lambda + a\sigma^2 + 2\gamma u_{01}u_{02} & -\alpha\sigma^2 + 2\gamma u_{02}^2 \end{pmatrix}. \tag{44}$$

The procedure of gaining the spectra λ from the necessary and sufficient condition $\det A = 0$ for (33) to have a nontrivial solution ${}^t(q_1, q_2, p_1, p_2) \neq \boldsymbol{0}$ is the same as before.

In the sequel, we regard the diffusion term as a perturbation, and retain its effect only to first order in a small parameter a/α, resulting in

$$\begin{aligned}\det A = \hat{\lambda}^4 + 2(p^2 + 4\kappa^2\alpha^2\sigma^2)\hat{\lambda}^2 - 32a\alpha\kappa^2\sigma^2(\alpha^2\sigma^2 - \gamma||\boldsymbol{u}_0||^2)\hat{\lambda} \\ +(p^2 - 4\kappa^2\alpha^2\sigma^2)^2 + O\left((a/\alpha)^2\right),\end{aligned} \tag{45}$$

where $\hat{\lambda} = \lambda + a\sigma^2$ and p is defined by (34). Correspondingly to this treatment, we construct the root of $\det A = 0$ in a power series in a/α. For $p > 0$, the root with positive imaginary part is calculated to be

$$\lambda = i(2\kappa\alpha\sigma \pm p) - a\sigma^2 \mp \frac{2a\kappa\sigma}{p}(\alpha^2\sigma^2 - \gamma||\boldsymbol{u}_0||^2) + O\left((a/\alpha)^2\right). \tag{46}$$

The roots with negative imaginary part are placed symmetrically with respect to the real axis. For the parameter region with $p > 0$, being free from the Benjamin-Feir instability, the diffusion effect pushes the roots, closer to the real axis, to the side of positive real part, being indicative of the exponential amplification. The corresponding eigenmodes have negative energy. The eigenmodes associated with their spectra located farther from the real axis which have positive energy are pushed to the side of negative real part, indicating decay in time by the diffusion effect.

The influence of dissipation effect ($b > 0, c > 0$) is similar to the above [3]. Krein's theory not only describes the bifurcation of the solution of a Hamiltonian system, but also has a rich implication on the effect of dissipation and diffusion. As a final remark, we emphases that Krein's theory is concerned with the system of a finite number of freedom. The topic under investigation is a system of an infinite number of freedom governed by partial differential equations.

8 Closing Remarks

We can list numerous examples of instabilities induced by dissipation or friction effects [7, 8]. A familiar example is the sleeping top, the vertically upright configuration of the axis of a heavy symmetrical top or Lagrange's top. In the absence of the friction at the end of axis in contact with a supporting plane, the upright configuration lasts permanently. When the friction comes into play, the rotation slows down, resulting in the tilting down of the axis. This is likened to the wake up. We reason that this is a manifestation of amplification of a negative energy mode. Recently we have found that unexpectedly the sleeping top does not necessarily follow this standard scenario [10]. In general, a disturbance superposed on the stable motionless state has only positive energy. Negative energy modes reside on nontrivial steady states, phenomena filling the nature surrounding us.

References

1. V.I. Arnol'd, *Mathematical Methods of Classical Mechanics*, 2nd ed. (Springer, New York, 1989)
2. G.K. Batchelor, *An Introduction to Fluid Dynamics* (Cambridge University Press, Cambridge, 1967)
3. T.J. Bridges, F. Dias, Enhancement of the Benjamin-Feir instability with dissipation. Phys. Fluids **19**, 104104 (2007)
4. A. Chabchoub, O. Kimmoun, H. Branger, N. Hoffmann, D. Proment, M. Onorato, N. Akhmediev, Experimental observation of dark solitons on the surface of water. Phys. Rev. Lett. **110**, 124101 (2013)
5. H. Hasimoto, H. Ono, Nonlinear modulation of gravity waves. J. Phys. Soc. Japan **33**, 805–811 (1972)

6. P. Huerre,M. Rossi, in *Hydrodynamic Instabilities in Open Flows*, ed. by C. Godrèche, P. Manneville, Hydrodynamics and Nonlinear Instabilities (Cambridge University Press, Cambridge, 1998)
7. O.N. Kirillov, *Nonconservative Stability Problems of Modern Physics*, (Walter de Gruyter, Berlin, 2013)
8. R. Krechetnikov, J.E. Marsden, Dissipation-induced instabilities in finite dimensions. Rev. Mod. Phys. **79**, 519–553 (2007)
9. R.S. MacKay, in *Stability of Equilibria of Hamiltonian Systems*, ed. by S. Sarkar. Nonlinear Phenomena and Chaos (Adam Hilger, Bristol, 1986) ,pp. 254–270.
10. A. Paerhati, Y. Fukumoto, An example exempted from Thomson-Tait-Chetayev's theorem. J. Phys. Soc. Japan **82**, 043002 (2013)
11. G.B. Whitham, *Linear and Non-Linear Waves* (Wiley, New York, 1974)

Financial Applications of Quasi-Monte Carlo Methods

Shu Tezuka

Abstract This article overviews major developments in the last two decades on the applications of quasi-Monte Carlo methods to financial computations.

Keywords Finance · Low-discrepancy sequences · Quasi-Monte Carlo methods

1 Introduction

In 1977, Boyle [1] applied Monte Carlo methods to finance problems, where he tried not only simple Monte Carlo but also variance reduction techniques, such as control variates and antithetic variates, for option pricing. For such applications, however, Monte Carlo approaches have a serious drawback of their notoriously slow convergence rate. According to the central limit theorem, the convergence rate is $O(N^{-1/2})$, where N is the number of sample points, even if we use variance reduction methods, by which the constant term included in the O notation can be (sometimes considerably) reduced. The more complex the stochastic models become, the more computing time is needed. Similarly, the more accurate the solution is required to be, the more computing time is needed. Speeding up the computation time is indispensable for these finance applications. Around 1990, even with the fastest parallel supercomputers, it was not possible to, for example, distinguish within a short time a highly profitable mortgage-backed security (MBS) from pools of collateralized mortgage obligations (CMOs) with poorer returns.

In the 1990s, we had witnessed a dramatic increase in the efficiency of Monte Carlo simulations for finance applications, in terms of both accuracy and speed [2, 5, 9, 11, 12, 23]. It was reported [9, 11] that speed-ups of as much as thousands

S. Tezuka (✉)
Institute of Mathematics for Industry, Kyushu University, 744, Motooka, Nishiku, Fukuoka 819-0395, Japan
e-mail: tezuka@imi.kyushu-u.ac.jp

R. Nishii et al. (eds.), *A Mathematical Approach to Research Problems of Science and Technology*, Mathematics for Industry 5,
DOI: 10.1007/978-4-431-55060-0_28,

of times relative to simple Monte Carlo simulations could be obtained for complicated fixed-income derivatives including CMOs. This made a tremendous impact not only on the finance industry but also on the computer software business [25, see articles such as in The New York Times]. The technology which this innovation was based on is called as quasi-Monte Carlo methods, which had been investigated by number theoreticians since Monte Carlo methods were devised by von Neumann and his colleagues in the 1940s. In fact, this technology has made marked progress and become a mainstay in high-dimensional numerical integrations, particularly over the last two decades [4, 6, 8, 15–18, 27].

The organization of this paper is as follows: In Sect. 2, we first describe what are quasi-Monte Carlo methods and also give some historical remarks. Next, we give the definitions of mathematical terms such as discrepancy and low-discrepancy sequences. Then, we introduce a general method of constructing low-discrepancy sequences, which we call $(t, \mathbf{e}, k)$-sequences, followed by a special realization of them called generalized Faure sequences. Lastly, we briefly summarize recent progress of theoretical research on explaining the success of quasi-Monte Carlo methods in finance problems. In Sect. 3, we give results from numerical experiments with financial problems related to computing the present values of MBS, along with some discussion of these results.

2 Speeding-Up by Quasi-Monte Carlo Methods

2.1 *What Are Quasi-Monte Carlo Methods?*

As was already mentioned earlier, the main drawback of Monte Carlo methods is their slow convergence rate $O(N^{-1/2})$ for N sample paths. Although we can improve the convergence speed by applying the variance reduction techniques, the improvement only reduces the constant factor included in the O term, still leaving the same rate in terms of N. Finance-related Monte Carlo problems, particularly those related to derivative pricing, can be formulated as problems of computing high-dimensional integration. The dimension of the integration is naturally equal to the number of time steps in the time period considered. Once a problem can be formulated as one of high-dimensional numerical integration, we have several "deterministic" approaches for computing it. However, the direct extensions of one-dimensional approaches, such as trapezoidal rules, to higher dimensional ones do not work. For example, if we apply the product rule to the trapezoidal rules for high-dimensional numerical integration, the error bound for the k-dimensional integration is known to be $O(N^{-2/k})$, which means that for a fixed error tolerance the computation time grows exponentially in dimension. Much more generally, we have a theoretical lower bound of computational complexity for high-dimensional numerical integration, which increases at the exponential rate in dimension. This is a well-known result in the field of Information-Based Complexity (see, e.g. [10, 23]), and today it is called *the curse of dimensionality* in numerical integration.

If we assume a certain smoothness for integrands, there is a much better approach, called quasi-Monte Carlo methods, which give the error bound $O((\log N)^k/N)$ for the k-dimensional integration problem. The idea is that by using deterministic sequences instead of *random numbers*, we can derive more precise (deterministic) error bounds for high-dimensional numerical integrations, and consequently achieve a significant improvement in the convergence rate. The extent to which the points are uniform has been mathematically defined as their *discrepancy*. The more uniformly distributed the points are, the lower the discrepancy. The so-called *low-discrepancy sequences* are to quasi-Monte Carlo methods as random numbers are to Monte Carlo methods. Here, we should remark that *pseudorandom numbers*, which we actually use in practice for Monte Carlo simulations, are also deterministic sequences, but are required to mimic truly random numbers. On the other hand, low-discrepancy sequences have no relevance with randomness at all. The practical advantage of low-discrepancy sequences is that for every $N > 1$, the first N points of a low-discrepancy sequence are uniformly distributed; in other words, we can add one point after another so that the entire set of points at any time remains very uniform throughout the domain. Therefore, we can use them sequentially until some stopping rule for the computation is met.

The application of low-discrepancy sequences to finance problems was triggered by the seminal paper of Woźniakowski [26] published in 1991. Paskov and Traub [12, 24] used Halton sequences and Sobol' sequences for pricing a ten-tranch CMO (Collateralized Mortgage Obligation), which they obtained from Goldman-Sachs, and reported that Sobol' sequences performed very well relative to simple Monte Carlo methods, as well as to antithetic Monte Carlo methods. Joy et al. [5] applied Faure sequences to several equity derivatives to obtain good performances. Notice that at that time Faure sequences were theoretically the best of all known low-discrepancy sequences, whereby "theoretically the best" we mean that the constant in front of the leading term $(\log N)^k/N$ in the discrepancy upper bound is the smallest asymptotically in dimension. Then, Ninomiya and Tezuka [9] reported that generalized Niederreiter sequences could provide further speed-up over Halton, Sobol', and Faure sequences, and that they had observed a speed-up of about 1,000 times over simple Monte Carlo for three pricing problems: a discount bond, an interest-rate lookback option, and MBS. Papageorgiou and Traub [11] reported that generalized Faure sequences, a special subset of generalized Niederreiter sequences, perform consistently better than their improved Sobol' sequences for pricing their CMO, and may allow speed-ups of about 1,000 times relative to simple Monte Carlo simulations in cases where high accuracy is desired. Thus, in what follows, we describe generalized Faure sequences in detail.

2.2 Generalized Faure Sequences

First, we recall the definition of discrepancy. For N points $X_0, X_1, \dots, X_{N-1}$ in $[0, 1]^k$, and a subinterval $J = \prod_{i=1}^{k}[0, u_i)$, where $0 < u_i \le 1$ for $1 \le i \le k$, we define the (star) discrepancy as

$$D_N^{(k)} = \sup_J \left| \frac{A(J;N)}{N} - \mathrm{Vol}(J) \right|,$$

where $A(J;N)$ is the number of n, $0 \le n < N$, with $X_n \in J$, and where $\mathrm{Vol}(J)$ is the volume of J, with the supremum extended over all subintervals J. If we employ the L_2-norm instead of the L_∞-norm in the above, we can define L_2-discrepancy, $T_N^{(k)}$, as

$$\left(T_N^{(k)}\right)^2 = \int\limits_{[0,1]^k} \left(\frac{A(J;N)}{N} - \mathrm{Vol}(J) \right)^2 \mathrm{d}u_1 \cdots \mathrm{d}u_k.$$

It is easy to see that $0 \le T_N^{(k)} \le D_N^{(k)} \le 1$.

The Koksma-Hlawka theorem describes the relation between discrepancy and numerical integration [3, 7]:

Theorem 1 *If the integrand f is of bounded variation $V(f)$ on the k-dimensional unit hypercube $[0,1]^k$ in the sense of Hardy and Krause, then for any $X_0, X_1, \ldots, X_{N-1} \in [0,1)^k$ we have*

$$\left| \frac{1}{N} \sum_{n=0}^{N-1} f(X_n) - \int\limits_{[0,1]^k} f(u_1, \ldots, u_k) \mathrm{d}u_1 \cdots \mathrm{d}u_k \right| \le V(f) D_N^{(k)}.$$

We also have another important result, the Woźniakowski theorem [26]:

Theorem 2 *Let $\mathcal{C}_k$ be the class of real continuous functions defined on $[0,1]^k$ equipped with the classical Wiener sheet measure w (that is, Gaussian with mean zero and covariance kernel*

$$R(\mathbf{s}, \mathbf{t}) \stackrel{\mathrm{def}}{=} \int\limits_{\mathcal{C}_k} f(\mathbf{s}) f(\mathbf{t})\, w(\mathrm{d}f) = \min(\mathbf{s}, \mathbf{t}) \stackrel{\mathrm{def}}{=} \prod_{i=1}^{k} \min(s_i, t_i)$$

for any vectors $\mathbf{s} = (s_1, \ldots, s_k)$ and $\mathbf{t} = (t_1, \ldots, t_k)$ in $[0,1]^k$). Then, for a given set of points $X_n = (x_{n1}, \ldots, x_{nk})$, $n = 0, 1, \ldots, N-1$, in $[0,1]^k$, we have

$$\int\limits_{\mathcal{C}_k} \left(\frac{1}{N} \sum_{n=0}^{N-1} f(X_n) - \int\limits_{[0,1]^k} f(u_1, \ldots, u_k) \mathrm{d}u_1 \cdots \mathrm{d}u_k \right)^2 w(\mathrm{d}f) = (\bar{T}_N^{(k)})^2,$$

where $\bar{T}_N^{(k)}$ is the L_2-discrepancy of the point set $\bar{X}_n = (1 - x_{n1}, \ldots, 1 - x_{nk})$, $n = 0, 1, \ldots, N-1$.

The theorem means that on average the integration error is dependent only on the discrepancy, not on the integrand f, and that low-discrepancy sequences are useful in integration. Both the above theorems tell us that the lower the discrepancy is, the smaller the integration error will be. We now introduce the definition of low-discrepancy sequences:

Definition 1 If a sequence $X_0, X_1, \ldots$ in $[0,1]^k$ satisfies the condition that for all $N > 1$, the discrepancy of the first N points is

$$D_N^{(k)} \leq C_k \frac{(\log N)^k}{N},$$

where C_k is a constant depending only on the dimension k, then we call it a low-discrepancy sequence.

Notice that the order of magnitude in terms of N on the right-hand side is believed to be the optimal upper bound.

In this article, we concentrate on the construction of low-discrepancy sequences based on $(t, \mathbf{e}, k)$-sequences. First, we need the following notions: Let $b \geq 2$ be an integer. An elementary interval in base b, which is a key concept of the net theory, is an interval of the form

$$E(j_1, \ldots, j_k) = \prod_{i=1}^{k} \left[\frac{a_i}{b^{j_i}}, \frac{a_i + 1}{b^{j_i}} \right), \quad (0 \leq a_i < b^{j_i},\ j_i \geq 0),$$

where a_i and j_i are integers for $i = 1, \ldots, k$.

Definition 2 Let $\mathbf{e} = (e_1, \ldots, e_k)$ and $\mathbf{j} = (j_1, \ldots, j_k)$ be integer vectors with $e_i \geq 1$ and $j_i \geq 0$ for $i = 1, \ldots, k$. Let t and m be integers with $0 \leq t \leq m$ such that $m - t \in M(\mathbf{e}) := \{(\mathbf{e}, \mathbf{j}) \mid \text{all } \ \mathbf{j}\ \}$, where $(\mathbf{e}, \mathbf{j}) := e_1 j_1 + \cdots + e_k j_k$. A $(t, m, \mathbf{e}, k)$-net in base b is a point set of b^m points in $[0,1]^k$ such that $A_{b^m}(E) = b^t$ for every elementary interval $E = E(e_1 j_1, \ldots, e_k j_k)$ in base b with $\mu(E) = b^{t-m}$ and $\mathbf{j}$ satisfying $(\mathbf{e}, \mathbf{j}) = m - t$.

Definition 3 A $(t, \mathbf{e}, k)$-sequence in base b is an infinite sequence, $X = (X_n)_{n \geq 0}$, of points in $[0,1]^k$ such that for all integers $\ell \geq 0$ and all $m > t$ satisfying $m - t \in M(\mathbf{e})$, the point set $\{[X_{\ell b^m}]_{b,m}, \ldots, [X_{(\ell+1)b^m - 1}]_{b,m}\}$ is a $(t, m, \mathbf{e}, k)$-net, where $[X_n]_{b,m}$ means the coordinate-wise b-ary m-digit truncation of a point X_n.

It is obvious that a (t, k)-sequence in base b is identical to a $(t, \mathbf{e}, k)$-sequence in base b with $\mathbf{e} = (1, \ldots, 1)$.

Tezuka [20, 21] proved the following theorem on the discrepancy of $(t, \mathbf{e}, s)$-sequences in an arbitrary integer base $b \geq 2$:

Theorem 3 *Let $b \geq 2$ be an arbitrary integer. The discrepancy for the first $N > b^t$ points of a $(t, \mathbf{e}, k)$-sequence in base b is bounded as follows:*

$$D_N \leq C_k \frac{(\log N)^k}{N} + O\left(\frac{(\log N)^{k-1}}{N} \right),$$

where $C_k = \frac{b^t}{k!} \prod_{i=1}^k \left(\frac{b^{e_i} - 1}{2e_i \log b} \right)$.

This means that if t, $\mathbf{e}$, and b are constant or depend only on k, then a $(t, \mathbf{e}, k)$-sequence in base b becomes a low-discrepancy sequence. Note that a smaller value of t gives a lower discrepancy.

The following is known as a general construction principle for (t, k)-sequences, which can be applied to $(t, \mathbf{e}, k)$-sequences as well. Let $k \geq 1$ and $b \geq 2$ and $B = \{0, 1, \ldots, b-1\}$. Accordingly, we define;

(i) A commutative ring R with identity and card$(R) = b$;
(ii) Bijections $\psi_r : B \to R$ for $r = 1, 2, \ldots,$ with $\psi_r(0) = 0$ for all sufficiently large r;
(iii) Bijections $\lambda_{im} : R \to B$ for $1 \leq i \leq k$ and $m = 1, 2, \ldots,$ with $\lambda_{im}(0) = 0$ for $1 \leq i \leq k$ and all sufficiently large m;
(iv) Elements $c_{mr}^{(i)} \in R$ for $1 \leq i \leq k$, $m \geq 1$, $r \geq 1$, where for fixed i and r we have $c_{mr}^{(i)} = 0$ for all sufficiently large m.

For $n = 0, 1, 2, \ldots,$ let $n = \sum_{r=1}^{\infty} a_r(n) b^{r-1}$ for $a_r(n) \in B$. Set the i-th coordinate of the point X_n as

$$X_n^{(i)} = \sum_{m=1}^{\infty} x_{nm}^{(i)} b^{-m},$$

for $1 \leq i \leq k$ and $n \geq 0$, with

$$x_{nm}^{(i)} = \lambda_{im} \left(\sum_{r=1}^{\infty} c_{mr}^{(i)} \psi_r(a_r(n)) \right) \in B$$

for $1 \leq i \leq k$, $m \geq 1$, and $n \geq 0$. We call $C^{(i)} = (c_{mr}^{(i)})$ the *generator matrix* of the i-th coordinate of a $(t, \mathbf{e}, k)$-sequence.

Hereafter, we assume that b is a prime power, and that R in the construction principle is the finite field $GF(b)$. Let

$$\mathbf{c}_m^{(i)}(l) = (c_{m,1}^{(i)}, \ldots, c_{m,l}^{(i)}) \in GF(b)^l,$$

and let

$$C(d_1, \ldots, \mathrm{d}_k; l) = \{\mathbf{c}_m^{(i)}(l) \mid 1 \leq m \leq \mathrm{d}_i, 1 \leq i \leq k\}.$$

We need construct the generator matrices, $C^{(i)}$, $1 \leq i \leq k$, so that $(t, \mathbf{e}, k)$-sequences become low-discrepancy sequences.

We now describe how to construct such generator matrices so that we obtain low-discrepancy sequences called generalized Niederreiter sequences [3, 14]. The construction is based on the formal Laurent series expansions over finite fields. Denote $S(z) \in GF\{b, z\}$ by

$$S(z) = \sum_{r=w}^{\infty} a_r z^{-r},$$

where all $a_r \in GF(b)$ and w is an arbitrary integer. Hereafter, we use the following notations: $[S(z)]$ denotes the polynomial part of $S(z) \in GF\{b, z\}$ and $[S(z)]_{p(z)} \stackrel{\text{def}}{=} [S(z)] \pmod{p(z)}$ with $0 \leq \deg([S(z)]_{p(z)}) < \deg(p(z))$.

Let polynomials $p_1(z), \ldots, p_k(z) \in GF[b, z]$ be pairwise coprime and let $e_i = \deg(p_i) \geq 1$ for $1 \leq i \leq k$. For $m \geq 1$, $1 \leq i \leq k$, and $j \geq 1$, consider the expansion

$$\frac{y_{im}(z)}{p_i(z)^j} = \sum_{r=w}^{\infty} a^{(i)}(j, m, r) z^{-r},$$

by which the elements $a^{(i)}(j, m, r) \in GF(b)$ are determined. Here $w \leq 0$ may depend on i, j, m, and each $y_{im}(z)$ is a polynomial such that the residue polynomials $[y_{im}(z)]_{p_i(z)}$, $(j-1)e_i \leq m - 1 < je_i$, are linearly independent over $GF(b)$ for any $j > 0$ and $1 \leq i \leq k$. Define

$$c_{mr}^{(i)} = a^{(i)}(m_i + 1, m, r),$$

for $1 \leq i \leq k$, $m \geq 1$, and $r \geq 1$, where $m_i = [(m-1)/e_i]$.

Tezuka [20] proved the following theorems:

Theorem 4 *A generalized Niederreiter sequence in base b is a* $(0, \mathbf{e}, k)$*-sequence in base b, where* $e_i = \deg(p_i)$ *for* $i = 1, \ldots, k$.

Theorem 5 *If* $p_i(z)$, $i = 1, 2, \ldots, k$ *are the first k irreducible polynomials over* $GF(2)$ *from the list sorted in nondecreasing order of degree, the leading constant in the upper bound of the discrepancy of generalized Niederreiter sequences in base 2 converges to 0 as the dimension k goes to infinity.*

Since Faure sequences are a special case of generalized Niederreiter sequences such that (1) $b \geq k$ is prime, (2) $p_i(z) = z - i + 1$ for $1 \leq i \leq k$, and (3) all $y_{im}(z) = 1$, we give the following definition:

Definition 4 Generalized Faure sequences are defined as $(0, \mathbf{e}, k)$-sequences in a prime base $b \geq k$, where $\mathbf{e} = (1, \ldots, 1)$, obtained from generalized Niederreiter sequences.

Theorem 3 implies that the leading constant in the upper bound of the discrepancy of generalized Faure sequences converges to 0 as the dimension k goes to infinity, if the base is the least prime $b \geq k$. In the matrix representation, for all generator matrices $C^{(i)}$, $1 \leq i \leq k$, we have

$$C^{(i)} = A^{(i)} P^{i-1}, \tag{1}$$

where $A^{(i)}$, $1 \le i \le k$, are nonsingular lower triangular matrices over $GF(b)$ and P is the Pascal matrix whose (i, j) element is equal to $\binom{j-1}{i-1}$. The original Faure sequences correspond to the case in which $A^{(i)} = I$ for all $1 \le i \le k$.

2.3 Theoretical Explanations of Speeding-Up

After remarkable success of quasi-Monte Carlo methods in financial computations, researchers have been interested in explaining the success theoretically [23, 27], because it seems to contradict the curse of dimensionality. There are two sides which we need take into consideration. One is an algorithm side, i.e., low-discrepancy sequences, and the other is a problem side, i.e., integrands.

For the algorithm side, Matoušek [7] pointed out that randomly chosen generalized Faure sequences can be viewed as a simplified version of Owen's scrambling scheme [2, 6]. His results imply that the expected integration error over all generalized Faure sequences is

$$O\left(\frac{(\log N)^{(k-1)/2}}{N^{3/2}}\right).$$

Note that this result can be interpreted as an existence theorem of a very good generalized Faure sequence for high-dimensional integration. Tezuka and Faure [22] proposed the so-called i-binomial scrambling as a partial derandomization of Owen's scrambling.

For the problem side, the notion of effective dimension was proposed by Paskov [12]. This notion is a measure to quantify the importance of each variable on the integrand with many variables. Caflisch, Morokoff, and Owen [2] defined it based on the ANOVA (Analysis of Variance) decomposition, which is defined as follows: Let $u \subseteq \{1, 2, \ldots, k\}$ be a subset of indecies and $\bar{u} = \{1, 2, \ldots, k\} - u$ be its complement. Also, $X = \{x_1, \ldots, x_k\}$ and $X^u = \{x_j; j \in u\}$. Then, the ANOVA decomposition of $f(x_1, \ldots, x_k)$ is defined by

$$f(x_1, \ldots, x_k) = \sum_{u \subseteq \{1,2,\ldots,k\}} \alpha_u(x_1, \ldots, x_k),$$

where $\alpha_u(x_1, \ldots, x_k)$ is given by

$$\alpha_\emptyset(x_1, \ldots, x_k) := I(f) \equiv \int_{[0,1)^k} f(z_1, \ldots, z_k) \mathrm{d}z_1 \ldots \mathrm{d}z_k,$$

and

$$\alpha_u(x_1, \ldots, x_k) := \int_{Z^u = X^u, Z^{\bar{u}} \in [0,1)^{\bar{u}}} (f(z_1, \ldots, z_k) - \sum_{v \subset u} \alpha_v(z_1, \ldots, z_k)) \prod_{j \in \bar{u}} \mathrm{d}z_j.$$

The meaning of $\alpha_u(x_1, \ldots, x_k)$ is the effect of the subset X^u on $f(x_1, \ldots, x_k)$ minus the effect of its proper subset X^v with $v \subset u$. These $\alpha_u(x_1, \ldots, x_k)$ have the following orthogonal property: Let $i \in u$. If we fix x_j, $j \neq i$,

$$\int_0^1 \alpha_u(x_1, \ldots, x_k) \mathrm{d}x_i = 0.$$

Thus, when $u \neq \emptyset \subset \{1, \ldots, k\}$,

$$\int_{[0,1)^k} \alpha_u(x_1, \ldots, x_k) \mathrm{d}x_1 \ldots \mathrm{d}x_k = 0.$$

When $u \neq v$,

$$\int_{[0,1)^k} \alpha_u(x_1, \ldots, x_k) \alpha_v(x_1, \ldots, x_k) \mathrm{d}x_1 \ldots \mathrm{d}x_k = 0.$$

Hence, the variance of $f(x_1, \ldots, x_k)$ is given by

$$\sigma^2 = \int_{[0,1)^k} (f(x_1, \ldots, x_k) - \alpha_\emptyset(x_1, \ldots, x_k))^2 \mathrm{d}x_1 \ldots \mathrm{d}x_k = \sum_{|u|>0} \sigma_u^2,$$

where $\sigma_u^2 = 0$ if $u = \emptyset$; otherwise

$$\sigma_u^2 := \sigma^2(\alpha_u) = \int_{[0,1)^k} \alpha_u(x_1, \ldots, x_k)^2 \mathrm{d}x_1 \ldots \mathrm{d}x_k.$$

The definition of effective dimension was introduced in two ways [2]:

- *Truncation sense*:

$$D_{\text{trunc}} := \min(1 \leq i \leq k \ \text{ such that } \sum_{u \subseteq \{1,2,\ldots,i\}} \sigma_u^2 \geq 0.99\sigma^2),$$

- *Superposition sense*:

$$D_{\text{super}} := \min(1 \leq i \leq k \ \text{ such that } \sum_{|u| \leq i} \sigma_u^2 \geq 0.99\sigma^2).$$

However, Tezuka [19] showed that low-effective dimension is not a necessary condition for quasi-Monte Carlo methods to perform better than simple Monte Carlo methods.

Sloan and Woźniakowski [13] took a different approach, which is now called the weighted discrepancy. Together with a notion of a weighted function space, they obtained the weighted version of the Koksma-Hlawka bound for the integration error in the weighted space. They proved that there exists a weighted space for which quasi-Monte Carlo methods run in $O(N^{-1})$ with the implied constant independent of the dimension. This direction of research has much advanced to what is today called the tractability theory [10], one major field of computational complexity. Recent progress in this field was surveyed by Woźniakowski [27], which also contains an interesting little history about the exciting early days of the tractability theory at Columbia University. It should be pointed out that the tractability of numerical integration using generalized Faure sequences is still open.

3 Numerical Experiments

In this section, we apply quasi-Monte Carlo methods to a practical problem related to pricing financial derivatives. It is numerical simulation originally described by Paskov [12], concerned with mortgage-backed securities (MBS), the most popular among fixed-income derivatives. Many people still remember that in 2008 the US subprime MBS caused the credit crisis in the international financial market, and shook the world. In the experiments, we used a randomly chosen generalized Faure sequence as a low-discrepancy sequence for quasi-Monte Carlo simulations. More precisely, nonsingular lower triangular matrices $A^{(i)}$, $1 \leq i \leq k$, in Eq. (1) for the generator matrices were chosen at random, and we omitted the first 100000 points; that is to say, we used the points $X_n, n = 100001, 100002, \ldots,$ of the sequence. Tezuka [14] describes an efficient implementation of generalized Faure sequences based on the b-ary Gray code. We used the random number generator CombTaus [14] for Monte Carlo simulation.

Mortgage-Backed Securities

Mortgage-backed securities (MBS) are a kind of interest-rate option, whose underlying asset is a pool of residential mortgage portfolios. They have a critical feature of prepayment privileges, because householders can prepay thier mortgages at any time. The integration problem associated with MBS is summarized as follows. We use the following notations:

r_k: The appropriate interest rate in month k
w_k: The percentage prepaid in month k
a_k: The remaining annuity after $361 - k$ months
C: The monthly payment on the underlying mortgage pool

for $k = 1, 2, \ldots, 360$, where $a_k = 1 + \mathrm{d}_1 + \cdots + \mathrm{d}_1^{k-1}$ is constant with $\mathrm{d}_1 = 1/(1 + r_0)$ and r_0 is the current monthly interest rate. C is also constant. The variable r_k follows the discrete-time version of the Rendleman-Bartter interest-rate model, which is mathematically equivalent to the Black-Scholes model:

$$\log r_k - \log r_{k-1} = (a - \frac{\sigma^2}{2})\Delta + \sigma \mathrm{d}B,$$

where $\Delta = 1$ and $\mathrm{d}B$ is the normal random variable with mean zero and variance Δ. Here, we assume zero drift (i.e., $a = 0$) in order to make $E(r_k) = r_0$ for $k = 1, \ldots, 360$. Thus,

$$r_k = K_0 \exp(\sigma z_k) r_{k-1}, \quad \text{for } k = 1, 2, \ldots, 360,$$

where z_k, $k = 1, 2, \ldots, 360$, are independent standard normally distributed random variables, and $K_0 = \exp(-\sigma^2/2)$.

The prepayment model for the variables w_k, $k = 1, 2, \ldots, 360$, depends on the interest rate r_k, $k = 1, 2, \ldots, 360$, as follows:

$$w_k = K_1 + K_2 \arctan(K_3 r_k + K_4),$$

where K_1, K_2, K_3, and K_4 are given constants. As is easily seen from our actual experience, in general, the lower the interest rate, the higher the prepayment rate. Thus, the cash flow in month k is

$$M_k(z_1, \ldots, z_k) = C(1 - w_1) \cdots (1 - w_{k-1})(1 - w_k + w_k a_{361-k}).$$

This is multiplied by the discount factor

$$\mathrm{d}_k(z_1, \ldots, z_{k-1}) = \prod_{i=0}^{k-1} \frac{1}{1 + r_i},$$

We have the following total present value of MBS:

$$PV(z_1, \ldots, z_{360}) = \sum_{k=1}^{360} \mathrm{d}_k(z_1, \ldots, z_{k-1}) M_k(z_1, \ldots, z_k).$$

What we want to compute is the expected value of the present value PV over all independent random variables z_k, $k = 1, \ldots, 360$. By using the inversion of the normal distribution, we can formulate this problem as one of computing a multivariate integration over $[0, 1]^{360}$:

$$\mathrm{E}(PV) = \int_{[0,1]^{360}} PV(u_1, \ldots, u_{360}) \mathrm{d}u_1 \cdots \mathrm{d}u_{360},$$

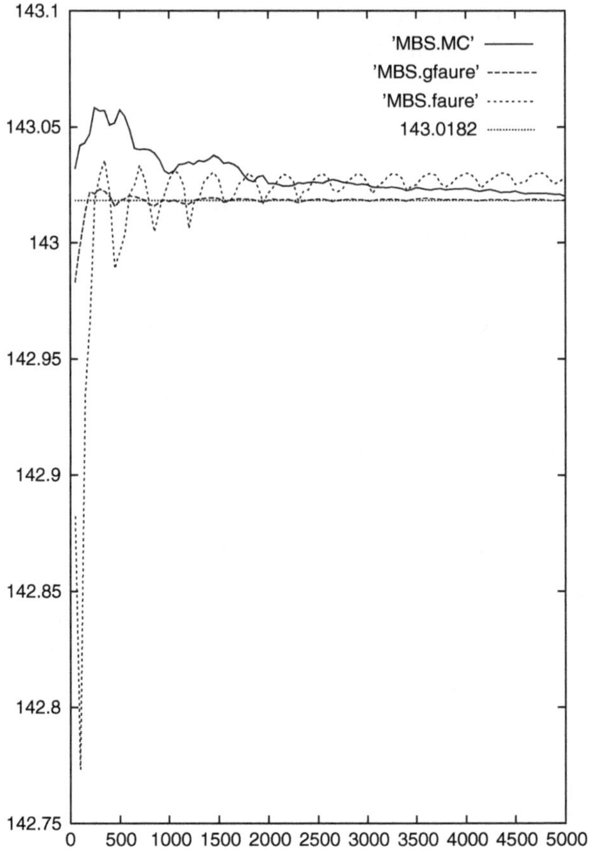

Fig. 1 Convergence of Monte Carlo and quasi-Monte Carlo methods for MBS pricing: The *vertical* and *horizontal lines* indicate the present value (PV) of MBS and the number of sample paths, respectively

where $u_k = \mathrm{N}(z_k)$ for $k = 1, \ldots, 360$.

In this experiment, we used the parameter set

$$(r_0, K_1, K_2, K_3, K_4, \sigma) = (0.00625, 0.24, 0.134, -261.17, 12.72, 0.2)$$

from [9, 15], where the expected value of PV is numerically computed as $143.0182 \times C$. Figure 1 shows the convergence performances of Monte Carlo and quasi-Monte Carlo methods. The solid line (MBS.MC) shows the result for Monte Carlo methods, while the dotted lines (MBS.faure and MBS.gfaure) show the results for quasi-Monte Carlo methods using the original Faure sequences and generalized Faure sequences, respectively. In this case, for example, with 1,000 samples quasi-Monte Carlo (MBS.gfaure) converges to the correct value within an accuracy of 10^{-5}. On the other hand, the standard deviation of PV computed from the first 1,000 sample values of the Monte Carlo simulation is about 0.276. Thus, the 99 % confidence interval is about [143.005, 143.042]. Therefore, we can say that about 250 times speed-up was gained by quasi-Monte Carlo for this problem. We should notice that generalized Faure sequences perform significantly better than the original Faure, though both are low-discrepancy sequences.

References

1. P. Boyle, Options: a Monte Carlo approach. J. Financ. Econ. **4**, 323–338 (1977)
2. R.E. Caflisch, W. Morokoff, A.B. Owen, Valuation of mortgage backed securities using Brownian bridges to reduce effective dimension. J. Computat. Financ. **1**, 27–46 (1997)
3. J. Dick, F. Pillichshammer, *Digital Nets and Sequences: Discrepancy Theory and Quasi-Monte Carlo Integration* (Cambridge University Press, Cambridge, 2010)
4. P. Jäckel, *Monte Carlo Methods in Finance* (Wiley, New York, 2002)
5. C. Joy, P. Boyle, K.S. Tan, Quasi-Monte Carlo methods in numerical finance. Manage. Sci. **42**, 926–938 (1996)
6. C. Lemieux, *Monte Carlo and Quasi-Monte Carlo Sampling* (Springer, New York, 2009)
7. J. Matoušek, *Geometric Discrepancy: An Illustrated Guide* (Springer, Berlin, 1999)
8. H. Niederreiter, *Low-Discrepancy Simulation*, Handbook of Computational Finance (Springer, New York, 2012)
9. S. Ninomiya, S. Tezuka, Toward real time pricing of complex financial derivatives. Appl. Math. Financ. **3**, 1–20 (1996)
10. E. Novak, H. Woźniakowski, *Tractability of Multivariate Problems: Standard Information for Functionals*, vol. 2 (European Mathematical Society, Zurich, 2010)
11. A. Papageorgiou, J.F. Traub, Beating Monte Carlo. RISK **9**, 63–65 (1996)
12. S.H. Paskov, *New Methodologies for Valuing Derivatives*, Mathematics of Derivative Securities (Cambridge University Press, Cambridge, 1997)
13. I.H. Sloan, H. Woźniakowski, When are quasi-Monte Carlo algorithms efficient for high dimensional integrals. J. Complex. **14**, 1–33 (1998)
14. S. Tezuka, *Uniform Random Numbers: Theory and Practice* (Kluwer Academic Publishers, Boston, 1995)
15. S. Tezuka, *Financial Applications of Monte Carlo and Quasi-Monte Carlo Methods, Random and Quasi-Random Point Sets*. Lecture Notes in Statistics, vol. 138 (Springer, New York, 1998), pp. 303–332
16. S. Tezuka, *Quasi-Monte Carlo Methods for Financial Applications* (ICIAM99) (Oxford University Press, Oxford, 2000), pp. 234–245
17. S. Tezuka, *Discrepancy Theory and Its Application to Finance* (IFIP TCS2000). Lecture Notes in Computer Science, vol. 1872 (Springer, New York, 2000), pp. 243–256
18. S. Tezuka, Quasi-Monte Carlo: *Discrepancy between Theory and Practice, Monte Carlo and Quasi-Monte Carlo Methods*, vol. 2000 (Springer, Berlin, 2002), pp. 124–140
19. S. Tezuka, On the necessity of low-effective dimension. J. Complex. **21**, 710–721 (2005)
20. S. Tezuka, On the discrepancy of generalized niederreiter sequences. J. Complex. **29**, 240–247 (2013)
21. S. Tezuka, Recent results on (t, e, s)-sequences, presented at the Oberwolfach workshop on uniform distribution theory and applications, Sept 30, 2013
22. S. Tezuka, H. Faure, I-binomial scrambling of digital nets and sequences. J. Complex. **19**, 744–757 (2003)
23. J.F. Traub, A.G. Werschulz, *Complexity and Information* (Cambridge University Press, Cambridge, 1998)

24. J.F. Traub, H. Woźniakowski, *Breaking Intractability* (Scientific American, New York, 1994), pp. 102–107
25. P. Truel, *From I.B.M., Help in Intricate Trading* (The New York Times, New York, 1995)
26. H. Woźniakowski, Average case complexity of multivariate integration. Bull. Am. Math. Soc. **24**, 185–194 (1991)
27. H. Woźniakowski, *Tractability of Multivariate Problems, Foundations of Computational Mathematics, Hong Kong 2008* (Cambridge University Press, Cambridge, 2009), pp. 236–276

Pure Mathematics and Applied Mathematics are Inseparably Intertwined: Observation of the Early Analysis of the Infinity

Masahito Takase

Abstract In this work I consider the connection between pure mathematics and applied mathematics from a historian's point of view and I conclude that pure mathematics and applied mathematics are inseparably intertwined.

Keywords Pure mathematics · Applied mathematics · Gauß's theory of errors · Folium of descartes · Cyclotomic equations · Proof of the impossibility · Analysis of the infinity

1 From Euler's Statements

The original idea of separating mathematical science into pure and applied mathematics can be found by looking back at the history of mathematics. Crelle's Journal founded by Crelle in Berlin at the beginning of the nineteenth century already had the name *Journal fur die reine und angewandte Mathematik*, in which we find the two phrases *pure mathematics* and *applied mathematics*. Leonhard Euler also compared pure mathematics and applied mathematics in his paper *De la controverse entre Mrs. Leibniz et Bernoulli sur les logarithmes des nombres negatifs et imaginaires*, where he writes;

> The views of mathematicians may be quite different on issues concerning applied mathematics, in which the real controversies may be caused by the different ways of conceiving objects and of referring them to precise ideas. On the other hand, the pure parts of mathematics are completely free from any item of dispute, and have always much proud of the fact that one will find there nothing of which we can't demonstrate either the truth or the falsity [4, Translation from the French by Stacy G. Langton].

M. Takase (✉)
Faculty of Arts and Science, Kyushu University, 744, Motooka, Nishiku,
Fukuoka 819-0395, Japan
e-mail: takase@artsci.kyushu-u.ac.jp

R. Nishii et al. (eds.), *A Mathematical Approach to Research Problems of Science and Technology*, Mathematics for Industry 5,
DOI: 10.1007/978-4-431-55060-0_29, © Springer Japan 2014

Prior to Euler's paper, there was a debate between Gottfried Wilhelm Leibniz and Johann Bernoulli over the logarithm $\log(-1)$ of a negative number -1 and the logarithm $\log(\sqrt{-1})$ of an imaginary number $\sqrt{-1}$. Euler spoke as above after having known the course of the debate. Determining the real nature of the logarithm of a negative number and of an imaginary number was a theme of pure mathematics, but because the Leibniz and Bernoulli debate dealt with only a vague, overall impression of the issue, it could not be said who was right and the matter gradually faded away. Euler claimed that it is impossible in pure mathematics for a falsehood to not be distinguished from the truth—indeed, he always boasted of it. In contrast, this is not so in applied mathematics, and so it seems to natural that debates on realism often occur. I have still never come across clear written or verbal distinction of applied mathematics and pure mathematics.

Even if it is not clear what Euler considers to be applied mathematics, I have recently begun to think that there might be an original mathematical science that predates both pure and applied mathematics. C. F. Gauß's is what got me thinking about such mathematics.

2 Gauß's Theory of Errors

Gauß's theory of errors is essentially a correction of the errors in astronomical observation. Gaußproposed the method of least squares to minimize errors and developed a probabilistic consideration for the decision of the orbits of celestial bodies. Probability theory is connected with this orbit decision because the orbit decision is a prediction of future position. Prediction techniques are the very essence of probability theory and comprised the bulk of Jacob Bernoulli's classic masterpiece on probability theory, *Ars Conjectandi*.

Pure mathematics, which has the reputation of *mathematics for mathematics*, is a theoretical system totally unrelated to physical natural phenomena. In this discussion, I reference facts that I have actually observed in the world of mathematics. Gauß's theory of numbers is a typical example of pure mathematics. Pierre de Ferma's theory of numbers also belongs to pure mathematics. Leibniz and Bernoulli pursued the real nature of the logarithm of negative and imaginary numbers, and Euler, several decades after in their contemplation, clarified the essential characteristic of the logarithm, i.e., that it has infinitely many values; Euler's investigation also feels like pure mathematics. In contrast, the single word *applied* in applied mathematics makes it seem more like the applications of theories of pure mathematics.

Here, I want to introduce my idea of a simple image of pure and applied mathematics. Gauß's theory of errors seems to be a typical example of applied mathematics since it is a device to rectify the errors of astronomical observations. I became greatly estranged from the general image of applied mathematics when I read Gauß's theory of errors. It makes sense to deduce from his theory that the idea of *new mathematics is rooted in a concept we call the correction of errors*. It seems more likely that probability theory is created from the theory of errors than that probability theory is applied to the theory of errors. This surprised me, as I had not assumed that I would have such an impression at all.

The mysterious impression I received from Gauß' theory of errors made me think I had felt a similar feeling somewhere else. Looking back at the history of mathematics, I found a concrete example in the analysis of the infinity of the days of creation. The analysis of the infinity of the European Continent begins in the two articles written by Leibniz: one, called *Nova methodus pro maximis et minimis, itemque tangentibus, quae nec fractas nec irrationales quantitates moratur, et singulare pro illis calculi genus* [9, 1684], is an article on differential calculus—and the other, *De geometria recondita et analysi indivisibilium atque infinitorum* [8, 1686], deals with integral calculus. The two papers were published in the scientific journal *Acta Eruditorum* founded by Otto Mencke in Leipzig in 1682. In those days, the Bernoulli brothers (Jacob and Johann) were based in Basel, Switzerland and seemed to be very much charmed by Leibnitz's papers; they cooperated with him and began decoding his work, writing him letters because his papers were so full of enigmatic words that they were quite difficult to understand. This was the scene in the early days of the analysis of the infinity.

Prior to the establishment of the analysis of the infinity, there was *the day of curves*, namely *the times when people developed an interest in curves*. In 1637, Descartes' *Discours de la methode pour bien conduire sa raison et chercher la verite dans les sciences* [3] was published and the Optics, the Meteorology, and the Geometry were discussed as concrete examples of his method discussed in the introduction. Here, we should focus on Descartes' Geometry, which includes three chapters. The second chapter, entitled *the properties of curves*, expresses the origin of today's concept of analytic geometry.

3 Folium of Descartes

The interest in curves is very old: curves such as the Conchoid of Nicomedes, the Cissoid of Diocles, the Spiral of Archimedes, and the conic sections of Apollonius have been found in Ancient Greece. This interest was inherited by the mathematical scholars of Modern Europe and many new curves were discovered, leading to an increased interest in the study of curves. According to an idea suggested by Descartes, curves are usually expressed by algebraic equations of the form $f(x, y) = 0$. In these equations, $f(x, y)$ denotes a polynomial of x, y, so the objects of contemplation are limited to *algebraic curves*. The original idea of *algebraic curves* did not include displaying them by using polynomials with $f(x, y) = 0$; rather, these curves were first expressed by algebraic equation, and then we were able to survey the whole algebraic curves.

The known algebraic curves were all displayed in algebraic equations and a curve drawned in various ways was displayed in an algebraic equation as far as it was algebraic. The truly essential point of Descartes' idea was to start from equations. First, we write down an algebraic equation $f(x, y) = 0$, and second, we think the displayed curve. *The all algebraic curves* were generated by this idea. In 1638, the year after Descartes' *Discours* was published, Descartes wrote the equation $x^3+y^3 = axy$ (a denotes a constant) in a letter addressed to Mersenne and pictured the rough shape of the curve that it expressed. This is the algebraic curve that came to be known as the *Folium of Descartes* . Descartes' point when doing this was to demonstrate the attitude of starting from an equation proactively by oneself.

4 From the Theory of Curves to the Analysis of the Infinity

As stated, in the seventeenth century, interest in curves was widespread and various methods were devised to draw tangent and normal lines for various kinds of curves, to determine the curvature of a curve, and to calculate the area of the domain surrounded by a curve. These methods were elaborated along the particular property associated with the individual curve. Then Leibniz proposed his new method of the analysis of the infinity, which was fine-tuned through his continued correspondence with the Bernoulli brothers during this time. In 1696, *Analyse des infiniment petits pour l'intelligence des lignes courbes* [10] by Marquis de L'Hopital was published. This is the first calculus text in the history of mathematics, and the title deserves attention because it suggests that the analysis of the infinity is the best theory to use to understand curves.

The power of the analysis of the infinity was extremely strong; for example, it led to the capability of drawing a tangent line to a curve freely in its arbitrary point. We do not require a specialized device to do this, and Leibniz's method is applicable not only to algebraic curves but also to transcendental curves. In fact, this led to the development of *the all-around tangent line method*. Descartes' *Discours* was published in 1637, and Lebniz's two papers on differential and integral calculus were published in 1684 and 1686, respectively. During this interval of approximately 50 years, curve theory was finalized by the birth of the analysis of the infinity. This remarkable scene closely resembles the situation surrounding the creation of Gauß's Theory of Errors.

Gauß created a new theory of mathematics in his efforts to correct the errors existing in astronomical observation. Mathematicians of the seventeenth century were fascinated with curves and the analysis of the infinity was generated largely from their passion. Being fascinated with a preexisting concept, and the situation of mathematics being born of a passion to pursue an ideal is identical to the case of Gauß's Theory of Errors. We do not obtain new knowledge of a scientific domain applying some kind of finished theory of mathematics. Rather, the creation of mathematics is intimately connected with the passion of wanting to know the unknown.

Mathematics is born of passion. Applied mathematics is born when people's passion turns to phenomena of nature, such as celestial bodies, and pure mathematics is born when we turn to phenomena of a mathematical nature, such as curves. Only *mathematics itself* exists without the distinction between pure and applied mathematics. This is my impression when I recollect the history of the formation of the analysis of the infinity and Gauß's Theory of Errors.

5 Euler's Analysis of the Infinity

Euler was aware of the distinction between *pure mathematics* and *applied mathematics* in mathematical science and wrote a few words about the distinction. It was quite difficult for me to approach the substance of what is called applied mathematics with only this as a clue. However, it suddenly occurred to me while studying *Gauß's Theory of Errors* that there is no distinction between pure mathematics and applied mathematics in *mathematics itself.* This realization caught me completely off guard.

Let me give another example, this one using Euler's analysis of the infinity. I mentioned above that the theory of curves was reached after Leibniz and the Bernoulli brothers had completed their analysis of the infinity. What kind of course would it make sense for the subsequent theory of the infinity to follow? Let us express it a little more concretely. Euler lived in the generation directly after Leibniz and the Bernoulli brothers. What kind of mathematical phenomena would Euler instinctually turn to in his analysis of the infinity? Since interest in the theory of curves had already peaked, it seems clear that he would have moved beyond that. Euler's books and papers from that era indicate he was trying to find a theme in the dynamics and the calculus of variations. In 1736, when he was 29-years old, Euler penned the two volumes of *Mechanica* [5, E15, E16], and in 1744, aged 37, a masterpiece entitled *Methodus inveniendi lineas curvas maximi minimive proprietate gaudentes, sive solutio problematis isoperimetrici lattissimo sensu accepti* [6, E65].

Since tracing a moving object creates a drawing of a curve, we should be able to use tracting to understand dynamics based on the theory of curves. Johann Bernoulli proposed the *Brachistochrone curve* problem, which led to the birth of the calculus of variations, and so we can say that the calculus of variations also came from the theory of curves. Euler surveyed a world opened up by the theory of curves and went on to explore the dynamics and the calculus of variations. He chose this route because Newton's theory of dynamics was on his mind. Not only *Mechanica* but also *Methodus inveniendi lineas curvas maximi minimive proprietate gaudentes, sive solutio problematis isoperimetrici lattissimo sensu accepti* were Euler's foundation stones in the study of dynamics. His mathematical objective was to understand the dynamics of Newton by analyzing Leibniz's and the Bernoulli brothers' concept of the infinity. The figure of the analysis of the infinity repeated big transformations in this flow.

The above is a common view. However, a question remains: Is the analysis of the infinity pure mathematics or applied mathematics? The essence of Euler's analysis of the infinity is a theory of solving various differential equations. Differential equations existed even in the times of Leibniz and the Bernoulli brothers, but they were equations that indicate the situation of a tangent line and the normal line of a curve rather than a differential equation because differential equations before Euler's time were considered the realm of the theory of curves. Leibniz and the Bernoulli brothers had been thinking about curves long before Euler came around, and they aimed to restore the perspective of a curve from local information on tangent and normal lines. This method, called *inverse method of tangent*, is the same as integral calculus from the viewpoint of the theory of differential equations.

6 Various Sources of Pure Mathematics

Euler's analysis of infinity is no longer considered a theory of curves. Euler introduced the concept of three functions into mathematics, established a viewpoint to consider a curve as the graph of a function, and moved the collective focus from curves to functions. Functions now play the leading role in the analysis of the infinity and the information on tangent and normal lines are shown in the form of differential equations while the former *inverse method of tangent* became modern integral calculus. In this case, integral calculus has the same meaning as the method of solving differential equations and integral calculus has the same meaning as the elucidation of the differential equation.

In the early days of curve theory, Descartes suggested calling a figure expressed by an equation a curve, and then Euler changed the concept of what a curve actually is once again. Because Euler's analysis of the infinity is essentially a theory of solving differential equations, it feels more like pure mathematics than applied mathematics. However, the source is dynamics. Both the common and the abstract theories of mathematics were born from serious interest in dynamics. Ultimately, the theory of differential equations is neither applied mathematics nor pure mathematics: it is just a theory of mathematics. It is pointless to wonder if the theory of differential equations is pure mathematics or applied mathematics because, in truth, it can be considered *mathematics created by dynamics* .

In the theory of algebraic equations, we search for a method of solving cubic and quartic equations by observing phenomena observed inside of mathematics. Various theories, including Gauß's theory of cyclotomic equations [7], Abel's *proof of the impossibility* [1], and Galois's *Galois theory* were born from this observation. There surely exists a *world of mathematics* in which Leibniz and the Bernoulli brothers' analysis of the infinity, Euler's analysis of the infinity, Gauß's theory of cyclotomic equations, Abel's *proof of the impossibility*, and Galois's *Galois theory* can coexist. It seems only logical to call *all* of these theories *pure mathematics*.

The source of Euler's analysis of the infinity is dynamics, but he did not apply the established theory of differential equations to the dynamics. Rather, he had the mathematical intention to understand the dynamics from the viewpoint of the theory of curves. The theory of curves resulted in the creation of the theory of differential equations.

I conclude that, if Euler had been around in modern times, he might have considered the analysis of the infinity to be applied mathematics. By this I mean that there was an opportunity for the formation of a theory outside of mathematics, and Euler's analysis of the infinity is an area of scientific mathematics, similar to number theory. This is why I believe that pure mathematics and applied mathematics are inseparably intertwined.

References

1. N.H. Abel, Démonstration de l'impossibilité de la résolution algébrique des équations générales qui passent le quatrieme degré, Ouvres 1, 1826, pp. 66–86 (German version published in Crelle's Journal, vol. 1, 1826)
2. J. Bernoulli, Ars conjectandi (1713)
3. R. Descartes, *Discours de la Methode Pour Bien Conduire sa Raison et Chercher la Verite dans les Sciences* (Maire, Leiden, 1637)
4. L. Euler, *De la Controverse Entre Mrs. Leibniz et Bernoulli sur les Logarithmes des Nombres Negatifs et Imaginaires*; [E168], 1749/51. Opera omnia: series 1, vol. 17, pp. 195–232 (English translation by Stacy Langton; Stacy Langton's Web Directory. http://home.sandiego.edu/langton/)
5. L. Euler, *Mechanica*; [E15][E16], 1736. Opera omnia: series 2, vol. 1, 2
6. L. Euler, *Methodus Inveniendi Lineas Curvas Maximi Minimive Proprietate Gaudentes, Sive Solutio Problematis Isoperimetrici Lattissimo Sensu Accepti*; [E65], 1744. Opera omnia: series 1, vol. 24
7. C.F. Gauß, Disquisitiones arithmeticae (1801)
8. G. Leibnitz, De geometria recondita et analysi indivisibilium atque infinitorum. Acta Eruditorum (1686)
9. G. Leibnitz, Nova methodus pro maximis et minimis, itemque tangentibus, quae nec fractas nec irrationales quantitates moratur, et singulare pro illis calculi genus. Acta Eruditorum (1684)
10. Marquis de L'Hôpital, *Analyse des Infiniment Petits Pour l'intelligence des Lignes Courbes* (Marquis de L'Hôpital, Paris, 1696)

High Performance Computing for Mathematical Optimization Problem

Katsuki Fujisawa

Abstract The semidefinite programming (SDP) problem is one of the central problems in mathematical optimization. The primal-dual interior-point method (PDIPM) is one of the most powerful algorithms for solving SDP problems, and many research groups have employed it for developing software packages. However, two well-known major bottlenecks, i.e., the generation of the Schur complement matrix (SCM) and its Cholesky factorization, exist in the algorithmic framework of the PDIPM. We have developed a new version of the semidefinite programming algorithm parallel version (SDPARA), which is a parallel implementation on multiple CPUs and GPUs for solving extremely large-scale SDP problems with over a million constraints. SDPARA can automatically extract the unique characteristics from an SDP problem and identify the bottleneck. When the generation of the SCM becomes a bottleneck, SDPARA can attain high scalability using a large quantity of CPU cores and some processor affinity and memory interleaving techniques. SDPARA can also perform parallel Cholesky factorization using thousands of GPUs and techniques for overlapping computation and communication if an SDP problem has over 2 million constraints and Cholesky factorization constitutes a bottleneck. We demonstrate that SDPARA is a high-performance general solver for SDPs in various application fields through numerical experiments conducted on the TSUBAME 2.5 supercomputer, and we solved the largest SDP problem (which has over 2.33 million constraints), thereby creating a new world record. Our implementation also achieved 1.713 PFlops in double precision for large-scale Cholesky factorization using 2,720 CPUs and 4,080 GPUs.

Keywords Mathematical optimization · Semidefinite programming · Interior-point method · Parallel computing · High performance computing

K. Fujisawa (✉)
Institute of Mathematics for Industry, Kyushu University, 744, Motooka, Nishiku, Fukuoka 819-0395, Japan
e-mail: fujisawa@imi.kyushu-u.ac.jp

R. Nishii et al. (eds.), *A Mathematical Approach to Research Problems of Science and Technology*, Mathematics for Industry 5,
DOI: 10.1007/978-4-431-55060-0_30,

1 Introduction

In the last two decades, semidefinite programming (SDP) problems have been intensively studied in both their theoretical and practical aspect in a wide range of fields. SDP has been regarded as one of the most important optimization problems for following several reasons.

1. SDP is used in a wide range of fields such as combinatorial optimization, structural optimization, control theory, economics, quantum chemistry, sensor network location, data mining, and machine learning [1, 2].
2. SDP theoretically includes a large number of convex programming such as linear programming (LP), convex quadratic programming (QP), and second-order cone programming (SOCP). For example, LP is a special case of SDP when all the matrices involved are diagonal. Therefore, we can convert typical convex programming problems to SDP problems.
3. SDP relaxation techniques can be used to obtain strong upper and lower bounds. This is true not only for convex optimization problems but also for nonconvex optimization problems, which are encountered in application areas such as control theory and smart grids and systems.
4. There exist several fast and stable theoretical algorithms [1]. For example, we can obtain an optimal solution in polynomial time using a primal-dual interior-point method (PDIPM).
5. Many state-of-the-art multithread-based parallel software packages for SDP, such as SDPA [3–5], CSDP [6], SeDuMi [7], and SDPT3 [8] have been developed. In addition, there exist MPI-based parallel software packages for supercomputers such as SDPARA [9–12], PCSDP [13], and PDSDP [14].

The standard form SDP has the following primal-dual form.

$$
\begin{array}{ll}
\mathcal{P} : & \text{minimize } \sum_{k=1}^{m} c_k x_k \\
& \text{subject to } X = \sum_{k=1}^{m} F_k x_k - F_0, \quad X \succeq O. \\
\mathcal{D} : & \text{maximize } F_0 \bullet Y \\
& \text{subject to } F_k \bullet Y = c_k \ (k = 1, \ldots, m), \quad Y \succeq O.
\end{array}
\tag{1}
$$

We denote by $\mathbb{S}^n$ the space of $n \times n$ symmetric matrices. The notation $X \succeq O$ ($X \succ O$) indicates that $X \in \mathbb{S}^n$ is a positive semidefinite (positive definite) matrix. The inner-product between $U \in \mathbb{S}^n$ and $V \in \mathbb{S}^n$ is defined by $U \bullet V = \sum_{i=1}^{n} \sum_{j=1}^{n} U_{ij} V_{ij}$.

In most SDP applications, it is common for the input data matrices $F_0, \ldots, F_m$ to share the same diagonal block structure $(n_1, \ldots, n_h)$. Each input data matrix F_k $(k = 1, \ldots, m)$ consists of submatrices in the diagonal positions as follows:

$$F_k = \begin{pmatrix} F_k^1 & O & O & O \\ O & F_k^2 & O & O \\ O & O & \ddots & O \\ O & O & O & F_k^h \end{pmatrix}$$

where $F_k^1 \in \mathbb{S}^{n_1}, F_k^2 \in \mathbb{S}^{n_2}, \ldots, F_k^h \in \mathbb{S}^{n_h}$.

Note that $\sum_{\ell=1}^{h} n_\ell = n$ and the variable matrices X and Y share the same block structure. We define $n_{\max}$ as $\max\{n_1, \ldots, n_h\}$. For the blocks where $n_\ell = 1$, the constraints of positive semidefiniteness are equivalent to the constraints of the non-negative orthant. Such blocks are sometimes called LP blocks.

The size of a given SDP problem can be approximately measured in terms of four metrics.

1. m: the number of equality constraints in the dual form $\mathscr{D}$ (which equals the size of the SCM)
2. n: the size of the variable matrices X and Y
3. $n_{\max}$: the size of the largest block of input data matrices
4. nnz: the total number of nonzero elements in all data matrices.

The SDP algorithm (SDPA) [4, 5] is one of the most well-known software packages of the PDIPM for solving the standard form SDP problem (formulation (1)). SDPA incorporates special data structures for handling block diagonal data matrices and efficient techniques for computing search directions when SDP problems become large and/or sparse [4]. Solving extremely large-scale SDP problems is considered to be important and challenging. In many applications, however, SDP problems become considerably large such that SDP software packages, including the SDPA, cannot solve them on a single node. There exist two well-known major bottlenecks in the algorithmic framework of the PDIPM. The first is the generation of the so-called Schur complement matrix (SCM). The second is the Cholesky factorization of the SCM. These two parts where bottlenecks occur are called ELEMENTS and CHOLESKY, respectively. We denote the time complexities of ELEMENTS and CHOLESKY by $O\left(mn^3 + m^2n^2\right)$ and $O\left(m^3\right)$, respectively.

We developed a new version of semidefinite programming algorithm parallel version (SDPARA) 7.6.0-G, which is a parallel implementation on multiple CPUs and GPUs for solving extremely large-scale SDP problems. SDPARA is designed to execute the PDIPM on parallel computers with distributed memory space. The speed-up achieved by SDPARA is essentially attributable to its use of parallel computation to overcome the computational bottlenecks of ELEMENTS and CHOLESKY. Each process reads the input data and keeps all the variables in the process memory space, whereas the SCM data are divided between the processes. We previously reported that SDPARA can compute each row of the SCM in parallel, and applied the parallel Cholesky factorization provided by ScaLAPACK[1] to the SCM [9, 10]. SDPARA is considerably faster than other parallel implementations, such as PCSDP [13] and

[1] http://www.netlib.org/scalapack/.

PDSDP [14] when solving large-scale sparse SDP problems [9–11]. In our previous work [11], we implemented SDPARA 7.5.0-G on TSUBAME 2.0, which is a high-performance GPU-accelerated supercomputer at the Tokyo Institute of Technology. We solved the largest SDP problem (which has over 1.48 million constraints), and created a new world record in 2012. In the same year, our implementation also achieved 533 TFlops in double precision for large-scale Cholesky factorization using 4,080 GPUs.

As mentioned above, SDP has many applications that involve SDP problems with special structures. We initiated the SDPA project,[2] whose objective is to develop high-performance software packages for SDP, and we have solved a large number of SDP problems since 1995; therefore, we can classify the various types of SDP problems into the following three cases:

1. Case 1: SDP problems are sparse and satisfy the property of correlative sparsity [15]; therefore, the SCM tends to become sparse (e.g., the sensor network location problem and the polynomial optimization problem). In this case, CHOLESKY constitutes the bottleneck in the PDIPM. We can perform parallel Cholesky factorization of a sparse SCM by utilizing MUMPS [16] and optimized BLAS libraries [10].
2. Case 2: m is less or not considerably greater than n and the SCM is fully dense (e.g., the quantum chemistry problem and the truss topology problem). In this case, ELEMENTS constitutes the bottleneck in the PDIPM, and therefore, we decrease the time complexity of ELEMENTS from $O\left(mn^3 + m^2n^2\right)$ to $O\left(m^2\right)$ by exploiting the sparsity of the data matrix [9]. ELEMENTS for large-scale SDP problems generally requires significant computational resources in terms of CPU cores and memory bandwidth. In our recent paper [12], we demonstrated that SDPARA can perform efficient parallel computation of ELEMENTS using a large quantity of CPU cores and some processor affinity and memory interleaving techniques in Case 2 (Sect. 4).
3. Case 3: m is considerably greater than n and the SCM is fully dense (e.g., the combinatorial optimization problem and quadratic assignment problem (QAP) [11]). In this case, CHOLESKY constitutes the bottleneck in the PDIPM. We accelerated CHOLESKY by using massively parallel GPUs with a computational performance much higher than that of CPUs. In order to achieve scalable performance with thousands of GPUs, we utilized a high-performance BLAS kernel together with optimization techniques to overlap computation, PCI-Express (PCIe) communication, and MPI communication [11]. In our recent paper [12], we improved the performance of CHOLESKY and verified the numerical results in Case 3 on the TSUBAME 2.5 supercomputer (Sect. 6).

We previously reported that SDPARA can certainly determine whether the SCM of an input SDP problem becomes sparse (Case 1) or not (Cases 2 and 3) [10]. In the recent study [12], we focused primarily on parallel computation of ELEMENTS and CHOLESKY in Cases 2 and 3, respectively. We also demonstrated through

[2] http://sdpa.sourceforge.net/.

numerical experiments on the TSUBAME 2.5 supercomputer that SDPARA is a high-performance general solver for SDPs in various application fields and solved the largest SDP problem (which has over 2.33 million constraints), thus creating a new world record. Our implementation also achieved 1.713 PFlops in double precision for large-scale Cholesky factorization using 2,720 CPUs and 4,080 GPUs.

Our assertion in this paper is that the advanced integration of optimization techniques that span various layers is essential in order to achieve a petascale performance. Such optimization techniques include efficient GPU utilization, hiding communication (Sect. 6), and proper NUMA allocation policy (Sect. 4).

2 Important Applications of SDP

As mentioned in Sect. 1, we can easily identify important applications of SDP in many research areas. Sets of SDP benchmark problems are helpful for the development and evaluation of SDP codes. There exist some standard benchmark test sets of SDP instances: SDPLIB,[3] the DIMACS test sets,[4] and benchmark sets maintained by Hans Mittelmann[5] and Renata Sotirov.[6] In this paper, we have selected two types of large-scale SDP instances associated with recent applications of SDP in truss combinatorial optimization and quantum chemistry; however, SDPARA can also achieve high performance when solving large-scale SDP problems in other important application areas.

2.1 *Quadratic Assignment Problems*

The QAP is a fundamental and important combinatorial optimization problem that involves finding a permutation for given flow and distance matrices, and it is formulated as follows:

$$\begin{array}{ll} \text{minimize} & \text{trace}\left(AXBX^{\mathrm{T}}\right) + \text{trace}(CX) \\ \text{subject to} & XX^{\mathrm{T}} = X^{\mathrm{T}}X = I_n, \end{array} \tag{2}$$

where A and B are $n \times n$ real symmetric matrices, C is an $n \times n$ real matrix, and I_n is an $n \times n$ identity matrix. For finding a lower bound or the exact value of the QAP in (2), we consider the following doubly nonnegative (DNN) relaxation problem [17]:

---

[3] http://euler.nmt.edu/~brian/sdplib/sdplib.html.

[4] http://dimacs.rutgers.edu/Challenges/Seventh/Instances/.

[5] http://plato.asu.edu/ftp/sparse_sdp.html.

[6] https://sites.google.com/site/sotirovr/library/.

$$
\begin{array}{ll}
\text{minimize} & \left\langle \begin{pmatrix} 0 & \mathrm{vec}(C)^{\mathrm{T}} \\ \mathrm{vec}(C) & B\mathrm{T} \otimes A \end{pmatrix}, \begin{pmatrix} y_{00} & y\mathrm{T} \\ y & Y \end{pmatrix} \right\rangle \\
\text{subject to} & Y = \left(Y^{ij}\right)_{1 \le i,j \le n} \in \mathbb{S}^{n^2}_{+}, \\
& Y^{ij} = \left(Y^{ij}_{k\ell}\right)_{1 \le k,\ell \le n} \in \mathbb{S}^{n} \quad (i, j = 1, \dots, n), \\
& y = \left(y^{i}\right)_{1 \le i \le n}, \ y^{i}_{k} = Y^{ii}_{kk} \quad (i, k = 1, \dots, n), \\
& y_{00} = 1, \\
& Y^{ij}_{k\ell} \ge 0 \quad (i, j, k, \ell = 1, \dots, n), \\
& \sum_{i=1}^{n} Y^{ii} = I_n, \\
& \left\langle I_n, Y^{ij} \right\rangle = \delta_{ij} \quad (1 \le i \le j \le n), \\
& \left\langle J_n, Y^{ij} \right\rangle = 1 \quad (1 \le i \le j \le n),
\end{array}
\tag{3}
$$

where δ_{ij} is a Kronecker delta, J_n is an $n \times n$ matrix of all ones, and $\mathrm{vec}(C)$ is the vector formed from the columns of matrix C. We can obtain the optimal value and solution of (3) by applying SDP solvers since (3) can be reformulated into an SDP problem. Before solving (3) with the SDP solvers, we apply a facial reduction algorithm based on the study of Zhao et al. [18] to (3) since (3) is too degenerate to get an accurate value and solution. Indeed, (3) does not have any interior feasible solutions, and thus, it is difficult to obtain an accurate value and solution by using SDP solvers in which PPDIMs are implemented. In addition, we rewrite the resulting DNN problem as a linear matrix inequality described by variables $Y^{ij}_{k\ell}$ to improve the numerical stability of the computation of SDP solvers. For instance, we can write $\left\langle I_n, Y^{ij} \right\rangle = \delta_{ij}$ and $\left\langle J_n, Y^{ij} \right\rangle = 1$, respectively, as follows:

$$
\sum_{k=1}^{n} Y^{ij}_{kk} = \delta_{ij} \text{ and } \sum_{k,\ell=1}^{n} Y^{ij}_{k\ell} = 1.
$$

2.2 *SDP Relaxation of Electronic Structure Problems for Atoms and Molecules*

The ultimate goal in chemistry is to know the exact wave function by solving the Schrödinger equation

$$
H\Psi = E\Psi, \tag{4}
$$

where H is the Hamiltonian of the system, E is the energy, and Ψ is the wave function [19]. We can know the Hamiltonian of the systems quite easily, so if we solve the Schrödinger equation, then we can obtain a lot of information: how superconductivity occurs, how a protein works as an enzyme, how CO_2 is converted to O_2, etc.

However, it is also known that solving the Schrödinger equation is difficult because it involves solving the eigenvalue problem in the form of a partial differential equation of second order. The task is difficult even if we employ localized atomic orbitals to discretize the problem, as it becomes an eigenvalue problem involving a huge matrix. Therefore, we are looking at an alternative method; we are interested in the direct determination of the second-order reduced density matrix

$$\Gamma^{i_1 i_2}_{j_1 j_2} = \frac{1}{2}\left\langle \Psi | a^{\dagger}_{i_1} a^{\dagger}_{i_2} a_{j_1} a^{\dagger}_{j_2} | \Psi \right\rangle.$$

If we restrict ourselves to the ground-state problem, in which the lowest eigenvalue and its eigenvector are found, the problem is further simplified to the minimization of the linear functional:

$$E_{\mathrm{g}} = \min_{\Gamma \in \mathscr{P}} \mathrm{Tr} H \Gamma. \tag{5}$$

This simplification is quite drastic as the number of variables reduces from $O(r!)$ to $O(r^4)$, where r is the number of localized atomic orbitals. The barter for such a drastic simplification is that we have to know about the N-representability condition [20]; otherwise, the calculated energy becomes much lower than the exact one. Therefore, for actual calculations, we just minimize under sufficient conditions like P, Q, and G conditions [20, 21].

Equation (4) can be cast as an SDP problem for the ground state.

$$\begin{cases} \inf \ \langle H, \Gamma^{\mathrm{full}} \rangle \\ \text{s.t. } \langle I, \Gamma^{\mathrm{full}} \rangle = N, \\ \quad 0 \preceq \Gamma^{\mathrm{full}}, \end{cases}$$

where N is the number of electrons in the system and Γ^{full} is the von Neumann or Landau density matrix. Equation (5) can also be cast as an SDP problem [22, 23].

$$\begin{cases} \inf \ \langle H, \Gamma \rangle \\ \text{s.t. } 0 \preceq \Gamma, \text{ etc. (``}N - \text{representabilityconditions}''\text{).} \end{cases}$$

Nakata et al. [22] formulated the problem as a primal SDP problem. They performed a direct variational calculation of the 2-reduced density matrix (2-RDM) by employing the P, Q, and G conditions and by using the well-established SDP solver known as SDPA [4, 5]. This approach was applied to many few-electron atoms and molecules. Nakata et al.'s [22] results obtained by using the P, Q, and G conditions were very encouraging; they obtained around 100–130 % of correlation energy and could produce a dissociation curve of the nitrogen dimer that was in good agreement with full CI results. Zhao et al. included the $T1$ and $T2$ conditions in addition to the P, Q, and G conditions in the above-mentioned approach in their calculations on small molecules [23].

3 Basic Framework of the Primal-Dual Interior-Point Method

In this section, we explain the basic framework of the PDIPM on which SDPA, SDPARA, and other software packages are based. The most important theoretical aspects of the PDIPM is that it solves both primal and dual forms simultaneously and finds their optimal solution in polynomial time.

The Karush–Kuhn–Tucker (KKT) conditions theoretically guarantee that a point (x^*, X^*, Y^*) satisfying the system in (6) below is an optimal solution of (1) when the so-called Slater's condition is satisfied.

$$\text{KKT} \begin{cases} X^* = \sum_{k=1}^{m} F_k x_k^* - F_0, \\ F_k \bullet Y^* = c_k \quad (k = 1, 2, \ldots, m), \\ \sum_{k=1}^{m} c_k x_k^* = F_0 \bullet Y^*, \\ X^* \succeq O, Y^* \succeq O. \end{cases} \tag{6}$$

As we have shown in Fig. 1, in the PDIPM, the algorithm starts from a feasible or infeasible point. In each iteration, it computes the search direction $(\mathrm{d}x, \mathrm{d}X, \mathrm{d}Y)$ from the current point toward the optimal solution, decides the step size, and advances by the step size in the search direction. If the current point reaches a small neighborhood of the optimal solution, the PDIPM terminates the iteration and returns the approximate optimal solution. The PDIPM is described in many papers (see [24–28]). The framework we use relies on the HRVW/KSH/M approach [25–27], and we will use the appropriate norms $||\cdot||$ for matrices and vectors.

3.1 Algorithmic Framework of PDIPM

Step 0: Choose a feasible or infeasible initial point (x^0, X^0, Y^0) such that $X^0 \succ O$ and $Y^0 \succ O$. Set the centering parameter $\beta \in (0, 1)$, the boundary parameter $\gamma \in (0, 1)$, the threshold parameter $\varepsilon > 0$, and the iteration number $s = 0$.

Step 1: Compute the residuals of the primal feasibility P, the dual feasibility d, and the primal-dual gap g:

$$\begin{cases} P = F_0 - \sum\limits_{k=1}^{m} F_k x_k^s + X^s, \\ d_k = c_k - F_k \bullet Y^s \quad (k = 1, \ldots, m), \\ g = \sum\limits_{k=1}^{m} c_k x_k^s - F_0 \bullet Y^s. \end{cases}$$

If $\max\{||P||, ||d||, |g|\} < \varepsilon$ (namely, all residuals above are sufficiently small), stop the iteration and output (x^s, X^s, Y^s) as an approximate optimal solution.

Step 2: Compute the search direction $(\mathrm{d}x, \mathrm{d}X, \mathrm{d}Y)$.

Step 2a: Compute the SCM B by the formula

$$B_{ij} = \left(\left(X^s \right)^{-1} F_i Y^s \right) \bullet F_j. \tag{7}$$

Step 2b: Apply the Cholesky factorization to B and obtain a lower triangular matrix L such that $B = LL^{\mathrm{T}}$.

Step 2c: Obtain a component of the search direction $\mathrm{d}x$ by solving the equations

$$L\,\widetilde{\mathrm{d}x} = r \text{ and } L^{\mathrm{T}}\,\mathrm{d}x = \widetilde{\mathrm{d}x}$$

for the right-hand-side vector r computed as

$$r_k = -d_k + F_k \bullet \left(\left(X^s \right)^{-1} \left(R + P Y^s \right) \right) \quad (k = 1, \ldots, m), \tag{8}$$

where $R = \beta\mu I - X^s Y^s$ with $\mu = X^s \bullet Y^s / n$.

Step 2d: Compute the remaining components of the search direction $(\mathrm{d}X, \mathrm{d}Y)$ as follows:

$$\begin{cases} \mathrm{d}X = -P + \sum\limits_{k=1}^{m} F_k\,\mathrm{d}x_k, \\ \widetilde{\mathrm{d}Y} = (X^s)^{-1}(R - \mathrm{d}X Y^s), \\ \mathrm{d}Y = \left(\widetilde{\mathrm{d}Y} + \widetilde{\mathrm{d}Y}^{\mathrm{T}} \right)/2. \end{cases}$$

Step 3: Maximize the step sizes (lengths) such that the following positive definiteness conditions are satisfied.

$$\begin{cases} \alpha_P = \max\{\alpha \in [0, 1] \colon X^s + \alpha\,\mathrm{d}X \succeq O\}, \\ \alpha_D = \max\{\alpha \in [0, 1] \colon Y^s + \alpha\,\mathrm{d}Y \succeq O\}. \end{cases}$$

Step 4: Update the current point as follows:

$$\left(x^{s+1}, X^{s+1}, Y^{s+1} \right) \leftarrow \left(x^s, X^s, Y^s \right) + \gamma(\alpha_P\,\mathrm{d}x, \alpha_P\,\mathrm{d}X, \alpha_D\,\mathrm{d}Y).$$

Set $s \leftarrow s + 1$, and return to Step 1.

Steps 2a and 2b correspond to the first and second bottleneck parts defined in Sect. 1, respectively. We shall define the computation in Steps 2a and 2b as ELEMENTS and CHOLESKY, respectively. As indicated in [9], ELEMENTS $\left(O\left(mn^3 + m^2n^2\right)\right)$ and CHOLESKY $\left(O\left(m^3\right)\right)$ have often accounted for 80–90 % of the total execution time of the PDIPM. Therefore, researchers have focused on reducing the time taken for these steps [5–8]. We have reduced the time complexity of ELEMENTS from $O\left(mn^3 + m^2n^2\right)$ to $O\left(m^2\right)$ when solving the sparse SDP problem [4].

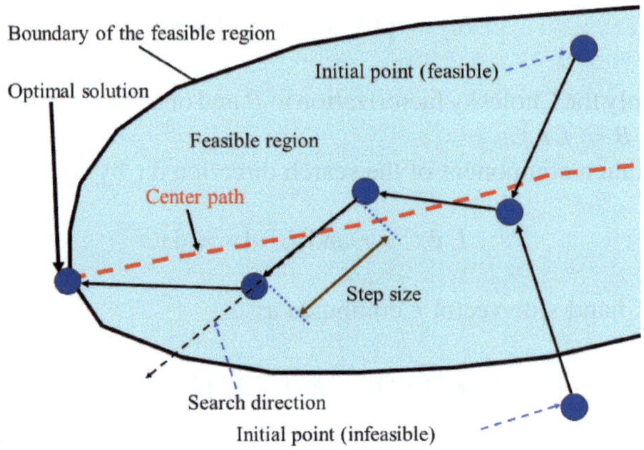

Fig. 1 Primal-dual interior-point method

4 Hybirid Parallel Computing for Computing the Schur Complement Matrix

4.1 Generic Algorithmic Framework for Hybrid Parallel Computing

As mentioned in Sect. 1, ELEMENTS forms a bottleneck in the PDIPM when solving SDP problems in Case 2. In this section, we explain the hybrid (MPI and OpenMP) parallel computation for ELEMENTS. SDPARA was the first parallel software package to incorporate a technique for exploiting the sparsity of ELEMENTS [4, 9]. In general, the elements of the SCM are evaluated by using $B_{ij} = \left((X^s)^{-1} F_i Y^s\right) \bullet F_j$ in Step 2 of the PDIPM. This formula requires two multiplications of two $n \times n$ matrices and one computation of the inner product of two $n \times n$ matrices. However, if we focus only on the nonzero elements of F_i and F_j, we can obtain another expression for the same formula, i.e.,

$$B_{ij} = \sum_{l=1}^{n} \sum_{\gamma=1}^{n} \sum_{\varepsilon=1}^{n} \left(\sum_{\alpha=1}^{n} \sum_{\beta=1}^{n} \left[\left(X^{-1} \right)^{l} \right]_{\gamma\alpha} \left[F_i^l \right]_{\alpha\beta} \left[Y^l \right]_{\beta\varepsilon} \right) \left[F_j^l \right]_{\gamma\varepsilon},$$

where $\left[F_i^l\right]_{\alpha\beta}$ is the (α, β) element of the l-th block of F_i.

Let nz $\left(F_i^l\right)$ be the set of indices of nonzero elements of the l-th block of F_i:

$$\mathrm{nz}(F_i^l) = \left\{ (\alpha, \beta) : \left[F_i^l \right]_{\alpha\beta} \neq 0 \right\}.$$

Futher, let $\#\mathrm{nz}(F_i^l)$ be its cardinality. The cost of B_{ij} based on the latter formula is almost proportional to $(2 \times \#\mathrm{nz}(F_i^l) + 1) \times \#\mathrm{nz}(F_j^l)$. If $\#\mathrm{nz}(F_i^l)$ and $\#\mathrm{nz}(F_j^l)$ are very small and $O(1)$ (namely, $\#\mathrm{nz}(F_i^l)$ and $\#\mathrm{nz}(F_j^l)$ are independent of n), the latter formula is less resource-intensive than the original formula. Depending on the sparsity of the input data matrices, SDPARA can automatically select the best formula from among the following three;

$$\mathscr{F}_1 : B_{ij}^l = \left(\left(X^{-1}\right)^l F_i^l Y^l\right) \bullet F_j^l$$

for the case when both F_i^l and F_j^l are dense,

$$\mathscr{F}_2 : B_{ij}^l = \sum_{(\gamma,\varepsilon)} \left(\sum_{\alpha=1}^{n} \left[\left(X^{-1}\right)^l\right]_{\gamma\alpha} \left[F_i^l Y^l\right]_{\alpha\varepsilon} \right) \left[F_j^l\right]_{\gamma\varepsilon},$$

for the case of dense F_i^l and sparse F_j^l,

$$\mathscr{F}_3 : B_{ij}^l = \sum_{(\gamma,\varepsilon)} \left(\sum_{(\alpha,\beta)} \left[\left(X^{-1}\right)^l\right]_{\gamma\alpha} \left[F_i^l\right]_{\alpha\beta} Y_{\beta\varepsilon}^l \right) \left[F_j^l\right]_{\gamma\varepsilon},$$

for the case when both F_i^l and F_j^l are sparse,

where $(\alpha, \beta) \in \mathrm{nz}(F_i^l)$, $(\gamma, \varepsilon) \in \mathrm{nz}(F_j^l)$, and $B_{ij} = \sum_{l=1}^{h} B_{ij}^l$.

Following the report of our proposed method [4], similar techniques for exploiting the sparsity of ELEMENTS were implemented in other software packages, such as CSDP [6] and SDPT3 [8].

The new version of SDPARA incorporates MPI+OpenMP-based hybrid parallel computing (Figs. 2 and 3). The most important property of $\mathscr{F}_1$, $\mathscr{F}_2$, or $\mathscr{F}_3$ from the viewpoint of parallel computation is that the computations on each column are completely independent of those of the other rows.

It should be noted that B is always symmetric in the PDIPM, and therefore, we have to evaluate only the upper triangular part. Suppose that u processes are available for the parallel computation of ELEMENTS. Then, within the framework of the column-wise distribution shown in Fig. 3, the p-th process evaluates the columns of B that have been assigned to it in a cyclic manner. In particular, the p-th process evaluates the columns in set $\mathscr{R}_p$ defined by

$$\mathscr{R}_p = \{j : (j - p)\%u = 0, \quad 1 \le j \le m\},$$

where $a\%b$ is the remainder of the division of a by b. If each processor on the parallel computer has multiple CPU cores, we can accelerate the computation of the columns in set $\mathscr{R}_p$ using the OpenMP library with the scheduling parameter *dynamic*. SDPARA can attain high scalability using a large quantity of CPU cores and some processor affinity and memory interleaving techniques [12].

set_mempolicy(*interleaving*)
$B = O$
for $l = 1, 2, \cdots, h$ **do**
  for $j \in \{j \mid F_j^l \neq O\}$ **parallel(MPI+thread) do**
    set_affinity(*scatter*)
    for $i \in \{i \mid F_i^l \neq O\}$ **do**
      Selects $\mathscr{F}_1, \mathscr{F}_2$ **or** $\mathscr{F}_3$ **and compute** B_{ij}^l
      $B_{ij} = B_{ij} + B_{ij}^l$
    end
    unset_affinity()
  end
end

Fig. 2 Algorithmic framework of MPI+OpenMP hybrid parallel computation for ELEMENTS

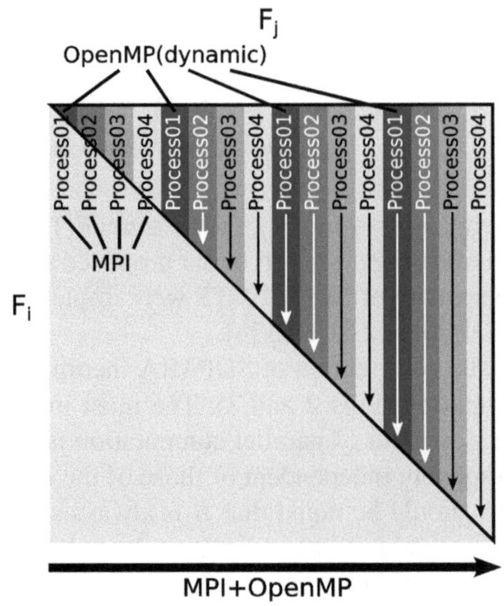

Fig. 3 Column-wise distribution and MPI+OpenMP hybrid parallel computation for ELEMENTS

5 TSUBAME 2.0 and 2.5: GPU-Accelerated Supercomputer

This section briefly describes the TSUBAME 2.0 and 2.5 supercomputers that were used for our evaluation of SDPARA, and the new techniques introduced in this paper. It is to be noted that SDPARA and the proposed techniques were developed for general GPU clusters.

TSUBAME 2.0, installed at Tokyo Institute of Technology in 2010, is a GPU-based heterogeneous supercomputer with a peak performance of 2.40 PFlops (for detailed information, refer to the technical paper [29]). The main part of the system

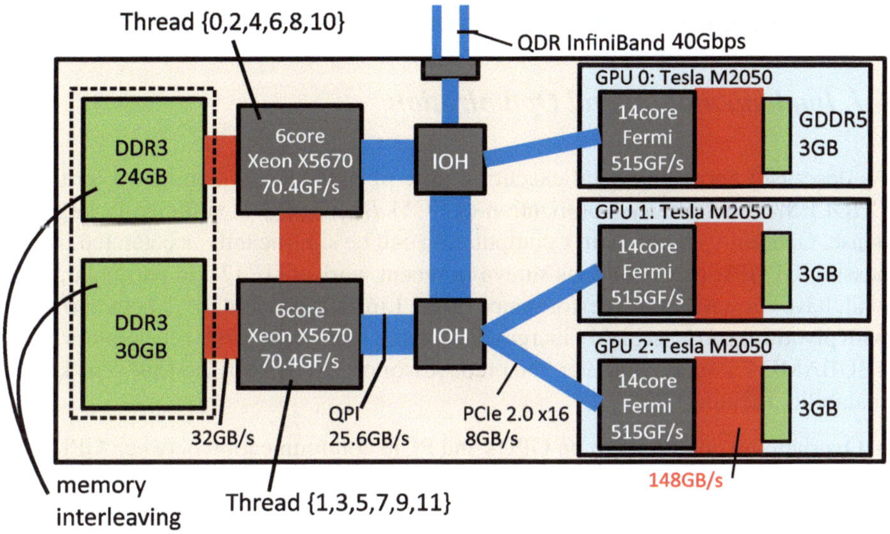

Fig. 4 Structure of each HP SL390s G7 node used in TSUBAME 2.0 and scatter-type affinity and memory interleaving. In TSUBAME 2.5, M2050 GPUs are replaced by K20X GPUs

consists of 1,408 HP Proliant SL390s G7 nodes. With these nodes, the system has 16,896 cores, 4,224 GPU devices, over 74 TB of host memory capacity, and interconnection with 200 Tbps of bisection bandwidth.

Figure 4 shows the structure of each node, which has two Intel Xeon X5670 2.93 GHz (six cores) CPUs, three NVIDIA Tesla M2050 GPUs, 54 GB (partly 96 GB) of DDR3 memory, and 120 GB SSDs as node local storage. Each node is connected to interconnect via two QDR 40 Gbps InfiniBand HCAs. The two CPUs share 54 GB of memory, and their total theoretical peak performance (double precision) is 140.8 GFlops. The CPUs and GPUs are connected via a PCIe 2.0 x16 bus.

As the interconnect, which connects all the compute nodes and the shared storage, we adopt a dual-rail interconnect with a full-bisection fat-tree topology. This design, which has many redundant routes, is adopted so that the cost of global communication is minimized, as compared to that in other topologies, such as the torus.

In September 2013, TSUBAME 2.0 was upgraded to a new version called TSUBAME 2.5, by replacing the M2050 GPUs with new-generation NVIDIA Tesla K20X GPUs. The peak performance (double precision) of each K20X GPU is 1.31 TFlops, which is 2.54 times faster than that of the M2050; the current total peak performance of TSUBAME 2.5 is 5.76 PFlops. No parts other than the GPUs, such as CPUs, host memory, main boards, and networks were changed.

6 Scalable Cholesky Factorization for Thousands of GPUs

6.1 Implementation and Optimization

As described above, the total execution time of SDPARA is dominated by that of CHOLESKY (Cholesky factorization of SCM B) when B is sufficiently large and dense. Generally, dense matrix computation can be significantly accelerated by harnessing GPGPU computing, as shown in recent works [11, 12]. In particular, Endo et al. have shown that the performance of the Linpack benchmark (LU factorization with pivoting) can be scalably increased by using more than 1,200 accelerators on the TSUBAME 1.2 supercomputer, the predecessor of TSUBAME 2.0 [30]. The keys to scalability include

- Overlapping computation on GPUs and PCIe communication between GPUs and CPUs.
- Configuring the *block size* nb so that data reuse is promoted and the amount of PCIe communication is reduced. To determine a better value of nb, we conducted some preliminary experiments. Table 1 lists the `DGEMM` kernel performance on a GPU. On an M2050 GPU, we see that the kernel performance improves with a larger value of nb, and reaches 343.8 GFlops with nb $=$ 1,024. Because we obtain only a slight benefit with an even larger value of nb, we chose to use nb $=$ 1,024 on TSUBAME 2.0. On the recently implemented K20X GPU, we see a three times better performance than on an M2050 GPU. However, the trade-off becomes more severe; nb $=$ 1,024 cannot hide the PCI-e costs sufficiently. Instead, we chose to use nb $=$ 2,048 on TSUBAME 2.5.

While the parallel Cholesky factorization algorithm and Linpack have many common points, the former poses new challenges because of the following differences.

- In each computation step, the part of the matrix to be updated is the lower triangle part, rather than a rectangle. Upon two-dimensional block-cyclic distribution, the shape of the updated part in each process becomes more complex.
- (Related to the above difference) The computation amount per step is halved compared to LU factorization. This makes the computation/communication ratio even worse.
- Cholesky factorization requires additional work called "panel transposition."

Hereafter, we call SDPARA whose CHOLESKY component is accelerated by GPUs "SDPARA (version 7.6.0-G) [12]." Our accelerated CHOLESKY component has properties similar to the `pdpotrf` function of ScaLAPACK. The dense matrix B, with a size of $m \times m$, is distributed among MPI processes in the two-dimensional block-cyclic distribution with block size nb. When we let mb $= \lceil m/\text{nb} \rceil$ be the number of blocks that are aligned in a row or a column, the CHOLESKY algorithm consists of mb steps. A single (k-th) step proceeds as follows:

- *Diagonal block factorization*: The k-th diagonal block is Cholesky-factorized locally. Then, the result block is broadcast to processes in the k-th process column.

Table 1 Performance of GPU DGEMM in GFlops, including PCI-Express communication costs

nb	M2050 in TSUBAME 2.0 (GFlops)	K20X in TSUBAME 2.5 (GFlops)
256	316.0	251.0
512	334.1	490.8
768	340.5	719.6
1,024	343.8 (*)	935.6
1,536	346.6	1,065
2,048	348.2	1,080 (*)
2,560	349.0	1,089
3,072	349.8	1,091

We use two matrices whose sizes are (16,384 × nb) and (nb × 16,384). In our SDPARA execution, we chose the value (nb) marked with (*)

- *Panel factorization*: The k-th block columns are called "panel" L, and the panel is factorized by using the `dtrsm` BLAS kernel.
- *Panel broadcast and transposition*: We need to broadcast L row-wise, obtain the transposition of L, and broadcast L^{t} column-wise.
- *Update*: This is the most computation-intensive part. Each process updates its own part of the rest matrix, taking the corresponding part of L and L^{t}. Let B' be the rest matrix. Then $B' = B' - L \times L^{\mathrm{t}}$ is computed. Thus, the `DGEMM` BLAS kernel dominates the execution time. Note that updating the lower triangular part is sufficient; thus, we can omit the computation of the unused upper part.

The basic approaches we applied to accelerate this algorithm are as follows.

1. We invoke one MPI process per GPU to drive it, and thus, three processes per node are invoked on TSUBAME 2.0 and 2.5 nodes.
2. On GPU clusters, we have to decide where the data structure is located since the GPU device memory is separated from the host memory. In order to accommodate larger sizes of B, we store it on the host memory.

Approaches (1) and (2) indicate that we need to divide the matrices into parts smaller than the device memory capacity and send the input matrices to the GPU via PCIe in order to perform partial computation. Of course, without overlapping computation and communication (as in "Version 1" in Fig. 5), the performance is strictly restricted. To ensure high performance, the following optimization and configuration are adopted.

- When the size of the partial matrix to be updated by a single GPU is $r \times s$, the computation cost in the "update" phase is $O(r \cdot s \cdot \mathrm{nb})$, while the communication cost is $O(r \cdot s + r \cdot \mathrm{nb} + s \cdot \mathrm{nb})$. To reduce the relative communication cost, the block size nb should be sufficiently large. Here, there is a trade-off since very large nb degrades load balancing. After a preliminary evaluation, we set nb to be 2,048.
- In order to reduce and hide the PCIe communication cost, we overlap GPU computation and PCIe communication.

In each computation step, each process uses its assigned GPU to accelerate the DTRSM and DGEMM kernels. In each process, its local part of B, which may be larger than the GPU memory size, is stored on the host memory. Therefore, we need to divide the matrices into small parts and send them to the GPU via PCI-e, and execute the DTRSM/DGEMM kernels. The PCI-e communication and GPU computation are conducted in a pipelined fashion. With these methods ("Version 2" in Fig. 5) we can enjoy the accelerated performance. However, we have noticed that inter-node MPI communication costs still restrict the scalability. Although each node in TSUBAME 2.5 has a wide injection bandwidth (8 GB/s = 40 Gbs), the communication costs increase relatively when the computation is accelerated. Here, we further promote the overlapping policy described above; we overlap all computations, PCIe communication, and MPI communication ("Version 3" in Fig. 5). For this, the "Panel broadcast and transposition" and "Update" phases are reorganized; the transposed panel L^t is divided into pieces before broadcasting, and each process transfers them to the GPU just after partial broadcast is completed.

The implementation ("Version 3" in Fig. 5) were effective and achieved good scalability for up to 4,080 GPUs; however, we observed that there remains a bottleneck in the "panel factorization" phase, which computes L. The computational cost of this phase is asymptotically smaller than that of the update phase; however, because this cost is in the critical path, hiding the cost is important. To achieve this, we improved the overlapping method ("Version 4 in Fig. 5). Here, the panel factorization phase and the following "panel broadcast" are overlapped by dividing the computation of the panel L into small parts. Using all the described techniques, the performance of our CHOLESKY was even improved, as shown below.

As future improvement, we could overlap MPI communication for L and computation by using an optimization technique called "lookahead," which has been introduced in High-Performance Linpack [31].

6.2 Numerical Results

This section demonstrates the performance evaluation results of CHOLESKY for large-scale SDP problems in Case 3 (Table 2). The largest problem is QAP10, where the SCM size is $m = 2{,}339{,}331$. Table 3 shows the system software used in the experimentation.[7]

Figure 6 and Table 2 show the speed of the CHOLESKY component in teraflops. The graph shows that the performance reaches 1.018 PFlops on TSUBAME 2.0 with the QAP10 problem. This result forms a world record in terms of the performance of an SDP solver.

The graph further shows the effects of the above described technique in which panel factorization and broadcast are overlapped. "New" corresponds to the more recent algorithm ("Version 4") in the Fig. 5, whereas "org" denotes the original

[7] "hpl-2.0_FERMI_v15" is distributed at https://nvdeveloper.nvidia.com/ to registered developers.

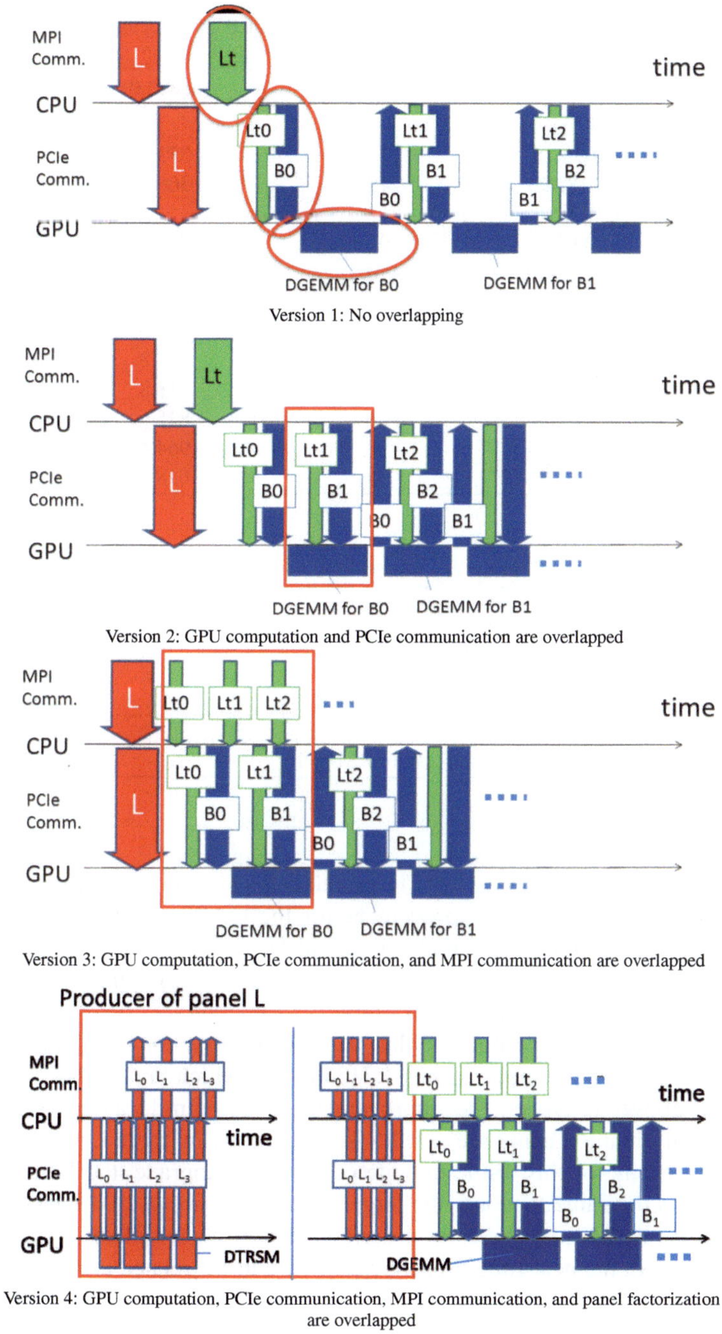

Fig. 5 Several versions of the Cholesky factorization algorithm

Table 2 Performance (teraflops) of GPU CHOLESKY obtained by using up to 1,360 nodes (4,080 GPUs) on TSUBAME 2.0 and 2.5

Name	*m*	org(2.0)	new(2.0)	new(2.5)
(a) 400 nodes (1,200 GPUs)				
QAP6	709,275	223.0	233.0	314.5
QAP7	1,218,400	248.8	306.2	505.8
(b)700 nodes (2,100 GPUs)				
QAP6	709,275	309.5	329.0	387.5
QAP7	1,218,400	440.0	470.0	707.1
QAP8	1,484,406	463.8	512.9	825.1
(c) 1,360 nodes (4,080 GPUs)				
QAP6	709,275	439.6	437.8	508.7
QAP7	1,218,400	695.2	718.8	952.0
QAP8	1,484,406	779.3	825.6	1186.4
QAP9	1,962,225	–	964.4	1526.5
QAP10	2,339,331	–	1018.5	1,713.0

Table 3 System software used in the experimentation

Name	TSUBAME 2.0	TSUBAME 2.5
Compiler	Intel 11.1.072	Intel 11.1.072
MPI	MVAPICH 1.5.1	MVAPICH 1.5.1
BLAS (CPU)	GotoBLAS2 1.13	GotoBLAS2 1.13
CUDA	4.0	5.0
BLAS (GPU)	hpl-2.0_FERMI_v15	CUBLAS 5.0

algorithm ("Version 3") in the Fig. 5. The graph shows that the "new" version achieves up to a 10.5 % performance improvement for the QAP8 problem. Although this improvement seems relatively small, it should be noted that the "org" version is already optimized by applying overlapping techniques of computation, PCI-e communication, and MPI communication.

On TSUBAME 2.5, we observe that the performance for the QAP10 problem becomes 1.7 times better and reaches 1.713 PFlops, which breaks the world record mentioned above. However, the improvement is rather mild as compared to the improvement obtained using the `DGEMM` kernel shown in Table 1. We suppose the following reasons for this. First, in spite of the GPU upgrade, the InfiniBand network remains the same. Second, the larger block size, nb $= 2{,}048$, degrades load balancing among MPI processes as described above. We are planning to propose a new algorithm in the near future that solves the trade-off related to nb.

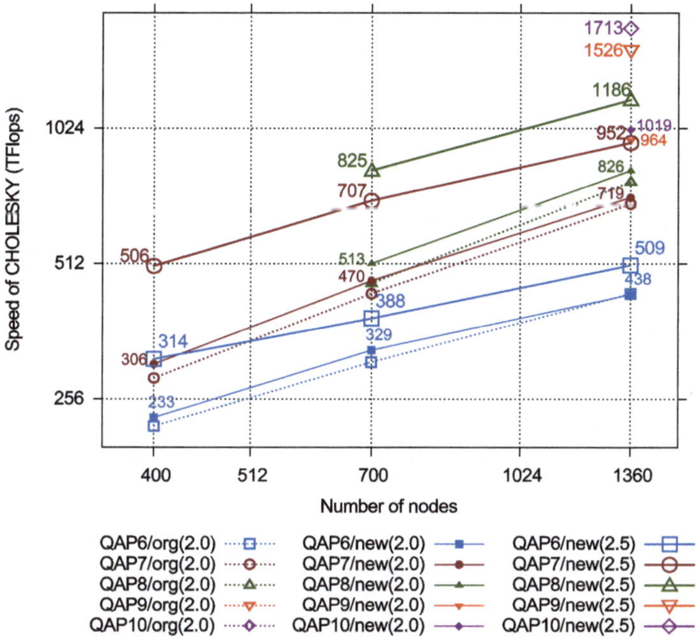

Fig. 6 Performance of GPU CHOLESKY obtained by using up to 1,360 nodes (4,080 GPUs) on TSUBAME 2.0 and 2.5

7 Conclusion

This paper described a high-performance solver, SDPARA 7.6.0-G, for large-scale SDP problems with over 2 million constraints. The key to the high performance of SDPARA is the acceleration of ELEMENTS and CHOLESKY by using thousands of CPUs and GPUs, respectively. SDPARA could attain high scalability using 16,320 CPU cores on the TSUBAME 2.0 supercomputer and some processor affinity and memory interleaving techniques when the generation of the SCM constituted a bottleneck. By using 4,080 NVIDIA Tesla K20X GPUs on the TSUBAME 2.5 supercomputer, our implementation achieved 1.713 PFlops in double precision for a large-scale problem with $m = 2{,}339{,}331$. We showed that SDPARA is a petascale general solver for real problems in Cases 2 and 3. Finally, we demonstrated that solving SDP problems with $m > 2 \times 10^6$ is now possible on modern accelerated supercomputers.

Acknowledgments This research was supported by the Japan Science and Technology Agency (JST), the Core Research of Evolutionary Science and Technology (CREST), and the TSUBAME 2.0 and 2.5 Supercomputer Grand Challenge Program at the Tokyo Institute of Technology.

References

1. H. Wolkowicz, R. Saigal, L. Vandenberghe (eds.), *Handbook of Semidefinite Programming (International Series in Operations Research and Management Science)*, (Kluwer Academic Publishers, Massachusetts, 2000)
2. M.F. Anjos, J.B. Lasserre (eds.), *Handbook on Semidefinite, Conic and Polynomial Optimization (International Series in Operations Research and Management Science)*. (Springer Science+Business Media, Dordrecht, 2011)
3. M. Yamashita, K. Fujisawa, M. Fukuda, K. Kobayashi, K. Nakata, M. Nakata, in *Latest Developments in the SDPA Family for Solving Large-Scale SDPs*, ed. by M.F. Anjos and J.B. Lasserre. Handbook on Semidefinite, Cone and Polynomial Optimization: Theory, Algorithms, Software and Applications. (Springer Press, New York, 2011), pp 687–713 (Chapter 24)
4. K. Fujisawa, M. Kojima, K. Nakata, Exploiting sparsity in primal-dual interior-point methods for semidefinite programming. Math. Prog. **79**, 235–253 (1997)
5. K. Fujisawa, K. Nakata, M. Yamashita, M. Fukuda, SDPA project: solving large-scale semidefinite programs. J. Oper. Res. Soc. Japan **50**, 278–298 (2007)
6. B. Borchers, CSDP, a C library for semidefinite programming. Optim. Meth. Softw. **11**, **12**, 613–623 (1999)
7. J.F. Sturm, SeDuMi 1.02, a MATLAB toolbox for optimization over symmetric cones. Optim. Meth. Softw. **11**, **12**, 625–653 (1999)
8. K.-C. Toh, M.J. Todd, R.H. Tütüncü, SDPT3—MATLAB software package for semidefinite programming, version 1.3. Optim. Meth. Softw. **11**, **12**, 545–581 (1999)
9. M. Yamashita, K. Fujisawa, M. Kojima, SDPARA: semidefinite programming algorithm parallel version. Parallel Comput. **29**, 1053–1067 (2003)
10. M. Yamashita, K. Fujisawa, M. Fukuda, K. Nakata, M. Nakata, Parallel solver for semidefinite programming problem having sparse Schur complement matrix. ACM Trans. Math. Softw. **39**(12), (2012)
11. K. Fujisawa, T. Endo, H. Sato, M. Yamashita, S. Matsuoka, M. Nakata, in *Proceedings of the 2012 ACM/IEEE Conference on Supercomputing, SC'12*. High-Performance General Solver for Extremely Large-Scale Semidefinite Programming Problems (2012)
12. K. Fujisawa, T. Endo, Y. Yasui, H. Sato, N. Matsuzawa, S. Matsuoka, H.Waki, in *The 28th IEEE International Parallel and Distributed Processing Symposium (IPDPS2014)*. Peta-scale General Solver for Semidefinite Programming Problems with over Two Million Constraints (2014)
13. I.D. Ivanov, E. de Klerk, Parallel implementation of a semidefinite programming solver based on CSDP in a distributed memory cluster. Optim. Meth. Softw. **25**(3), 405–420 (2010)
14. S. Benson, Parallel computing on semidefinite programs. (Mathematics and Computer Science Division Argonne Science Laboratory, IL, 2002) (Preprint ANL/MCS-P939-0302)
15. K. Kobayashi, S. Kim, M. Kojima, Correlative sparsity in primal-dual interior-point methods for LP, SDP and SOCP. Appl. Math. Optim. **58**, 69–88 (2008)
16. P.R. Amestoy, I.S. Duff, J.-Y. L'Excellent, Multifrontal parallel distributed symmetric and unsymmetric solvers. Comput. Methods in Appl. Mech. Eng. **184**, 501–520 (2000)
17. S. Burer, On the copositive representation of binary and continuous nonconvex quadratic programs. Math. Prog. **120**, 479–495 (2009)
18. Q. Zhao, S.E. Karisch, F. Rendl, H. Wolkowicz, Semidefinite programming relaxations for the quadratic assignment problem. J. Comb. Optim. **2**, 71–109 (1998)
19. P.A.M. Dirac, The quantum theory of the electron. Roy. Soc. A **123**, 714 (1929) (London)
20. A.J. Coleman, Structure of fermion density matrices. Rev. Mod. Phys. **35**, 668–687 (1963)
21. C. Garrod, J.K. Percus, Reduction of the N-particle variational problem. J. Math. Phys. **5**, 1756–1776 (1964)
22. M. Nakata, H. Nakatsuji, M. Ehara, M. Fukuda, K. Nakata, K. Fujisawa, Variational calculations of fermion second-order reduced density matrices by semidefinite programming algorithm. J. Chem. Phys. **114**, 8282–8292 (2001)

23. Z. Zhao, B.J. Braams, M. Fukuda, M.L. Overton, J.K. Percus, The reduced density matrix method for electronic structure calculations and the role of three-index representability conditions. J. Chem. Phys. **120**, 2095–2104 (2004)
24. F. Alizadeh, J.-P.A. Haeberly, M.L. Overton, Primal-dual interior-point methods for semidefinite programming: convergence rates, stability and numerical results. SIAM J. Optim. **8**, 746–768 (1998)
25. C. Helmberg, F. Rendl, R.J. Vanderbei, H. Wolkowicz, An interior-point method for semidefinite programming, SIAM J. Optim. **6**, 342–361 (1996)
26. M. Kojima, S. Shindoh, S. Hara, Interior-point methods for the monotone semidefinite linear complementarity problems. SIAM J. Optim. **7**, 86–125 (1997)
27. R.D.C Monteiro, Primal-dual path following algorithms for semidefinite programming. SIAM J. Optim. **7**, 663–678 (1997)
28. Y.E. Nesterov, M.J. Todd, Primal-dual interior-point methods for self-scaled cones. SIAM J. Optim **8**, 324–364 (1998)
29. S. Matsuoka, T. Endo, N. Maruyama, H. Sato, S. Takizawa: The Total Picture of TSUBAME2.0. TSUBAME e-Science Journal, GSIC, Tokyo Institute of Technology, Vol. 1, 2–4 (2010).
30. T. Endo, A. Nukada, S. Matsuoka, N. Maruyama, in *Proceedings of IEEE IPDPS10*. Linpack Evaluation on a Supercomputer with Heterogeneous Accelerators, pp. 1–8 (2010)
31. A. Petitet, R.C. Whaley, J. Dongarra, A. Cleary, HPL—A Portable Implementation of the High-Performance Linpack Benchmark for Distributed-Memory Computers, http://www.netlib.org/benchmark/hpl/

Part VI
Application of Mathematics

Modeling of Head-Disk Interface for Magnetic Recording

Kanzo Okada

Abstract All the existing models of thin film gas lubrication developed for designing head sliders of hard disk drives are chronologically reviewed so as to show how each model was improved and finally generalized to treat gas lubrication flows for arbitrary Knudsen numbers. Each model is compared with the others using specific examples for benchmarking purposes. A possible approach to the modeling of head-disk interface is also proposed for further consideration that has the potential of addressing one of the extreme operations in which the reader/writer element of the head slider surfs through the lubricant on the disk.

Keywords Magnetic recording · Hard disk drives · Head-disk interface · Thin film gas lubrication · Flying height · Head slider · Knudsen numbers · Modeling

1 Introduction

The hard disk drive is an integral and essential component of the modern computer system. The head-disk interface technology is one of the key technologies of hard disk drives and has been regarded as most instrumental in driving the ever-increasing recording density trend. The frontiers of the head-disk interface technology are already well into the regime of atomic and molecular phenomena.

As the physics of magnetic recording dictates, the head-to-disk separation should be as little as possible to provide a sufficiently large signal for reading/writing a small magnetic bit. Most of the head sliders commercialized 20 years ago broke the flying height barrier of 100 nm and today it goes below 10 nm at the laboratory level, an

K. Okada (✉)
Institute of Mathematics for Industry, Kyushu University, 744 Motooka, Nishi-ku, Fukuoka 819-0395, Japan
e-mail: okada@imi.kyushu-u.ac.jp

R. Nishii et al. (eds.), *A Mathematical Approach to Research Problems of Science and Technology*, Mathematics for Industry 5,
DOI: 10.1007/978-4-431-55060-0_31, © Springer Japan 2014

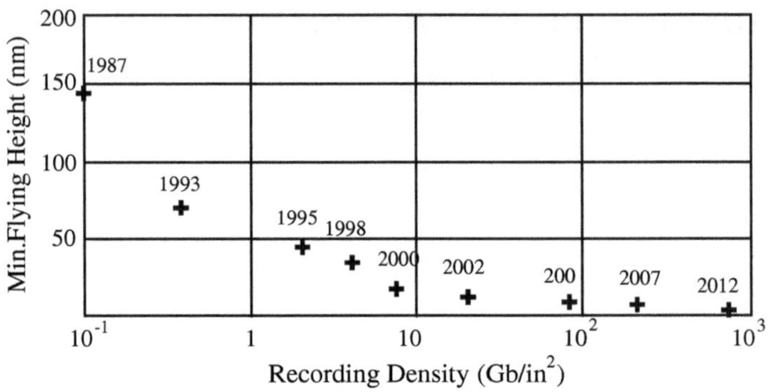

Fig. 1 Recording density versus minimum flying height

order of magnitude lower [1]. Quite amazingly, 10 nm flying height of the standard 0.85 mm long head slider is comparable, by analogy, to flying the latest Boeing 787 aircraft steadily at an altitude of less than 1 mm. As shown in Fig. 1, due to the rapidly decreasing flying height in the seemingly never-ending pursuit for higher recording densities, the air-bearing film thickness became comparable to and soon surpassed the mean free path of air molecules (~64 nm under the condition of 1 atmospheric pressure and 0 °C), inducing the evolution of flying head slider design methodologies as we know it today.

The ratio of the mean free path of air molecules λ to the film thickness h is a measure of the degree of molecular rarefaction and known as the Knudsen number $K_n = \lambda/h$. The Knudsen number generally classifies gas flows into the following four types: (1) $K_n \to 0$ for continuum flow, (2) $K_n \ll 1$ for slip flow, (3) $K_n \sim 1$ for transition flow, and (4) $K_n \gg 1$ for free molecular flow. As such, modeling gas lubrication flows involves scale effects.

Our plan is to chronologically review all the existing models of thin film gas lubrication developed for designing head sliders of hard disk drives so as to show how each model was improved and finally generalized to treat gas lubrication flows for arbitrary Knudsen numbers. In Sect. 2 we provide the readers with a basic derivation of each model, covering from the classical Reynolds equation to the generalized Reynolds equation of Fukui and Kaneko [2, 3] derived from the Boltzmann equation of molecular gas dynamics, with the modified Reynolds equation of Burgdorfer [4] and the higher order approximation model due to Hsia and Domoto [5] reviewed in the transition. At the end of Sect. 2, we compare one model against another by using specific examples for benchmarking purposes. In Sect. 3, in consideration for the stringent magnetic spacing requirement of 2–3 nm for the next-generation ultrahigh density hard disk drives, we conclude this article by making a brief comment, for further consideration, on a possible approach to the modeling of head-disk interface, which has the potential of addressing one of the extreme operating conditions where the reader/writer element of the head slider surfs through the lubricant on the disk [1].

2 Thin Film Gas Lubrication Equations

In Sect. 2.1, we introduce the classical Reynolds equation based on the continuum theory of viscous fluids since it is the foundation of hydrodynamic lubrication theory. In Sect. 2.2 the modified Reynolds equation and its higher order version are derived by adopting the first order slip flow condition and extending it up to the second order slip flow condition based on the kinetic theory of gases for improved approximation. In Sect. 2.3 the generalized Reynolds equation valid for arbitrary Knudsen numbers is derived from the linearized Boltzmann equation. Section 2.4 is provided to benchmark all the four models reviewed in this article.

2.1 Reynolds Equation

We consider a general curved solid surface (head slider) floating in a steady state over a planar solid surface (disk) moving in the direction parallel to the xy-plane as shown in Fig. 2. For convenience, we let the planar surface be in the xy-plane of the Cartesian coordinate system. We also let the local separation between the surfaces be a prescribed function $z = h(x, y)$. The solid surfaces are assumed to be sufficiently rigid so that any deformations of the surfaces due to hydrodynamic pressure and surface forces are negligible. The fluid flow at low Reynolds number in the region between the sliding surfaces can be described by what is known as the Reynolds lubrication approximation if the characteristic length l of the head slider in relative motion is large compared to the minimum flying height h_0, that is, under the condition of $h_0/l \ll 1$. In this approximation, it is assumed that locally the flow is similar to that between the parallel plates, namely, the lateral components of the flow velocity field are large and derivatives in the direction normal to the disk are dominant. By applying this approximation to the compressible Navier-Stokes equations for a viscous fluid of shear viscosity μ, we obtain the following steady state problem:

$$\mu \frac{\partial^2 \boldsymbol{u}}{\partial z^2} = \nabla p \tag{1}$$

Here $\boldsymbol{u} = (u, v)$ is the flow velocity field, $\nabla = (\partial/\partial x, \partial/\partial y)$, and the pressure p is only a function of x and y.

Throughout this article, without loss of generality, we assume an infinite disk along the y-direction. The disk is forced to move with a constant velocity U along the x-direction. Under these conditions, the problem can be treated as one in two-dimensions. Using the non-slip flow conditions:

$$u = U \quad \text{at } z = 0 \tag{2}$$

$$u = 0 \quad \text{at } z = h \tag{3}$$

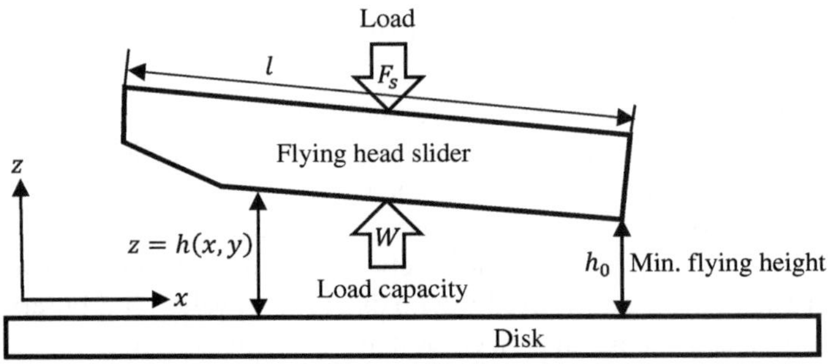

Fig. 2 A schematic description of head-disk interface

Equation (1) can be integrated with respect to z to give

$$u(x, z) = -\frac{1}{2\mu}\frac{dp}{dx}\left(hz - z^2\right) + U\left(1 - \frac{z}{h}\right) \tag{4}$$

It is to be noted from the two terms on the right-hand side of Eq. (4) that thin film gas lubrication flows are composed of the two fundamental parallel plate flows, namely, the first term corresponds to the pressure flow (Poiseuille flow) due to the pressure gradient, while the second term to the shear flow (Couette flow) due to the disk motion. The equation of continuity in differential form can be replaced by the condition that the volume of flow in every section must be constant, namely,

$$\frac{\mathrm{d}}{\mathrm{d}x}\left(\rho\int_0^h u\mathrm{d}z\right) = 0 \tag{5}$$

Substituting Eq. (4) into Eq. (5) and integrating it with respect to z under the nonslip flow conditions (2) and (3), we obtain the Reynolds equation in the following form:

$$\frac{\mathrm{d}}{\mathrm{d}x}\left(\rho h^3\frac{dp}{dx}\right) = 6\mu U\frac{\mathrm{d}}{\mathrm{d}x}(\rho h) \tag{6}$$

If the temperature distributions on the boundary surfaces are uniform and approximately the same, it can be proven that the flow field at the head-disk interface is isothermal. Under the isothermal condition, the pressure can be taken to be proportional to the density:

$$p \propto \rho \tag{7}$$

For compressible fluids such as air, replacing ρ in Eq. (6) by p through the relation (7), the Reynolds equation can be expressed entirely in terms of the pressure distribution $p(x)$ with an arbitrary height function $h(x)$ given.

$$\frac{d}{dx}\left(ph^3\frac{dp}{dx}\right) = 6\mu U\frac{d}{dx}(ph) \tag{8}$$

The pressure at the perimeter of the air-bearing surface is equal to the ambient pressure p_a. Projecting the perimeter on the domain of definition ($0 \leq x \leq l$), we can set the pressure boundary condition by

$$p_{\mathrm{bdry}} = p_a \tag{9}$$

Thus, the Reynolds equation of lubrication can be put in the final forms of Eqs. (8) and (9).

For ease of comparison in what follows, we write both the Reynolds equation and the boundary condition in nondimensional forms by dividing the position coordinate x, the pressure p, and the height function h by the head slider length l, the ambient pressure p_a, and the minimum flying height h_0, respectively.

$$\frac{d}{dX}\left[PH^3\frac{dP}{dX}\right] = \Lambda\frac{d}{dX}(PH) \tag{10}$$

$$P_{\mathrm{bdry}} = 1 \tag{11}$$

Here, the dimensionless quantity $\Lambda(=6\mu Ul/p_a h_0^2)$ is known as the bearing number and gives the ratio of the mass flow rate of the Couette flow to that of the Poiseuille flow. A brief discussion will be made later in Sect. 2.4 on how the bearing number characterizes the lubrication flow.

2.2 Modified Reynolds Equations

As already remarked in Sect. 1, the film thickness became no longer negligible in comparison to the mean free path of air molecules as the former decreased rapidly in response to the demand for higher and higher recording densities. Clearly the Reynolds equation was not adequate to model such thin film gas lubrication flows. It was Burgdorfer [4] who first introduced from the kinetic theory of gases an ad hoc slip boundary condition (by expanding the velocity in a Taylor series about the wall position and retaining the zeroth and first order terms as shown in Eqs. (12) and (13)) to account for molecular effects, although still limited to the continuum flow region of $K_n \ll 1$.

$$u = U + a\lambda\frac{\partial u}{\partial z} - \frac{(a\lambda)^2}{2}\frac{\partial^2 u}{\partial z^2} + \ldots \quad \text{at} \quad z = 0 \tag{12}$$

$$u = -a\lambda\frac{\partial u}{\partial z} - \frac{(a\lambda)^2}{2}\frac{\partial^2 u}{\partial z^2} - \ldots \quad \text{at} \quad z = h \tag{13}$$

Here, a is defined by $a = (2-\alpha)/\alpha$ where α is called the accommodation coefficient of the boundary surface and $0 < \alpha < 1$. There are mainly three modes of gas particle reflection at the boundary: (1) specular reflection ($\alpha = 0$), (2) diffuse reflection ($\alpha = 1$), and (3) Maxwell-type reflection (a linear interpolation between specular and diffuse reflection modes). In this article, we set $\alpha = 1$ for simplicity.

In the same way as the Reynolds equation was derived in Sect. 2.1, Eq. (1) is solved for the flow velocity u by using the first order slip flow conditions (12) and (13) in place of the nonslip flow conditions (2) and (3) and it is given by

$$u(x, z) = -\frac{1}{2\mu}\frac{dp}{dx}\left(\lambda h + hz - z^2\right) + U\left(1 - \frac{z+\lambda}{h+2\lambda}\right) \tag{14}$$

Following exactly the same procedure as employed in Sect. 2.1, we can finally write down the nondimensional form of the modified Reynolds equation of Burgdorfer as in Eq. (15). Note that there is an additional term $6K_{n0}/PH$ on the left-hand side of the equation, which is associated with the mass flow rate of the Poiseuille flow and accounts for the first order effect due to the mean free path λ of gas molecules.

$$\frac{d}{dX}\left[PH^3\left(1 + \frac{6K_{n0}}{PH}\right)\frac{dP}{dX}\right] = \Lambda\frac{d}{dX}(PH) \tag{15}$$

Here $K_{n0}(= \lambda/h_0)$ is the Knudsen number evaluated at the minimum flying height h_0.

By retaining terms up to the second derivative of u with respect to z in the velocity boundary conditions (12) and (13), Hsia and Domoto [5] proposed a higher order approximation model to the Reynolds equation and it reads as follows:

$$\frac{d}{dX}\left[PH^3\left(1 + \frac{6K_{n0}}{PH} + \frac{6K_{n0}^2}{P^2H^2}\right)\frac{dP}{dX}\right] = \Lambda\frac{d}{dX}(PH) \tag{16}$$

We will take a look, later in Sect. 2.4, at the problem of applicability of the modified Reynolds equation (15) and its higher order version (16) to those Knudsen numbers on the order of $K_n \sim 1$.

2.3 Generalized Reynolds Equation

Around 1987, when Fukui and Kaneko [2] first reported the derivation of a generalized Reynolds equation valid for arbitrary Knudsen numbers, the minimum flying heights of head sliders used then were around 150 nm (see Fig. 1). We had been knocking on the door of $K_n \sim 1$, and therefore, a more advanced model other than the slip flow Reynolds equations was in order. This provided the researchers with impetus to extend or generalize the slip flow Reynolds equations, which are corrections to the classical Reynolds equation to one degree or another, so that it becomes valid for arbitrary Knudsen numbers. As the Boltzmann equation of molecular gas dynamics describes the behavior of gas molecules for arbitrary Knudsen numbers, and therefore, has a definite advantage over the then existing gas lubrication models [6], Fukui and Kaneko took a good look at the linearized Boltzmann equation to derive an ultrathin film gas lubrication model valid for arbitrary Knudsen numbers exactly in the same form as the Reynolds equation. Given below is an essential part of their derivation of the generalized Reynolds equation [2, 3].

Essential to molecular gas dynamics is the probabilistic and statistical description of the behavior of gas molecules by means of a molecular velocity distribution function $f(\boldsymbol{x}, \boldsymbol{\xi}, \boldsymbol{t})$ where $\boldsymbol{x}$ is the position vector, $\boldsymbol{\xi}$ is the molecular velocity vector, and $\boldsymbol{t}$ is the time. It gives the probability of finding a gas molecule at a given phase coordinates of $\boldsymbol{x}$ and $\boldsymbol{\xi}$ at time $\boldsymbol{t}$ and is the solution of the celebrated integro-differential equation known as the Boltzmann equation:

$$\frac{\partial f}{\partial t} + \boldsymbol{\xi} \cdot \nabla f = Q(f, f) \tag{17}$$

Here, $Q(f, f)$ is an integral operator quadratic in f, representing the gains and losses through intermolecular collisions which cause changes of f in space and time through Eq. (17). In general, the collision operator $Q(f, f)$ is highly complex and difficult to treat.

However, we can approximate it by using the well-known BGK model proposed by Bhatnagar, Gross and Krook [7] without losing the essence of the problem under consideration. The BGK model is most frequently used for practical problems and takes the form

$$\frac{\partial f}{\partial t} + \boldsymbol{\xi} \cdot \nabla f = \nu (f_e - f) \tag{18}$$

Here, ν is the mean collision frequency and f_e is the Maxwellian distribution function, which is one of the exact solutions of the Boltzmann equation, representing a uniform and steady state of a gas, namely an equilibrium state independent of space and time.

$$f_e(\boldsymbol{\xi}) = \frac{\rho}{(2\pi RT)^{3/2}} \exp\left\{ -\frac{|\xi - u|^2}{2RT} \right\} \tag{19}$$

Since it can safely be assumed for the actual head-disk interface that the magnitude of the flow velocity field is appreciably small compared to the molecular counterpart, namely, the Mach number is small compared to unity and that the flow field in the region of lubrication is maintained in an isothermal state, we can regard the deviation of the velocity distribution function f from the velocity distribution function f_0 of the static equilibrium state as small. This provides us with the reasonable ground for using the BGK model equation linearized about f_0 as the fundamental equation of ultrathin film gas lubrication. The nondimensional form of the linearized BGK model equation is given by

$$\varepsilon \zeta_X \frac{\partial \phi}{\partial X} + \zeta_Z \frac{\partial \phi}{\partial Z} = \frac{1}{k_0} (-\phi + \omega + 2\zeta_X V_X) \tag{20}$$

The nondimensional variables used above are defined as follows:

$$X = \frac{x}{l}, \quad Z = \frac{z}{h_0}, \quad \varepsilon = \frac{h_0}{l}$$
$$\zeta_X = \frac{\xi_x}{\sqrt{2RT}}, \quad \zeta_Z = \frac{\xi_z}{\sqrt{2RT}},$$
$$V_X = \frac{u}{\sqrt{2RT}} = \iiint_{-\infty}^{\infty} (\zeta_X \phi E)\, \mathrm{d}\zeta \tag{21}$$
$$\phi = f f^{-1}|_{u=0} - 1$$
$$\omega = \frac{\rho}{\rho_0} - 1 = \iiint_{-\infty}^{\infty} (\phi E)\, \mathrm{d}\zeta$$

Here, X and Z are the nondimensional coordinates of x and z, ζ_X and ζ_Z are the nondimensional molecular velocity components of ξ_x and ξ_z, V_X is the flow velocity in the X-direction, and h_0 is the minimum flying height. Those quantities with a subscript 0 refer to the static equilibrium state. ϕ and ω are the non-dimensional perturbations of the velocity distribution function f and density ρ, respectively. R and T are gas constant and reference temperature. E and k_0 are given by

$$E = \pi^{-3/2} \exp\left(-\zeta^2\right) \tag{22}$$
$$k_0 = \frac{\sqrt{\pi}}{2}\left(\frac{\lambda}{h_0}\right) = \frac{\sqrt{\pi}}{2} K_{n0} \tag{23}$$

Under the isothermal condition and the assumption of diffuse reflection mode($\alpha = 1$), the boundary conditions of the governing equation (20) are simplified to yield

$$\phi = 2V_W \quad \text{at} \quad Z = 0, \zeta_Z > 0 \tag{24}$$
$$\phi = 0 \quad \text{at} \quad Z = H, \zeta_Z < 0 \tag{25}$$

Here, V_W is the nondimensional disk speed $\left(= U/\sqrt{2RT}\right)$.

Next, we consider the following similarity solution to the fundamental equation (20).

$$\phi = \left(\frac{X}{\varepsilon}\right)\phi_0\left(\zeta^2\right) + \zeta_X \phi_1\left(Z, \zeta_Z, \zeta^2\right) \tag{26}$$

Substituting it into Eq. (20) and arranging the resultant in terms of X and ζ_X, we get the following relations for ϕ_0 and ϕ_1:

$$\phi_0 = \beta \tag{27}$$

$$\zeta_Z \frac{\partial \phi_1}{\partial Z} = \frac{1}{k_0} \left(-\phi_1 + 2V_X\right) - \beta \tag{28}$$

Here, β is the nondimensional pressure gradient $(= \varepsilon \, \mathrm{d}P/\mathrm{d}X)$.

Solving Eq. (28) with the boundary conditions (24) and (25), we obtain the following analytical expressions for ϕ_1 with V_X yet to be determined.

$$\begin{aligned} \phi_1 &= 2V_W \exp\left(-\frac{Z}{k_0\zeta_Z}\right) \\ &+ \frac{1}{\zeta_Z} \exp\left(-\frac{Z}{k_0\zeta_Z}\right) \int_0^Z \left(\frac{2V_X(Z')}{k_0} - \beta\right) \exp\left(\frac{Z'}{k_0\zeta_Z}\right) \mathrm{d}Z' \quad \text{for} \quad \zeta_Z > 0 \end{aligned} \tag{29}$$

$$\phi_1 = \frac{1}{\zeta_Z} \exp\left(-\frac{Z}{k_0\zeta_Z}\right) \int_H^Z \left(\frac{2V_X(Z')}{k_0} - \beta\right) \exp\left(\frac{Z'}{k_0\zeta_Z}\right) \mathrm{d}Z' \quad \text{for} \quad \zeta_Z < 0 \tag{30}$$

Substituting Eqs. (29) and (30) into Eq. $(21)_3$ with ϕ replaced by the second term on the right-hand side of Eq. (26) while taking an appropriate care of the sign of ζ_Z, we get the integral equation for V_X.

$$V_X(Z) = \frac{1}{\sqrt{\pi}} \left\{ V_W T_0 \left(\frac{Z}{k_0}\right) + \frac{1}{k_0} \int_0^H T_{-1} \left(\frac{|Z - Z'|}{k_0}\right) \left[V_X(Z') - \frac{k_0\beta}{2}\right] \mathrm{d}Z' \right\} \tag{31}$$

Here, $T_n(x)$ is the Abramowitz function [8] defined by

$$T_n(x) = \int_0^\infty t^n \exp\left(-t^2 - \frac{x}{t}\right) \mathrm{d}t \tag{32}$$

Fukui and Kaneko wrote $V_X(Z)$ as a linear sum of those two terms in the following equation for the reason to become clear shortly.

$$V_X(Z) = \frac{k_0\beta}{2} \left(1 - \psi_p\right) + V_W \psi_c \tag{33}$$

Putting Eq. (33) into Eq. (31) and arranging the resultant in terms of $k_0\beta/2$ and V_W, we obtain the following integral equations for ψ_p and ψ_c.

$$\psi_p(Z) = 1 + \frac{1}{\sqrt{\pi} k_0} \int_0^H T_{-1} \left(\frac{|Z - Z'|}{k_0}\right) \psi_p(Z') \, \mathrm{d}Z' \tag{34}$$

$$\psi_c(Z) = \frac{1}{\sqrt{\pi}}\left[T_0\left(\frac{Z}{k_0}\right) + \frac{1}{k_0}\int_0^H T_{-1}\left(\frac{|Z - Z'|}{k_0}\right)\psi_c(Z')\,\mathrm{d}Z'\right] \quad (35)$$

It was already known then that the solutions ψ_p and ψ_c of Eqs. (34) and (35) give the velocity profiles V_{Xp} and V_{Xc} of the Posiseuille and Couette flows through the first and second terms on the right-hand side of Eq. (33) [9, 10]. This is the reason why Fukui and Kaneko adeptly expressed $V_X(Z)$ as in Eq. (33) so that their lubrication equation could conveniently be put in the same form as all the other prior lubrication equations.

$$V_X = V_{Xp} + V_{Xc} \quad (36)$$

where

$$V_{Xp} = \frac{k_0\beta}{2}(1 - \psi_p) \quad (37)$$

$$V_{Xc} = V_W\psi_c \quad (38)$$

Accordingly, the total mass flow rate q can be written as a linear sum of the corresponding mass flow rates q_p and q_c.

$$q = q_p + q_c \quad (39)$$

Since the velocity profile of the Couette flow is symmetric, it is clear that the Couette flow rate q_c is always given by the same quantity $\rho Uh/2$ as for the continuum case, regardless of the Knudsen number. According to Eq. (39), the part of the total mass flow rate q that is dependent on the Knudsen number must come from the Poiseuille flow. Therefore, the final form of the generalized Reynolds equation can take the same form as all the other prior lubrication equations, the only difference being in the expression of the Poiseuille flow rate.

In order to obtain the non-dimensional form of the generalized Reynolds equation, we make the Poiseuille flow rate q_p non-dimensional by

$$Q_p(D) = -\frac{q_p}{h^2\left(\frac{\mathrm{d}p}{\mathrm{d}x}\right)/\sqrt{2RT}} \quad (40)$$

Here, D is the inverse Knudsen number and defined by

$$D = \frac{\sqrt{\pi}}{2K_n} \quad (41)$$

Normalizing $Q_p(D)$ by the continuum Poiseuille flow rate $Q_{con}(D)$, we can finally express the nondimensional form of the generalized Reynolds equation as

$$\frac{d}{dX}\left[PH^3\left(\frac{Q_p(D)}{Q_{con}(D)}\right)\frac{dP}{dX}\right] = \Lambda\frac{d}{dX}(PH) \tag{42}$$

Looking at Eq. (42), we realize that the history of generalizing the Reynolds equation was indeed the history of generalizing the Poiseuille flow rate $Q_p(D)$. In fact, the Poiseuille flow rates $Q_p(D)$ for Reynolds' original theory of hydrodynamic lubrication, Burgdorfer's modified Reynolds equation, and its higher order approximation model are given, respectively, by $D/6$, $D/6\left(1+3\sqrt{\pi}/D\right)$ and $D/6\left(1+3\sqrt{\pi}/D+3\pi/2D^2\right)$.

We end this section by providing the readers with a brief sketch of the numerical procedure of solving the generalized Reynolds equation. The Poiseuille flow rate can be obtained by first solving Eq. (34) for the velocity profile ψ_p numerically and then by integrating the result along every section in the separation gap according to the formula of mass flow rate by way of Eq. (37). Since the generalized Reynolds equation differs from all the prior lubrication equations only in the Poiseuille flow rate, one can easily modify any of the available simulation codes for the head-disk interface by computing Q_p with the Poiseuille flow rate database available in the literature [11]. In the same paper [11], there is also proposed an effective interpolation scheme to speed up the process of computing Q_p using the database so that one can solve the generalized Reynolds equation as fast as the slip flow Reynolds equations.

2.4 Benchmark Results

Since it boils down to the comparison of mass flow rates to distinguish all the four models reviewed in Sects. 2.1 through 2.3, the mass flow rate Q_p of each model was evaluated as a function of the inverse Knudsen number D [2]. Figure 3 reveals that the first order slip model underestimates the actual mass flow rate, whereas the second order slip approximation overestimates it. Note that the solution obtained from the generalized Reynolds equation describes its asymptotic behavior in the region of $D \gg 1$.

As mentioned in Sect. 2.1, the bearing number Λ is the ratio of Couette flow rate to Poiseuille flow rate. As it increases, the Couette flow becomes dominant over the Poiseuille flow, diminishing the effect of the Knudsen number. This is why there is no significant difference in mass flow rate among the models for a certain range of Knudsen numbers. Figure 4, which shows the relation between the bearing number Λ and the load capacity W, verifies that as the bearing number increases with the inverse Knudsen number held at a fixed value, all the load capacities asymptotically approach a constant value. However, it needs to be stressed that, under those experimental conditions in which the Poiseuille flow becomes dominant, the numerical solutions of the generalized Reynolds equation are demonstrated to be in good agreement with the experimental results [12].

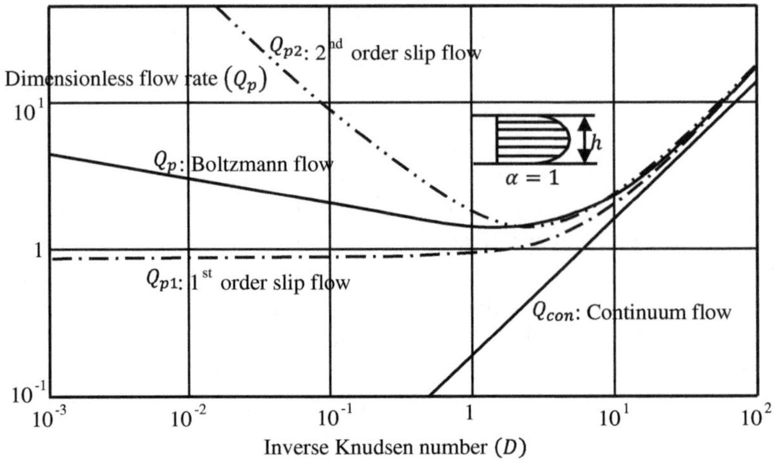

Fig. 3 Mass flow rates of plane Poiseuille flow (*Source* [2])

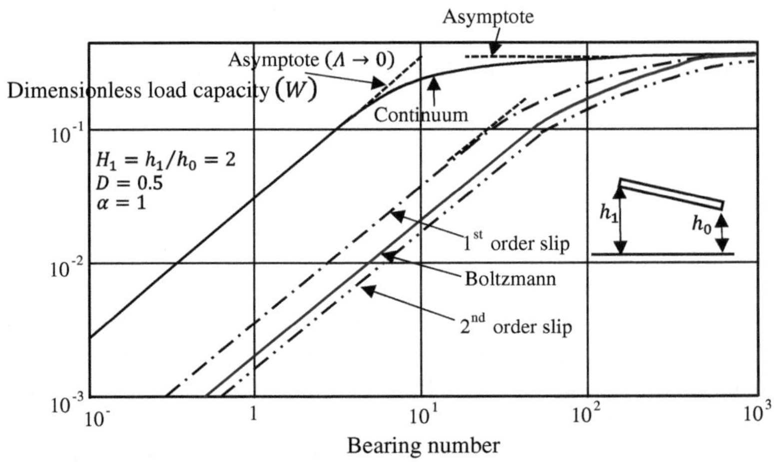

Fig. 4 Load carrying capacity versus bearing number (*Source* [2])

3 On a New Head-Disk Interface Design

Apparently even the generalized Reynolds equation is limited in its application to the future head-disk interface where the magnetic spacing is projected to be around 2–3 nm. It should be pointed out that the magnetic spacing referred to earlier is no longer a physical clearance, but should be understood in the sense that the reader/writer element can sense it only magnetically. In fact, a new concept called the "lube-surfing head-disk interface scheme" was proposed a few years ago as part of the efforts to develop hard disk drives capable of storing data at the recording

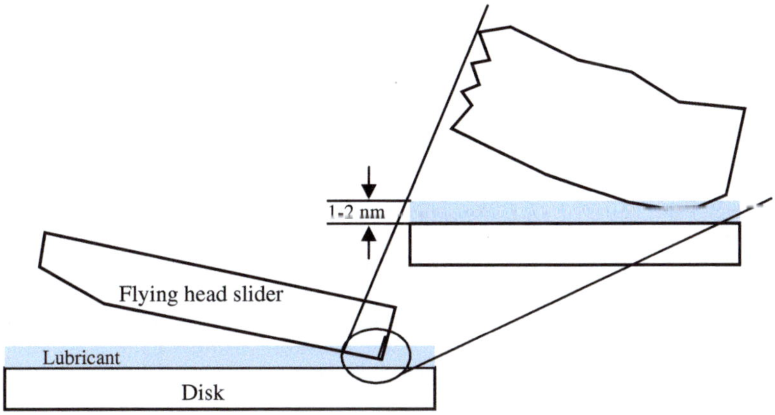

Fig. 5 A schematic description of lube-surfing head-disk interface (*Source* [1])

density of 10 Tb/in^2 (see Fig. 5) [13]. In this newly proposed scheme, the head slider steadily flies over the disk so that it can support its reader/writer element surfing through the lubricant layer of a thickness of a few nanometers. In order to design such head sliders, it would be necessary to take into consideration the lubricant film flow on a rotating disk and its interaction with the head slider in flight as well. It is well-established that in very thin liquid films of less than several molecules thick near a solid wall, the classical Reynolds equation breaks down. For the flow behavior of a collection of molecules near a solid wall, microstructural effects such as the micro-inertia of the liquid and the nonlocal surface stress play relevant roles.

Eringen and Okada [14, 15] developed a nonlocal theory of lubrication capable of describing the physics of dynamic processes in very thin liquid films at the molecular level and it can predict the forces of interaction between the head slider and the lubricant film. We believe that, leveraging on the nonlocal lubrication theory, it should be possible to develop a new head-disk interface design methodology for the likes of the lube-surfing head-disk interface scheme, whereby one can address questions regarding the ways in which an appropriate head-disk interface can be designed by combining it with the generalized Reynolds equation.

Acknowledgments I should like to express my appreciation to Prof. Shigehisa Fukui of Tottori University for kindly providing me with useful references and comments.

References

1. B. Liu et al., Air-bearing design towards highly stable head-disk interface at ultra-low flying height. IEEE Trans. Magn. **43**(2), 715–720 (2007)
2. S. Fukui, R. Kaneko, Analysis of ultra-thin gas film lubrication based on linearized Boltzmann equation: first report—derivation of a generalized lubrication equation including thermal creep flow. Trans. ASME J. Tribol. **110**, 253–262 (1988)

3. S. Fukui, R. Kaneko, in *Molecular Gas Film Lubrication (MGL)*, ed. by B. Bushan. Handbook of Micro/Nanotribology, (CRC Press, Boca Raton, 1995) pp. 559–604
4. A. Burgdorfer, The influence of the molecular mean free path on the performance of hydrodynamic gas lubricated bearings. Trans. ASME J. Basic Eng. **81**, 94–100 (1959)
5. Y.H. Hsia, G.A. Domoto, An experimental investigation of molecular rarefaction effects in gas lubricated bearings at ultra-low clearances. Trans. ASME J. Lubr. Technol. **105**, 120–130 (1983)
6. C. Cercignani, *Rarefied Gas Dynamics—From Basic Concepts to Actual Calculations* (Cambridge University Press, Cambridge, 2000)
7. P.L. Bhatnagar, E.P. Gross, M. Krook, A model for collision processes in gases. Phys. Rev. **94**, 511–525 (1954)
8. M. Abramowitz, I.A. Stegan, in *Handbook of Mathematical Functions* (Dover, New York, 1968)
9. C. Cercignani, A. Daneri, Flow of a rarefied gas between two parallel plates. J. Appl. Phys. **34**(12), 3509–3513 (1963)
10. D.R. Willis, Comparison of kinetic theory analyses of linearized Couette flow. Phys. Fluids **5**(2), 127–135 (1962)
11. S. Fukui, R. Kaneko, A database for interpolation of Poiseuille flow rates for high Knudsen number lubrication problem. Trans. ASME J. Tribol. **112**, 78–83 (1990)
12. S. Fukui, R. Kaneko, Experimental investigation of externally pressurized bearings under high Knudsen number conditions. Trans. ASME J. Tribol. **110**, 144–147 (1988)
13. B. Liu et al., Lube-surfing recording and its feasibility exploration. IEEE Trans. Magn. **45**(2), 899–904 (2009)
14. A.C. Eringen, *Nonlocal Continuum Field Theories* (Springer, New York, 2002)
15. A.C. Eringen, K. Okada, A lubrication theory for fluids with microstructures. Int. J. Eng. Sci. **33**(15), 2297–2308 (1995)

Nonstationary Analysis of Blast Furnace Through Solution of Inverse Problem and Recurrence Plot

Junichi Nakagawa

Abstract A recurrence plot is a method for directly visualizing, on a two-dimensional surface, information regarding the proximal points and distance between two points created using time delay coordinates from temporal sequence data This method qualitatively captures the nonstationary nature of temporal sequence data. However, as the complexity of phenomena increases, the plotted structure of visualizations using two-dimensional surfaces also becomes complex. This can cause problems when trying to interpret changes in the plotted structure. This paper proposes a method of recurrence plot structural analysis, that is, on the basis of deterministic principles. Furthermore, because the proposed method is applied to a blast furnace, which involves the handling of enormous quantities of high-temperature molten iron as well as complex phenomena accompanying reactions in the gas, liquid and solid phases, direct measurement of the internal states when they are treated as spatial distributions is an extremely difficult process. Thus, this study undertakes an analysis of the principles of the determinable properties underlying the temperature shifts in the thermoelectric couples embedded in the brickwork of the furnace floor.

Keywords Nonstationary · Recurrence plot · Blast furnace · Inverse heat conduction problem

1 Introduction

Blast furnaces are a typical complex process, involving all the phenomena accompanying reactions between the gas, liquid, and solid phases. Also, since blast furnaces must handle enormous quantities of high-temperature molten iron, any direct

J. Nakagawa (✉)
Nippon Steel and Sumitomo Metal Corporation, 20-1 Shintomi, Futtsu, Chiba 293-8511, Japan
e-mail: nakagawa.q9p.junichi@jp.nssmc.com

R. Nishii et al. (eds.), *A Mathematical Approach to Research Problems of Science and Technology*, Mathematics for Industry 5,
DOI: 10.1007/978-4-431-55060-0_32,

measurement of the internal states when they are to be treated as spatial distributions is an extremely difficult process.

One problem with blast furnaces is the abnormal states believed to occur when a malfunction causes furnace conditions to diverge from a stationary state. Since furnaces are usually controlled so as to avoid abnormal states as much as possible, it is rare for these abnormal states to continue in a stationary manner, and the ones we see are transition processes. Accordingly, in order to respond to this sort of problem, it is important to discover and quantify universal attributes from the conditions present in transition processes.

One of the main factors in determining the longevity of blast furnaces is the thickness of the brick of the furnace floor. The furnace floor is usually covered by a coagulated layer comprised mainly of a compound of molten metal and coke, called the "furnace upheaval," that protects it from direct contact with molten metal which can cause damage the brick. However, if for any reason a sudden dynamic change occurs inside the furnace, a local increase in the flow rate of molten iron accompanying the rise or fall of the furnace core, called the "deadman" may cause the furnace upheaval to disappear, causing the brick to suffer damage from the heat of the molten metal [1].

When a furnace is in operation, it is desirable to detect as early as possible any sudden dynamic changes within it that might cause damage to the furnace floor brick, and to take the appropriate action to bring these internal changes under control. However, since direct internal measurement of the furnace is extremely difficult, as outlined above, we are forced to estimate any changes in the furnace's internal state from the extremely limited temperature measurement information coming from the thermocouples embedded in the brick of the furnace floor.

This paper applies the inverse problem to the heat transfer phenomenon of blast furnaces, on the basis of temporal sequence data consisting of thermocouple measurement values. Here, the inverse problem has a twofold meaning. One is the inverse heat conduction problem to reconstruct a heat flux on an inaccessible boundary. The other is the identification of the system's characteristic nonstationary principles.

In the inverse heat conduction problem, for the nonstationary heat conduction equation, we are required to reconstruct a heat flux on an inaccessible boundary from measurements made near the accessible boundary. There are a number of numerical methods for solving the inverse heat conduction problem. However, most of them require some data to serve as the initial conditions. On the other hand, in many practical situations such as blast furnaces, we cannot know the initial condition because we have to estimate the problem for a process that has already started. One of the main purposes of this paper is to propose a numerical method which does not need initial data for reconstructing the boundary values and is stable against the intrinsic instability of the inverse heat conduction problem.

On a reconstructed attractor created to bring about a one-to-one corresponding relationship with the original dynamic system, we attempt to analyze the information regarding internal furnace changes contained within the temporal sequence data. One challenge to identifying the nonstationary states of dynamic systems with comparatively large degrees of freedom and split-second timing is to discover the system's characteristic nonstationary and nonlinear principles. In order to do this, it is necessary to analyze attributes related to transitions over time in the dynamic structure of transitional dynamic trajectories and to visualize transitions over time in the structure of attractors. We use the recurrence plot method, as it displays excellent capabilities in this area. A recurrence plot [2] is a way of directly visualizing, on a flat two-dimensional surface, information regarding the proximal points and distance between two points created using time delay coordinates from temporal sequence data. This method is effective for qualitatively capturing the nonstationary nature of temporal sequence data [3, 4]. However, as the complexity of a phenomenon increases, the plot structure of a visualization using a two-dimensional surface also becomes complex. This can cause problems when trying to interpret changes in the plot structure.

Thus, in this paper, we extract detection functions from the recurrence plot by using the principles of the measure of determinism and we pose the problem of processing of temporal sequence data as an inverse heat conduction problem and numerically index the recurrence plot structure. Then, we discuss the nonlinear principles applicable to bringing back under control a furnace experiencing a sudden increase in heat flux that potentially can damage its floor.

2 Formulation of the Inverse Heat Conduction Problem

Let us consider the one-dimensional heat equation:

$$\partial_t u = \alpha \partial_x^2 u(x,t), \quad 0 < x < l, \ t > 0 \tag{1}$$

Here, $\alpha > 0$ is a given constant which represents the thermal diffusion coefficient. The inverse heat conduction problem for (1) is to determine

$$f(t) \equiv \partial_x u(0,t), \quad t > 0 \tag{2}$$

and

$$u_0(x) \equiv u(x,0), \quad 0 < x \leq l \tag{3}$$

from

$$\partial_x u(l,t) \equiv g(t) = u(l - \Delta x) - u(l), \quad t > 0 \tag{4}$$

and

$$u(l,t) \equiv h(t), \quad t > 0. \tag{5}$$

Let $0 < t_1 < \cdots < t_M$ be given. Our task is to reconstruct the values of $f(t)$ and $u_0(x)$ from discrete noisy values of $g(t_j)$ and $h(t_j)$, $j = 1, 2, \ldots, M$.

We refer the reader to [5] in which the algorithm for solving the inverse heat conduction problem is shown.

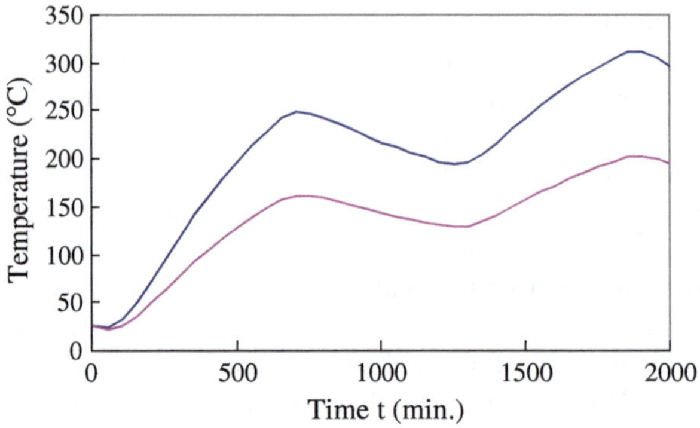

Fig. 1 Time history of temperature used in numerical experiment

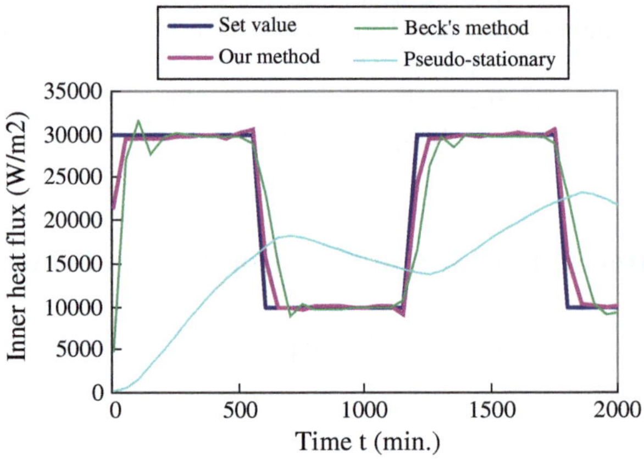

Fig. 2 Results of numerical experiment

3 Results of Numerical Experiment

In this section, we will give an example to test the algorithm described in the previous section.

Figure 1 displays temperature values from two thermocouples set at different depths in a refractory material. The thermocouples are, respectively, positioned at 0.85 and 0.95 m from the inner surface. We set the heat flux at the inner boundary as in Fig. 2 and reconstructed it using the temperatures shown in Fig. 1. Figure 2 is the numerical result. From it, we can see that the calculated value is quite close to the set value.

For the sake of comparison, other numerical results are shown in Fig. 2. Beck's method [6] is based on the variational principle. Since this method requires an initial value, its numerical performance is poor for small t due to a long time required to converge to a stable answer. Although the initial values were set to be constant (25 °C) at all positions of the material, there is a subtle temperature gradient.

The pseudo-stationary condition is calculated with Formula (6).

$$q_{st} = \frac{\lambda \Delta T}{l} \tag{6}$$

Here, λ represents the thermal conductivity of the furnace floor brick, which is here set as 21 W/m · K. In addition, l represents the distance between the two thermocouples and ΔT represents the difference in temperature values at any given time. The numerical performance is worse for all t.

4 Reconstruction of Heat Flux in Blast Furnace

Figure 3 displays temperature measurement values from two thermocouples set at different depths in the brick of the floor of a blast furnace. The distance between them is 0.1 m.

Figure 4 displays the change in pseudo-stationary heat flux, calculated with Formula (7) based on the temperature measurements made with the two thermocouples.

$$q_{st} = \frac{\lambda \Delta T}{l} \tag{7}$$

Here, λ represents the thermal conductivity of the furnace floor brick (14 W/m · K). The distance between the thermocouples is denoted by l, and the difference in temperature measurement values at any given time between the thermocouples is denoted by ΔT. The dotted line in the diagram is the shutdown period; it represents the times when the operation of the blast furnace is suspended for scheduled repair.

Figure 5 displays the change in nonstationary heat flux calculated with the inverse heat conduction problem method.

We can see that the changes in the heat flux estimated by inverse heat conduction problem method in Fig. 5 are large in comparison with those in Fig. 4. This difference is due to a heat transfer delay, caused by heat transfer resistance occurring between the positions of the thermocouples in the blast furnace brick and the surface of the brick. Another reason for why the change in heat flux is larger in Fig. 5 is that the measurement errors in the original temperature data from the thermocouples have been amplified by the inverse problem. However, if this amplification of measurement errors has occurred its influence would extend across all time periods to the same extent, but we can see that the scope of the changes in heat flux in Fig. 5 depends on the period of time. Also, around the vicinity of 1,650 h, the scope of changes in the heat flux is almost to the same extent and at the same time as in Fig. 3 showing the change in stationary heat flux (where there is no amplification of measurement errors). This leads us to judge that the influence from amplification of measurement

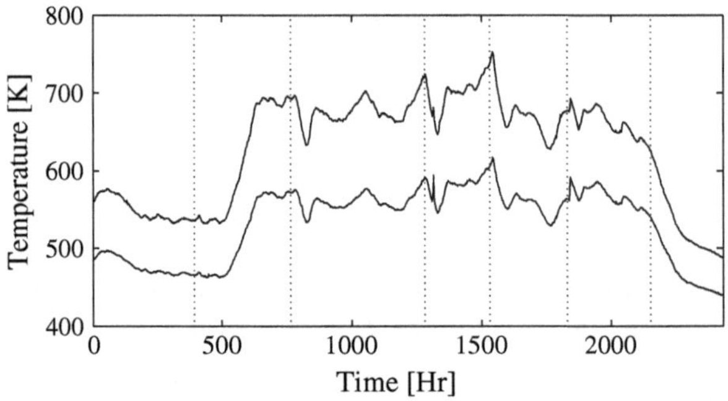

Fig. 3 Change in thermocouple measurement temperatures

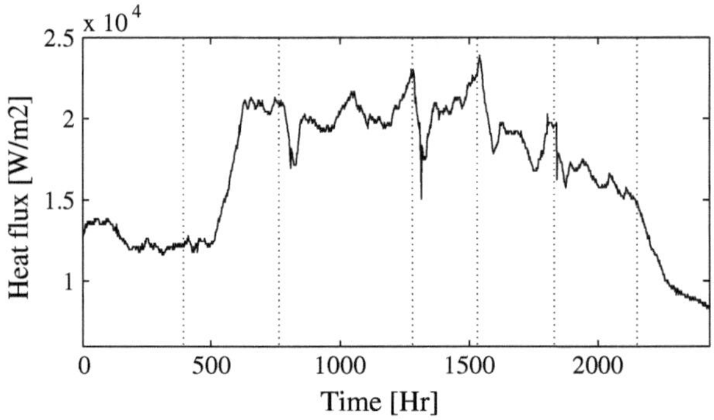

Fig. 4 Change in stationary heat flux

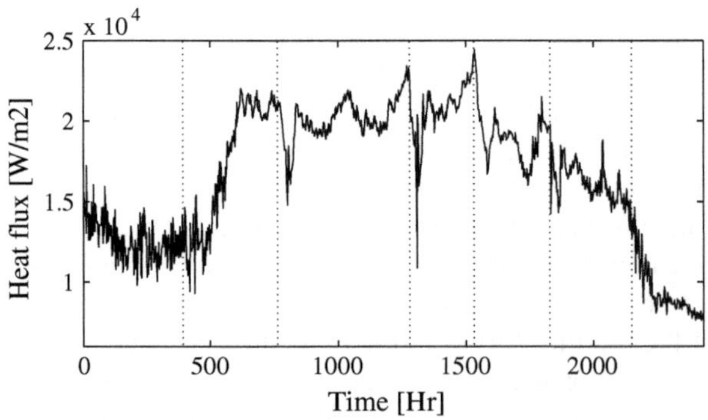

Fig. 5 Change in nonstationary heat flux

errors was not extensive. In the next section, taking Figs. 3 and 4 as the objects of analysis, we will develop our argument by focusing on discrepancies in the results of analysis between those cases in which the heat transfer delay has been considered and cases in which it is not.

5 Analysis of Recurrence Plot

In order to create a recurrence plot methodology [4], we first must imagine an image formed of $N \times N$ pixels. Then, we must consider an index i of temporal sequence data in the x direction of this image and an index j of temporal sequence data in the y direction, and obtain both the points where these meet $\{i, j\}$ within the distance between the two points $D\{i, j\}$.

$$D(i, j) = |X(i) - X(j)| \tag{8}$$

Here, $X(i)$ and $X(j)$ are reconstructed vectors formed using the observed temporal sequence data $\{x(t), t = 0, 1, 2, \ldots, N\}$:

$$X(i) = (x(i), x(i + \tau), \ldots, x(i + (m - 1)\tau) \tag{9}$$

$$X(j) = (x(j), x(j + \tau), \ldots, x(j + (m - 1)\tau) \tag{10}$$

The 14 h maximum value for the response time of 6–14 h [7, 8] reported as the response characteristics of the blast furnace is used here as the time delay τ. The dimension m of the space of the reconstructed states was set at 7, in consideration of the embedding theorem [9, 10] and the values of the correlation dimensions d of the stationary heat flux and nonstationary heat flux (calculated using the correlation dimensional analysis presented in [8]), which were respectively 2.2 and 2.9. Also, the Euclidean distance was used in calculating the distance between the two points $X(i)$ and $X(j)$. When the threshold ε is appropriately stipulated,

$$D(i, j) < \varepsilon \tag{11}$$

a black point is plotted at the pixel lying at point (i, j).

The threshold value ε is set for optimum visual clarity of the recurrence plot. If ε is too small, the plot becomes too sparse; conversely if it is too large, the plot becomes too dense, making it difficult to ascertain the plot structure. The value of ε is often set somewhere between 0.01 times and 0.28 times the maximum value of $D(i, j)$ [3, 11, 12].

Figures 6 and 7 display the recurrence plots of the temporal sequence data from the stationary heat flux of Fig. 4 and the non-stationary heat flux of Fig. 5. Here, ε is set at 0.065 of the maximum value of $D\ (i,j)$.

The patterns of the recurrence plots depicted in Figs. 6 and 7 resemble each other to a certain extent. This suggests that the macro structure of the reconstructed attractors between the stationary heat flux and the nonstationary heat flux is conserved. Also, in comparison with Fig. 6, the thick shading is preserved in the plot pattern in Fig. 7. This is nearly identical to the macro structure of reconstructed attractors outlined above, but fine differences are observable. The following section analyzes the mechanisms that are expressed in this shading structure.

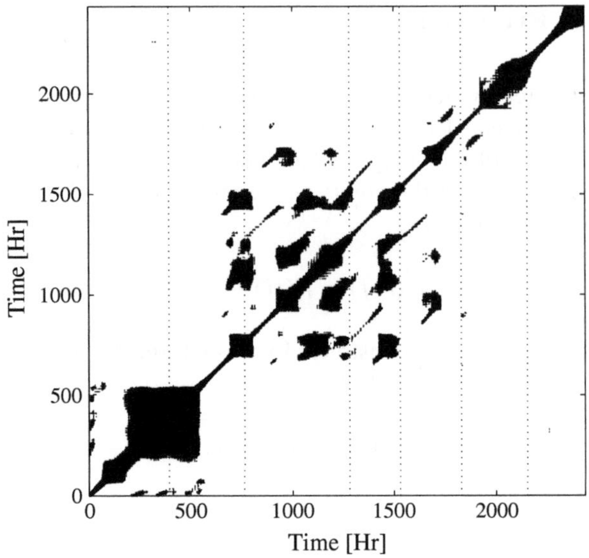

Fig. 6 Recurrence plot of stationary heat flux

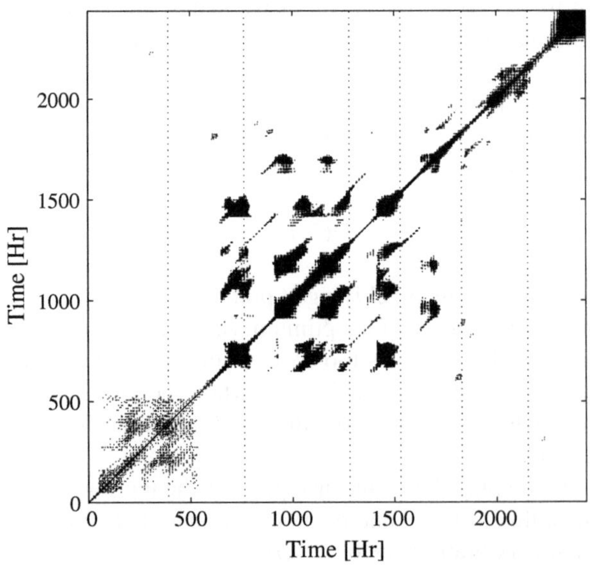

Fig. 7 Recurrence plot of nonstationary heat flux

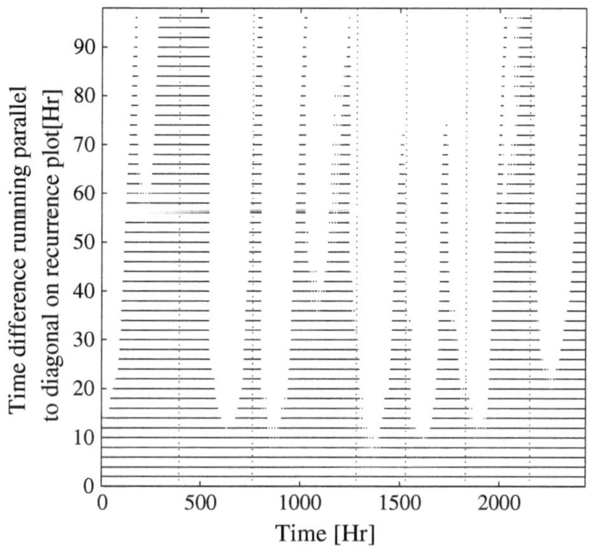

Fig. 8 Detection of determinism from stationary heat flux

6 Detecting Determinism from a Recurrence Plot

In order to analyze the shading structure in the recurrence plot described above from the perspective of deterministic principles, we took the diagonal line of the recurrence plot as the transverse axis, and assigned the plot running parallel to the diagonal line on the transverse axis of the graph. The re-depicted results of Figs. 6 and 7 are displayed as Figs. 8 and 9. Here, the time component until $r \times m = 98$ h is displayed on the vertical axis of the graph.

In some places on Fig. 8, some whited-out sections can be seen, but it can be seen that apart from these whited-out sections the parallel lines mostly continue along the horizontal axis of the graph. On the other hand, in Fig. 9 some upsets in the continuity of the parallel lines can be seen around the starting point in time where the heat flux values in Fig. 5 increased (in the vicinity of 520 h). This appears in the recurrence plot pattern from Fig. 7 as the shaded section. This suggests that the level of determinability decreases before a state changes greatly. We believe that this decrease in determinability can be seen in the shading pattern of the recurrence plot in Fig. 7.

7 Quantitative Evaluation Index for Determinism

As outlined above, a numerical index should be developed for the determinability of the continuity of the parallel lines in Figs. 8 and 9. The plotted points for the times t listed in Figs. 8 and 9 are defined as a numerical index for the principle of

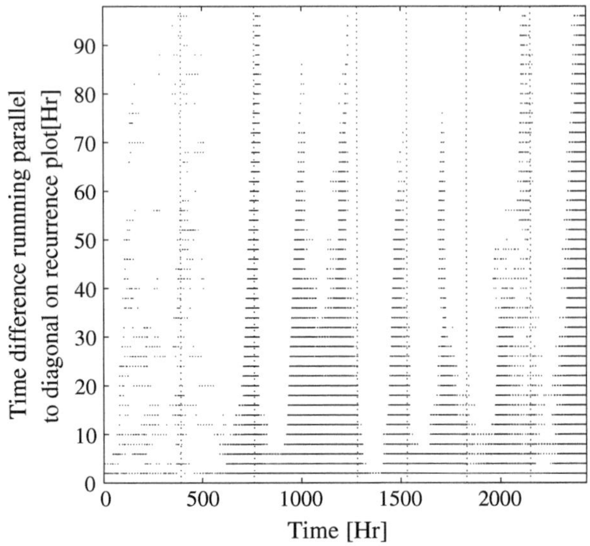

Fig. 9 Detection of determinism from nonstationary heat flux

determinability for the ratio of the recurrence points forming parallel lines to the diagonal line in the recurrence plot. Here, Δt represents the sampling time for the original temporal sequence data.

Here, the concept of *%determinism* defined by Webber Jr, and Zbilut [13] is identical to the ratio of the recurrence points forming parallel lines to the diagonal line in the recurrence plot, so we shall refer to this numerical index as *%determinism*. Figs. 10 and 11 represent the temporal changes in *%determinism* of Figs. 8 and 9. In Fig. 11, the *%determinism* values decreased right before 520 and 2,150 h (the times at which large changes in heat flux began).

8 Discussion

The "shutdown period" refers to the period, usually every day or two, where the production of molten iron in the blast furnace is suspended for conservation and maintenance purposes. We have observed some cases where the shutdown operation can trigger changes in the internal state of the blast furnace. Accordingly, just as Fig. 12 indicates the changes in the internal state of the blast furnace, Figs. 13 and 14 plot the relationship between level change Q in heat flux value before and after the shutdown operation and the *%determinism* value directly before shutdown D. Figure 13 represents the case of stationary heat flux and is based on the results of Figs. 4 and 10. On the other hand, Fig. 14 represents the case of nonstationary heat flux, based on the results of Figs. 5 and 11.

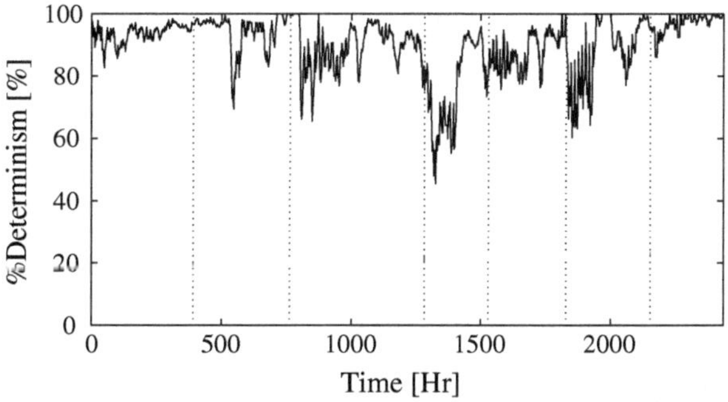

Fig. 10 % Determinism of stationary heat flux

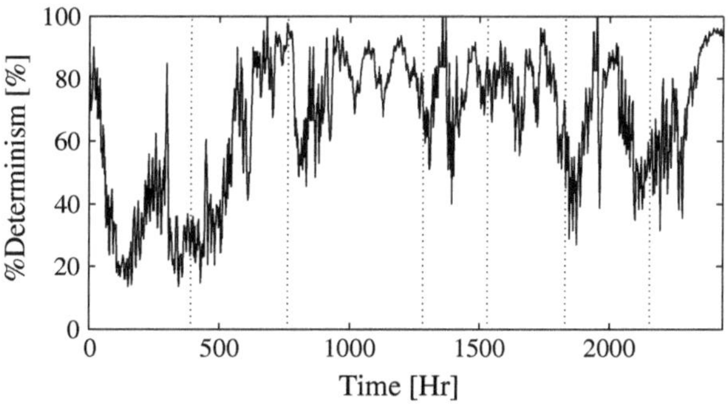

Fig. 11 % Determinism of nonstationary heat flux

In Fig. 14, a mutual relationship can be seen; as the % *determinism* value D directly before the shutdown period decreases, the level change Q in heat flux values before and after shutdown increases. This figure illustrates that when the %*determinism* value is small, operations such as the "shutdown period" will stimulate the system, leading to a greater chance of shifting the system from its current state to another state. Accordingly, we believe that this means the %*determinism* value is a kind of numerical index expressing the stability of the system.

On the other hand, the mutual relationship outlined above cannot be seen in Fig. 13. This suggests that, in the same fashion as the heat transfer phenomenon, in systems where the diffusion effect of a physical quantity becomes a problem, signal deterioration due to diffusion resistance has a large influence on the microstructure of the recurrence plot.

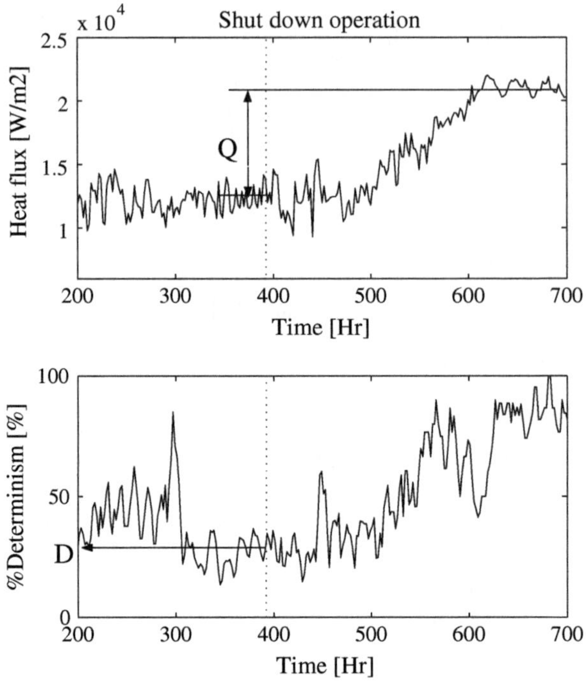

Fig. 12 %Determinism value D just before shutdown of operations and heat flux change Q before and after shutdown

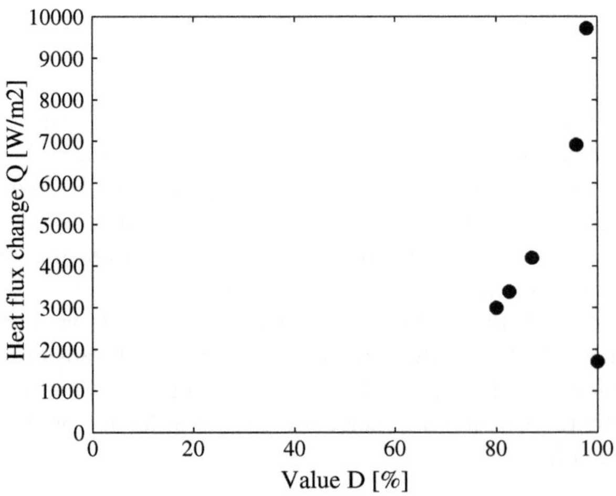

Fig. 13 Relationship between *%determinism* value D and heat flux change Q (Case of stationary heat flux)

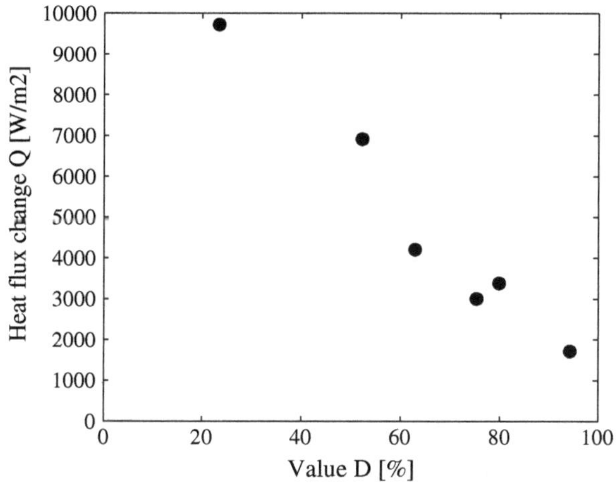

Fig. 14 Relationship between *%determinism* value D and heat flux change Q (Case of nonstationary heat flux)

9 Conclusion

We applied recurrence plotting methods to heat transfer phenomena in blast furnace floors by analyzing the determinable properties underlying rises or falls in temperature of the thermocouples embedded in the furnace floor brickwork.

Due to the heat transfer resistance of the furnace floor brick, signals from the thermocouples deteriorate, but it was shown that an inverse analysis of signals sent from the thermocouples compensates the influence of heat transfer resistance by calculating the nonstationary heat flux at the inner surface. Also, in order to extract a measure of determinability from the recurrence plots, we proposed taking the diagonal lines of recurrence plots as the transverse axis and assigning plots running parallel to these lines on the transverse axis of the graph as away of re-depicting recurrence plots. In the same diagram, for the points plotted at a given time *t*, by calculating the ratio of the recurrence points forming parallel lines to the diagonal line in the recurrence plot, we were able to obtain the *% determinism* value, which is a numerical index for principles of determinability.

Furthermore, we discovered a mutual relationship between the *% determinism* value directly before shutdown and the change in heat flux before and after the shutdown period. It was shown that when the *% determinism* value is small, shutdown operations can stimulate the system, possibly causing it to shift from its current state into other states. It is believed that this is one of the main factors in rises or falls in signal levels from thermocouples.

Recurrence plot methods based on the derivation of the measure of determinability illustrated in this paper can be applied to many kinds of phenomena besides blast furnaces. Verifying the effectiveness and universality of this method remains an issue for the future work.

Acknowledgments This research is supported by the Aihara Innovative Mathematical Modeling Project, the Japan Society for the Promotion of Science (JSPS), through the "Funding Program for World-leading Innovative R & D on Science and Technology (FIRST Program)", initiated by the Council for Science and Technology Policy (CSTP).

References

1. Akihiko Shinotake, Satoshi Nakamura, Hajime Otsuka, Nozomi Sasaki, Yasushi Kurita, Concept of longevity of blast furnaces under the operation of high-powered metal. Mater. Process **14**(4), 750–753 (2001)
2. J.P. Eckmann, S.O. Kamphorst, D. Ruelle, Recurrence plots of dynamical systems. Europhys. Lett. **4**(9), 973–977 (1987)
3. T. Yamada, K. Aihara, Recurrence plot and non-stationary sequence analysis of distribution of distance between 2 points. J. Inst. Electr. Inf. Commun. Eng. (Shingakuron) **J82-A**(7), 1016–1028 (1999)
4. K. Aihara, *Introduction to Chaos Theory*. (The Society for the Promotion of the University of the Air, 2001)
5. Y. Wang, J. Cheng, M. Yamamoto, J. Nakagawa, A numerical method for solving the inverse heat conduction problem without initial value. Inverse Prob. Sci. Eng. **18**, 655–671 (2010)
6. J.V. Beck, Nonlinear estimation applied to the non-linear inverse heat conduction problem. Int. J. Heat Mass Transfer **13**, 703–716 (1970)
7. Michiharu Hatano, Keisuke Misaka, Yoshiyuki Matoba, Kouichi Otsuka, The blast furnaces mathematical formula for the molten metal temperature control. Iron. Steel **67**(3), 518–527 (1981)
8. Y. Otsuka, M. Konishi, T. Maki, Stabilization method for hot metal temperature in operation change of blast furnace. ISIJ Int. **40**(4), 324–347 (2000)
9. K. Aihara (others), *Basis and Application of Chaos Temporal Sequence Analysis*. (Sangyo zusho, Tokyo, 2000)
10. F. Takens, Detecting strange attractors in turbulence, in *Dynamical Systems of Turbulence*, ed. by D.A. Rand, B.S. Young, Lecture Notes in Mathematics, vol. 898. (Springer, New York, 1981), pp. 366–381
11. M.C. Casdagli, Recurrence plots revisited. Physica D **108**, 12–44 (1977)
12. M. Koebbe, G. Mayer-Kress, Use of recurrence plots in the analysis of time-series data, in *Nonlinear Modeling and Forecasting*, ed. by M. Castagli, S. Eubank. (Addison-Wesley, Redwood, 1992), pp. 361–378
13. C.L. Webber Jr, J.P. Zbilut, Dynamical assessment of physiological systems and states using recurrence plot strategies. J. Appl. Phys. **76**(2), 366–381 (1994)

Time-Periodic Nonlinear Steady Field Analysis

Kenji Miyata

Abstract Error correction Time interval Flexible (ETF) method is presented to fastly obtain time-periodic nonlinear fields in the presence of extremely slow decay fields. The analysis variables are corrected by using the steady-state condition with respect of time variations of the fundamental components. The ETF method is classified into Self ETF and Mutual ETF methods. The time interval of error corrections is flexibly selected, and then step-by-step continuous error corrections are available by using the Mutual ETF method. The ETF method improves the convergence properties of the conventional method like the simplified Time-Periodic Explicit Error Correction (TP-EEC) and the simplified polyphase TP-EEC methods. The presented methods were verified in three-variable simultaneous equations as a simple linear example problem and a nonlinear magnetic field simulation of a synchronous motor by the finite element method as a multivariable problem.

Keywords Time-periodic solution · Steady field · Correction · Transient field · Magnetic field

1 Introduction

A highly accurate time-periodic solution of nonlinear time-differential equation can be derived in several ways. As a standard technique, the shooting method [1, 2] is more commonly used in a case of a small-scale system like an electric circuit. The time-periodic finite element method [3, 4] can be powerfully used in the two-dimensional finite element analysis of electromagnetic field. The harmonic balance finite element method [5] uses nonlinear analysis in frequency domain where a large-scale matrix equation must be analyzed.

K. Miyata (✉)
Hitachi Research Laboratory, Hitachi, Ltd., 7-1-1, Omika, Hitachi, Ibaraki 319-1292, Japan
e-mail: kenji.miyata.fv@hitachi.com

R. Nishii et al. (eds.), *A Mathematical Approach to Research Problems of Science and Technology*, Mathematics for Industry 5,
DOI: 10.1007/978-4-431-55060-0_33,

A step-by-step solution in the transient analysis is corrected toward a steady-state one-half periodic solution every one-half period in Time-Periodic Explicit Error Correction (TP-EEC) method [6, 7]. The correction toward a steady-state is executed every one-sixth of period in three-phase AC TP-EEC method [8, 9] proposed by Tokumasu as the polyphase AC TP-EEC method in a general form. The TP-EEC and polyphase AC TP-EEC methods are very powerful techniques to obtain secure periodic solutions in several corrections. In other words, however, the TP-EEC and polyphase AC TP-EEC methods require one-half and one-sixth period of calculations for one correction, respectively. The Error correction Time interval Flexible (ETF) method [10] as a new correction method requires only transient calculation in a time interval much shorter than one-half of period. The ETF method improves the convergence property of the conventional methods like the simplified TP-EEC and the simplified three-phase AC TP-EEC, and so on.

2 Principle and Formulations of the ETF Method

In the ETF method, all or a part of variables, where their time differentials affect the transient behavior, are corrected by using the steady-state condition with respect of time variation of a fundamental wave component. The ETF method is categorized into self and mutual types, i.e., Self ETF and Mutual ETF methods. The Self ETF method uses self field only, while the Mutual ETF method uses plural fields with phases different from each other in the polyphase AC system.

2.1 Self ETF Method

The variables are conveniently represented in complex notation. The complex variable z_1 becomes z_1', z_2, and z_2', after rotation by θ, ϕ, and $\theta + \phi$, respectively, in electrical angle (See Fig. 1).

Here, we define Δz_1 and Δz_2 as the time variations from z_1 to z_1' and from z_2 to z_2', respectively. Supposing that alternating field $z_1$reaches the steady state, z_2' can be expressed with the time variations Δz_1 and Δz_2. When the complex variable z_1 is normalized to 1, the complex variables of z_1', z_2, and z_2' can be written as follows:

$$z_1' = e^{j\theta},\ z_2 = e^{j\phi},\ z_2' = e^{j(\phi+\theta)} \tag{1}$$

Then the time variations Δz_1 and Δz_2 become

$$\Delta z_1 = z_1' - z_1 = e^{j\theta} - 1, \tag{2}$$

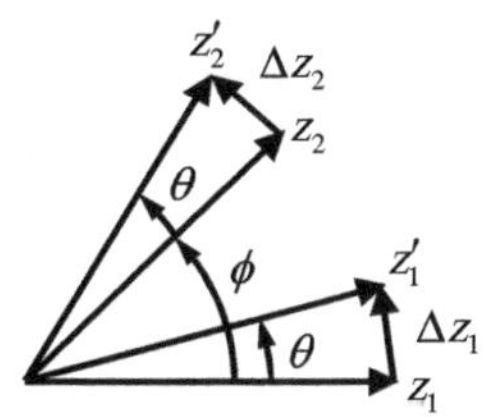

Fig. 1 Phasor diagram to derive a correction formula based on the Self ETF method

$$\Delta z_2 = z_2' - z_2 = e^{j\phi}\left(e^{j\theta} - 1\right). \tag{3}$$

Therefore we obtain the following equation

$$z_2' = a\Delta z_1 + b\Delta z_2, \tag{4}$$

$$a = \frac{1}{2}\left(\frac{\sin\phi}{\sin\theta} + \frac{\cos\phi}{1-\cos\theta}\right),\quad b = \frac{1}{2}\left(\frac{\sin\phi}{\sin\theta} - \frac{\cos\phi}{1-\cos\theta}\right). \tag{5}$$

Equation (4) can be used as an error correction formula to get a steady-state solution, and is rewritten in the following form,

$$z_2^{\prime new} = a\Delta z_1 + b\Delta z_2 \tag{6}$$

In the case of $\theta = \phi$, Eq. (6) becomes

$$z_2^{\prime new} = \Delta z_2 - \left(\frac{\Delta z_2 - \Delta z_1}{4\sin^2(\theta/2)}\right). \tag{7}$$

Especially, in the case that $\phi = \theta = \pi/2$, Eq. (7) leads to the correction formula based on the simplified TP-EEC method,

$$z_2^{\prime new} = \frac{1}{2}\left(\Delta z_1 + \Delta z_2\right) = \frac{1}{2}\left(z_2' - z_1\right) \tag{8}$$

Furthermore, in the case that $\phi = \theta << 1$, Eq. (7) is approximated to

$$z_2^{\prime new} \cong \Delta z_2 - \left(\frac{\Delta z_2 - \Delta z_1}{\theta^2}\right) \cong \Delta z_2 - \frac{\partial^2 z_2}{\partial\theta^2} \tag{9}$$

The correction formula approximately reduces to the following error correction formula of z_2 based on the TDC method [11] without time averaging process,

$$z_2^{\prime new} = \Delta z_2^{new} + z_2^{new},\quad z_2^{new} \cong -\frac{\partial^2 z_2}{\partial\theta^2}, \tag{10}$$

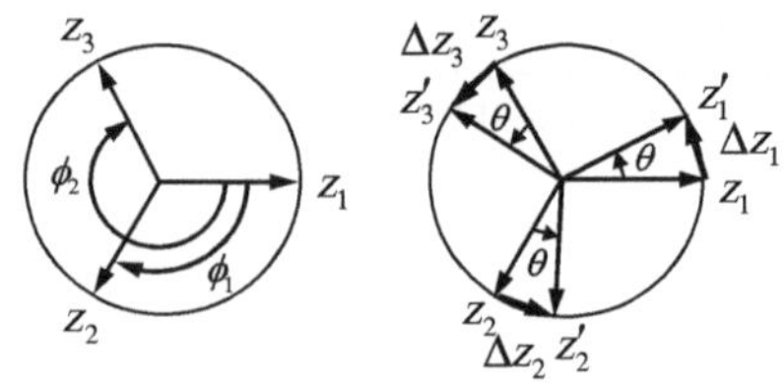

Fig. 2 Phasor diagram to derive a correction formula based on the Mutual ETF method

where Δz_2^{new} indicates the time variation $z_2' - z_2$ after z_2 is corrected. As mentioned above, the Self ETF method contains both the simplified TP-EEC and the approximate TDC method.

2.2 Mutual ETF Method

In the polyphase system, the field can be quickly corrected by using the information of another AC field with the different phase. The variables are also represented in the complex notation. We use the three complex variables z_1, z_2, and z_3 displayed in Fig. 2. The variables z_2 and z_3 have lagging angles of ϕ_1 and ϕ_2, respectively, to the variable z_1. The variables z_1, z_2 and z_3 become z_1', z_2' and z_3', respectively, after rotating by electrical angle θ. The time variations of z_1, z_2 and z_3 during the time interval are described as Δz_1, Δz_2 and Δz_3. When the complex variable z_1 is normalized to be 1, the complex variables of z_2, and z_3 can be written as follows;

$$z_2 = e^{-j\phi_1}, \quad z_3 = e^{-j\phi_2}, \tag{11}$$

then we obtain

$$\begin{aligned} \Delta z_1 &= e^{j\theta} - 1, \quad \Delta z_2 = \left(e^{j\theta} - 1\right) e^{-j\phi_1}, \\ \Delta z_3 &= \left(e^{j\theta} - 1\right) e^{-j\phi_2}. \end{aligned} \tag{12}$$

The variable z_1' can be expressed in the style of a linear combination of Δz_1, Δz_2 and Δz_3 as follows:

$$z_1' = e^{j\theta} = \alpha \Delta z_1 + \beta \Delta z_2 + \gamma \Delta z_3. \tag{13}$$

By substituting Eq. (12) into Eq. (13), the parameters β and γ can be written using the parameter α as follows;

$$\beta = \frac{1}{\sin(\phi_1 - \phi_2)} \left[\left(\alpha - \frac{1}{2}\right) \sin\phi_2 + \frac{1}{2} \cot\left(\frac{\theta}{2}\right) \cos\phi_2 \right], \tag{14}$$

$$\gamma = \frac{1}{\sin(\phi_2 - \phi_1)} \left[\left(\alpha - \frac{1}{2}\right) \sin\phi_1 + \frac{1}{2} \cot\left(\frac{\theta}{2}\right) \cos\phi_1 \right]. \tag{15}$$

Then we obtain the following correction formula;

$$z_1^{\text{new}} = \alpha \Delta z_1 + \frac{1}{\sin(\phi_2 - \phi_1)} \left[\left(\alpha - \frac{1}{2} \right) F + \frac{1}{2} \cot \left(\frac{\theta}{2} \right) G \right] \tag{16}$$

where F and G are given by

$$F = \sin \phi_1 \Delta z_3 - \sin \phi_2 \Delta z_2, \tag{17}$$

$$G = \cos \phi_1 \Delta z_3 - \cos \phi_2 \Delta z_2. \tag{18}$$

The special case with α=0 leads to the following correction formula;

$$z_1^{\text{new}} = \frac{\sin \phi_2 \Delta z_2 - \sin \phi_1 \Delta z_3 + \cot(\theta/2)(\cos \phi_1 \Delta z_3 - \cos \phi_2 \Delta z_2)}{2 \sin(\phi_2 - \phi_1)} \tag{19}$$

with the self time variation Δz_1not used. Another special case with $\gamma = 0$ leads to the following correction formula;

$$z_1^{\text{new}} = \frac{1}{2} \Delta z_1 + \frac{\cot(\theta/2)}{2 \sin \phi_1} (\Delta z_2 - \Delta z_1 \cos \phi_1) \tag{20}$$

by using the time variations Δz_1 and Δz_2.

The correction equation based on three-phase AC ETF method can be derived from Eq. (16) with the three-phase condition ($\phi_1 = 2\pi/3$ and $\phi_2 = 4\pi/3$). Taking into account that the variables z_1, z_2 and z_3 compose a cyclic system, we obtain

$$z_1^{\text{new}} = \alpha_1 \Delta z_{\text{sum}} - \frac{1}{2} (\Delta z_2 + \Delta z_3) + \frac{\cot(\theta/2)}{2\sqrt{3}} (\Delta z_2 - \Delta z_3), \tag{21}$$

$$z_2^{\text{new}} = \alpha_2 \Delta z_{\text{sum}} - \frac{1}{2} (\Delta z_3 + \Delta z_1) + \frac{\cot(\theta/2)}{2\sqrt{3}} (\Delta z_3 - \Delta z_1), \tag{22}$$

$$z_3^{\text{new}} = \alpha_3 \Delta z_{\text{sum}} - \frac{1}{2} (\Delta z_1 + \Delta z_2) + \frac{\cot(\theta/2)}{2\sqrt{3}} (\Delta z_1 - \Delta z_2), \tag{23}$$

where $\Delta z_{\text{sum}} = \Delta z_1 + \Delta z_2 + \Delta z_3$. The summation of Eqs. (21)–(23) is

$$z_1^{\text{new}} + z_2^{\text{new}} + z_3^{\text{new}} = (\alpha_1 + \alpha_2 + \alpha_3 - 1) \Delta z_{\text{sum}}, \tag{24}$$

and therefore the three-phase balance condition $\left(z_1^{\text{new}} + z_2^{\text{new}} + z_3^{\text{new}} = 0\right)$is identically satisfied when $\alpha_1 + \alpha_2 + \alpha_3 = 1$. Furthermore, Eqs. (21)–(23) under the condition $\theta = \pi/3$ reduce to the following correction formulae based on the three-phase AC TP-EEC method,

$$z_1^{\text{new}} = \alpha_1 (\Delta z_1 + \Delta z_2 + \Delta z_3) - \Delta z_3, \quad (25)$$

$$z_2^{\text{new}} = \alpha_2 (\Delta z_1 + \Delta z_2 + \Delta z_3) - \Delta z_1, \quad (26)$$

$$z_3^{\text{new}} = \alpha_3 (\Delta z_1 + \Delta z_2 + \Delta z_3) - \Delta z_2, \quad (27)$$

where the condition $\alpha_1 = \alpha_2 = \alpha_3 = 1/2$ is usually used because of good correcting performance over all higher order time-harmonic wave sources.

2.3 Discussion of the ETF Method

Both the Self and Mutual ETF methods mentioned above allow flexible time intervals between successive error corrections. The computational cost of error corrections is very low comparably to the simplified TP-EEC method. Step-by-step error corrections become available without higher order time-harmonic wave source. When the source terms include time-harmonic wave components, successive corrections by the ETF method work effectively only for the special time interval of half period or 1/(2n) of one period as the simplified TP-EEC and the simplified polyphase AC TP-EEC methods, because the correction formula of the ETF method utilizes the steady-state condition of the fundamental wave component.

The correction performance of the long time interval ETF method like the simplified TP-EEC or the simplified polyphase AC TP-EEC methods can be improved by the initial corrections based on the short time interval ETF method. Thus, it is very useful to use the hybrid ETF method, i.e., the serial usage of the short and long time-interval ETF methods. The simple case of three variables is shown in the three-phase linear AC system and the electromagnetic nonlinear field analysis as following sections.

3 Simple Example of Transient Analysis in a Three-Phase AC System

Several error correction methods including the ETF method are tested in the steady-state problem written by the following linear simultaneous equations of three variables;

$$\begin{bmatrix} 3 & -1 & -1 \\ -1 & 3 & -1 \\ -1 & -1 & 3 \end{bmatrix} \begin{Bmatrix} \mathrm{d}U/\mathrm{d}\theta \\ \mathrm{d}V/\mathrm{d}\theta \\ \mathrm{d}W/\mathrm{d}\theta \end{Bmatrix} + \begin{Bmatrix} U \\ V \\ W \end{Bmatrix} = \sum_n b_n \begin{Bmatrix} \cos n\theta \\ \cos n(\theta - 2\pi/3) \\ \cos n(\theta - 4\pi/3) \end{Bmatrix}, \quad (28)$$

where the angle θ is used as a time variable. The theoretical steady-state solution is

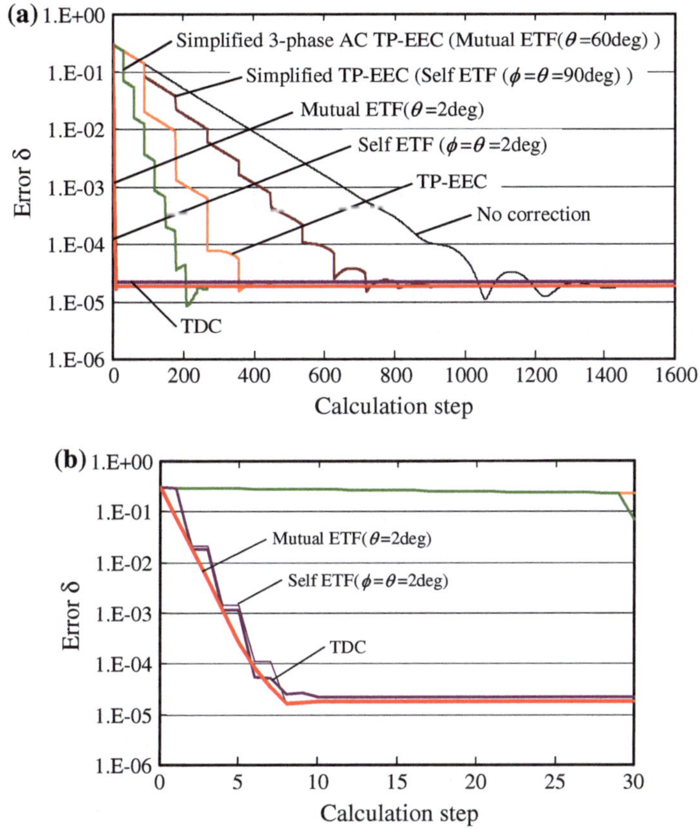

Fig. 3 Comparison among several correction methods with a source including only a fundamental wave. **a** Overall view. **b** Initial-stage view

$$\begin{aligned} U_{\rm th} &= \textstyle\sum_n a_n \cos(n\theta + \varphi_n), \quad a_n = b_n/\sqrt{1+g_n^2}, \\ V_{\rm th} &= \textstyle\sum_n a_n \cos\left[n\left(\theta - \frac{2\pi}{3}\right) + \varphi_n\right], \quad \varphi_n = -\tan^{-1} g_n, \\ W_{\rm th} &= \textstyle\sum_n a_n \cos\left[n\left(\theta - \frac{4\pi}{3}\right) + \varphi_n\right], \quad g_n = n\left[3 - 2\cos\left(\frac{2n\pi}{3}\right)\right]. \end{aligned} \tag{29}$$

When the source term in the right hand side of Eq. (28) has only a fundamental wave component; i.e., $b_1 = 1$, and $b_n = 0$ for n > 1, the computational result is shown in Fig. 3. Figure 3b indicates the initial stage of Fig. 3a in a scaled-up horizontal axis. The free parameters α_1, α_2 and α_3 are all set to be 1/2 in Eqs. (21)–(23) and Eqs. (25)–(27) because the simplified three-phase AC TP-EEC method indicates a maximum performance of error corrections. The error of correction δ is defined as follows;

$$\delta = \sqrt{(U - U_{\rm th})^2 + (V - V_{\rm th})^2 + (W - W_{\rm th})^2}. \tag{30}$$

The number of time divisions per one period is 180 so that the time width of one time step corresponds to an angle of 2°. Self ETF ($\theta = \phi = 2°$) indicates corrections every 2 time steps, while Mutual ETF ($\theta = 2°$) indicates corrections every time step.

The TDC method presents the following correction formula,

$$Z_{\text{new}}\,(\theta - \alpha) = -\left(\frac{\alpha}{\sin\alpha}\right)\frac{\mathrm{d}^2\,\langle z\rangle}{\mathrm{d}\theta^2}, \tag{31}$$

where <z> indicates an time average of z, α is expressed in an angle as one-half of time span for time averaging, and the angle θ is used as a time variable. The time average process is required to reduce harmful effects due to several higher order time-harmonic modes. This example has no higher order time-harmonic modes, and so TDC has no time average process.

Figure 3 indicates that the correction performances of Self ETF ($\theta = 2°$), Mutual ETF ($\theta = 2°$), and TDC exceed those of the other correction methods.

Next, Fig. 4 shows the results in the case where third, fifth, and seventh order time-harmonic modes are included in the right hand side source term with $b_1 = 1$ and $b_3 = b_5 = b_7 = 0.01$. Successive corrections are executed for TP-EEC, Self ETF ($\phi = \theta = 90°$) (simplified TP-EEC) and Mutual ETF ($\theta = 60°$) (simplified three-phase AC TP-EEC), while only initial two times of corrections are executed for TDC and Self ETF ($\phi = \theta = 2°$) and only initial eight times of corrections are executed for Mutual ETF ($\theta = 2°$). Figure 4 indicates that the correction performances degrade for TDC and Self ETF ($\phi = \theta = 2°$) and Mutual ETF ($\theta = 2°$). The higher order time-harmonic source terms do not allow the correction methods like TDC and short time interval ETF methods to execute successive corrections. However, the correction performance is improved by the initial short time interval ETF method followed by the simplified TP-EEC or the simplified polyphase AC TP-EEC methods. The result of example calculations is shown in Fig. 5. The successive corrections by Mutual ETF ($\theta = 60°$) (i.e., the simplified three-phase AC TP-EEC method) follow the initial strong corrections; the initial two times of corrections by TDC and Self ETF methods and eight times of corrections by Mutual ETF ($\theta = 2°$). The combination of Mutual ETF ($\theta = 2°$) and Mutual ETF ($\theta = 60°$) presents a maximum performance of corrections.

4 Nonlinear Example of Magnetic Field Analysis in a Synchronous Motor Coupled with an External Power Supply Circuit

The ETF method is applied to a numerical simulation of magnetic field of a permanent magnet synchronous motor coupled with an external power supply circuit to verify the performance by comparing with the conventional methods. The transient nonlinear magnetic field is analyzed in the edge-element based finite element method.

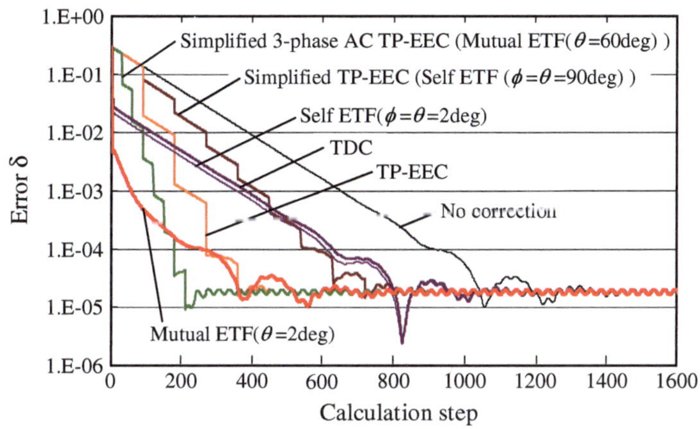

Fig. 4 Comparison among several correction methods with a source including third, fifth, and seventh order harmonic waves in addition to a fundamental wave ($b_1 = 1, b_3 = b_5 = b_7 = 0.01$)

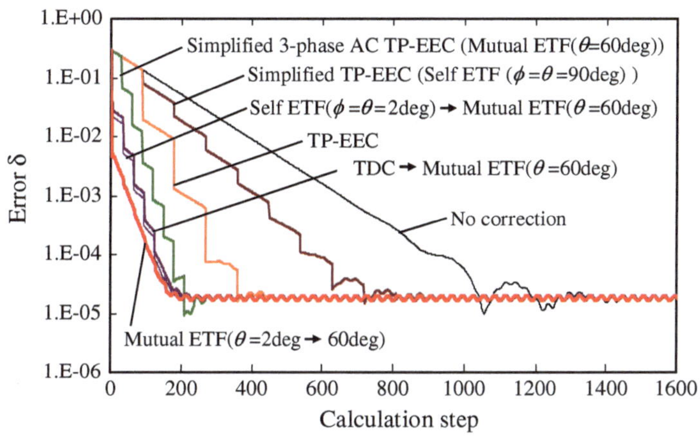

Fig. 5 Comparison among several correction methods including corrections followed by long time interval Mutual ETF method with a source including third, fifth and seventh order harmonic waves in addition to a fundamental wave ($b_1 = 1, b_3 = b_5 = b_7 = 0.01$)

The discretized model of the synchronous motor is shown in Fig. 6, where the number of elements is 8,682. The motor with four poles and six slots has a doubly periodic structure which allows numerical analysis in a half cut model. The specifications of the motor are listed in Table 1. The slide surface between the rotor and the stator is equally divided by 180 in the circumferential direction. In the rotor moving simulation, the rotor is stepwisely rotated with the step width of the minimum mesh size, and so the number of time divisions in one electrical period is 180. The rotation speed is 3,000 min^{-1} and the maximum voltage between two windings is 200 V.

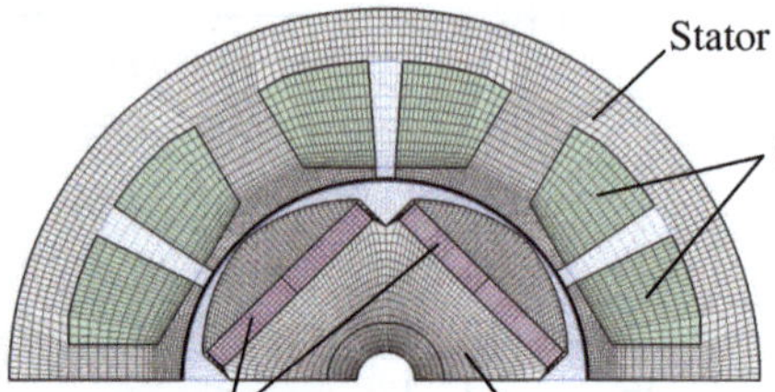

Fig. 6 2D finite element model of a synchronous motor (number of elements: 8,682)

Table 1 Specifications of the synchronous motor

Maximum diameter of rotor	54.8 mm
Inner diameter of stator	56.0 mm
Outer diameter of stator	103 mm
Core length	55 mm
Remanent magnetization of permanent magnet	1.315 T
Conductivity of permanent magnet	6.94×10^5 S/m

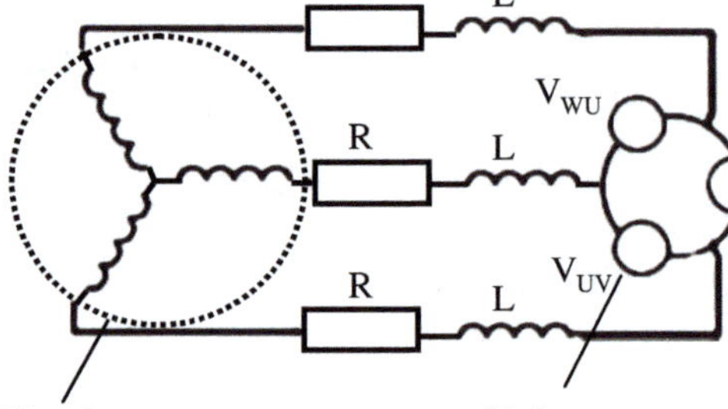

Fig. 7 External power supply circuit connected with coils embedded in the FEM region

The coupled system of the motor and the external power supply is schematically shown in Fig. 7. The three stator windings are connected in star configuration with electric resistances of 0.2 Ω including coil resistances and no external inductance.

The three-phase coil currents are converged in 4,000 time steps with no correction in the transient magnetic field analysis coupled with the external power supply circuit. Figure 8 indicates the comparison of computational results by using several types of correction methods showing waveforms of U-phase coil current, torque and eddy current loss. Corrections by the simplified TP-EEC method are executed every 90 time steps corresponding to one-half of period, while corrections by the simplified three-phase AC TP-EEC method are executed every 30 time steps corresponding to one-sixth of period. Both the simplified TP-EEC and the simplified three-phase AC TP-EEC methods give successive corrections, and TDC method with no time averaging process gives 10 corrections in the initial stage.

In Fig. 8, Mutual ETF ($\theta = 2° \Rightarrow 60°$) indicates that 10 times of corrections by the Mutual ETF method with $\theta = 2°$ are followed by the successive Mutual ETF

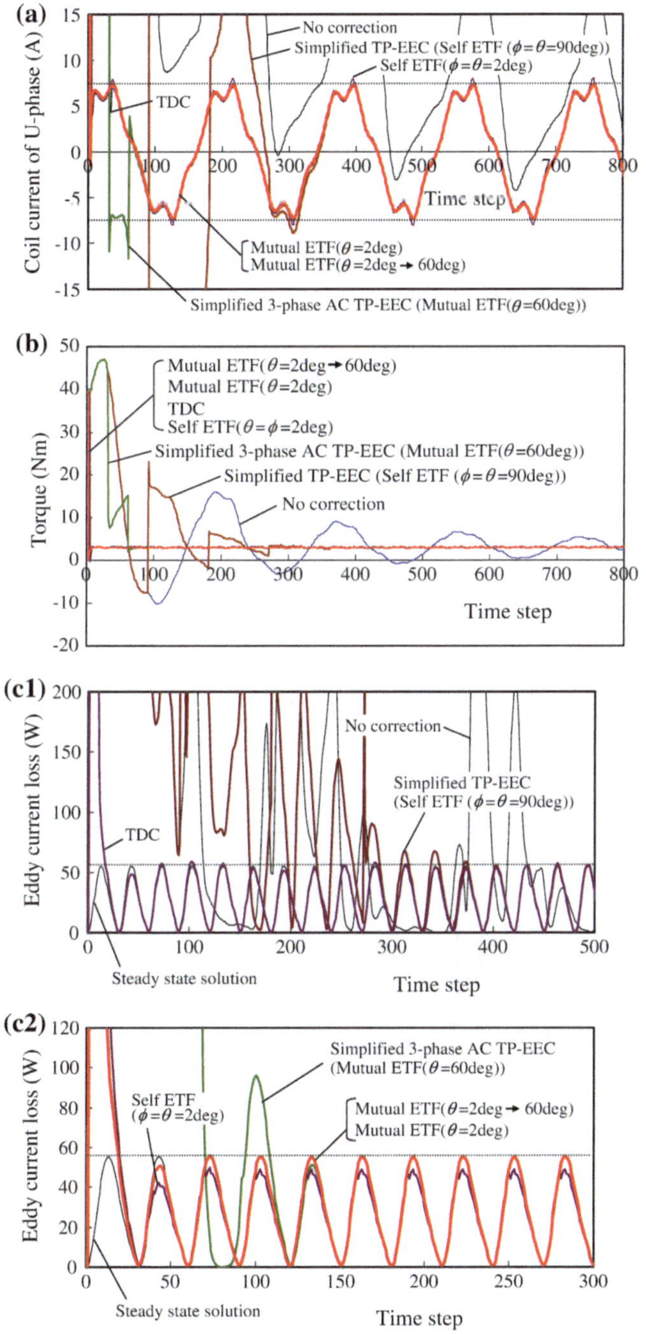

Fig. 8 Comparison among several correction methods in the analyses of torque, eddy current loss, and coil current of the synchronous motor model. **a** Coil current waveform of U-phase. **b** Waveform of torque. **c.1** and **c.2** Waveform of eddy current loss

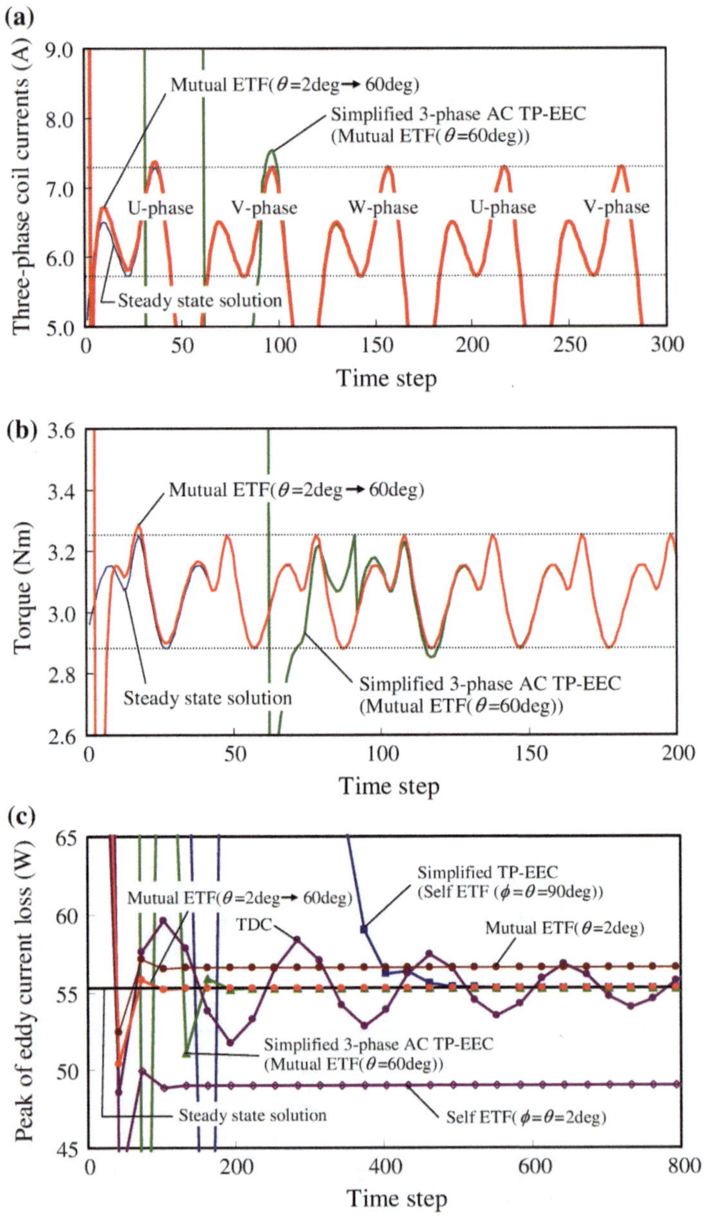

Fig. 9 Comparison of correction precise in the analyses of coil current, torque, and peak of eddy current loss of the synchronous motor model. **a** Current waveforms of U-, V-, and W-phases. **b** Waveform of torque. **c** Peak of eddy current loss

method with $\theta = 60°$ (i.e., the simplified three-phase AC TP-EEC method). The waveforms of eddy current losses are separately shown in Fig. 8c.1, c.2 together with the real steady-state solution obtained as a calculation result after 50,040th time step corresponding to 278 periods.

In order to evaluate easily the precision of converged steady solution, Fig. 9 shows the scaled-up view of waveforms of coil current and torque, and a transient behavior of peak of eddy current loss together with the real steady solutions. and indicates the high performance of Mutual ETF ($\theta = 2° \Rightarrow 60°$). As shown in Fig. 9a, b, waveforms of three-phase coil currents and torque are converged to steady state in 120 time steps by using Mutual ETF ($\theta = 60°$) (i.e., the simplified three-phase AC TP-EEC method), while converged to steady state in 40 time steps by using Mutual ETF ($\theta = 2° \Rightarrow 60°$) (i.e., initial ten times step-by-step successive corrections by Mutual ETF ($\theta = 2°$) followed by Mutual ETF ($\theta = 60°$), that is, the simplified three-phase AC TP-EEC method). The number of time steps required to obtain a steady solution is reduced to one-third by using Mutual ETF ($\theta = 2° \Rightarrow 60°$) compared with the simplified three-phase AC TP-EEC method as a conventional one. On the other hand, the steady-state solution of eddy current loss requires comparatively larger number of time steps. As shown in Fig. 9c, the number of time steps required to obtain a steady-state solution of eddy current loss within the precision of 0.2 % is 190 for the simplified three-phase AC TP-EEC method, while 100 (about one-half of the former case) for Mutual ETF ($\theta = 2° \Rightarrow 60°$). Generally, TDC does not allow successive corrections and indicates the convergence property like a damped wave in Fig. 9c. Self ETF ($\theta = \phi = 2°$) and Mutual ETF ($\theta = 2°$) allow successive corrections, and the steady solutions of eddy current loss deviate -11 and 2.4 %, respectively, from the real solution. Therefore, the process of switching to the simplified TP-EEC or the simplified three-phase AC TP-EEC is necessary in the same way as Mutual ETF ($\theta = 2° \Rightarrow 60°$). In this example, the high performance correction methods are Mutual ETF ($\theta = 2° \Rightarrow 60°$), Mutual ETF ($\theta = 60°$) and Self ETF ($\phi = \theta = 90°$) in the order of convergence speed.

5 Conclusions

We have presented the ETF method as an effective correction method to obtain a steady-state solution of nonlinear field with a long time constant. The ETF method is classified into two categories of Self ETF and Mutual ETF methods. Self ETF method uses self field only, while Mutual ETF method uses plural fields with phases different from each other in the polyphase AC system. In the ETF method, transient fields are effectively corrected toward the steady fields by using the relationship between the fundamental wave components of the nonlinear field and their time variations in the steady state. The simplified TP-EEC and TDC methods belong to the Self ETF method, while the simplified AC TP-EEC method belongs to the Mutual ETF method. In the ETF method, the time interval to obtain time variations can be flexibly selected, and then a shortest time interval is selected to realize step-by-step corrections in the Mutual ETF method. The short time interval ETF method has a powerful performance in an initial stage, but its single use is restricted due to harmful effects of higher order time-harmonic modes. The convergence property of the long time interval ETF method, like the simplified TP-EEC and the simplified three-phase AC TP-EEC, is improved by several corrections in an initial stage by the short time interval ETF. The hybrid ETF method, i.e., the serial usage of the short and long time-interval ETF methods, is verified by the test example calculation of the nonlinear magnetic field analysis coupled with a power supply circuit in a synchronous motor with eddy current fields.

References

1. T.J. Aprille Jr., T.N. Trick, Steady-state analysis of nonlinear circuits with periodic inputs, Proc. IEEE **60**(1), 108–114 (1972)
2. S. Li, H. Hofmann, Numerically efficient steady-state finite-element analysis of magnetically saturated electromechanical devices. IEEE Trans. Magn. **39**(6), 3481–3484 (2003)
3. T. Hara, T. Naito, J. Umoto, Field analysis of corona shield region in high voltage rotating machines by time-periodic finite element method: I numerical calculation method, Trans. IEE Jpn. **102-B**(7), 423–430 (1982) (in Japanese)
4. T. Nakata, N. Takahashi, K. Fujiwara, K. Muramatsu, H. Ohashi, H.L. Zhu, Practical analysis of 3-D dynamic nonlinear magnetic field using time-periodic finite element method. IEEE Trans. Magn. **31**(3), 1416–1419 (1995)
5. S. Yamada, J. Lu, K. Bessho, T. Yoshimoto, Alternating-current magnetic field analysis including magnetic saturation by a harmonic balance finite element method. Trans. IEE Jpn. **109-D**(10), 756–762 (1989) (in Japanese)
6. T. Tokumasu, M. Fujita, T. Ueda, Problems remained in practical usage of 2 dimensional electromagnetic analyses (3), The papers of joint technical meeting on static apparatus and rotating machinery. IEE Japan, SA-08-62/RM-08-69 (2008) (in Japanese)
7. Y. Takahashi, T. Tokumasu, M. Fujita, S. Wakao, T. Iwashita, M. Kanazawa, Improvement of convergence characteristics in nonlinear transient eddy-current analyses using the error correction of time integration based on the time-periodic FEM and the EEC method, IEEJ Trans. PE, **129**(6), 791–798 (2009) (in Japanese)
8. T. Tokumasu, M. Fujita, T. Ueda, Problems remained in practical usage of 2 dimensional electromagnetic analyses (4), The papers of joint technical meeting on static apparatus and rotating machinery. IEE Japan, SA-09-6/RM-09-6 (2009) (in Japanese)
9. T. Tokumasu, M. Fujita, T. Ueda, Problems remained in practical usage of 2 dimensional electromagnetic analyses (6), The papers of joint technical meeting on static apparatus and rotating machinery. IEE Japan, MAG-10-1/SA-10-1/RM-10-1 (2010) (in Japanese)
10. K. Miyata, Convergence improvement of time-periodic nonlinear field analysis by using step-by-step continuous error corrections, The papers of joint technical meeting on static apparatus and rotating machinery. IEE Japan, SA-13-3/RM-13-3 (2013) (in Japanese)
11. K. Miyata, Fast analysis method of time-periodic nonlinear fields. J. Math. Ind. **3**, 131–140 (2011) (JMI2011B-7)

Mathematical Models in First-Principles Calculations for Materials Science

Hajime Kobayashi

Abstract The mathematical models used in first-principles calculations are summarized, and the uses of mathematics in various industries are introduced. The different approximations used to obtain the Hartree-Fock equation from the Schrödinger equation for multi-atom systems are summarized, and difficulties in solving the Hartree-Fock equation in a self-consistent way are presented. Novel algorithms are needed in order to reduce computational costs of large systems.

Keywords First-principles calculations · Hartree-Fock · molecular orbital · SCF · DFT · computer simulation · materials science · electron correlation

1 Introduction

The diverse features of materials are determined by their electron states, which are in turn described by quantum mechanics. All material properties subject to a non-relativistic limit, such as total energy, energy levels, electron density, electrostatic potential, dielectric moments, frequencies and elastic moduli, can be understood by solving the Schrödinger equation for multi-atom systems. The time-independent Schrödinger equation for multi-atom systems is as follows:

$$H\Psi = E\Psi \tag{1}$$

H. Kobayashi (✉)
Advanced Materials Laboratories, Sony Corporation, 4-14-1 Asahicho,
Atsugi 243-0014, Japan
e-mail: HajimeA.Kobayashi@jp.sony.cpm

R. Nishii et al. (eds.), *A Mathematical Approach to Research Problems of Science and Technology*, Mathematics for Industry 5,
DOI: 10.1007/978-4-431-55060-0_34,

$$H = \sum_{A}^{N_n}\left(-\frac{\hbar^2}{2M_A}\nabla_A^2\right) + \sum_{i}^{n}\left(-\frac{\hbar^2}{2m}\nabla_i^2\right) - \sum_{i}^{n}\sum_{A}^{N_n}\frac{Z_A e^2}{r_{iA}} + \sum_{i<j}^{n}\frac{e^2}{r_{ij}} + \sum_{A<B}^{N_n}\frac{Z_A Z_B e^2}{R_{AB}} \quad (2)$$

Here, H is the Hamiltonian and Ψ is the wave function of the system. N_n is the number of nuclei; n is the number of electrons; M_a and Z_A are the mass and electric charge of nucleon A, respectively; R_{AB} is the distance between nucleons A and B; m is the electron mass; r_{iA} is the distance between the ith electron and the Ath nucleon; and r_{ij} is the distance between the ith and jth electrons. The terms on the right side in Eq. (2) are the kinetic energy of the nuclei, the kinetic energy of the electrons, the Coulomb energy between the nuclei and electrons, the Coulomb energy of the electrons, and the Coulomb energy of the nuclei, respectively. This equation dates from the 1920s, just after quantum mechanics was established. However, only small systems, such as hydrogen atoms, can be exactly solved because it is impossible to analytically solve three body systems. Moreover, even with powerful computers, it quickly becomes computationally too expensive to calculate Eq. (2) as the number of atoms included in the calculation is increased. P. A. Dirac noted this difficulty, writing as follows:

> The underlying physical laws necessary for the mathematical theory of a large part of physics and the whole chemistry are thus completely known, and the difficulty is only that the exact application of these laws leads to equations much too complicated to be soluble [3].

Thus, approximations are required to overcome this difficulty. The term "first-principles calculation" means not to use any empirical parameters; however, it does use approximations. Various mathematical models have been used to reduce the calculation costs while keeping accuracy as high as possible [5, 7, 12].

2 Mathematical Models Used in First-Principles Calculations

2.1 Born-Oppenheimer Approximation

Velocities of nuclei are considerably smaller than those of electrons because of their larger masses. Therefore, the Born-Oppenheimer approximation assumes the nuclei to be in the resting state. Under this approximation, the first term in Eq. (2) becomes negligible and the fifth term becomes a constant. Therefore, the remaining terms, which are related to electrons, can be separated, and the Schrödinger equation for the electrons can be written as

$$H_e \Psi_e = E_e \Psi_e, \quad (3)$$

$$H_e = \sum_{i}^{n} \left(-\frac{\hbar^2}{2m} \nabla_i^2 \right) - \sum_{i}^{n} \sum_{A}^{N_n} \frac{Z_A e^2}{r_{iA}} + \sum_{i<j}^{n} \frac{e^2}{r_{ij}}. \tag{4}$$

2.2 Molecular Orbital Method

The one-electron Hamiltonian, h_i, and the one-electron wave function, ϕ_i, satisfy the following equation:

$$h_i \phi_i = \varepsilon_i \phi_i, \tag{5}$$

where ε_i is the eigenvalues of the energy. ϕ_i is called the molecular orbital. When h_i is the result of taking the mean field approximation, the Hamiltonian of such an n electron system can be expressed by the sum of h_i:

$$H_e = \sum_{i}^{n} h_i. \tag{6}$$

Suppose that the total wave function, Ψ_0, is the product of n electron wave function, ϕ_i. Then Ψ_0 is an eigenvalues of H_e:

$$\Psi_0 = \phi_1 \phi_2 \cdots \phi_n \tag{7}$$

$$H_e \Psi_0 = E \Psi_0. \tag{8}$$

Ψ_0 is called the Hartree product. The total energy, *E*, is the sum of the eigenvalues corresponding to the one-electron energies, ε_1:

$$E = \sum_{i}^{n} \varepsilon_i. \tag{9}$$

The one-electron wave function, ϕ_I, can be expressed as a superposition of *N* basis functions:

$$\phi_i = \sum_{\nu=1}^{N} C_{\nu i} \chi_\nu. \tag{10}$$

Here, $C_{\nu i}$ are called molecular orbital coefficients. The atomic orbitals of real atoms have known shapes, and they are used as the basis functions in the linear combination of atomic orbitals (LCAO) approximation. Moreover, Gaussian functions can be used as atomic orbitals in the LCAO method. Molecular orbitals satisfy orthonormality:

$$\int \phi_i^* \phi_j \mathrm{d}r = \delta_{ij}. \tag{11}$$

Thus, the n electrons problem becomes a one-electron Schrödinger equation. Note that the mean field assumption has been used in this approximation. The effect of the approximation will be described in the Sect. 2.8.

2.3 Pauli Exclusion Principle

The Pauli exclusion principle requires that the wave function should be antisymmetric when two electrons are exchanged. Since the Hartree product in Eq. (7) does not satisfy this requirement, the Slater determinant is used instead:

$$\Psi = \frac{1}{\sqrt{n!}} \begin{vmatrix} \phi_1(1) & \phi_2(1) & \cdots & \phi_n(1) \\ \phi_1(2) & \phi_2(2) & \cdots & \phi_n(2) \\ \vdots & \vdots & \ddots & \vdots \\ \phi_1(n) & \phi_2(n) & \ldots & \phi_n(n) \end{vmatrix} . \tag{12}$$

The Slater determinant, *Y*, is expressed by using the Hartree product *Y*0 and the anti-symmetrical operator $\hat{A}$:

$$\Psi = \sqrt{n!}\hat{A}\Psi_0 \tag{13}$$

$$\hat{A} = \frac{1}{n!}\sum_{p_n}^{n!}(-1)^{p_n}\hat{P}_n, \tag{14}$$

where $\hat{P}_n$ is the permutation operator and p_n is the number of permutations. Ψ satisfies orthonormality:

$$\int \Psi^*\Psi \mathrm{d}r = 1. \tag{15}$$

In order to derive a convenient relationship, the expectation value of $\hat{S}$, which is a symmetric operator under the exchange of two coordinates, is taken:

$$\int \Psi^*\hat{S}\Psi \mathrm{d}r = n! \int \hat{A}\Psi_0^*\hat{S}\hat{A}\Psi_0 \mathrm{d}r. \tag{16}$$

Since $\hat{A}$ is Hermitian,

$$\int \Psi^*\hat{S}\Psi \mathrm{d}r = n! \int \Psi_0^*\hat{A}\hat{S}\hat{A}\Psi_0 \mathrm{d}r. \tag{17}$$

Moreover, since $\hat{S}$ is symmetric, $\hat{A}\hat{S} = \hat{S}$. Therefore,

$$\int \Psi^* \hat{S} \Psi \mathrm{d}r = n! \int \Psi_0^* \hat{S} \hat{A} \Psi_0 \mathrm{d}r$$

$$\text{and} \quad \int \Psi^* \hat{S} \Psi \mathrm{d}r = \sum_{p_n}^{n!} (-1)^{p_n} \int \Psi_0^* \hat{S} \hat{P}_n \Psi_0 \mathrm{d}r. \tag{18}$$

The next section shows that a calculation of physical properties that would otherwise involve Slater determinant can be performed using Hartree products.

2.4 The Hartree-Fock Method

Equation (4) can be separated into one-electron and two-electron parts, as follows:

$$H_e = \sum_i^n h_i + \sum_{i<j}^n g_{ij}, \tag{19}$$

$$h_i = -\frac{\hbar^2 \nabla_i^2}{2m} - \sum_A^{N_n} \frac{Z_A e^2}{r_{iA}}, \tag{20}$$

$$g_{ij} = \frac{e^2}{r_{ij}}. \tag{21}$$

The expectation value of the energy is

$$\begin{aligned} E_e &= \int \Psi^* H \Psi \mathrm{d}r \\ &= \sum_i^n \int \Psi^* h_i \Psi \mathrm{d}r + \sum_{i<j}^n \int \Psi^* g_{ij} \Psi \mathrm{d}r. \end{aligned} \tag{22}$$

After transformation with Eq. (18),

$$E_e = \sum_i^n \sum_{p_n}^{n!} (-1)^{p_n} \int \Psi_0^* h_i \hat{P}_n \Psi_0 \mathrm{d}r + \sum_{i<j}^n \sum_{p_n}^{n!} (-1)^{p_n} \int \Psi_0^* g_{ij} \hat{P}_n \Psi_0 \mathrm{d}r. \tag{23}$$

The orthonormality of the orbitals means that the first term on the right side is not zero only when $\hat{P}_n$ is the identity permutation.

$$\text{The first term in Eq. (23)} = \sum_i^n \int \phi_i^* h_i \phi_i \mathrm{d}r. \tag{24}$$

Moreover, when $\hat{P}_n$ is the identity permutation, or i and j are exchanged, the second term is not zero.

$$\text{The second term in Eq. (23)} = \sum_{i<j}^{n} \int\int \phi_i^* \phi_j^* g_{ij} \phi_i \phi_j \mathrm{d}r_1 \mathrm{d}r_2 - \sum_{i<j}^{n} \int\int \phi_i^* \phi_j^* g_{ij} \phi_j \phi_i \mathrm{d}r_1 \mathrm{d}r_2 \tag{25}$$

$$= \frac{1}{2}\left(\sum_{i,j}^{n} \int\int \phi_i^* \phi_j^* g_{ij} \phi_i \phi_j \mathrm{d}r_1 \mathrm{d}r_2 - \sum_{i,j}^{n} \int\int \phi_i^* \phi_j^* g_{ij} \phi_j \phi_i \mathrm{d}r_1 \mathrm{d}r_2 \right). \tag{26}$$

Thus,

$$E_e = \sum_{i}^{n} h_{ii} + \frac{1}{2} \sum_{i,j}^{n} (J_{ij} - K_{ij}). \tag{27}$$

Here,

$$h_{ii} = \int \phi_i(1)^* h_1 \phi_i(1) \mathrm{d}r_1, \tag{28}$$

$$J_{ij} = \int\int \phi_i^*(1) \phi_j^*(2) g_{12} \phi_i(1) \phi_j(2) \mathrm{d}r_1 \mathrm{d}r_2, \tag{29}$$

$$K_{ij} = \int\int \phi_i^*(1) \phi_j^*(2) g_{12} \phi_j(1) \phi_i(2) \mathrm{d}r_1 \mathrm{d}r_2. \tag{30}$$

Since the electrons in the Slater products are distinguishable, h_{ii} is usually expressed by the coordinates of the first electron, and J_{ij} and K_{ij} are expressed by the coordinates of the second electron.

From the variational principle, the wave function, Ψ, that makes the total energy a minimum, is the solution in the real system. This indicates that the set of ϕ_i should be obtained which minimizes E_e under the condition $\int \phi_i^* \phi_j \mathrm{d}r = \delta_{ij}$ and should be found in the set of $C_{\nu i}$ in Eq. (10) corresponding to them. The condition which $C_{\nu i}$ should satisfy can be obtained by using the Lagrange multiplier method:

$$\sum_{\nu=1}^{N} (F_{\mu\nu} - \varepsilon_i S_{\mu\nu}) C_{\nu i} = 0. \tag{31}$$

Here,

$$F_{\mu\nu} = h_{\mu\nu} + \sum_{\lambda,\sigma}^{N} P_{\lambda\sigma} \{2(\mu\nu\,|\lambda\sigma) - (\mu\sigma\,|\lambda\nu)\}\,, \tag{32}$$

$$h_{\mu\nu} = \int \chi_\mu^*(1) h_1 \chi_\nu(1) \mathrm{d}r_1, \tag{33}$$

$$(\mu\nu\,|\lambda\sigma) = \int\int \chi_\mu^*(1)\chi_\nu(1) g_{12} \chi_\lambda^*(2)\chi_\sigma(2) \mathrm{d}r_1 \mathrm{d}r_2, \tag{34}$$

$$P_{\lambda\sigma} = \sum_i^{n/2} C_{\lambda i}^* C_{\sigma i}, \tag{35}$$

$$S_{\mu\nu} = \int \chi_\mu^*(1)\chi_\nu(1) \mathrm{d}r_1. \tag{36}$$

Equation (31) is named the Hartree-Fock-Roothaan equation. It can be written in matrix form:

$$\mathbf{FC} = \mathbf{SC}\varepsilon. \tag{37}$$

Now, the problem of solving the Schrödinger equation of multi-atom systems Eq. (1) becomes tone of solving the Hartree-Fock-Roothaan Equation (37). This problem is named the Hartree-Fock method. Since F contains P and P contains C, this is a nonlinear equation.

2.5 Procedure for Solving the Equation

Since Eq. (37) is a nonlinear equation, it is usually solved using the self-consistent field (SCF) procedure. Generally $\mathbf{S} \neq \mathbf{I}$ because atomic orbitals are not usually orthogonal. Therefore, Eq. (37) is a generalized eigenvalue problem. In computer calculations, S is orthonormalized before solving the eigenvalue problem, to increase computing performance. A unitary matrix exists to orthonormalize **S,** since **S** is Hermitian:

$$\mathbf{U}^{+}\mathbf{SU} = \mathbf{I}. \tag{38}$$

$\mathbf{F}'$ and $\mathbf{C}'$ are obtained by using $\mathbf{U}$, as follows:

$$\mathbf{F}' = \mathbf{U}^{+}\mathbf{FU}, \tag{39}$$

$$\mathbf{C}' = \mathbf{U}^{-1}\mathbf{C}. \tag{40}$$

By using $\mathbf{F}'$ and $\mathbf{C}'$, Eq. (37) becomes

$$\mathbf{F}'\mathbf{C}' = \mathbf{C}'\varepsilon. \tag{41}$$

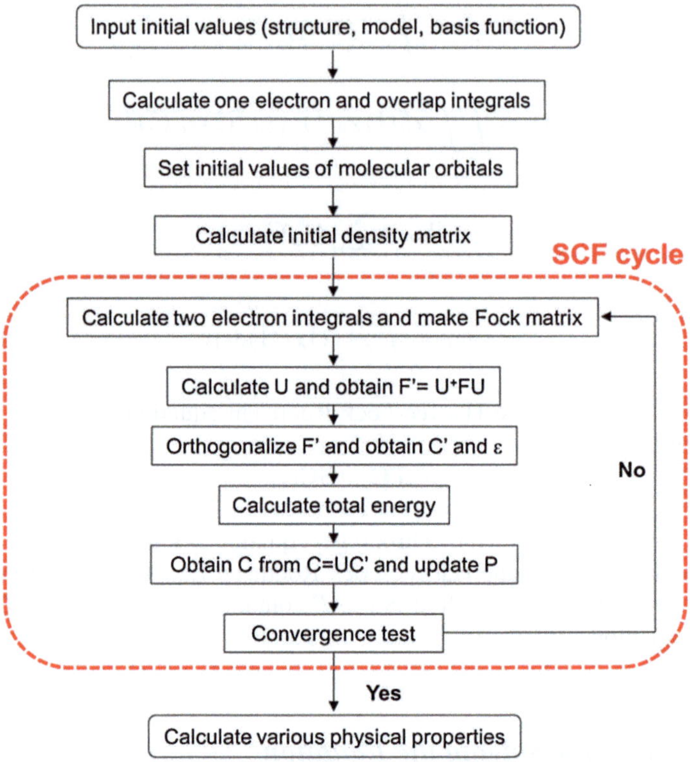

Fig. 1 Flowchart of SCF calculation

$\mathbf{C}'$ is obtained after orthonormalization of $\mathbf{F}'$, and $\mathbf{C}$ is obtained from the following equation:

$$\mathbf{C} = \mathbf{U}\mathbf{C}'. \tag{42}$$

Figure 1 shows a flowchart of the SCF calculation.

2.6 Occupied and Unoccupied Orbitals

N molecular orbitals and energy levels are obtained from the Hartree-Fock-Roothaan Equation (10) when N basis functions are used. Taking account of spin, two electrons are allowed to occupy an orbital, each orbital is filled up with two electrons starting from the lowest energy level in the ground state. Therefore, $n/2$ orbitals are occupied in an n electron system (occupied orbitals). The remaining $N - n/2$ orbitals are unoccupied (Fig. 2). The unoccupied orbitals are important when excited states or

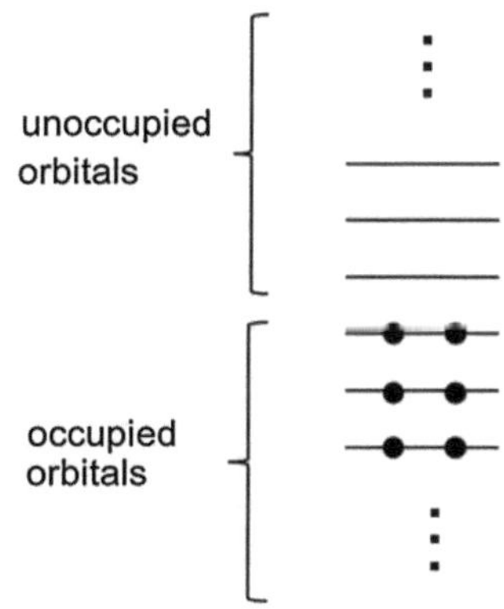

Fig. 2 Occupied and unoccupied orbitals

chemical reactions induced by thermal energy or light are studied. They also play an important role in electron correlations, as described later.

2.7 Difficulty of Calculations

The outline of the Hartree-Fock method, which is a basic mathematical model of first-principles calculation, was reviewed in the previous section. Calculations can be performed by following the flow chart in Fig. 1. However, difficulties arise during actual calculations. Though it is desirable that the total energy of the system converge rapidly, as shown in Fig. 3a, this becomes hard to guarantee as the system size increases. The difficulties typically fall into one of the two situations below.

(a) Total energy does not converge.

The total energy does not always converge; it may oscillate around a higher energy than the convergence condition, as shown in Fig. 3b. In this case, the selection of basis functions or structure (atom configuration) is sometimes important. Convergence is also dependent on the solvers of the SCF calculation. Numerous methods have been proposed, such as SD, CG, DM [11], QC [1], DIIS [9] and EDIIS [6]. However, each of these methods has its problems, and the appropriate method depends on the system to be solved. Though first-principles calculations are the nonempirical methods, at present, the SCF calculation requires a lot of know-how based on experience. Novel algorithms, which are versatile enough to be used on any system, must be developed to make first-principles calculations easy to use.

(b) Converging, but time-consuming.

Figure 3c shows an example where the SCF calculation converges but consumes a lot of time. Calculation of the two-electron integrals is the most time-consuming process; here parallelization algorithms [8] can be used to shorten the calculation time. The data involved in the two-electron integrals becomes large as the system size increases, and critically affects the calculation time. It should be considered whether

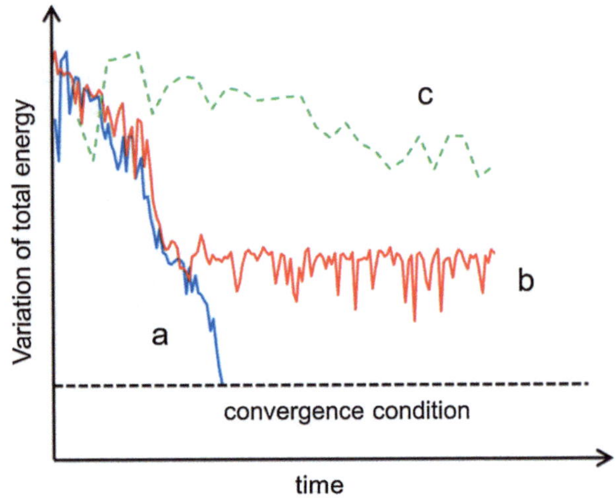

Fig. 3 Convergence issue affecting SCF calculations

the data are on memories, on hard disks or the integrals are calculated every time without storage. In practice, one has to make optimizations taking account of the size of the calculation and the performance of the hardware, such as CPUs, memories, hard disks and I/O.

2.8 Electron Correlation

The Hartree-Fock method is the basis of other first-principles calculations. However, it does not include the effect of electron correlation, such as electron-electron scattering or reorganization of the electron configuration, because it uses the mean field approximation, which assumes one electron moves in an averaged potential generated by the other electrons. Therefore, the Hartree-Fock method is inaccurate for strong correlated electron systems. In such cases, post Hartree-Fock methods are used. There are essentially three such methods, i.e., the configuration interaction method (CI), the coupled cluster method (CC), and the perturbation method. All methods use the unoccupied orbitals for reorganization of the electron configuration induced by the electron correlation. The Slater determinant from the Hartree-Fock method is used as a ground state σ_0, and excited states, wherein electrons are excited into unoccupied orbitals, are added in order to increase the degree of freedom of the electron configuration (Fig. 4).

In the CI method, the excited states are superimposed linearly, as follows:

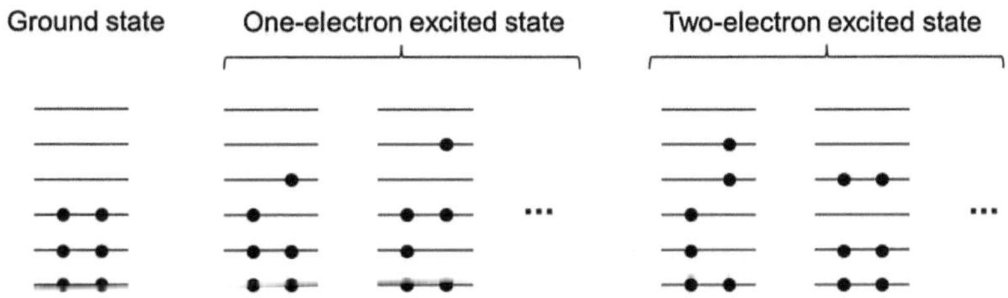

Fig. 4 Excited states for taking account of electron correlations

$$\Psi^{CI} = \Phi_0 + \left(\sum_i C_i^{CI} \hat{T}_i \right) \Phi_0, \tag{43}$$

where $\hat{T}_i$ is the excitation operator to create i electron excited states from the ground state σ_0. Ψ^{CI} satisfies the Pauli exclusion principle by using Slater determinants for the excited states. The wave functions including electron correlations can be obtained by using the variational principle to determine C_i^{CI}. When all excited states on n electrons are considered, an exact electron correlation is obtained (full CI).

In the CC method, the following wave function is used:

$$\Psi^{CC} = \exp\left(\sum_i C_i^{CC} \hat{T}_i \right) \Phi_0. \tag{44}$$

A Taylor expansion yields

$$\Psi^{CC} = \Phi_0 + \left\{ \left(\sum_i C_i^{CC} \hat{T}_i) \right) + \frac{1}{2} \left(\sum_i C_i^{CC} \hat{T}_i \right)^2 + \ldots \right\} \Phi_0. \tag{45}$$

The difference from the CI method is the term of $\frac{1}{2} \left(\sum_i C_i^{CC} \hat{T}_i \right)^2 + \ldots$. Thus, the CC method contains higher-order excited states despite using the same number of coefficients as the CI method. Therefore, the CC method is more accurate than the CI method.

In the perturbation method, the effects of well-known weak interactions perturbing the main interaction, which are calculated successively:

$$H = H_0 + V. \tag{46}$$

Here, H_0 is the main nonperturbative interaction and V is the weak perturbative interaction. Once the wave function and eigenvalues of H_0 are obtained,

$$H_0 \Psi_i^{(0)} = E_i^{(0)} \Psi_i^{(0)}. \tag{47}$$

The superscripts in parentheses express the degree of perturbation. When there is no degenerate energy levels, the true wave function Ψ_i and eigenenergies E_i are written as follows:

$$H\Psi_i = E_i\Psi_i, \tag{48}$$

$$\Psi_i = \Psi_i^{(0)} + \Psi_i^{(1)} + \Psi_i^{(2)} + \cdots \tag{49}$$

$$= \Psi_i^{(0)} + \sum_{n\neq} \frac{\int \Psi_n^{(0)} V \Psi_i^{(0)} dr}{E_i^{(0)} - E_n^{(0)}} \Psi_n^{(0)} + \cdots, \tag{50}$$

$$E_i = E_i^{(0)} + E_i^{(1)} + E_i^{(2)} + \cdots \tag{51}$$

$$= E_i^{(0)} + \int \Psi_i^{(0)} V \Psi_i^{(0)} \mathrm{d}r + \sum_{n\neq} \frac{\left| \int \Psi_i^{(0)} V \Psi_n^{(0)} \mathrm{d}r \right|^2}{E_i^{(0)} - E_n^{(0)}} + \cdots. \tag{52}$$

The perturbation method is useful when V is considerably smaller than H_0. The Møller-Plesset (MP) method uses the ground state (Hartree-Fock configuration) as H_0. Since $E_i^{(0)}$ is usually greater than 99 % of the total energy, this is a good approximation. In the calculation procedure, H_0 is calculated with a standard SCF first, and then successive perturbation terms are calculated after that. The perturbation part does not have convergence problem because it is not SCF calculation. Principally, accuracy increases as the degree of perturbation increases. These post-Hartree-Fock methods require longer calculation times than the Hartree-Fock method, which calculates only the ground state, and hence, novel computing techniques need to be developed in order to reduce this time.

There is another way to consider electron correlations: the density functional method (DFT). Like the Hartree-Fock method, the DFT method uses the mean field approximation; however, it uses only electron densities and does not use wave functions. The DFT method is generally more accurate than the Hartree-Fock method but has similar calculation costs.

2.9 Accuracy and Calculation Cost

The accuracy of first-principles calculations is determined by the size of the basis functions and the theoretical model of electron correlation. Extremely high accuracy is obtained when a large basis set and the CC method with three-electron excited states are used. For example, the energy obtained by such method is consistent with the experimental value within 0.02 eV. However, a tradeoff exists in that any increase in accuracy is at the expense of a higher calculation cost (Fig. 5). For this reason, relevant basis functions and theoretical models should be carefully chosen in order to reduce calculation time while keeping a certain level of accuracy.

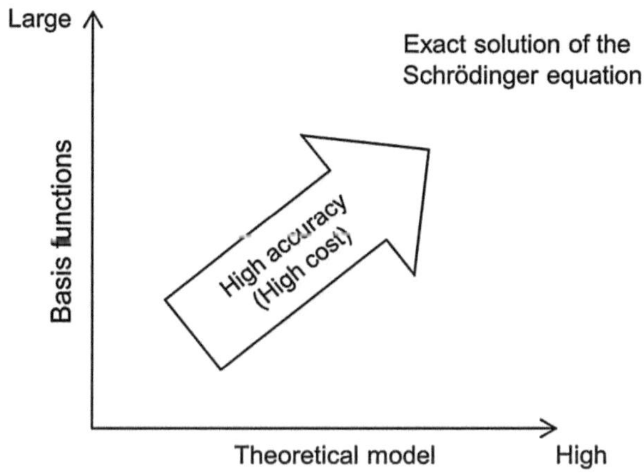

Fig. 5 Relationship between basis functions, theoretical model, and accuracy

2.10 Structure Optimizations

The structure of the system under study must be known before starting the calculations. It is preferable to determine the structure accurately by experiment. However, this is not always possible. In such cases, the structure can be numerically optimized; first, an SCF calculation is performed by taking account of the fifth term in Eq. (2), which is the Coulomb energies of the nuclei and was ignored in Eq. (4). Then, forces at the position of each atom are obtained from the first-order derivative of the potential energy, and the atoms are displaced according to the forces. After that, a new SCF calculation is performed on the new structure. This procedure is iterated until a convergence condition is reached, e.g., the variation of the average force of all atoms falls below a certain value. More stable (lower energy) structures are found in this way, but the search for the minimum potential energy is conducted in the 3N-dimensional space formed by N nuclei (Fig. 6). That means one must try to find the global minimum amidst possibly many local minima in a large space. This method entails large calculation costs, because it contains two loops: one for the SCF calculation and one for the structure search. In order to make the search efficient, a rough search is used at the beginning and it is successively refined. Various search algorithms have been proposed, including the Berny [10], GDIIS [4] and eigenvalue-following [2]. However, faster algorithms would be welcomed.

3 Summary

Various approximations are used in first-principles calculations because of the difficulty in exactly solving the Schrödinger equation of multi-atom systems. It is important to know the contribution of these approximations to the final accuracies. It is desirable to calculate a system on which there is experimental data first in order

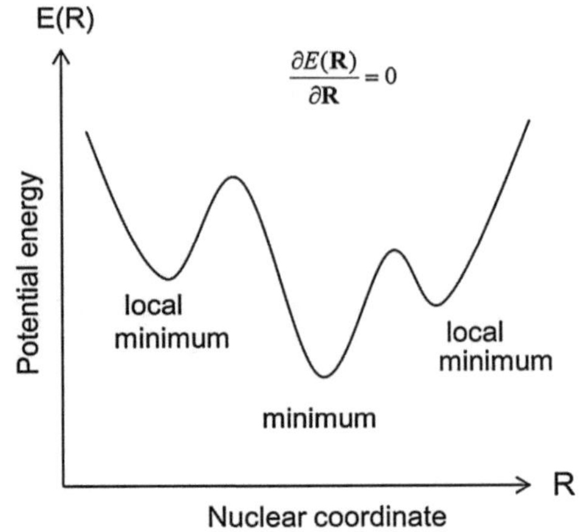

Fig. 6 Structure optimization by searching for the global minimum of potential energy

to confirm the accuracy of a method. If there is no experimental data, it is possible to estimate the accuracy by performing a higher level calculation with larger basis functions and a more detailed theoretical model, because the higher level calculation guarantees higher accuracy, as shown in Fig. 5. Novel algorithms are needed in order to reduce the costs of the SCF calculation and structure optimization. Supercomputers are used for large-scale simulations, and efficient algorithms for large-scale parallel computing are necessary in order to use them efficiently. The advent of new mathematical methods will benefit computational materials science.

References

1. G.B. Bacskay, A quadratically convergent Hartee-Fock (QC-SCF) method. Application to closed shell systems. Chem. Phys. **61**, 385 (1981)
2. C.J. Cerjan, W.H. Miller, On finding transition states. J. Chem. Phys. **75**, 2800 (1981)
3. P.A.M. Dirac, Quantum mechanics of many-electron systems. Proc. Roy. Soc. **A123**, 714 (1929)
4. Ö. Farkas, H.B. Schlegel, Methods for optimizing large molecules. II. Quadratic search. J. Chem. Phys. **111**, 10806 (1999)
5. J.B. Foresman, Æ. Frisch, *Exploring Chemistry with Electronic Structure Methods*, 2nd edn. (Gaussian Inc, Pittsburgh, 1996)
6. K.N. Kudin, G.E. Scuseria, A black-box self-consistent field convergence algorithm: one step closer. J. Chem. Phys. **116**, 8255 (2002)
7. T. Nakajima, *Ryoshi kagaku* (Shokabo, Japan, 2009) (in Japanese)
8. S. Obara, A. Saika, Efficient recursive computation of molecular integrals over Cartesian Gaussian function. J. Chem. Phys. **84**, 3963 (1986)
9. P. Pulay, Improved SCF convergence acceleration. J. Comput. Chem. **3**, 556 (1982)
10. H.B. Schlegel, Optimization of equilibrium geometries and transition structures. J. Comput. Chem. **3**, 214 (1982)
11. R. Seeger, J.A. Pople, Self-consistent molecular orbital methods. XVI. Numerically stable direct energy minimization procedures for solution of Hartree-Fock equations. J. Chem. Phys. **65**, 265 (1976)
12. A. Szabo, N.S. Ostlund, *Modern Quantum Theory* (Macmillan, New York, 1982)

Mathematics and Manufacturing: The Symbolic Approach

Ryusuke Masuoka and Hirokazu Anai

Abstract This chapter discusses applications of the symbolic approach to the manufacture of hardware and software. Two example applications, one hardware and the other software, are illustrated. The first example is the design of a hard disk drive (HDD) head by using quantifier elimination (QE), and the other is software validation using symbolic execution. Both examples demonstrate the strengths of the symbolic approach over conventional numerical approaches. While there are, of course, challenges facing the symbolic approach such as faithful modeling and the need for abstraction, it is an extremely powerful and game-changing technology.

Keywords Manufacturing · Mathematics · Quantifier elimination · Software validation · Symbolic execution · Symbolic optimization

---

R. Masuoka (✉)
Center for International Public Policy Studies, Mitsui Main Building, 5th Floor, 2-1-1, Nihonbashi Muromachi, Chuo-ku, Tokyo 103-0022, Japan
e-mail: masuoka@cipps.org; masuoka.ryusuke@jp.fujitsu.com

R. Masuoka · H. Anai
Fujitsu Laboratories Limited, 4-1-1, Kamikodanaka, Nakahara-ku, Kawasaki, Kanagawa 211-8588, Japan
e-mail: anai@jp.fujitsu.com; anai@imi.kyushu-u.ac.jp

H. Anai
Institute of Mathematics for Industry, Kyushu University, 744, Motooka, Nishi-ku, Fukuoka 819-0395, Japan

R. Nishii et al. (eds.), *A Mathematical Approach to Research Problems of Science and Technology*, Mathematics for Industry 5,
DOI: 10.1007/978-4-431-55060-0_35,

1 Introduction

This chapter discusses manufacturing and mathematics, in particular, applications of symbolic approach to manufacturing.

A number of mathematical approaches, not just symbolic, but also numerical analysis and optimization, are used in various aspects of manufacturing hardware and software. The methodologies and associated challenges of applying mathematics to manufacturing can be generally described as follows:

1. Modeling: To model the target system
2. Analysis: To simulate and analyze the target system's behaviors in its environment
3. Design:
   - To formulate the design as a mathematical problem according to the given design goals
   - To solve the above mathematical problem
4. Validation:
   - To show that the system design is correct and defect-free
   - To make sure the system does not fall into undesirable states

In the case of the hard disk drive (HDD) head discussed in Sect. 2, the steps above can be described as follows:

1. Modeling: To build a mathematical model for the actuator's control system. This model takes parameters for the arm's actuator controller and ouputs the calculated position of the head (which is attached to the arm)
2. Analysis: To simulate the arm's actuator model with the head defined by design parameters within the actual environment (considering air-resistance and other interactions with its environment) and analyze its movements and states
3. Design: To build target functions, which evaluates the head's states such as its hovering height and angle. Then to find a set of design parameters to optimize the target functions
4. Validation: To verify that the design based on the parameters obtained from (1), (2), and (3) is actually within the given specifications

This paper describes two applications of the mathematical symbolic approach to hardware and software manufacture. The design (Step 3 above) of an HDD head and software validation (Step 4 above) are illustrated in Sects. 2 and 3.

Unlike numerical approaches, the symbolic approach supports formulas without instantiating their variables, and this is one of its strengths. Let us see how this can be true by examining a simplified version of the HDD design problem presented in Sect. 2.

Let us assume, due to design constraints, that the hovering height, x^2+bx+a, of the HDD head needs to be more than a given d, where x represents the arm position (of HDD head) and a and b are design parameters. If we put $c=a-d$, this condition can be written as $x^2+bx+c>0$ for $\forall x$.

We need to determine appropriate values for the design parameters, *b* and *c* (*a* to be exact, but we will use *c* for the sake of simplicity). If we were to take a numerical approach, we would assign random values to *b*, *c*, and *x* and see if the conditions hold. Or we would change the values incrementally in order to let them reach the solutions gradually. Such numerical approaches have merits; for instance, they are independent of the kind of formula and elements involved in the problem. On the other hand, they also depend on "luck" and more importantly, one cannot tell how good the obtained solution is.

If we use a symbolic approach (that is, using a discriminant for the quadratic equation in this simplified example), the range for *b* and *c* can be strictly expressed as $b^2 - 4c < 0$. (Note that the variables, *b* and *c*, remain variables, i.e., symbols.) This range can be plotted in a two-dimensional coordinate system (it would be one of the two regions separated by the quadratic function). As such, one can obtain a very good understanding of the solution space for the design parameters, (*b*, *c*), and pick any set of *b* and *c* in the region. This is of critical importance in the manufacture of hardware where manufacturing variances are unavoidable. If (*b*, *c*) is picked within the solution space, but too close to the boundary, there will be a very large chance that the end result will be out of specification because of manufacturing variances. Numerical approaches do not reveal whether the solution is well within the solution space, which makes it difficult for them to provide enough of a buffer against manufacturing variances. Symbolic approaches provide a solution space and allow the design parameters to be picked within the region in order to avoid the effects of manufacturing variances and gain a higher yield.

What are the merits of symbolic approaches in software manufacturing?

Let us assume that we want to make a website safer by performing SQL injection penetration tests. An SQL injection attack causes the back-end database of the website to take an unexpected action through the entry of special strings on the site's input fields and/or URLs. The special strings are passed through a set of programs from the front-end Web server to the back-end database, where the SQL commands are formed and passed to the database. For example, if an SQL command includes within it the substring, ";SHUTDOWN;", an unexpected termination of the database occurs.

In order to avoid such vulnerabilities, human testers input many different strings at the website to see whether unexpected and dangerous SQL commands can be formed and passed to the database. This is not a numerical approach, per se, but it is conceptually very close to one, as its uses specific values for variables (i.e., it instantiates variables instead of keeping them as symbols). It would be more efficient if we could use computers instead of human testers to input special strings automatically for such tests, but it makes no difference to the approach's essence. The issue is that there may be infinitely many strings that could be input and that it would be impossible to test all of them. Hackers are very creative and they are always coming up with new and unexpected attack methods.[1] Testing with specific values only nominally improves a website's safety as it tries only previously known strings.

---

[1] In one case, they bypassed the check by entering ";SHUTDOWN;" in HEX and turning it back into the standard coding just before the SQL command was sent to the database.

In symbolic execution, a symbolic approach for software validation, the string for an input field, s, for example, does not get instantiated (i.e., it does not get assigned a specific value) in the program. Instead, the program is executed with s being a variable (symbol) all the way. When the program reaches a conditional branching, it may not be able to decide which branch to take. In such cases, it continues by executing each branch with the condition for the corresponding branch added. When it reaches a statement in which an SQL command string is formed, the SQL command string should be expressed as a function, $SQL(s)$, of the input string, s, along with the conditions collected on the path to the statement. Then, one solves the proposition, "$SQL(s)$ includes ';SHUTDOWN;' as its substring." If the proposition has no solution, it means there is no s which poses a SHUTDOWN threat to the database. If the proposition has a solution, s_0, it means that the database will shutdown unexpectedly if s_0 is in the input field. In that case, code to reject s_0 in the input field can be added to make the website safer.

The first approach of testing with specific values can cover only a finite (and very limited) set of points in an infinitely large input space. The second approach of symbolic execution finds solutions in an infinitely large input space, if they exist. If there is no solution in the symbolic execution, we can assert that the website does not have such a vulnerability.

As described above, symbolic approaches are very powerful for both hardware and software. However, they are not a panacea. In particular, their modeling capabilities are weak. Many details can be lost when you try to break down reality into manipulatable symbolic expressions. A real situation consists of numerous elements and can constitute an extremely complex system. It is naive to think the whole of physical reality can be expressed as a set of symbolic expressions. The situation is similar for software. It is possible to execute the program symbolically and solve the proposition at the end if the program is relatively small. Unfortunately, recent software programs often consist of hundreds of thousand or sometimes even millions of lines of code. In order to make software validation applicable to large pieces of software, abstraction of the parts irrelevant to the validation becomes a necessity.

Therefore, in applying the symbolic approach to real-world manufacturing problems, it is imperative to abstract out or omit irrelevant parts of the problem. Another possibility is to use high-performance computers or distributed computing, such as cloud computing, to extend the symbolic approach's scalability.

In what follows, we describe how the symbolic approach can be applied to real-world manufacturing problems. Section 2 describes the problem of HDD head design, and Sect. 3 describes the problem of software validations. The last Sect. 4 includes a summary and discusses future prospects of mathematical symbolic approaches in manufacturing and other applications.

2 Hardware Design

In manufacturing, there is a compelling need for mathematical optimization technologies to make the design process more efficient, cut costs, and create products with high added value. Current mathematical optimization technologies are mainly composed of various numerical algorithms that are broadly used in science and engineering. However, there are many issues when it comes to solving practical problems numerically. For example, many optimization problems in industry include "nonconvexities," which make obtaining a globally optimal solution difficult. Moreover, multi-objective design in manufacturing requires repeated simulations on a wide variety of parameters in order to find the optimal design.

Symbolic approaches are promising ways of tackling such issues. They are based on *symbolic and algebraic computations* that handle mainly polynomials with indeterminate elements and parameters as they are. We call the optimization methods based on the symbolic approach "symbolic optimizations". A key tool of symbolic optimization is *quantifier elimination (QE)* (see [1, 2] for a description of QE). Symbolic optimization enables us to solve nonconvex optimization problems exactly and parametric optimization problems directly. In other words, we can obtain feasible parameter regions in a parameter space for given specifications. This brings with it a deeper understanding of design problems and systematic flows for multi-objective design. Recently, symbolic optimization has been applied to design and verification problems in many fields (see [1]).

In the following, we show a symbolic optimization accomplished by QE and its application to a problem of designing a HDD.

2.1 Optimization Using Symbolic Approaches

QE is a symbolic and algebraic algorithm that deals with first-order formulas. First-order formulas consist of polynomial equations, inequalities, quantifiers ($\forall$, $\exists$), and boolean operations ($\wedge$, $\vee$, $\Rightarrow$, $\neg$, etc.). QE outputs an equivalent quantifier-free formula for a given first-order formula. For example, QE derives an equivalent quantifier-free formula $b^2 - 4c < 0$ for the input formula $\forall x(x^2 + bx + c > 0)$ over the field of real numbers. If all variables in a given first-order formula are quantified, QE returns true or false for the input formula. In this paper, we consider real QE.

Now let us show how we can solve optimization problems by using QE.

$$\begin{aligned} &\text{Objective function: } f(x_1, \ldots, x_n) \to \min \\ &\text{Constraints: } g_1(x_1, \ldots, x_n)\rho_1 0, \ldots, \quad g_k(x_1, \ldots, x_n)\rho_k 0 \end{aligned} \tag{1}$$

where $\rho_i \in \{\neq, <, >, \leq, \geq\}$. Optimization problem (1) can translated into a QE problem (first-order formula) as follows:

$$\exists x_1 \ldots \exists x_n (k - f(x_1, \ldots, x_n) = 0 \wedge \varphi(x_1, \ldots, x_n)) \tag{2}$$

where $\varphi(x_1, \ldots, x_n) \equiv \bigwedge_i (g_i(x_1, \ldots, x_n)\rho_i 0)$ and k is a newly introduced variable assigned to the objective function $f(x_1, \ldots, x_n)$. Performing QE on the first-order formula (2), we obtain a formula $\psi_1(k)$ that shows the possible range of k, i.e., f. Thus, the minimal value of k in $\psi_1(k)$ is the minimum of the objective function f and it is the globally minimal value.

We briefly summarize the properties of symbolic optimization below.

- Optimization problems can be solved parametrically, meaning all feasible regions of the parameters in the parameter space can be obtained.
- Nonconvex optimization problems can be solved exactly.
- The feasibility of the optimization problem can be exactly verified.

Effective exploitation of these properties leads to efficiency and performance advantages.

2.1.1 Example

Here, we will consider the following optimization problem.

$$\begin{array}{ll} \text{Objective function:} & 2x_1 + x_2 \to \min \\ \text{Constraints:} & 4x_1 + x_2 \leq 9, x_1 + 2x_2 \geq 4, 2x_1 - -3x_2 \geq -6 \end{array} \tag{3}$$

This problem reduces to the following QE problem.

$$\exists x_1 \exists x_2 (k - (2x_1 + x_2) = 0 \wedge 4x_1 + x_2 \leq 9 \wedge x_1 + 2x_2 \geq 4 \wedge 2x_1 - -3x_2 \geq -6) \tag{4}$$

Performing QE on (4), we obtain the feasible region of k:

$$2 \leq k \leq 6.$$

This implies that the maximum of k is 6 and the minimum is 2; that is, the maximum of the objective function $2x_1 + x_2$ is 6, and the minimum is 2.

Next, we consider the following *multi-objective optimization (MOO)* problem, which simultaneously optimizes more tha one objective function. For a nontrivial multiobjective optimization problem, there is no single solution that simultaneously optimizes each objective. In fact, most MOO problems appearing in real-world problems have trade-offs between two or more conflicting objective functions. In that case, there exists (a possibly infinite number of) Pareto optimal solutions. A solution is called Pareto optimal, if none of the objective functions can be improved in value without degrading some of the other objective values. We note that all Pareto optimal solutions are considered equally good, so we need additional subjective preference information to order them. Hence, solving an MOO problem entails computing Pareto optimal solutions. (See [3] for a description of MOO.)

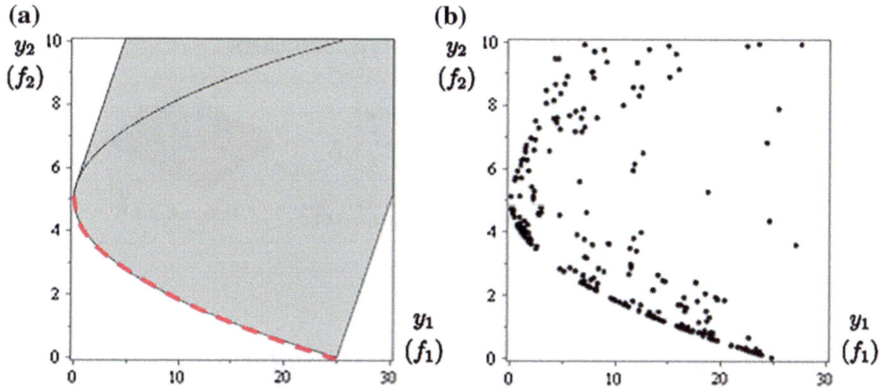

Fig. 1 Feasible region of objective functions for MOO problem (5). **a** Feasible region by QE. **b** Feasible solutions by GA

Let us demonstrate how MOO problems are solved by using QE. Consider the following MOO problem.

$$\begin{aligned}\text{Objective functions: } & f_1(x_1, x_2),\ f_2(x_1, x_2), \rightarrow \min \\ & f_1 = x_1^2 + x_2^2,\ f_2 = 5 + x_2^2 - x_1, \\ \text{Constraints: } & -5 \leq x_1 \leq 5, -5 \leq x_2 \leq 5.\end{aligned} \tag{5}$$

We first translate the MOO (5) into the following QE problems.

$$\exists x_1 \exists x_2 (y_1 = x_1^2 + x_2^2 \wedge y_2 = 5 + x_2^2 - x_1 \wedge -5 \leq x_1 \leq 5 \wedge -5 \leq x_2 \leq 5) \tag{6}$$

where y_1 and y_2 are newly introduced variables assigned to the objective functions f_1 and f_2, respectively. By performing QE on the formula, we obtain the following formula that describes the feasible regions of y_1, y_2 (i.e., f_1, f_2) in the f_1-f_2 objective space.

$$\begin{aligned}&(y_2 - y_1 + 25 \geq 0 \ \wedge \ y_2^2 - 60y_2 - y_1 + 925 \geq 0 \wedge \\ &y_2 \leq 30 \ \wedge \ y_1 \geq 25) \vee \\ &(4y_2 - 4y_1 - 21 \leq 0 \wedge \ y_2 \geq 30 \ \wedge \ 4y_1 \leq 101) \vee \\ &(y_2 - y_1 + 15 \geq 0 \wedge \ y_2^2 - 60y_2 - y_1 + 925 \leq 0) \vee \\ &(y_2 - y_1 + 25 \geq 0 \wedge \ y_2^2 - 10y_2 - y_1 + 25 \leq 0) \vee \\ &(4y_2 - 4y_1 - 21 \leq 0 \wedge \ y_2 \geq 5 \ \wedge \ y_1 \leq 25 \ \wedge \ 4y_1 \geq 1).\end{aligned}$$

This region is shown in gray in Fig. 1a. The Pareto optimal solutions are obtained as a Pareto optimal front (the dashed red line in the figure). Figure 1b shows the Pareto optimal solutions obtained by a numerical method based on the genetic algorithm (GA). This is a typical output of a numerical approach; it is not easy to see the Pareto optimal front or the feasible regions of the objective functions.

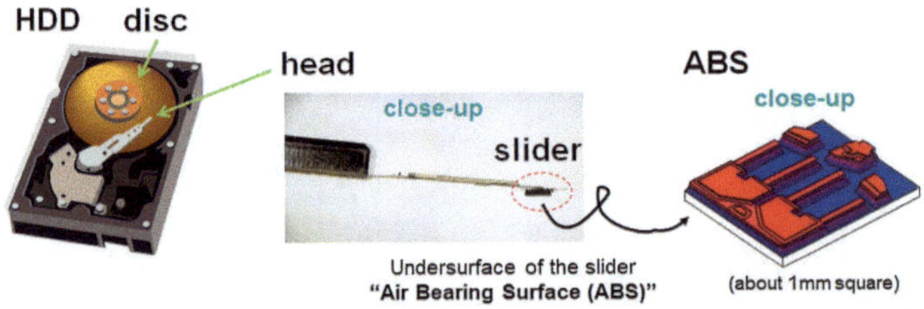

Fig. 2 Slider of HDD and its air bearing surface (ABS)

2.2 *Application: Design of Shape of HDD Slider Air Bearing Surface*

We applied symbolic optimization to the design of the optimal shape of an HDD slider (Fig. 2). The slider is a thin, nearly square, flat part attached to the actuator arm of the HDD. On the top of the slider is a magnetic head that writes (or reads from) binary information on the disk. The surface of the slider, designed on the nanometer scale, is shaped so as to stabilize the head. The design problem is to determine the shape, or the pattern, of the slider's surface. When the disk rotates, the slider lifts or hovers by air pressure like a glider. To ensure high read/write performance and durability at the same time, it is very important to control the relative position between the slider head and the disk. For example, the distance between them, called the flying height, is one of the most important indicators that affect the quality of hard disk drives because contact between the head and disk might cause a system crash. The relative angles between the slider and disk, such as pitch and roll angles, are also to be controlled. These performance indicators form a set of objective functions in the design problem of the hard disk slider. Environmental changes are also taken into consideration. Reduced atmospheric pressure at high altitude changes the relative position. In our case, we have nine objective functions in total to be optimized, indicating that this is a multiobjective optimization problem.

A typical way of designing an air bearing surface (ABS) is as follows. A designer first draws a basic geometrical shape of the surface and chooses a set of parameters to be optimized as well as their ranges. For a set of real values for the parameters, a simulator computes various physical values related to the slider's relative position to the disk. We treat the simulator as a black box. Conventionally, the design problem is solved by using numerical methods for multiobjective optimization on the basis of metaheuristic approaches.

Instead, we shall take the symbolic approach explained in Sect. 2.1. First, we need to make an approximate model (a model expressed in terms of polynomials) for the objective functions. After that, we apply the symbolic method using QE. The result of the symbolic optimization is the feasible region of the objective functions. For example, Fig. 3 shows the feasible region for two of nine objective functions (say, f_1 and f_2).

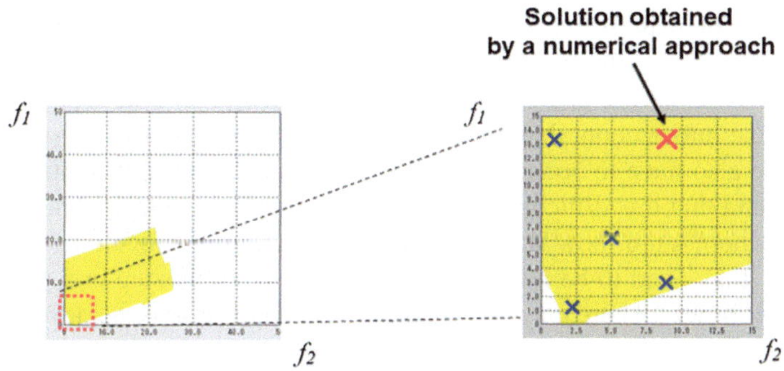

Fig. 3 Feasible region of objective functions f_1 and f_2

By taking a symbolic approach, we can obtain a Pareto optimal front with all feasible regions in the objective space. It follows that it greatly reduces the total time to determine the design solution. For example, it can reduce the time required for completing the ABS design from 14 days to one day.

3 Software Validation

3.1 Background: Why the Technology is Important for Today's Society

Software permeates every aspect of society. It is especially apparent that software plays a huge role in PCs, smart phones, and tablets that people use every day. Software has also become crucial in places not so visible to everyone. From social infrastructure like financial systems and transportation systems to everyday appliances and modern cars with more than 200 computer chips, software continues to become more and more prevalent and critical to their functionalities.

Concepts like the Smart (Power) Grid, Smart City, and Smart Home have started being implemented, and this surely means the pace of software penetration into society can only accelerate.

Not only can one lose one's money to bugs (i.e., "defects") in software; software bugs can also precipitate dire consequences for society, including loss of life. For example, at least six patients lost their lives due to a bug in one model of medical accelerator. This and other cases, which exemplify what sorts of disaster a software bug is capable of causing, can be found in [4].

Unfortunately, software bugs will always be there, and there are several reasons why. We list two reasons here. First is that software systems have become too large and complex, and the entirety of the system states is simply beyond human comprehension. A huge system is divided into modules and the specifications for each module are written as "what" the module is supposed to do. Since there can be many modules interacting with one particular module, it would be an extremely difficult, if not impossible, task to give a correct set of specifications. This is particularly true when we take into account that huge systems tend to be maintained for long period of time. We also have to consider that specifications are usually written in natural language[2] and that it is difficult to remove ambiguity completely from them.

Second, there is the "how" issue. Even if the program specifications (the "whats") are perfect, they need to be interpreted and written step by step in a programming language as "hows". There may be a gap between "what" is to be achieved and "how" it is implemented, and this is one more place where bugs can slip in.

Software is tested to check if it functions as it is supposed to before it goes into production. However, it is virtually impossible to cover all possible execution paths. When the program size grows, the resources necessary for testing grow exponentially compared to the resources used to write the program. Testing is currently quite a labor-intensive task[3] and an attempt at being comprehensive may incurs a tremendous amount of time and money. In today's cost-conscious world, system integrators cannot afford such (comprehensive and somewhat complete) testing or else the added costs would make their services or products too expensive. System integrators have to limit their resources for testing to keep costs manageable. Therefore, software systems are likely distributed with undiscovered and potentially critical bugs still in them.

This is where software validation comes in. Software validation can automate some of software testing processes to make them more cost-efficient, and it enables a larger test coverage for possible software states.[4]

3.2 History of Software Validation

Software validation originated in the 1960s. It initially took a very rigorous approach in which the "what" of a program is described in a formal language and is compared with the actual program to "prove" whether the "what" and the program match exactly.

Eventually, people realized that this approach is not of much use in real-world programming. A programmer needs to learn a formal language, which is often very difficult and different from ordinary programming languages, and writing "what" in a formal language can entail as much, if not more, effort as writing the actual working program. Because of this, the rigorous approach has remained applicable

---

[2] Only a very few people write (can write) program specifications in formal language.

[3] For example, human testers may have to manually input data on Web browsers to test a website software system.

[4] [5, 6] are general introductions to software validation. Please refer to [7] for a detailed description of the symbolic execution (outlined in Sect. 3.3) that is being developed at Fujitsu Laboratories of America.

Fig. 4 Mini-program

```
foo(a, b, c) {
   int a,b,c;
   c = a + b;
   if (c > 0) {
     c++;
   }
   return c;
}
```

only to small-scale programs, and it hasn't reached a critical mass of users within the programming community. In other words, the approach is inadequate for dealing with most software that has a sizeable amount of code.

A change came around 1998. Instead of pursuing stringent proofs of exact matches, researchers started focusing on ways of finding more bugs even if they sacrificed rigorousness and completeness in doing so. Prior efforts were not in vain, but there is no denying that it was research in the ivory tower. The industry asked academia what it is useful for and academia changed its direction so that programs could be validated directly without specifications given in a formal language. Since then, universities and corporations such as CMU, Stanford University, Microsoft, NASA, UC Berkeley, NEC, and Fujitsu have been instrumental in making software validation applicable to ever growing realms of software.

3.3 Symbolic Execution

Here, we will use a simple example to illustrate how symbolic execution works and discuss its merits, limits, and future prospects.

Consider the mini-program in Fig. 4 to be validated. Suppose it needs to always satisfy the following property[5]:

$$(a > 1) \wedge (b > 0) \rightarrow (c > 4) \tag{7}$$

We proceed as follows to see if the program always satisfies this condition.

First, we negate (7) and obtain the following negated property:

$$(a > 1) \wedge (b > 0) \rightarrow (c <= 4) \tag{8}$$

If this negated property has a solution (a_0, b_0, c_0) at the end of execution, the solution is a counter example and this implies that the program does *NOT* always satisfy the original property, (7). On the other hand, we can conclude that the program *DOES*

---

[5] We use the term, "property" for a condition in symbolic execution.

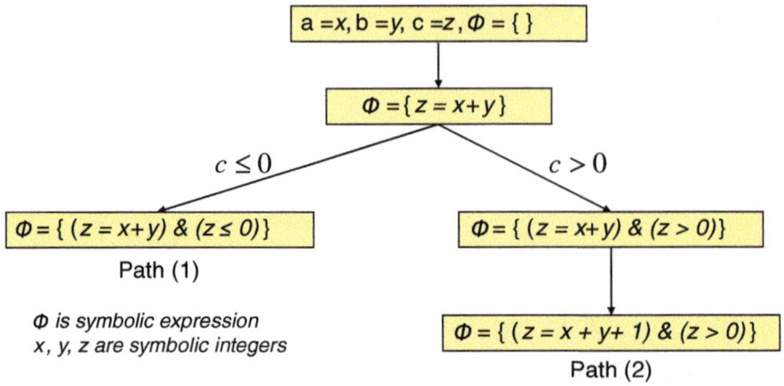

Fig. 5 Symbolic execution flow

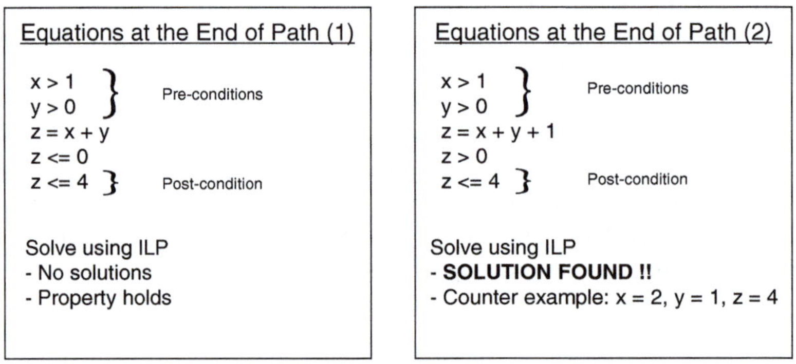

Fig. 6 Set of equations to be solved at the end of each path

always satisfy the original property, (7), if the negated property, (8), has no solution at the end of program execution.

In order to see which is the case, we execute the program symbolically. Figure 5 depicts the process. In the first step, we substitute variables, a, b, and c in the program with corresponding symbolic variables, x, y, and z, as shown in the top box of Fig. 5. At the assignment statement, $c = a+b$, we add $z = x+y$ to the symbolic expression, Φ. As we cannot fix its boolean value at the conditional statement in the next step of the program, we go down both paths separately, each path with the corresponding condition, $z \leq 0$ or $z > 0$, added to Φ. Eventually, Path (1) and Path (2) reach their last statements and Φ has the condition for the function execution to follow each path.

As in Fig. 6, we add the negated property, (8), to Φ at the end of each path. (The negated property consists of the precondition and the postcondition.) We solve the set of conditions thus obtained for each path using, for example, integer linear programming (ILP). For Path (1), there is no solution, meaning this path has the original property, (7). For Path (2), there exist solutions (for example, $x = 2$, $y = 1, z = 4$) and this path does *NOT* have the original property, (7). The solu-

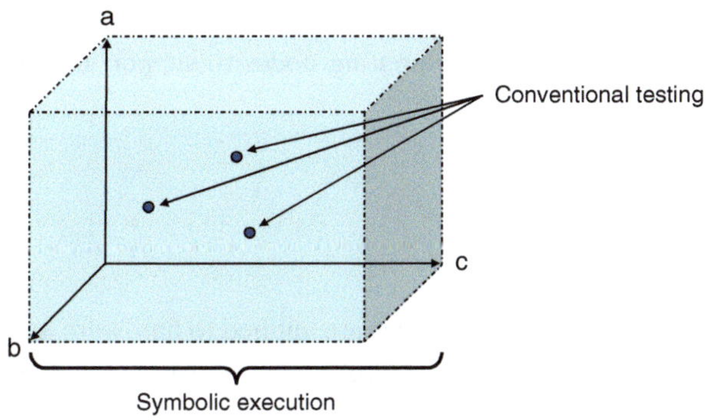

Fig. 7 Conventional testing and symbolic execution

tions for Path (2) serve as test data, which cause errors (i.e. do not satisfy the given condition, (7)), and we can use them to find bugs.

As Fig. 7 shows, conventional testing covers only a finite number of points in the variable space to check if they have a certain property. However, symbolic execution can cover the whole variable space at once to see if there is any possible variable set in the space that violates a condition.

This characteristic of symbolic execution makes it extremely powerful, especially for checking terminal conditions[6] and paths rarely traversed in ordinary executions of the program.[7]

Symbolic execution is an extremely powerful tool, but it has limitations. The number of paths can explode. When there are too many conditionals in a program, the number of paths can increase uncontrollably and make the state space (i.e. the set of paths that the symbolic execution traverses) explode. The same happens when too many symbolic variables are used. In both cases, a computer's processing power and memory capacity can be easily exceeded. Numerous methods and techniques have been devised to make symbolic execution applicable to large programs by placing weaker or fewer restrictions on things like the number of conditionals and symbolic variables. Such efforts have borne fruit and there are cases where symbolic execution was applied to programs with hundreds of thousands or even millions of lines of code.

There are mainly two fronts to explore. One is appropriate abstraction of the program in order to contain state space explosions. For example, symbolic execution could be strictly applied to only the modules in question and the rest of the program is abstracted out. Another example is to use only three states, zero, positive, and negative, for integer and real number variables. There are many more ways to reduce complexity, and this is currently an active research area.

---

[6] Conditions like $c > 545$ for a real number, c. The conditions cannot be checked completely with only specific numbers like $c = 546, 545.1, 545.01, \ldots$.

[7] Conventional testing requires specific values for variables in order for the function executions to traverse rarely traversed paths, and finding such values is quite difficult. On the other hand, symbolic execution follows all the paths in the same manner.

The other front is to use distributed computing such as cloud computing. Here, tasks are distributed to multiple computing nodes to support the larger state space for symbolic execution.

4 Summary

We illustrated how symbolic approaches are applied to hardware and software manufacturing processes. This is part of a bigger trend in which mathematics is making manufacturing technologies more effective.

Recently, model-based development (MBD) has been introduced to make the design process more efficient, to reduce costs, and to create higher value products. The symbolic approaches described in this paper, we believe, are a very good fit for MBD and that continued efforts in applying mathematics to manufacturing will be critical to the advancement of MBD.

For wider acceptance of mathematical approaches in manufacturing, easy-to -understand-and-use software tools and/or push-button solutions are crucial. In the future, research results from the field of artificial intelligence (e.g., automated mathematical problem solving [8]) will be employed to make them possible.

We also foresee that accumulated knowledge about mathematical applications in manufacturing will become important in other systems, including socio-economic systems and energy management systems.

References

1. H. Anai, K. Yokoyama, Algorithms of Quantifier Elimination and Their Applications: Optimization by Symbolic and Algebraic Methods (University of Tokyo Press, Tokyo, 2011) (In Japanese)
2. B.F. Caviness, J.R. Johnson (ed.), Quantifier elimination and cylindrical algebraic decomposition. *Texts and Monographs in Symbolic Computation* (Springer, New York, 1998)
3. M. Yoon, Y. Yun, H. Nakayama, *Sequential Approximate Multiobjective Optimization Using Computational Intelligence* (Springer, New York, 2009)
4. Matt Lake, Epic failures: 11 infamous software bugs. ComputerWorld, Sept 2010. http://www.computerworld.com/s/article/9183580
5. C. Cadar, P. Godefroid, S. Khurshid, K.C. Pasareanu, Sen, N. Tillmann, Symbolic execution for software testing in practice: preliminary assessment, in *ICSE '11* (2011)
6. V. D'Silva, D. Kroening, G. Weissenbacher, A survey of automated techniques for formal software verification, in *IEEE Transactions on Computer-Aided Design of Integrated Circuits and Systems, TCAD* (2008)
7. G. Li, I. Ghosh, S.P. Rajan, Klover: a symbolic execution and automatic test generation tool for c++ programs, in *CAV 2011* (2011)
8. T. Matsuzaki, H. Iwane, H. Anai, N. Arai, The complexity of math problems—linguistic, or computational? in *The 6th International Joint Conference on Natural Language Processing (IJCNLP13)*, pp. 73–81 (2013)

Error Correcting Codes Based on Probabilistic Decoding and Sparse Matrices

Hironori Uchikawa

Abstract These days we encounter many digital storage and communication devices in our daily lives. They contain error correcting codes that operate when data is read from storage devices or received via communication devices. For example, you can listen to music on a compact disc even if its surface is scratched. This article introduces low density parity check (LDPC) codes and the sum-product decoding algorithm. LDPC codes, one class of error correcting codes, have been used for practical applications such as hard disk drives and satellite digital broadcast systems because their performance closely approaches the theoretical limit with manageable computational complexity. In particular, it is shown that an optimal decoding algorithm from the viewpoint of probabilistic inference can be derived with LDPC codes.

Keywords LDPC · Error correcting · Probabilistic decoding · Sparse · Sum-product algorithm · MAP

1 Communication System Model and Error Correcting Codes

A communication system model for transmitting information from a source to a destination through a channel is shown in Fig. 1. Let us define a transmitted message of length k from the source as $\mathbf{m} = m_1 m_2 \dots m_k$. The message $\mathbf{m}$ is mapped to a length-n codeword $\mathbf{x} = x_1 x_2 \dots x_n$, where $n > k$ at the encoder to protect the message from noise in the channel. The codeword $\mathbf{x}$ is transmitted through the channel. The resulting output of the channel is a received word $\mathbf{y} = y_1 y_2 \dots y_n$, which is fed into the decoder. Our model of the channel is that of the discrete probabilistic channel. More precisely, the channel is described as a conditional probability distribution

H. Uchikawa (✉)
Center for Semiconductor Research and Development, Toshiba Corporation, Semiconductor & Storage Products Company, 2-5-1, Kasama, Sakae-Ku, Yokohama 247-8585, Japan
e-mail: hironori.uchikawa@toshiba.co.jp

R. Nishii et al. (eds.), *A Mathematical Approach to Research Problems of Science and Technology*, Mathematics for Industry 5,
DOI: 10.1007/978-4-431-55060-0_36, © Springer Japan 2014

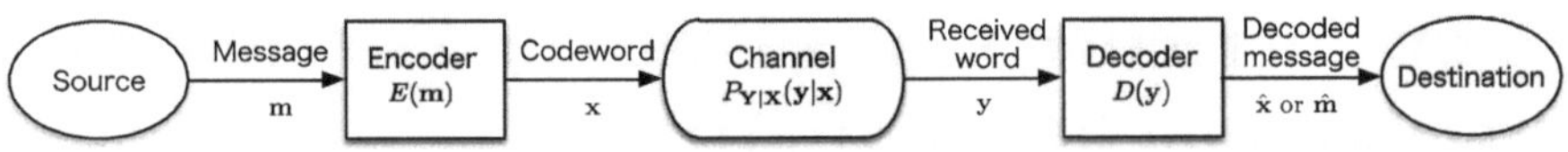

Fig. 1 Communication system model

$P_{\mathbf{Y}|\mathbf{X}}(\mathbf{y}|\mathbf{x})$. The decoder produces a decoded codeword $\hat{\mathbf{x}}$ or a decoded message $\hat{\mathbf{m}}$. Note that, from the viewpoint of communication systems, the decoder should produce a decoded message $\hat{\mathbf{m}}$ instead of decoded codeword $\hat{\mathbf{x}}$. Since mapping at the encoder is one to one, the decoder usually produces a decoded codeword $\hat{\mathbf{x}}$ in the model on coding theory studies.

For the sake of simplicity, the alphabets of message symbol m_i $(1 \le i \le k)$, codeword symbol x_i $(1 \le i \le n)$, and received word symbol y_i $(1 \le i \le n)$ are assumed to be binary, $\{0, 1\}$, in this article. The channel is also assumed to be a memoryless binary symmetric channel with an independent and identically distributed conditional probability

$$P_{\mathbf{Y}|\mathbf{X}}(\mathbf{y}|\mathbf{x}) = \prod_{i=1}^{n} P_{Y|X}(y_i|x_i),$$

where

$$P_{Y|X}(y_i|x_i) = \begin{cases} 1-p & x_i = y_i \\ p & x_i \neq y_i \end{cases}$$

where the probability p in the range $0 \le p \le 1$ denotes the crossover probability of the channel.

Now we define the information rate as $R = k/n$. Shannon showed that there exists an error correcting code that can lower the decoding error probability as much as possible if the rate R is less than the channel capacity C derived from the probability distribution of the channel [5]. Constructing error correcting codes with a rate closely approaching the capacity with low computational encode and decode operations is one of the ultimate goals for researchers in coding theory.

2 Probabilistic Decoding

The decoding of error correcting codes can be regarded as an inference problem: to infer the most likely $\mathbf{x}$ from the received word $\mathbf{y}$. Here, we use probability as a tool for dealing with the likelihood.

Assume that a codeword $\mathbf{x} = x_1 x_2 \cdots x_n$ is uniformly chosen from a codebook and transmitted through the channel; then the receiver gets the message $\mathbf{y} = y_1 y_2 \cdots y_n$. The optimal decoding rule to minimize the symbol error probability is expressed as

$$\hat{x}_i = \underset{x_i \in \{0,1\}}{\mathrm{argmax}}\, P_{X|\mathbf{Y}}(x_i|\mathbf{y}), \tag{1}$$

where the notation $\mathrm{argmax}_{x_i \in \{0,1\}}$ returns the argument of x_i to maximize the right term in (1). Since x_i is a binary symbol, $\mathrm{argmax}_{x_i \in \{0,1\}}$ returns 0 or 1, whichever has the highest probability, by comparing $P_{X|\mathbf{Y}}(x_i = 0|\mathbf{y})$ and $P_{X|\mathbf{Y}}(x_i = 1|\mathbf{y})$. The decoding algorithm in (1) is called the *maximum a posteriori probability* (MAP) decoding because it gives $\hat{x}_i$ that maximizes the *a posteriori probability* (APP) $P_{X|\mathbf{Y}}(x_i|\mathbf{y})$.[1]

In order to make Eq. (1) computable by using the conditional probabilities of the channel $P_{Y|X}(y|x)$, Eq. (1) can be transformed as follows.

$$
\begin{aligned}
\hat{x}_i &= \operatorname*{argmax}_{x_i \in \{0,1\}} P_{X|\mathbf{Y}}(x_i|\mathbf{y}) \\
&= \operatorname*{argmax}_{x_i \in \{0,1\}} \sum_{\sim x_i} P_{\mathbf{X}|\mathbf{Y}}(\mathbf{x}|\mathbf{y}) && (2) \\
&= \operatorname*{argmax}_{x_i \in \{0,1\}} \sum_{\sim x_i} \frac{P_{\mathbf{Y}|\mathbf{X}}(\mathbf{y}|\mathbf{x}) P_{\mathbf{X}}(\mathbf{x})}{P_{\mathbf{Y}}(\mathbf{y})} && (3) \\
&= \operatorname*{argmax}_{x_i \in \{0,1\}} \sum_{\sim x_i} P_{\mathbf{Y}|\mathbf{X}}(\mathbf{y}|\mathbf{x}) P_{\mathbf{X}}(\mathbf{x}) && (4) \\
&= \operatorname*{argmax}_{x_i \in \{0,1\}} \sum_{\sim x_i} \left(\prod_{j=1}^{n} P_{Y|X}(y_j|x_j) \right) \mathbb{I}[\mathbf{x} \in C] && (5)
\end{aligned}
$$

where $\sum_{\sim x_i}$ indicates the summation over all the alphabets of x except for x_i, i.e., $\sum_{\sim x_1}$ means

$$\sum_{x_2 \in \{0,1\}} \sum_{x_3 \in \{0,1\}} \cdots \sum_{x_n \in \{0,1\}} .$$

Equation (2) is derived from marginalization[2] and the transformation from (2) to (3) is given by the Bayes rule:

$$
\begin{aligned}
P_{\mathbf{Y},\mathbf{X}}(\mathbf{y}, \mathbf{x}) &= P_{\mathbf{X}|\mathbf{Y}}(\mathbf{x}|\mathbf{y}) P_{\mathbf{Y}}(\mathbf{y}) \\
&= P_{\mathbf{Y}|\mathbf{X}}(\mathbf{y}|\mathbf{x}) P_{\mathbf{X}}(\mathbf{x}).
\end{aligned}
$$

Since $P_{\mathbf{Y}}(\mathbf{y})$ does not affect the operation $\mathrm{argmax}_{x_i \in \{0,1\}}$, Eq. (4) is derived. In the last step, we the fact that transmitted codewords are chosen uniformly at random

[1] Sometimes called symbol MAP decoding to distinguish it from the MAP algorithm that maximizes the APP of the codeword $\mathbf{X}$.

[2] More precisely, it is the opposite way to marginalization. Marginalization is the computation to obtain the probability distribution with fewer variables from a multivariate probability distribution, e.g. the marginal distribution of the random variable A is given by $P_A(a) = \sum_{b \in \mathscr{B}} P_{AB}(a, b)$, where $\mathscr{B}$ is the domain of the random variable B.

Fig. 2 Distributive law: eliminating 1 multiplication

Distributive law

$$ax + ay \rightarrow a(x + y)$$

2 multiplications
1 addition

1 multiplication
1 addition

and the channel is memoryless. The indicator function $\mathbb{I}$[condition] returns 1 if the condition is satisfied; otherwise, it returns 0.

Can we calculate Eq. (5) when the code length is large? The answer is no because the scale of the summation $\sum_{\sim x_i}$ increases exponentially with code length n. Thus, we need a more efficient algorithm.

3 Computation Reduction by Using the Distributive Law

The computation of the APP has a sum-product form in (5). If we can take common factors out from the summation, i.e., by applying the distributive law, the number of computations decreases. For example, we can eliminate 1 multiplication by using the distribution law in Fig. 2.

In the following, we demonstrate APP computation reduction by using the code

$$C_1 = \{\mathbf{x} \in \{0, 1\}^7 | H_1 \mathbf{x}^\mathsf{T} = \mathbf{0}\},$$

where the parity-check matrix H_1 is defined as

$$H_1 = \begin{bmatrix} 1\,0\,0\,1\,1\,0\,0 \\ 1\,1\,0\,0\,0\,1\,0 \\ 0\,1\,1\,0\,0\,0\,1 \end{bmatrix} \tag{6}$$

From (5), the APP of the transmitted symbol x_1 is given as

$$P_{X|\mathbf{Y}}(x_1|\mathbf{y}) = \sum_{\sim x_1} \mathbb{I}[H_1 \mathbf{x}^\mathsf{T} = \mathbf{0}] \prod_{j=1}^{7} P_{Y|X}(y_j|x_j), \tag{7}$$

where $\mathbf{x}^\mathsf{T}$ denotes the transposition of $\mathbf{x}$. Since $H_1 \mathbf{x}^\mathsf{T} = \mathbf{0}$ can be written as the product of each parity equation, Eq. (7) is transformed to

$$P_{X|\mathbf{Y}}(x_1|\mathbf{y}) = \sum_{\sim x_1} \mathbb{I}_1 \mathbb{I}_2 \mathbb{I}_3 \prod_{j=1}^{7} P_{Y|X}(y_j|x_j), \tag{8}$$

where

$$\mathbb{I}_1 = \mathbb{I}[x_1 + x_4 + x_5 = 0],$$
$$\mathbb{I}_2 = \mathbb{I}[x_1 + x_2 + x_6 = 0],$$
$$\mathbb{I}_3 = \mathbb{I}[x_2 + x_3 + x_7 = 0].$$

Each indicator function corresponds to the respective parity equation. Then Eq. (8) is further factorized as

$$P_{X|\mathbf{Y}}(x_1|\mathbf{y}) = P_{Y|X}(y_1|x_1) \sum_{\sim x_1} \mathbb{I}_1\mathbb{I}_2\mathbb{I}_3 \prod_{j=2}^{7} P_{Y|X}(y_j|x_j) \tag{9}$$

$$= P_{Y|X}(y_1|x_1) \left(\sum_{x_4x_5} \mathbb{I}_1 \prod_{j' \in \{4,5\}} P_{Y|X}(y_{j'}|x_{j'}) \right) \times \left(\sum_{x_2x_3x_6x_7} \mathbb{I}_2\mathbb{I}_3 \prod_{j'' \in \{2,3,6,7\}} P_{Y|X}(y_{j''}|x_{j''}) \right) \tag{10}$$

$$= P_{Y|X}(y_1|x_1) \left(\sum_{x_4x_5} \mathbb{I}_1 \prod_{j' \in \{4,5\}} P_{Y|X}(y_{j'}|x_{j'}) \right) \times \left(\sum_{x_2x_6} \mathbb{I}_2 \prod_{j'' \in \{2,6\}} P_{Y|X}(y_{j''}|x_{j''}) \left(\sum_{x_3x_7} \mathbb{I}_3 \prod_{j''' \in \{3,7\}} P_{Y|X}(y_{j'''}|x_{j'''}) \right) \right), \tag{11}$$

where $\sum_{x_a x_b}$ denotes the summation of all the alphabets of both x_a and x_b, i.e. $\sum_{x_4x_5}$ shows $\sum_{x_4 \in \{0,1\}} \sum_{x_5 \in \{0,1\}}$.

Since the $P_{Y|X}(y_1|x_1)$ is common for all terms, we obtain Eq. (9). Then Eqs. (10) and (11) are derived by factorizing equations according to the form of each indicator function.

Equation (11) has fewer computations, 20 multiplications and 6 additions, compared with 96 multiplications and 14 additions in (7). The reader might think that the benefit of the distributive law is limited because the number of computations is already small in (7). Since the number of computations grows exponentially with the code length, we cannot compute in the form of (7) when the code length is large. On the other hand, the form of (11) can be computed efficiently by using the sum-product algorithm described in the following section.

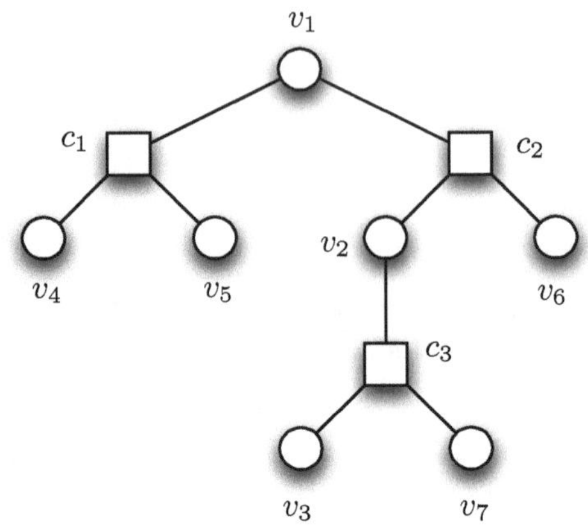

Fig. 3 Bipartite graph of the parity-check equation in (6)

4 Tree and Sum-Product Algorithm

In the previous section, we saw that the APP equation is factorized by each indicator function with the distributive law. It is preferred that the factorization is available for any code. However, the transformation is limited only if the corresponding bipartite graph forms a tree.

A bipartite-graph representation of the parity-check matrix in (6) is shown in Fig. 3. The bipartite graph consists of variable nodes (circle) corresponding to code bits and check nodes (rectangular) corresponding to parity-check equations. An edge (v, c) between the variable node v and the check node c denotes the relation between the corresponding code bit and parity-check equation. For example, the variable node v_1 corresponding to x_1 connects the check nodes c_1 and c_2 corresponding to the first and second parity-check equations, respectively.

If there is only one path from a node a to a node b in a graph, i.e., there is no cycle in the graph, such a graph is regarded as a tree. You can see that the graph in Fig. 3 has no cycle; thus, the graph is a tree. When a graph is a tree, the APP can be computed efficiently by using the sum-product algorithm.

The sum-product algorithm is also called the message passing algorithm because the sum-product algorithm is a sequence of message passing between nodes. The algorithm sets the variable node whose APP we want to compute as the root node, e.g., v_1 in Fig. 3. Toward the root node, messages computed by (12) and (13) at each node are sent from leaf nodes, which are at the tree's boundaries, e.g., $v_2, v_3, \ldots, v_7$ in Fig. 3. After the root node has received all messages through its edges, the APP of the root node is obtained by (14).

Variable node operation

$$M_{v_j \to c_i}(x_j) = P_{Y|X}(y_j|x_j) \prod_{i' \in \mathscr{I}(j) \setminus \{i\}} M_{c_{i'} \to v_j}(x_j) \tag{12}$$

Check node operation

$$M_{c_i \to v_j}(x_j) = \sum_{\sim x_j} \mathbb{I}_i \prod_{j' \in \mathscr{J}(i) \setminus \{j\}} M_{v_{j'} \to c_i}(x_{j'}) \tag{13}$$

Tentative estimation

$$g(x_j) = P_{Y|X}(y_j|x_j) \prod_{i \in \mathscr{I}(j)} M_{c_i \to v_j}(x_j) \tag{14}$$

where $\mathscr{I}(j)$ denotes the set of check-node indices connected to the variable node v_j (rows whose parity-check equation includes the symbol x_j), $\mathscr{J}(i)$ denotes the set of variable-node indices connected to the check node c_i (symbol indices involved in the ith parity-check equation), and $\mathscr{S} \setminus \{i\}$ denotes the subset of $\mathscr{S}$ except the set $\{i\}$.

Since the leaf node v_3 is a variable node, a message M_{v_3} is set as

$$M_{v_3 \to c_3}(x_3) = P_{Y|X}(y_3|x_3),$$

and then transmitted to the check node c_3. Since every leaf node has one edge, the corresponding message equation does not have the product in (12). Although, due to the space limit, the only example we showed was the computation for the variable node v_1, we can compute other variable nodes in the same way.

5 Sum-Product Decoding Algorithm

Individually computing the APP of each symbol $x_j (1 \leq j \leq n)$ referred to as a root node is not computationally efficient. Thus, all nodes operate simultaneously and iteratively until all of the parity-check equations are satisfied. Such an iterative decoding algorithm is called the sum-product decoding algorithm and shown below.

Sum-product Decoding Algorithm

Step 1 INITIALIZATION
Set the maximum number of iterations to K and $k = 1$. Also set $M_{c_i \to v_{j''}}(x_{j''}) = 1$ for all $1 \leq i \leq m$, $j'' \in \mathscr{J}(i)$, $x_{j''} \in \{0, 1\}$.

Step 2 VARIABLE NODE OPERATION
For each variable node v_j $(1 \leq j \leq n)$, proceed with the variable node operation (Eq. (12)).

Step 3 CHECK NODE OPERATION
For each check node c_i $(1 \leq i \leq m)$, proceed with the check node operation (Eq. (13)), where m denotes the number of rows in the parity-check matrix.

Step 4 PARITY-CHECK OPERATION
For each variable node v_j $(1 \leq j \leq n)$, proceed with the tentative estimation (Eq. (14)); then make a tentative decision for $1 \leq j \leq n$,

$$\tilde{x}_j = \begin{cases} 0 & (g(x_j = 0) \geq g(x_j = 1)), \\ 1 & \text{(otherwise)}, \end{cases}$$

and a tentative codeword $\tilde{\mathbf{x}} = \tilde{x}_1 \tilde{x}_2 \ldots \tilde{x}_n$. If $H\tilde{\mathbf{x}}^{\mathsf{T}} = \mathbf{0}$, then output $\tilde{\mathbf{x}}$ as the decoded codeword and stop; otherwise, go to **Step 5**.

Step 5 TERMINATION
If $k = K$, then declare a decoding failure and stop; otherwise, set $k = k + 1$ and go to **Step 2**.

We discuss only the case in which a bipartite graph forms a tree; however, bipartite graphs corresponding to parity-check matrices used in practice are not trees. When a bipartite graph is not a tree, the sum-product decoding algorithm computes the approximate APP. However, it has been observed that we can obtain reasonable error correction performance when a subgraph for a variable node as a root with *depth* of at least 6 has a tree form, where depth denotes the number of edges in the path from a root node.

How do we design a parity-check matrix so that a subgraph for each node forms a tree? In the next section, it will be revealed that the sparsity of LDPC codes has an important role for the answer.

6 Low Density Parity Check Codes

LDPC codes were invented by Gallager in his dissertation [1], and are a class of linear block codes defined by sparse parity-check matrices. Sparsity means that the number of nonzeros ("1" for binary codes) is relatively small. In general, the average number l of nonzeros in columns is $2 \leq l \leq 8$, but it is not proportional to the code length n. You can see that the number is much smaller when you compare with that for Hamming codes, known as 1 bit correction codes. The number of nonzeros in columns for Hamming codes is at least $\dfrac{\log_2 n}{2}$ on average.

Since parity-check matrices are sparse, the number of edges in corresponding bipartite graphs becomes small so it is hard to induce short cycles. Thus, a subgraph for a variable node as a root easily forms a tree of depth more than 4 so that a good approximate APP can be obtained by the sum-product algorithm.

Here, we show a construction introduced by Gallager [1]. This construction produces a parity-check matrix $H^{(n,l,r)}$ that has n columns, $\frac{ln}{r}$ rows, l nonzeros in each column, and r nonzeros in each row. For simplicity, we assume that n can be divided by r.

Construct the sub matrix $H_0^{(n,l,r)}$,

$$H_0^{(n,l,r)} := \begin{bmatrix} \mathbf{h}(r) \\ s^{(r)}(\mathbf{h}(r)) \\ s^{(2r)}(\mathbf{h}(r)) \\ \vdots \\ s^{(n-r)}(\mathbf{h}(r)), \end{bmatrix}$$

where $\mathbf{h}(r)$ denotes a length-n row vector having non zeros at the first r entries and zeros at remaining entries, and $s^{(r)}(\mathbf{h})$ denotes a cyclic shift function that shifts $\mathbf{h}$ by r entries to the right in a cyclic manner.

By using $H_0^{(n,l,r)}$, we can construct a parity-check matrix $H^{(n,l,r)}$ as follows.

$$H^{(n,l,r)} = \begin{bmatrix} H_0^{(n,l,r)} \\ \pi_1(H_0^{(n,l,r)}) \\ \vdots \\ \pi_{l-1}(H_0^{(n,l,r)}) \end{bmatrix}, \tag{15}$$

where $\pi_i(H_0^{(n,l,r)})$ $(1 \le i \le l-1)$ is a permutation function that permutes columns of $H_0^{(n,l,r)}$ randomly.

For example, $H^{(8,2,4)}$ is constructed as follows.

$$H^{(8,2,4)} = \begin{bmatrix} 1\,1\,1\,1\,0\,0\,0\,0 \\ 0\,0\,0\,0\,1\,1\,1\,1 \\ 1\,0\,0\,1\,1\,0\,1\,0 \\ 0\,1\,1\,0\,0\,1\,0\,1 \end{bmatrix}$$

We demonstrate the decoding performance of an LDPC code over the binary symmetric channel in Fig. 4. The parity-check matrix used in the simulation was constructed from (15). The size of the parity-check matrix is 510×1020 and the numbers of nonzeros in each row and column are $l = 3$ and $r = 6$, respectively. We used the sum-product decoding algorithm with a maximum of 500 iterations. For comparison, we also show the decoding performance of a Bose-Chaudhuri-Hocquenghem (BCH) code [2] in Fig. 4. The length of the BCH code is 1023 and the design distance is 103, which means the code can correct up to 51 errors in each codeword so that the lengths and rates of both codes are almost equivalent.

In Fig. 4, the decoding error probability of the BCH code is almost 1 at the crossover probability of 0.06. By contrast, the LDPC code achieves a decoding error probability of $\frac{1}{100}$.

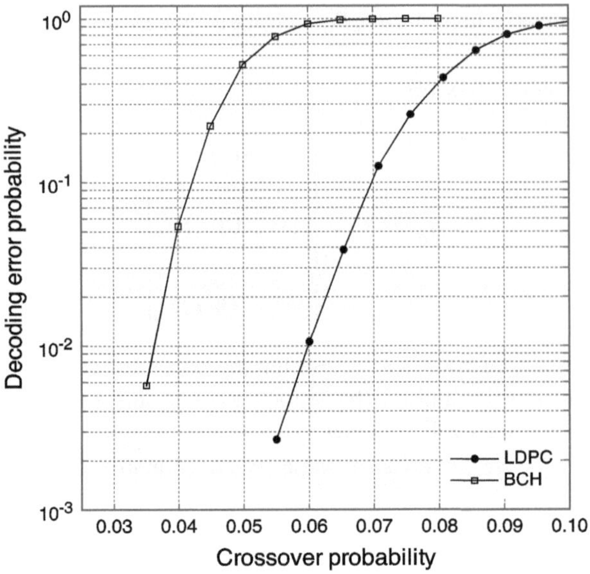

Fig. 4 Simulation results for a memoryless binary symmetric channel

7 Conclusions

In this article, we showed that a computationally efficient form for MAP decoding, the optimal with respect to probabilistic inference, can be derived from the distributive law. We also discussed computation of the derived equation by using the sum-product decoding algorithm. Finally, we showed that sparse parity-check matrices for LDPC codes are essential to compute good approximate APPs by using the sum-product decoding algorithm.

To clarify the concepts of this study, we explained ideas using toy examples instead of describing theorems and proofs. Readers interested in the mathematical background of this study should see [3]. And readers interested in practical implementations and code designs should read [4].

References

1. R.G. Gallager, in *Low-Density Parity-Check Codes* (MIT Press, Cambridge, 1963)
2. S. Lin, D.J. Costello, in *Error Control Coding*, 2nd edn. (Prentice Hall, Englewood Cliffs, 2004)
3. T.J. Richardson, R. Urbanke, in *Modern Coding Theory* (Cambridge University Press, Cambridge, 2008)
4. W.E. Ryan, S. Lin, in *Channel Codes: Classical and Modern* (Cambridge University Press, Cambridge, 2009)
5. C.E. Shannon, A mathematical theory of communication. Bell Syst. Tech. J. (1948)

Index

R. Nishii et al. (eds.), *A Mathematical Approach to Research Problems of Science and Technology*, Mathematics for Industry 5,
DOI: 10.1007/978-4-431-55060-0, © Springer Japan 2014

MIX
Papier aus verantwortungsvollen Quellen
Paper from responsible sources
FSC® C105338

If you have any concerns about our products,
you can contact us on
ProductSafety@springernature.com

In case Publisher is established outside the EU,
the EU authorized representative is:
Springer Nature Customer Service Center GmbH
Europaplatz 3, 69115 Heidelberg, Germany

Printed by Libri Plureos GmbH
in Hamburg, Germany